INTRODUCTORY APPLICATIONS OF PARTIAL DIFFERENTIAL EQUATIONS

INTRODUCTORY APPLICATIONS OF PARTIAL DIFFERENTIAL EQUATIONS

With Emphasis on Wave Propagation and Diffusion

G. L. Lamb, Jr.
The University of Arizona

A WILEY-INTERSCIENCE PUBLICATION

JOHN WILEY & SONS, INC.

New York • Chichester • Brisbane • Toronto • Singapore

Library of Congress Cataloging in Publication Data:

Lamb, G. L. (George L.), 1931–
     Introductory applications of partial differential equations with
emphasis on wave propagation and diffusion / G. L. Lamb, Jr.
          p.   cm.
     Includes bibliographical references and index.
     ISBN 0-471-31123-5
     1. Differential equations, Partial.   I. Title.
QA347.L32   1995
531′.1133′01515353—dc20   94-33111

Printed in the United States of America

10 9 8 7 6 5 4 3 2 1

To Joan

CONTENTS

PREFACE

This book has evolved from a one-semester undergraduate course in applied partial differential equations that the author has given over the past decade. Students in this course are rarely encountering partial differential equations for the first time but rather are looking for the background to and synthesis of the scattered exposure to the subject already received in previous physics and/or engineering courses. Frequently they are beginning graduate students who find that such a course is actually a prerequisite for some of their graduate engineering courses. Usually they have a working knowledge of Fourier series and Laplace transform methods. These topics, along with a brief summary of Sturm-Liouville theory and solutions of the Bessel and Legendre equations, are summarized here in appendices. The course provides a number of opportunities for the student to see the usefulness of various topics that are often introduced in an unmotivated way in a previous course in ordinary differential equations. For instance, the Cauchy-Euler equation is now found to provide a very simple means for examining diffusion and vibration in inhomogeneous systems. Lord Rayleigh's use of this solution to examine a model for the reflection of a wave from an inhomogeneous medium (*Theory of Sound*, Dover, 1945, Vol. 1, pp. 235–239) can certainly be read with profit by students at this stage in their education.

Although simple derivations of the diffusion equation and the wave equation are included, some prior knowledge of heat conduction and wave motion is implicitly assumed. I have attempted to develop a presentation that interweaves the physics and mathematics so as to indicate their interdependence. The book may thus be classified as neither physics nor mathematics or on the other hand, as a combination of both, depending upon one's own convictions and prejudices.

A course at the level of this text is taken by students whose mathematical abilities and physical understanding are changing quite rapidly. This fact, coupled with the varied interests and backgrounds of the students, makes it difficult to choose an appropriate level and mode of presentation. The physical settings have thus been drawn from the fields of mechanical vibration and heat conduction since an understanding of these topics requires only the simplest physical intuition and is almost immediately available to students in all engineering disciplines. I have attempted to begin at the beginning and gradually include topics that require more extensive calculation. In keeping with the approach that engineering students usually find most natural, the development proceeds from the particular to the general. Thus, maximum principles, mean value theorems, and uniqueness of solutions as well as the general classification of partial differential equations are not considered until Chapter 8. In a number of texts the solutions of partial differential equations are left in a ponderous form that adds little to one's intuitive understanding of the physical processes being analyzed. Wherever possible, I have tried to *use* the solutions to obtain some specific information about the physical system being considered. It is my feeling that an important aspect of a student's mathematical training is the ability to translate physical situations into mathematical terms. Thus, the wording of many of the problems included here, while perhaps lengthy in some instances, is designed to provide opportunities for the student to carry out this process of translating a physical situation into a corresponding mathematical formulation. In addition, it has been found that students with less than an optimal background can learn considerably from being led through the subsections into which many problems at this level may be decomposed.

I am indebted to the many students who, through their questions and comments, have aided significantly in my attempt to develop a pedagogic exposition of this subject. Finally, I wish to thank Gary Geernaert for providing the opportunity for me to assimilate the wonders of modern-day word processing and to my wife Joan for implementing these possibilities so efficiently.

G. L. LAMB, JR.

Tuscon, Arizona
January 1995

INTRODUCTORY APPLICATIONS OF PARTIAL DIFFERENTIAL EQUATIONS

1 One-Dimensional Problems— Separation of Variables

Quantities such as temperature, the displacement of a vibrating string or membrane, or the air pressure in a sound wave are examples of scalar quantities that may be evaluated at a specific location in space (of one, two, or three dimensions) as well as at some given time. They thus require more than one independent variable for their specification. The differential equations that must be solved to determine these quantities, for example, the temperature, are relations among the partial derivatives of the temperature with respect to these independent variables and are thus referred to as partial differential equations. The most commonly occurring partial differential equations describe relations among the same relatively small number of physical quantities such as density, displacement, and the change of these quantities that occur in various physical contexts. Hence the same equations recur in various fields, and to obtain a satisfactory introductory perspective of the subject, relatively few partial differential equations need to be considered. In fact, we shall find that there are only three basic types of equations.

Success in solving partial differential equations by analytical means has been due in large measure to the fact that the solution can be expressed in terms of solutions of one or more ordinary differential equations. This is accomplished by the method of separation of variables, to be discussed in this chapter, by the method of characteristics, or by the use of various integral transform techniques. These latter two topics will be considered in subsequent chapters.

1.1 INTRODUCTION

Before taking up any detailed consideration of the methods of solution, we first introduce the three basic types of equations.

1. We will show later (Section 1.8) that the simplest examples of wave motion on a string can be described by solutions of the equation

$$\frac{\partial^2 u}{\partial x^2} - \frac{1}{c^2} \frac{\partial^2 u}{\partial t^2} = 0 \qquad (1.1.1)$$

where $u(x, t)$ represents the displacement of the string from its equilibrium position along an x axis, c is a constant (shown later to be the velocity at which waves propagate along the string), and t is the time. Henceforth, partial derivatives will frequently be indicated by a subscript notation, that is, $u_{xx} = \partial^2 u/\partial x^2$, $u_{xy} = \partial^2 u/\partial x \partial y$, etc. The equation for the string given above is then written $u_{xx} - (1/c^2)u_{tt} = 0$.

2. The simplest examples of diffusion of heat along a beam will be shown in Section 1.2 to be described by solutions of the equation

$$u_{xx} - \gamma^{-2} u_t = 0 \qquad (1.1.2)$$

where $u(x, t)$ now refers to the temperature on the beam and γ^{-2} is a constant that is related to the specific heat and thermal conductivity of the medium.

One's intuitive familiarity with wave motion and diffusion leads to the expectation that the two equations listed above describe two fundamentally different types of processes. For instance, a motion picture of the type of wave motion described by (1.1.1) could be run backward and the result would represent an equally likely form of wave motion (the wave merely proceeding in the opposite direction). A typical diffusion process would look fundamentally different, however, when the film is run in reverse. The "undiffusing" would not conform to any phenomenon observed in nature. This fundamental difference is evident in the governing equations as well, since the second time derivative in the wave equation (1.1.1) is unchanged when t is replaced by $-t$ while in the diffusion equation (1.1.2) the sign of the first time derivative is reversed. This seemingly minor change in the equation accounts for the completely different type of behavior exhibited by solutions of the time-reversed diffusion equation.

3. The third type of partial differential equation usually occurs in problems that contain no time dependence. As we shall see later in Section 3.4, the equation for the steady displacement of a membrane sagging under its own weight is

$$u_{xx} + u_{yy} = -\rho g/T \qquad (1.1.3)$$

Here $u(x, y)$ is the static displacement below an xy plane of a membrane that is stretched with a tension T while ρg is the force of gravity per unit area on a membrane having density ρ. The combination of derivatives appearing in (1.1.3) is usually abbreviated $\nabla^2 u$ and is referred to as the Laplacian of u. For $g = 0$ we obtain the equation $\nabla^2 u = 0$, known as Laplace's equation. Without any force of gravity acting on the membrane ($g = 0$) we would expect no sagging of the membrane to occur. The satisfaction of Laplace's equation is a statement of this fact. A nonzero value of the Laplacian $\nabla^2 u$ is thus a measure of the sagging or bulginess of the surface of the membrane, $u(x, y)$. In three space dimensions the Laplacian $\nabla^2 u = u_{xx} + u_{yy} + u_{zz}$ frequently has the interpretation of the "lumpiness" of some density function $u(x, y, z)$. The universality of the physical concepts associated with the Laplacian leads one to expect this quantity to appear quite frequently when processes are described by partial differential equations. A more analytical discussion of these three types of equations as well as techniques for determining how other equations may be reduced to one of these three categories will be considered in Chapter 8.

Another classification of equations, essentially according to the difficulty of effecting their solution, is the classification into equations for which the coefficients of the various partial derivatives are (1) constants, (2) functions of the independent variables, or (3) functions of the dependent variable. The first two categories are linear equations while the third category yields a nonlinear equation. This classification is also used for ordinary differential equations, and since these equations are presumably already familiar to the reader, we first review these three categories in this context.

The ordinary differential equation that describes the oscillation of a frictionless mass-spring system is

$$m \frac{d^2 y}{dt^2} = -ky \qquad (1.1.4)$$

where y is the displacement of the system from equilibrium, m is the mass of the oscillator, and $-ky$ is the restoring force acting on the system. If m and k are constant, we have a differential equation with constant coefficients and a familiar example of simple harmonic oscillation. If, on the other hand, $k = \alpha t$, that is, the spring "constant" k varies with time (e.g., the small amplitude oscillation of a pendulum for which the length varies inversely with time), then the equation for the displacement is $m\ddot{y} + \alpha t y = 0$ where the dots indicate time derivatives. The variable coefficient makes this equation more difficult to solve than the equation with constant coefficients. If the term k has a dependence upon the displacement itself, such as $k = \beta y$ so that $m\ddot{y} = \beta y^2$, then the equation for y is nonlinear. Nonlinear ordinary differential equations are in general the most challenging to solve. Nonlinear partial differential equations present an even greater challenge and, except in a few isolated instances, will not be treated in this book.

The categories outlined above for ordinary differential equations are readily taken over to describe partial differential equations. As we shall see in Section 1.8, the vibration of a string of density ρ that is stretched to a uniform tension T and is encased in a rubberlike material that provides an elastic restoring force $-ky$ is governed by the equation

$$\rho y_{tt} - T y_{xx} = -ky \qquad (1.1.5)$$

where y is the displacement of the string from equilibrium. The three categories of equations mentioned above arise once again if we set the restoring force term k equal to any of the three possibilities: a constant, $-\alpha(x, t)$, or $-\beta(y)$. One could, of course, have the completely general situation $-\beta(x, t, y) y$. The solution of partial differential equations for these latter three categories can provide very formidable tasks and, except for a few special cases, will not be taken up in this text.

From prior experience with ordinary differential equations the reader should already have noted that the solution of an equation may be relatively simple, while satisfaction of boundary or initial conditions may be somewhat cumbersome. Indeed, an advantage of the Laplace transform technique is that it expedites this process for initial-value problems. The role played by boundaries is even more dominant in partial differential equations. The shape of the boundary and the conditions imposed along the boundary impose severe restrictions upon our ability to solve partial differential equations analytically.

1.2 ONE-DIMENSIONAL HEAT CONDUCTION

As a first example of a partial differential equation associated with a simple physical system, we consider heat conduction along a beam of constant cross-sectional area A. We assume that the temperature is uniform over each cross section but may vary along the beam. We also assume that heat does not radiate out from the sides of the beam, that is, we assume that the sides are perfectly insulated.

In the most elementary situation, the heat H (in some unit such as the calorie) contained in a material of mass M that is at a uniform temperature u is expressed as $H = cMu$, where c is a constant (the specific heat). If the linear density of the beam, $\rho(x)$, varies along the beam, then the mass between $x = a$ and $x = b$ may be written $M = A\int_a^b \rho(x)\,dx$. Similarly, if the temperature of the beam varies both with position and time, then the heat on the beam between $x = a$ and $x = b$ is

$$H_{ab}(t) = A \int_a^b c\rho(x)u(x, t)\,dx \qquad (1.2.1)$$

If we ignore any dependence of ρ (or c) upon either x or u and assume that they are constant, the change in $H_{ab}(t)$ with time is given by

$$\frac{dH_{ab}}{dt} = \rho cA \int_a^b u_t(x, t)\,dx \qquad (1.2.2)$$

The change in H_{ab} can take place because of a flow of heat along the beam into or out of the region $a \le x \le b$ as well as because of source terms (such as a radioactive source embedded in the beam) that introduce additional heat into the region between a and b.

Heat flow along the beam, $J(x, t)$, measured, say, in calories per square centimeter per second, is most easily analyzed by considering flow in the positive x direction. Then heat enters the region at $x = a$ and leaves at $x = b$. For a beam of area A we have

$$\frac{dH_{ab}}{dt} = A(J_{in} - J_{out}) + \text{sources}$$

$$= A[J(a, t) - J(b, t)] + A \int_a^b s(x, t)\, dx$$

$$= A \int_a^b \frac{\partial J(x, t)}{\partial x}\, dx + A \int_a^b s(x, t)\, dx \qquad (1.2.3)$$

where $s(x, t)$ is a heat source density in calories per second per cubic centimeter. The replacement of the difference $J(a, t) - J(b, t)$ by an integral over $J_x(x, t)$ is the one-dimensional version of the divergence theorem that would be used if we were deriving the equation for heat diffusion in three dimensions.

Equating the definition of dH_{ab}/dt from (1.2.3) with the expression for it [Eq. (1.2.2)] we have

$$\int_a^b (\rho c u_t + J_x - s)\, dx = 0 \qquad (1.2.4)$$

Since the interval $a \leq x \leq b$ is arbitrary, the vanishing of the integral implies the vanishing of the integrand itself,[1] and we obtain

$$\rho c u_t + J_x = s(x, t) \qquad (1.2.5)$$

This is the equation for the conservation of heat on the beam. To obtain a partial differential equation for the temperature $u(x, t)$, we must introduce an assumption concerning the dependence of the heat flow $J(x, t)$ upon the temperature. Since heat flows from hotter to cooler regions, we can expect J to depend upon the gradient of the temperature, $u_x(x, t)$, in some way. The simplest relation (due to Fourier and substantiated by considerable experimental evidence) is the linear one

$$J(x, t) = -K u_x(x, t) \qquad (1.2.6)$$

The proportionality parameter K, known as the conductivity, may depend upon both x and u. Combining this assumption for $J(x, t)$ with the conservation

---

[1] To conclude otherwise would lead to a contradiction. Since $u(x, t)$ and the entire integrand in (1.2.4) is presumably a continuous function of x, a nonzero, say positive, value of the integrand at some point x_0 would imply an interval about x_0 in which the integrand is also positive. By placing the limits of the integrand a and b in this interval, we would have a nonzero value for the integrand in (1.2.4), thus providing a contradiction.

equation (1.2.5) we have

$$\rho c u_t - (K u_x)_x = s(x, t) \tag{1.2.7}$$

This result is the partial differential equation for one-dimensional heat conduction. If there are no sources, so that $s(x, t), = 0$ and K is a constant, we obtain

$$u_{xx} - \gamma^{-2} u_t = 0, \qquad \gamma^2 = K/\rho c \tag{1.2.8}$$

which is the equation quoted in (1.1.2).

1.2.1 Boundary and Initial Conditions

As with ordinary differential equations, the association of a partial differential equation with a specific experimental situation requires the imposition of certain additional conditions. For the heat conduction situation just described we will be required to give information concerning what happens at each end of the beam at all times, the boundary conditions, as well as to specify an initial condition, $u(x, 0)$, the temperature distribution along the entire beam at some time labeled $t = 0$. The most common boundary conditions are that either u or J, where $J = -K u_x$, be specified at each end. If an end is perfectly insulated, then $J = 0$ at that end and thus $u_x = 0$ there. If the heat flow at an end, say at $x = 0$, is proportional to the temperature at the end, then a linear combination of u and u_x is specified there, that is, $u(0, t) = \alpha J(0, t) = -\alpha K u_s(0, t)$ where α is a proportionality constant. The two cases mentioned previously, that is, u or $u_x = 0$, than arise in the limits $\alpha \to 0$ and $\alpha \to \infty$, respectively.

Note that a background or reference temperature may always be ignored in problems governed by the diffusion equation (1.2.8). If the ends of the beam are maintained at temperature u_0, for instance, then we may set $u = u_0 + \vartheta(x, t)$ and determine the function $\vartheta(x, t)$, which is zero at each end. Since only *derivatives* of u appear in the diffusion equation (1.2.8), the temperature increment ϑ satisfies the same diffusion equation as that satisfied by u. After obtaining ϑ we need only add the constant u_0 to obtain the temperature $u(x, t)$.

1.3 STEADY STATE SOLUTIONS

Whenever a new equation is encountered, it is usually instructive to examine that equation and its solutions in some limiting cases that are readily understood. A simple limit of this sort for the heat equation is the so-called steady state situation in which there is no time dependence, that is, $u_t = 0$. When we consider the case of constant conductivity K, the heat equation (1.2.7) reduces, in the steady state situation, to the form

$$-K u_{xx} = s(x) \tag{1.3.1}$$

which is merely an ordinary differential equation. Since a source might be expected to heat the beam and thus render any steady state situation impossible, some restriction must obviously be satisfied in order to have a steady state temperature distribution on the beam. Requirements that apply to an entire system may frequently be obtained by integrating the governing equations over the extent of the system. If we consider the steady state heat equation (1.3.1) as it applies to a beam of length L and integrate the equation over the length of the beam, we obtain

$$-K \int_0^L u_{xx} \, dx = \int_0^L s(x) \, dx = H$$

$$-K[u_x(L) - u_x(0)] = H \qquad (1.3.2)$$

where H (in calories per second) is the rate at which heat is being deposited on the entire beam. According to the definition of heat flow J given in (1.2.6), the result obtained in (1.3.2) may be written

$$J(L) - J(0) = H \qquad (1.3.3)$$

We thus obtain the physically reasonable result that for a steady state situation to prevail, the rate of the heat flow out the ends of the beam must equal the rate at which heat is supplied to the entire system by the source $s(x)$. Note that since heat is assumed to flow off of the beam, $J(0)$ represents a heat flow in the *negative* x direction and thus $-J(0)$ is a positive quantity for heat flow out of the beam at $x = 0$. If there is no source term $s(x)$, then $H = 0$ in (1.3.3) and the steady state situation requires that the boundary condition satisfy $J(0) = J(L)$. We thus find that for steady state problems the boundary values of the derivative of u *may not be imposed arbitrarily*. If we consider a steady state problem in which the temperature itself is specified at each end rather than the heat flow, then the condition on J given in (1.3.3) will be automatically satisfied by the steady state solution that is obtained.

Example 1.1. The end points of a beam of constant conductivity K are located at $x = 0$ and L and are maintained at temperatures 0 and u_0, respectively. An amount of heat Q calories per second is continually supplied over the length of the beam with a density $s(x) = (\pi Q/2L) \sin \pi x/L$. [Note that $\int_0^L s(x) \, dx = Q$.] Determine the steady state temperature of the beam.

From (1.3.1) we require the solution

$$u_{xx} = -\alpha \sin \pi x/L, \qquad \alpha = \pi Q/2KL \qquad (1.3.4)$$

subject to the boundary conditions $u(0) = 0$, $u(L) = u_0$. This ordinary differential equation is sufficiently simple that it may be solved by direct integration to yield

$$u = \alpha(L/\pi)^2 \sin \pi x/L + ax + b \qquad (1.3.5)$$

where a and b are constants of integration. (We may also think of the first term as a particular solution and the last two terms as the solution of the homogeneous equation $u_{xx} = 0$.) Applying the boundary conditions we find $u(0) = b = 0$ and then $u(L) = aL = u_0$. Recalling the definition of α from (1.3.4) we obtain the solution

$$u(x) = (LQ/2\pi K)\sin \pi x/L + u_0 x/L \qquad (1.3.6)$$

Determination of the derivative of this result and evaluation at $x = 0, L$ show that the steady state condition (1.3.3) is satisfied.

As a final result, note also that if there is no external source ($Q = 0$), the steady state temperature due to the boundary conditions for this problem is the linear relation $u(x) = u_0 x/L$.

Example 1.2. The boundary conditions on both ends of the beam considered in Example 1.1 are changed. At $x = L$ the insulated end condition $u_x(L) = 0$ is imposed. If the source term remains the same, that is, $s(x) = (\pi Q/2L)\sin \pi x/L$, *determine* the value that must be imposed upon $u_x(0)$ so that a steady state condition is maintained. Then show that with these values of the u_x imposed at the ends, the temperature of the beam is only determined to within a constant.

For the steady state condition (1.3.3) to be satisfied with $u_x(L) = 0$ and $H = \int_0^L s(x)\,dx = Q$, we require $u_x(0) = -Q/K$.

To determine $u(x)$ we integrate (1.3.4) once and obtain $u_x = (Q/2K)\cos \pi x/L + c_1$. The integration constant c_1 is found to equal $Q/2K$ when we impose either of the boundary conditions $u_x(0) = Q/K$ or $u_x(L) = 0$. A second integration yields

$$u(x) = (LQ/2K)\sin \pi x/L + Qx/2K + c_2 \qquad (1.3.7)$$

The constant of integration c_2 is thus left undetermined.

Problems

1.3.1 (a) Determine the steady state temperature on a beam if $u(0) = u(L) = 0$, and the source is

$$s(x) = \begin{cases} s_0, & 0 \le x < L/2 \\ 0, & L/2 < x \le L \end{cases}$$

   (b) Determine the ratio of the heat flow off the beam at $x = 0$ to that at $x = L$, that is, $|J(0)|/J(L)$.

   (c) Where is the temperature a maximum?

   (d) Sketch the temperature profile on the beam.

1.3.2 Reconsider the previous problem when the boundary at $x = L$ is insulated, that is, $u_x(0) = 0$.

   (a) Show that all heat flows off the end at $x = 0$, that is, $J(0) = -Ku_x(0) = s_0 L/2$.

   (b) Sketch $u(x)$.

1.3.3 A beam of length L is located between $x = a$ and $x = b = a + L$. The conductivity of the beam varies linearly along the beam and is given by $K(x) = K_0 x/L$. The sides of the beam are insulated and the ends are maintained at zero temperature. The beam is subjected to a steady, spatially uniform heat source Q_0 calories per second over its entire length. The equation for the temperature on the beam is thus

$$\frac{d}{dx}\left[K(x)\frac{du}{dx} \right] = -\frac{Q_0}{L}$$

(a) Use the relation $J = -K(x)\,du/dx$ to determine the amount of heat flowing out each end of the beam and show that the total heat flow out both ends is Q_0.

(b) Show that if $b = 2a$ so that $K(b)/K(a) = 2$,

$$\frac{J(b)}{|J(a)|} = \frac{2\ln 2 - 1}{\ln 2 - 1} \cong 1.26$$

1.3.4 The ends of a beam of length $2L$ are maintained at zero temperature at $x = \pm L$. A source deposits Q_0 calories per second over the length of the beam with a heat density

$$s(x) = \frac{Q_0/\pi}{\sqrt{L^2 - x^2}}$$

Show that the steady state temperature on the beam is

$$u(x) = \frac{Q_0 L}{\pi K}\left[\frac{\pi}{2} - \frac{x}{L}\sin^{-1}\frac{x}{L} - \sqrt{1 - \left(\frac{x}{L}\right)^2} \right]$$

(This problem is a limiting case of a time dependent problem considered in Section 9.1.)

1.3.5 A beam of length L is maintained at zero temperature at $x = 0$. At $x = L$ there is a constant heat flux J_0 onto the beam. At each point along the beam there is a heat loss $\nu u(x)$ where ν is a constant and $u(x)$ is the temperature on the beam at that point. This loss term may be interpreted as a negative heat source $s(x) = -\nu u(x)$. Hence the steady state temperature on the beam is governed by

$$u_{xx} = (\nu/K)u$$

Show that the heat flux off of the beam at $x = 0$ is equal to J_0 sech $(L\sqrt{\nu/K})$.

1.3.6 A beam of length L is kept at zero temperature at the ends located at $x = \pm L$. The beam is heated by the steady source $Q(x) = (H_0/2a)$ $\text{sech}^2 (x/a)$.

(a) Show that the steady state temperature on the beam is

$$u(x) = \frac{H_0 a}{2K} \ln \frac{\cosh L/a}{\cosh x/a}$$

(b) Show that in the limit $a/L \ll 1$, in which the source may be described as a "point source," the temperature becomes .

$$u(x) = \begin{cases} \dfrac{H_0}{2K} (L - x), & 0 \le x \le L \\[2ex] \dfrac{H_0}{2K} (L + x), & -L \le x \le 0 \end{cases}$$

(c) Sketch the temperature $u(x)$ obtained in (b) as well as the function $Q(x)$ in the limit $a/L \ll 1$.

1.4 TIME DEPENDENT HEAT FLOW—SEPARATION OF VARIABLES

The one-dimensional steady state examples of the previous section involved a single independent variable and thus required only the solution of an ordinary differential equation. When the temperature on the beam depends upon both space and time, a partial differential equation of the type given in (1.2.7) or (1.2.8) must solved. Such equations may frequently be solved by relating them to ordinary differential equations in each of the independent variables. One technique for carrying out this simplification is known as the method of separation of variables. In adopting this approach to solving a partial differential equation we put aside, temporarily, the goal of obtaining a complete solution that satisfies both the initial and boundary conditions of a specific problem. We merely attempt to obtain some sort of solution in terms of functions $X(x)$ and $T(t)$, each of which depends upon only one of the two independent variables x and t. How $u(x, t)$ is to be decomposed into two such functions is a combination of guesswork and experience. For the *linear* equations that we shall encounter in this text, the decomposition is usually made in terms of the *product* expression $u(x, t) = X(x) T(t)$. For *nonlinear* equations the combination may be quite different (cf. Problem 1.4.10).

An advantage of the separation-of-variables method is that partial derivatives of $u(x, t)$ are replaced by ordinary derivatives of $X(x)$ and $T(t)$, for example, the partial derivatives that appear in the diffusion equation (1.2.8) become $u_t(x, t) = X \, dT/dt$ and $u_{xx}(x, t) = (d^2 X/dx^2) T$. We have no a priori expectation that such solutions can be combined so as to satisfy the boundary and initial conditions of any given problem. How Fourier series techniques are to be used to carry out the superposition of these elementary solutions so that initial and boundary conditions can be satisfied for *linear* partial differential equations is the main topic of this chapter.

1.4.1 Elementary Solutions

Using the forms of u_{xx} and u_t that result from the product representation for $u(x, t)$, the diffusion equation

$$u_{xx} - \gamma^{-2} u_t = 0 \qquad (1.4.1)$$

can be written as

$$X'' T = \gamma^{-2} X T' \qquad (1.4.2)$$

where, as noted above the primes indicate a derivative of a function with respect to the associated independent variable. On dividing this equation by XT we obtain

$$\frac{X''}{X} = \frac{1}{\gamma^2} \frac{T'}{T} \qquad (1.4.3)$$

We have now derived an equation that separates the spatial and temporal aspects of the problem. Since x and t are independent variables, we may assign values to each of them at will. For equality of both sides of the equation to persist for any choice of x and t, each side must be a constant,[2] the so-called separation constant.

Although there are some instances in which useful solutions are obtained by allowing the separation constant α^2 to be imaginary (cf. problem 1.4.9), we shall assume in the following that it is real and write it in the form $\pm \alpha^2$. The proper choice of sign will be determined subsequently. Equating each side of (1.4.3) to this constant, we obtain the pair of relatively simple ordinary differ-

---

[2]An alternate way to arrive at this result is to differentiate (1.4.3) with respect to one of the variables, say x, and obtain

$$\frac{\partial}{\partial x} \frac{X''}{X} = \frac{1}{\gamma^2} \frac{\partial}{\partial x} \frac{T'}{T} = 0$$

since X is a function of x alone, $\partial/\partial x(X''/X) = d/dx\,(X''/X)$ and integration with respect to x yields $X''/X = $ const. Similar usage of a time derivative leads to $T'/T = $ const, and by (1.4.3) these two integration constants must be equal.

ential equations

$$X'' \mp \alpha^2 X = 0 \tag{1.4.4}$$
$$T' \mp (\alpha\gamma)^2 T = 0$$

For the two possible choices of sign as well as $\alpha^2 = 0$, the solutions of these equations are

$$-\alpha^2: \quad X = Ae^{\alpha x} + Be^{-\alpha x}, \qquad T = Ce^{(\alpha\gamma)^2 t}$$
$$+\alpha^2: \quad X = A\cos\alpha x + B\sin\alpha x, \qquad T = Ce^{-(\alpha\gamma)^2 t} \tag{1.4.5}$$
$$\alpha^2 = 0: \quad X = Ax + B, \qquad T = C$$

where A, B, and C are integration constants.[3]

For each of these three cases, the function $u = XT$ is now a very specialized solution of the diffusion equation (1.4.1). We note in passing that the possibility of dividing by zero in the separated form (1.4.3) is taken care of by the fact that X'' vanishes whenever X does. Thus, we merely obtain an indeterminate form for the ratio X''/X.

The choice of sign associated with α^2 as well as specific values of α are determined when we impose the further restriction that the elementary solution $u(x, t) = X(x)T(t)$ satisfy the boundary conditions for a given problem. In the product form for $u(x, t)$, this boundary information is incorporated into the function $X(x)$. If, for instance, we are considering diffusion of heat on a beam of length L that is maintained at zero temperature at the end points $x = 0$ and L, then we impose the requirements $X(0) = X(L) = 0$.

Let us turn immediately to the second of the three solutions listed in (1.4.5). This choice will be seen to be the proper one. The reason why the other two must be discarded will be given below. For the second solution we have $X(0) = A = 0$ and $X(L) = A\cos\alpha L + B\sin\alpha L = 0$. From the condition at $x = 0$ we have $A = 0$. The condition at $x = L$ then reduces to $X = B\sin\alpha L = 0$ and we thus see that either $B = 0$ or $\sin\alpha L = 0$. With the first choice we would have a zero value for both A and B and thus $X = 0$. The whole solution $u = XT$ would then be zero, an uninteresting result. The second choice, namely $\sin\alpha L = 0$, does allow for nonvanishing solutions since it is satisfied by $\alpha L = n\pi$, where $n = 0, \pm1, \pm2, \ldots$. For each value of n we then obtain a different possible expression for $X(x)$ as well as a corresponding expression for $T(t)$, namely, $T(t) = C\exp[-n\pi\gamma/L)^2 t]$. The solution for $u(x, t)$ is then

$$u_n(x, t) = X_n T_n = b_n e^{-(n\pi\gamma/L)^2 t} \sin(n\pi x/L) \tag{1.4.6}$$

---

[3] Since $-\alpha^2 = (i\alpha)^2$, the first two forms of the solution may be transformed into each other by using $e^{i\alpha x} = \cos\alpha x + i\sin\alpha x$ and then redefining the constants of integration. Also, if we set $B = B'/\alpha$ in the second solution and then let $\alpha \to 0$, we obtain $X = A + B'x$, $T = C$, which has the form of the third solution.

where we have labeled the solutions by the value of n and have combined the integration constants BC into a single constant b_n that can be different from each value of n. The value $n = 0$ may be discarded since it merely yields the solution $u = 0$. Negative values of n may also be discarded. They provide no solution that could not be obtained by merely changing the sign of b_n. The specific values of α that enable us to obtain the nonzero solutions X_n (i.e., $\alpha = \alpha_n = n\pi/L$ in the present instance) are frequently referred to as *eigenvalues*. The solutions X_n themselves are referred to as *eigenfunctions*.

We now briefly consider the other two possible solutions that were listed in (1.4.5). When the first solution is required to vanish at $x = 0$, we obtain $A + B = 0$. To vanish at $x = L$ the solution must also satisfy $Ae^{-\alpha L} + Be^{-\alpha L} = 0$. This pair of homogeneous algebraic equations for A and B has no solution for real α other than $A = B = 0$. As noted in footnote 3, solutions may be obtained for purely imaginary values of α but they are merely the same as those obtained from the second solution in (1.4.5). Similarly, for the third solution in (1.4.5) we obtain $X(0) = B = 0$ and $X(L) = AL + B = 0$. Here again the two constants A and B must vanish. Hence, only the second solution in (1.4.5) is useful in solving this problem.

It is worth examining the solution in (1.4.5) a bit further and noticing that it conforms to our intuition. Since heat conduction is a diffusion process, we expect that any temperature variations will tend to smooth out as time progresses. The more fine grained the variations in the temperature, the more rapidly the variation should smooth out. In the solutions to the diffusion equation listed in (1.4.6), the larger the value of n, the more rapidly the spatial term oscillates. As expected, the associated time factor $\exp[-(n\pi\gamma/L)^2 t]$ decays more rapidly in time the larger the value of n.

1.4.2   Synthesis of Elementary Solutions—Fourier Series

The solutions given in (1.4.6) describe very specialized situations. They show that an initial temperature distribution of the form $u_n(x, 0) = b_n \sin n\pi x/L$ decays in time as $u_n(x, t) = b_n \exp[-(n\pi\gamma/L)^2 t] \sin n\pi x/L$. To approach the more interesting problem of describing the temporal variation of essentially *arbitrary* types of initial temperature distributions, we exploit the fact that the partial differential equation (1.4.1) is *linear*. As with linear ordinary differential equations, a sum of solutions to a given equation is still a solution to that equation. In the case of a linear partial differential equation we can have an infinite number of solutions such as those given in (1.4.6) for $n = 1, 2, 3, \ldots$. We can thus write a solution to the original partial differential equation as a sum of all of these solutions. This procedure yields

$$u(x, t) = \sum_{n=1}^{\infty} u_n(x, t) = \sum_{n=1}^{\infty} b_n e^{-n^2\nu t} \sin \frac{n\pi x}{L} \qquad (1.4.7)$$

where

$$\nu = \left(\frac{\gamma\pi}{L}\right)^2 \tag{1.4.8}$$

In so doing, we have written the solution in the form of a Fourier series. Each term in the series has the time-varying coefficient $b_n \exp(-n^2 \nu t)$. At $t = 0$ the coefficients are just b_n and, as will now be shown, can be determined by using the techniques available from the theory of Fourier series and summarized in Appendix A.

When the solution (1.4.7) is considered at $t = 0$, we have

$$u(x, t) = \sum_{n=1}^{\infty} b_n \sin \frac{n\pi x}{L} \tag{1.4.9}$$

and from the results given in Eq. (A.23) we have

$$b_n = \frac{2}{L} \int_0^L u(x, 0) \sin \frac{n\pi x}{L} \, dx \tag{1.4.10}$$

We now consider two examples of the use of Fourier series methods for solving an initial-value problem in heat conduction.

Example 1.3. Determine the temperature distribution at any time on a beam of length L if the initial temperature distribution is $u(x, 0) = u_0 \sin^3 \pi x/L$ and the ends are kept at zero temperature. Also show that after a long time (i.e., as $t \to \infty$) half of the initial heat on the beam will have flowed off at each end of the beam.

The coefficients b_n are obtained by evaluating the integral in (1.4.10) for the given form of $u(x, 0)$. The calculation is particularly simple in the present case when use is made of the trigonometric identity $4 \sin^3 \vartheta = 3 \sin \vartheta - \sin 3\vartheta$. The two integrations required in evaluating (1.4.10) are in the form of orthogonality integrals given in Eq. (A.2) with $d = -L$ and $p = L$. Since the integrand is even in x we have

$$b_1 = \frac{2}{L} \int_0^L u_0 \sin^3 \frac{\pi x}{L} \sin \frac{\pi x}{L} \, dx$$

$$= \frac{2u_0}{L} \int_0^L \left(\frac{3}{4} \sin \frac{\pi x}{L} - \frac{1}{4} \sin \frac{3\pi x}{L}\right) \sin \frac{\pi x}{L} \, dx$$

$$= \frac{2u_0}{L} \frac{3}{4} \frac{L}{2} = \frac{3}{4} u_0 \tag{1.4.11}$$

and similarly

$$b_3 = \frac{2}{L} \int_0^L u_0 \sin^3 \frac{\pi x}{L} \sin \frac{3\pi x}{L} \, dx$$

$$= -\frac{1}{4} u_0 \tag{1.4.12}$$

while all other b_n vanish due to orthogonality. In this example, then, the solution (1.4.7) is not an infinite series at all but merely

$$u(x, t) = u_0 \left(\frac{3}{4} e^{-\nu t} \sin \frac{\pi x}{L} - \frac{1}{4} e^{-3\nu t} \sin \frac{3\pi x}{L} \right) \qquad (1.4.13)$$

This solution is readily shown to predict results that are in conformity with our intuition. As an example, note that the symmetry of the initial temperature distribution about the midpoint of the beam suggests that half of the initial heat on the beam will flow off each end of the beam. From the definition given in (1.2.1), the total amount of heat initially on the beam is[4]

$$H(0) = \rho c u_0 A \int_0^L \sin^3 \frac{\pi x}{L} \, dx = \frac{4}{3\pi} \rho c u_0 LA \qquad (1.4.14)$$

The total amount of heat flowing off one end, say at $x = L$, is obtained by integrating the expression for $J(L, t)$ given by (1.2.6) over all time. Using the solution just obtained in (1.4.13) to calculate $u_x(L, t)$, we have

$$\int_0^\infty J(L, t) \, dt = KA \int_0^\infty u_x(L, t) \, dt$$

$$= \frac{3\pi K A u_2}{4L} \int_0^\infty (e^{-\nu t} - e^{-3\nu t}) \, dt \qquad (1.4.15)$$

$$= \frac{2}{3\pi} \rho c u_0 LA$$

which is just one-half of the total heat initially on the beam.

An initial temperature profile that leads to a Fourier series containing an infinite number of terms is provided by the problem $u(0, t) = u(L, t) = 0$, $u(x, 0) = u_0$ which can be used to describe a beam initially at a uniform temperature u_0 that has its ends suddenly set to zero temperature. From (1.4.10) the coefficients b_n are given by

$$b_n = \frac{2u_0}{L} \int_0^L \sin \frac{n\pi x}{L} \, dx = \frac{2u_0}{n\pi} (1 - \cos n\pi)$$

$$= \begin{cases} \dfrac{4u_0}{\pi}, & n = 1, 3, 5, \dots \\ 0, & n = 2, 4, 6, \dots \end{cases} \qquad (1.4.16)$$

The vanishing of the terms with even values of n is evident on the basis of simple symmetry considerations. Graphs of $u(x, 0)$ and the functions $\sin n\pi x/L$

[4]Set $\zeta = \pi x/L$ and use the result $\int_0^\pi \sin^3 \zeta \, d\zeta = \frac{4}{3}$.

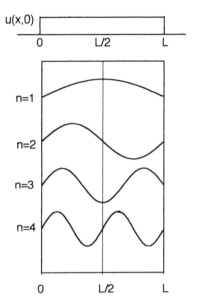

Figure 1.1. Initial temperature profile and first few mode shapes.

for the first few values of n are shown in Figure 1.1. The even values of n are seen to correspond to functions that are odd with respect to the midpoint of the beam at $x = L/2$. Since the initial temperature distribution is even with respect to the midpoint, the product $u(x, 0)\sin n\pi x/L$ is odd with respect to the midpoint $x = L/2$ for $n = 2, 4, 6, \ldots$. The integrals in (1.4.16) over the range $0 \le x \le L$ thus vanish as a result of this antisymmetry.

The first few terms in the series for $u(x, t)$ are thus

$$u(x, t) = \frac{4u_0}{\pi}\left(e^{-\nu t}\sin\frac{\pi x}{L} + \frac{1}{3}e^{-3\nu t}\sin\frac{3\pi x}{L} + \frac{1}{5}e^{-5\nu t}\sin\frac{5\pi x}{L} + \ldots\right)$$

$$(1.4.17)$$

where again $\nu = (\pi\gamma/L)^2$. The exponential time dependence of the coefficients in this series provide rapid convergence for $\nu t > 1$. For $\nu t \ll 1$, however, the series converges only as n^{-1}, which is quite slow.[5] In Section 2.2 we will develop a completely different representation of the solution that does converge rapidly at these early times.

According to the form of the solution given in (1.4.17), the temperature at

---

[5]At $x = L/2$ and as t approaches zero, the solution in (1.4.17) approaches $u(L/2, 0) = 4u_0/\pi(1 - \frac{1}{3} + \frac{1}{5} - \cdots) = 4u_0/\pi$, $\pi/4 = u_0$. The series representation for π that is encountered here is known to be very slowly convergent. In identifying the value of the series evaluated at $t = 0$ with the initial condition of the problem, it should be noted that we are assuming the validity of the interchange $\lim \Sigma u_n(x, t) = \Sigma \lim u_n(x, t)$. In the present instance this interchange may be justified by using Abel's test (cf. Section A.3).

any given point along the beam is represented by a sum of terms each of which has its own characteristic rate of decay. Note in particular that at the locations $x = L/3$ and $x = 2L/3$ along the beam the second term in the series vanishes at *all* times. These two points on the beam would thus be the most appropriate locations at which to obtain experimental information. Since the second term in the series is absent at either of these locations, the decrease in temperature would be given in terms of the single exponential term $4(u_0/\pi)e^{-\nu t}\sin \pi x/L$ after the brief initial time beyond which the term proportional to $e^{-25\nu t}$ could be neglected.

1.4.3 Changing the Boundary Conditions—Insulated Ends

If an end of the beam is covered with some insulating material (asbestos was customarily allowed at one time), then no heat flows out of the beam at that end and $J = -Ku_x = 0$ there. We thus have the boundary condition $u_x = 0$ at a perfectly insulated end. To incorporate this information into the solution of a heat diffusion problem, we must return to the solution of the ordinary differential equation satisfied by $X(x)$ in the second of (1.4.5), namely, $X(x) = A \cos \alpha x + B \sin \alpha x$, and apply the new boundary condition. In particular, if both ends are insulated, we set $X'(0) = X'(L) = 0$. For the given form of $X(x)$ we obtain

$$X'(0) = \alpha B = 0$$

$$X'(L) = \alpha(-A \sin \alpha L + B \cos \alpha L) \qquad (1.4.18)$$

which yields $B = 0$ and $\alpha = n\pi/L$ where $n = 0, 1, 2, \ldots$. The solution is thus $X(x) = A \cos \pi x/L$. Note that the solution $n = 0$ must be retained in the present instance since the cosine does not vanish for this value of n. The solution obtained for $n = 0$ is also provided by the third of the three solutions listed in (1.4.5). The elementary solutions for a beam with two insulated ends are thus

$$u_n(x, t) = a_n e^{-n^2\nu t} \cos \frac{n\pi x}{L}, \qquad n = 0, 1, 2, \ldots \qquad (1.4.19)$$

with $\nu = (\pi\gamma/L)^2$. The term $n = 0$, since it is time independent, has an important physical significance. The insulated ends on the beam prevent heat from escaping from the system. As $t \to \infty$, any initial temperature distribution will thus have smoothed out to a constant value over the entire beam. The value of that constant temperature is determined by the conservation of heat on the beam. This result is evident physically and also from an examination of the coefficients in the Fourier series. From Eq. (A.24), these coefficients are given by

$$a_0 = \frac{1}{L} \int_0^L u(x, 0) \, dx = \langle u(x, 0) \rangle_{\text{ave}}$$

$$a_n = \frac{2}{L} \int_0^L u(x, 0) \cos \frac{n\pi x}{L} \, dx, \qquad n = 1, 2, 3 \ldots \qquad (1.4.20)$$

The coefficient a_0 is seen to have the interpretation of being the average temperature on the beam.

Problems

1.4.1 Determine the eigenvalues and eigenfunctions for the boundary value problem $y'' + k^2 y = 0$, $y'(0) = y(L) = 0$. Sketch the first three eigenfunctions.

1.4.2 A beam of length L has the initial temperature distribution

$$u(x, 0) = \begin{cases} u_0, & 0 \le x < L/2 \\ 0, & L/2 < x \le L \end{cases}$$

If the ends are placed at zero temperature, show that the temperature at any later time is given by

$$u(x, t) = \frac{2u_0}{\pi} \sum_1^\infty \frac{1}{n} \left(1 - \cos \frac{n\pi}{2}\right) e^{-n^2 \nu t} \sin \frac{n\pi x}{L}$$

Show that three-fourths of the heat on the beam flows out through the end at $x = 0$.

1.4.3 The ends of the beam at $x = 0$ and L are kept at zero temperature. The initial temperature distribution is $u(x, 0) = u_0 \sin^2 \pi x/L$. Determine the temperature on the beam at any later time.

1.4.4 The ends of the beam at $x = 0$ and L are insulated. The initial temperature distribution is $u(x, 0) = u_0 \sin^2 \pi x/L$. Determine the temperature on the beam at any later time.

1.4.5 Determine elementary solutions of the equation $u_{xx} - \gamma^{-2} u_t + \alpha u = 0$ that satisfy the boundary conditions $u(0, t) = u(L, t) = 0$. Discuss the two cases $\alpha \gtrless (\pi/L)^2$ and comment on the physical significance of the results.

1.4.6 The initial temperature on a beam ($0 \le x \le L$) varies linearly along the beam so that $u(x, 0) = u_0(x/L)$. The ends are then set to zero temperature. Determine the temperature on the beam at any later time and determine the fraction of the initial heat on the beam that flows out through each end.

1.4.7 The initial temperature on a beam of length L is

$$u(x, 0) = \begin{cases} u_0(1 - 2x/L), & 0 \le x < L/2 \\ u_0(2x/L - 1), & L/2 < x \le L \end{cases}$$

The ends of the beam are placed at zero temperature.

(a) Determine the temperature on the beam at any later time.

(b) The temperature at the midpoint, initially equal to zero, rises to a maximum value and then returns to zero at large times. Use the first two terms in the series expansion obtained in part (a) to estimate the maximum temperature and the time at which it occurs. Can you justify using only the first two terms in the series?

1.4.8 A beam of length L has insulated ends and an initial temperature distribution $u(x, 0) = u_0 \sin \pi x/L$.

(a) Determine the temperature on the beam at any later time.

(b) Show that $u(x, \infty)$ is the average of the initial temperature on the beam.

(c) Use the series to show that $u(L/2, 0) = u_0$ (cf. Table F.1).

1.4.9 Assume that the separation constant in (1.4.4) is purely imaginary, that is, $\alpha^2 = i\mu^2$.

(a) Show that the product solution $u(x, t) = X(x)T(t)$ that vanishes as $x \to +\infty$ is

$$u(x, t) = ce^{-(x/y)\sqrt{\Omega/2} - i(\sqrt{\Omega/2}\, x/\gamma - \Omega t)}$$

where $\Omega = (\mu\gamma)^2$.

(b) Show that either the real or the imaginary part of this solution is also a solution of the diffusion equation.

(c) Show that the real part describes a damped temperature wave in a semi-infinite region $x \ge 0$ having the boundary condition $u(0, t) = u_0 \cos \Omega t$. Show also that similar considerations apply to the imaginary part.

(d) Show that if the heat flow at $x = 0$ is given by $J(0, t) = Ku_x(0, t) = J_0 \cos \Omega t$, then

$$u(x, t) = \frac{J_0\gamma}{K\sqrt{\Omega}} e^{-(x/\gamma)\sqrt{\Omega/2}} \cos\left(\Omega t - \frac{\sqrt{\Omega/2}\, x}{\gamma} - \frac{\pi}{4}\right)$$

1.4.10 A more elaborate example of the separation-of-variables procedure is provided by the *nonlinear* equation

$$u_{xx} - u_{yy} = \sin u$$

(a) Introduce a solution in the form

$$u = 4 \tan^{-1} \frac{X(x)}{Y(y)}$$

and use the identity $\sin 4a = 4 \tan a (1 - \tan^2 a)/(1 + \tan^2 a)^2$ to show that $X(x)$ and $Y(y)$ are related according to

$$XX'' + YY'' + \frac{X^2 Y''}{Y} + \frac{Y^2 X''}{X} = 2(X'^2 + Y'^2) + Y^2 - X^2$$

(i)

(b) Differentiate this result with respect to both x and y to obtain the *separated* equations

$$\frac{(Y''/Y)'}{(Y^2)'} = -\frac{(X''/X)'}{(X^2)'} = m$$

(ii)

(c) Integrate these equations to obtain

$$(X')^2 = -\tfrac{1}{2}mX^4 + \alpha X^2 + \beta$$

$$(Y')^2 = \tfrac{1}{2}mY^4 + aY^2 + b$$

(iii)

where α, β, a, and b are constants of integration.

(d) Substitute (ii) and (iii) into (i) to obtain $b = -\beta$ and $a = \alpha - 1$.

(e) Show that the special case $m = 0$, $\alpha = 1$, $\beta = -A^2$ yields

$$u = 4 \tan^{-1} \frac{\cosh (x + c_1)}{y + c_2}$$

$$= 2\pi - 4 \tan^{-1} [(y + c_2) \operatorname{sech} (x + c_1)]$$

where c_1 and c_2 are constants of integration.

1.5 STEADY STATE HEATING BY A LOCALIZED SOURCE— DELTA FUNCTION

When analyzing a system, in either a steady state or a time dependent situation, it is frequently instructive to be able to describe how the system responds to a source that is *highly localized* in space, time, or both. In particular, if the source only extends over a region that is small compared to any length of interest in the problem, it is possible to introduce the notion of a *point* source. The idea will be employed here in the context of a heat source, but as will be evident later, the notion of a point source has application to other situations as well. It will be shown that the temperature distribution due to this extreme version of a localized source is particularly easy to determine.

A convenient way to describe a point source is as a limit of some sharply peaked function. Three frequently used examples are

$$
1. \qquad \Delta_1(x) = \lim_{d \to 0} \begin{cases} 0, & x < -d/2 \\ d^{-1}, & -d/2 < x < d/2 \\ 0, & x > d/2 \end{cases}
$$

$$
2. \qquad \Delta_2(x) = \lim_{d \to 0} \frac{1}{\sqrt{\pi} d} e^{-(x/d)^2}
$$

$$
3. \qquad \Delta_3(x) = \lim_{d \to 0} \frac{d}{\pi} \frac{1}{x^2 + d^2} \tag{1.5.1}
$$

In each case, the constants have been chosen so that $\int_{-\infty}^{\infty} \Delta_n(x)\, dx = 1$. If a heat source density is now written as $s(x) = Q_0 \Delta_n(x)$, then as long as the system, say of length L and located within $-L/2 < x \le L/2$, is much larger than the width of the source as characterized by d, we may write $\int_{-L/2}^{L/2} s(x)\, dx \approx \int_{-\infty}^{\infty} s(x)\, dx = Q_0$; that is, the total amount of heat supplied to the system is independent of d.[6] In the limit that $d \to 0$ in each case, the source becomes more localized and more intense. The strength of the source, given by the integral of $s(x)$ over all values of x for which $s(x)$ is nonzero, remains constant at the value Q_0. The source thus approaches a point source. It is customary to express the result of this limiting process in the form

$$
\lim_{d \to 0} \Delta_n(x) = \delta(x) \tag{1.5.2}
$$

Such a source is usually referred to as a delta function source.

If the source is located at $x = x_0$ rather than at the origin, then limiting

---

[6]Since $\Delta_1(x)$ is equal to zero for $|x| > d$, the integrals are exactly equal in this case as long as $L > d$.

forms of expressions such as $\Delta_2(x - x_0) = (1/d\sqrt{\pi})\exp[-(x - x_0)/d]^2$ would be used and the result would be a point source located at x_0 and written $\delta(x - x_0)$.

As a result of the above considerations, we arrive at the first important property of the delta function $\delta(x - x_0)$, namely,

$$\int_{-\infty}^{\infty} \delta(x - x_0)\, dx = 1 \tag{1.5.3}$$

Since the limiting form of the various functions used to construct $\delta(x - x_0)$ go to zero for all values of x not equal to x_0, we may even write

$$\int_{x_0 + \epsilon}^{x_0 + \epsilon} \delta(x - x_0)\, dx = 1 \tag{1.5.4}$$

where ϵ is a small positive quantity. In other words, it is only necessary to integrate across the peak of the delta function to obtain the value of unity.

Another important property of $\delta(x - x_0)$ follows from a consideration of the integral $\int_a^b f(x)\delta(x - x_0)\, dx$ where $a < x_0 < b$ and $f(x)$ is smoothly varying in the vicinity of x_0. Since $\delta(x - x_0)$ is zero for $x \neq x_0$, we expect that only values of $f(x)$ in the vicinity of x_0 will contribute to this integral. To analyze the situation in detail, we expand $f(x)$ about x_0 and again consider the limiting procedure. To simplify the expressions that occur, we set $x_0 = 0$ and consider the form $\Delta_2(x)$. We then have the expression

$$\int_a^b \Delta_2(x)f(x)\, dx = \int_a^b \frac{1}{\sqrt{\pi}d} e^{-(x/d)^2} [f(0) + f'(0)x + \frac{1}{2}f''(0)x^2 + \cdots]$$

$$\tag{1.5.5}$$

Integrating term by term we obtain

$$\int_a^b \Delta_2(x)f(x)\, dx = f(0) + f'(0)\frac{1}{\sqrt{\pi}d}\int_a^b e^{-(x/d)^2} x\, dx$$

$$+ \frac{1}{2}f''(0)\frac{1}{\sqrt{\pi}d}\int_a^b e^{-(x/d)^2} x^2\, dx + \cdots \tag{1.5.6}$$

Recalling that as $d \to 0$ we may replace $\int_a^b$ by $\int_{-\infty}^{\infty}$, we see that the integrand in the second term is an odd function about 0 and thus the integral of this term vanishes. (Note that the form Δ_3 cannot be used for this purpose since the subsequent integrals in the series would not converge.) The integral in the third term is available from tables of integrals,[7] and we have

$$\int_a^b \Delta_2(x) f(x) \, dx = f(0) + \tfrac{1}{4} f''(0) d^2 + \cdots \qquad (1.5.7)$$

In the limit $d \rightarrow 0$ the second and higher terms in this result, which are all proportional to higher powers of d, will vanish and we finally obtain $\int_a^b f(x) \delta(x) \, dx = f(0)$. For the more general situation in which the delta function is peaked about x_0 we obtain

$$\int_a^b f(x) \delta(x - x_0) \, dx = f(x_0) \qquad (1.5.8)$$

that is, the sharply peaked expression $\delta(x - x_0)$ picks out the value of $f(x)$ at $x = x_0$.

There are a number of identities involving delta functions. The following list contains the more commonly occurring ones:

$$\delta(-x) = \delta(x)$$
$$x\delta(x) = 0$$
$$x\delta'(x) = -\delta(x)$$
$$\delta(ax) = \delta(x)/|a|$$
$$\delta(x^2 - a^2) = [\delta(x - a) + \delta(x + a)]/2\,|a|$$
$$\delta[f(x)] = \sum_{i=1}^n \frac{\delta(x - x_i)}{|f'(x_i)|}, \qquad f(x_i) = 0, \qquad f'(x_i) \neq 0 \quad (1.5.9)$$

It should be emphasized that whenever a calculation is done with a delta function, the calculation is always performed under an integral. For instance, the validity of each of the relations in (1.5.9) is shown by multiplying it by an arbitrary function $f(x)$ and integrating each side across the singularity in the delta function. The second entry follows by noting that the result of such a procedure yields $xf(x)$ evaluated at $x = 0$ which is zero. The third entry requires a somewhat more elaborate calculation. Again multiplying by a continuous function $f(x)$ and integrating as before, we perform an integration by parts to obtain

$$\int_a^b dx\, f(x) x \delta'(x) = xf(x) \delta(x) \big|_a^b - \int_a^b dx\, \frac{d}{dx} [xf(x)] \delta(x), \qquad a < 0 < b$$

$$(1.5.10)$$

[7]$\int_0^\infty x^2 \exp(-x^2/d^2)\, dx = \sqrt{\pi}\, d^3/4.$

The integrated term vanishes since $\delta(x) = 0$ at the end points and the differentiation under the integral sign yields

$$\int_a^b dx\, f(x)x\delta'(x) = -\int_a^b dx\,[xf'(x) + f(x)]\delta(x) \qquad (1.5.11)$$

The term involving $xf'(x)$ vanishes by the second property listed in (1.5.9) and we have

$$\int_a^b dx\, f(x)x\delta'(x) = -\int_a^b dx\, f(x)\delta(x) \qquad (1.5.12)$$

which is the third relation given in (1.5.9).

Integration over the delta function is conveniently summarized by writing

$$\int_a^x \delta(x' - x_0)\, dx' = \begin{cases} 1, & x > x_0 \\ 0, & x < x_0 \end{cases} \qquad (1.5.13)$$

where $a < x_0$. This result is usually represented by the unit step function $H(x - x_0)$, which is defined as

$$H(x) = \begin{cases} 1, & x > 0 \\ 0, & x < 0 \end{cases} \qquad (1.5.14)$$

As an example of the use of the delta function to treat a localized source, consider a beam of length L that is kept at zero temperature at ends located at $x = 0$ and L. A steady source of strength Q_0 calories per second is located at the midpoint. Such a source density can be written

$$s(x) = Q_0\delta(x - L/2) \qquad (1.5.15)$$

From (1.3.1), the steady state equation to be solved is

$$-Ku_{xx} = Q_0\delta(x - L/2) \qquad (1.5.16)$$

which, for $x \neq L/2$, reduces to the *homogeneous* equation $u_{xx} = 0$. We proceed by solving this homogeneous equation in the two regions $0 \leq x < L/2$, where the solution is represented by $u_<(x)$, and in $L/2 < x \leq L$, where the solution is represented by $u_>(x)$. To determine $u_<$ and $u_>$, we first apply the boundary condition to the appropriate solution at $x = 0$ and L and subsequently match these two solutions at the source point at $x = L/2$. For $0 \leq x < L/2$ the solution is $u_<(x) = ax + b$, where a and b are constants of integration. For $L/2 < x \leq L$ the solution is $u_>(x) = cx + d$, where c and d are also integration constants. Since $u_<(0) = 0$, we must have $b = 0$. Similarly, the condition $u_>(L) = 0$ requires $d = -cL$. The two remaining constants, a and c, are determined by matching $u_<(x)$ and $u_>(x)$ at $x = L/2$. Since the temperature can be expected to be continuous at $x = L/2$, we set $u_<(L/2) = u_>(L/2)$. This yields $a = c$ and we have $u_<(x) = ax$, $u_>(x) = a(L - x)$. Finally, a condition on $u_x(x)$ is obtained by integrating (1.5.16) across the delta function. We have

$$\int_{L/2-\epsilon}^{L/2+\epsilon} \frac{\partial^2 u}{\partial x^2} \, dx = -\frac{Q_0}{K} \int_{L/2-\epsilon}^{L/2+\epsilon} \delta(x - L/2) \, dx \qquad (1.5.17)$$

Both of these integrations are especially easy to perform. The left-hand side is a perfect derivative while on the right-hand side we merely use the property of the delta function given in (1.5.4). We thus obtain

$$\left.\frac{\partial u}{\partial x}\right|_{L/2-\epsilon}^{L/2+\epsilon} = -\frac{Q_0}{K} \qquad (1.5.18)$$

Since $u(x) = u_>(x)$ at $x = L/2 + \epsilon$ and $u(x) = u_<(x)$ at $x = L/2 + \epsilon$, we have

$$u_{>x}(L/2) - u_{<x}(L/2) = -Q_0/K \qquad (1.5.19)$$

Note that we may now ignore the increment ϵ since its purpose has been to distinguish the two sides of the discontinuity and this information is now contained in the subscripts $>$ and $<$. Using the expressions already developed for $u_>(x)$ and $u_<(x)$, (1.5.19) yields

$$-2a = -Q_0/K \qquad (1.5.20)$$

With the last constant in $u(x)$ now determined, we may write the final solution as

$$u(x) \begin{cases} u_<(x) = \dfrac{Q_0}{2K_0} x, & 0 \le x < \dfrac{L}{2} \\[4mm] u_>(x) = \dfrac{Q_0}{2K_0} (L - x), & \dfrac{L}{2} < x \le L \end{cases} \qquad (1.5.21)$$

Note that when we calculate the heat flow out each end, we obtain

$$J(0) = -Ku_{<x}(0) = -\tfrac{1}{2}Q_0$$

$$J(L) = -Ku_{<x}(L) = \tfrac{1}{2}Q_0 \qquad (1.5.22)$$

As expected on the basis of symmetry, an equal amount of heat flows out each end.

Problems

1.5.1 Show that the fourth property of the delta function listed in (1.5.9) follows from the fifth property.

1.5.2 Show that $\int_{-\infty}^{\infty} \sin(x + b)\delta(x^2 - a^2)\, dx = (1/|a|)\cos a \sin b$.

1.5.3 Use Eqs. (A.23) and (A.24) to show that $\delta(x - x_0)$ has the Fourier series expansions

$$\delta(x - x_0) = \frac{2}{L} \sum_{n=1}^{\infty} \sin \frac{n\pi x_0}{L} \sin \frac{n\pi x}{L}$$

$$\delta(x - x_0) = \frac{1}{L} + \frac{2}{L} \sum_{n=1}^{\infty} \cos \frac{n\pi x_0}{L} \cos \frac{n\pi x}{L}$$

Although these expansions lack the convergence expected of a Fourier series, they are useful, as the next problem shows, in solving problems by Fourier series methods.

1.5.4 Consider the point source problem described in (1.5.16). Introduce a Fourier series for $u(x)$ in the form $u(x) = \Sigma\, a_n \sin n\pi x/L$ as well as the Fourier sine series for $\delta(x - L/2)$ obtainable from problem 1.5.3. Equate coefficients of $\sin n\pi x/L$ and show that

$$u(x) = \frac{2Q_0 L}{\pi^2 K} \sum_{n=1}^{\infty} \frac{1}{n^2} \sin \frac{n\pi}{2} \sin \frac{n\pi x}{L}$$

The sum of the series obtained here is listed in Table F.1, entry 30, and is the Fourier series for the form of the solution given in (1.5.21).

1.5.5 A homogeneous beam of length L is kept at zero temperature at $x = 0$ and is insulated at $x = L$. The initial temperature distribution on the beam is taken to be $u(x, 0) = (H_0/\rho c)\delta(x - a)$ where $0 < a < L$.
(a) Determine the temperature on the beam at any later time.
(b) Show that all the heat initially on the beam (H_0) ultimately flows off the beam at $x = 0$.

1.5.6 A beam of length L with ends at zero temperature has a localized steady source of strength Q_0 at $x = L/N$ so that the equation governing the temperature on the beam is

$$-Ku_{xx} = Q_0\delta(x - L/N), \qquad u(0) = u(L) = 0$$

Show that $1/N$ of the heat input to the beam flows off the end at $x = L$.

1.5.7 The nonuniform beam described in Problem 1.3.3 is subjected to a localized heat source u_0 at its midpoint. The equation for the temperature on the beam is then

$$\frac{d}{dx} K_0 \frac{x}{L} \frac{du}{dx} = -Q_0 \delta \left(x - \frac{a + b}{2} \right)$$

(a) Determine the temperature on the beam.

(b) Show that if $b = 2a$, the ratio of heat flow at the ends is

$$\frac{J(b)}{|J(a)|} = \frac{\ln \frac{3}{2}}{\ln \frac{4}{3}} \approx 1.41$$

1.5.8 Consider the function $\exp[-f(x)/\epsilon]$ where $f(x)$ is always positive and $\epsilon \ll 1$. The exponential will take on its largest value when $f(x)$ is at a minimum, that is, at values of $x = x_n$ such that $f'(x_n) = 0$, $f''(x_n) > 0$. In the vicinity of any such point, say x_0, we may approximate $f(x)$ by using just the first few terms in the Taylor expansion of $f(x)$ about x_0. Retain the first three terms and use the second representation of the delta function given in (1.5.1) to show that

$$\lim_{\epsilon \to 0} e^{-f(x)/\epsilon} = e^{f(x_0)/\epsilon} \sqrt{\frac{2\pi}{|f''(x_0)|}}\, \delta(x - x_0), \qquad f''(x_0) \neq 0$$

Thus, for $\epsilon \ll 1$ the function $\exp[-f(x)/\epsilon]$ acts like a delta function and we may write

$$\int_a^b \varphi(x) e^{-f(x)/\epsilon}\, dx \approx e^{-f(x_0)/\epsilon} \sqrt{\frac{2\pi}{|f''(x_0)|}}\, \varphi(x_0)$$

for $a < x_0 < b$ and $f''(x_0) > 0$. If there is more than one such value of x_n, the dominant contribution can be expected to come from the value of x_n that corresponds to the smallest value of $f(x_n)$.

The approximate result for the evaluation of the integral over $\varphi(x)$ that has been obtained here is the same as that obtained by a method known as the saddle point method.

1.5.9 The steady state temperature distribution on a beam of length L is governed by

$$-K \frac{d^2 u}{dx^2} = \frac{Q_0/\pi}{\sqrt{x(x - L)}}, \qquad 0 < x < L$$

(a) Show that the total source strength is equal to Q_0.

(b) Introduce a Fourier series for both $u(x)$ and the source term to show that

$$u(x) = \frac{2Q_0 L}{\pi^2 K} \sum_{n=1}^{\infty} \frac{1}{n^2} J_0\left(\frac{n\pi}{2}\right) \sin \frac{n\pi}{2} \sin \frac{n\pi x}{L}$$

The coefficients in the expansion for the source term have been evaluated by using the integral representation for the zero order Bessel function given in Eq. (D.50).

1.6 INHOMOGENEOUS BOUNDARY CONDITIONS

Thus far we have only considered the Fourier series method when the boundary conditions were either $u = 0$ or $u_x = 0$. Any solution of the diffusion equation that satisfies either of these boundary conditions will still be a solution that satisfies these boundary conditions after the solution has been multiplied by an arbitrary constant. Consequently, such boundary conditions are usually said to be *homogeneous*. A third homogeneous boundary condition $u + \alpha u_x = 0$ will also be considered in later sections (cf. Problem 1.8.4).

Time dependent situations are frequently encountered in which the ends of the beam are maintained at different temperatures, for example, $u(0, t) = 0$, $u(L, t) = u_0$, or have nonvanishing heat fluxes $u_x = -J/K$ at one or both ends as was considered for time independent situations in Section 1.3. The above-mentioned multiplication property now no longer applies and such boundary conditions are referred to as being *inhomogeneous*.

Note that the nonzero temperature or heat flux at the end of the beam is actually a *source* of heat for the problem. In general, inhomogeneities in the boundary conditions, initial conditions, and/or differential equation itself, such as in (1.2.7), play the role of source terms. Later we will see how the same source term from a physical standpoint may be introduced into the analysis as an inhomogeneous term in more than one way. (cf. Problem 2.2.5.)

Maintenance of different end temperatures frequently leads, after an initial transient period, to a steady state temperature distribution on the beam that can be determined by solving the steady state heat equation $u_{xx} = 0$ subject to the given inhomogeneous boundary conditions. The last result mentioned in Example 1.1 provides just such a situation. If any other temperature variation is specified initially, it will evolve to this final steady state profile.

In problems having inhomogeneous boundary conditions it is convenient to consider the temperature as composed of two parts, the steady state part $u_{SS}(x)$ and the transient part $\theta(x, t)$. If we then write the temperature $u(x, t)$ as

$$u(x, t) = \theta(x, t) + u_{SS}(x) \tag{1.6.1}$$

the boundary conditions satisfied by $\theta(x, t)$ may be obtained from those imposed upon the complete solution $u(x, t)$. For example, if the boundary conditions are $u(0, t) = 0$ and $u(L, t) = u_0$, then

$$u(0, t) = \theta(0, t) + u_{SS}(0) = 0$$

$$u(L, t) = \theta(L, t) + u_{SS}(L) = u_0 \qquad (1.6.2)$$

Since $u_{SS}(0) = 0$ and $u_{SS}(L) = u_0$, the boundary conditions satisfied by $\theta(x, t)$ are thus the *homogeneous* ones $\theta(0, t) = \theta(L, t) = 0$. The initial condition for $\theta(x, t)$ is obtained from $u(x, 0)$ in a similar way. Writing

$$\theta(x, 0) = u(x, 0) - u_{SS}(x) \qquad (1.6.3)$$

and noting that $u(x, 0)$ is given while $u_{SS}(x) = u_1 x/L$, the initial condition $\theta(x, 0)$ is immediately available to us from (1.6.3). The equation satisfied by $\theta(x, t)$ is obtained by substituting the decomposition (1.6.1) into the heat equation (1.2.8), that is

$$(\theta + u_{SS})_{xx} - \gamma^{-2}(\theta + u_{SS})_t = 0 \qquad (1.6.4)$$

Since u_{SS} is independent of time and $(u_{SS})_{xx} = 0$, we have

$$\theta_{xx} - \gamma^{-2}\theta_t = 0 \qquad (1.6.5)$$

which shows that $\theta(x, t)$ satisfies the same diffusion equation as that satisfied by $u(x, t)$. For large times $\theta(x, t)$ approaches zero and therefore $u(x, t)$ approaches $u_{SS}(x)$ as expected.

Example 1.4. The temperature distribution on a beam of length L is initially $u(x, 0) = u_1 \sin \pi x/L$. For $t > 0$ the end of the beam at $x = 0$ is maintained at zero temperature while the end at $x = L$ is raised to the higher constant temperature u_0. As noted above in the use of (1.6.3), the final steady state temperature of the beam will be $u_{SS}(x) = u_0 x/L$. Solve the heat equation to determine the temperature at any intermediate time.

Decomposing the temperature into two contributions as suggested above, with $u_{SS}(x) = u_0 x/L$, we have $u(x, t) = \theta(x, t) + u_0 x/L$. The term $\theta(x, t)$ satisfies the diffusion equation (1.6.5). The initial condition is $\theta(x, 0) = u(x, 0) - u_{SS}(x) = u_1 \sin \pi x/L - u_0 x/L$ and the homogeneous boundary conditions are $\theta(0, t) = \theta(L, t) = 0$. According to (1.4.7) the solution for $\theta(x, t)$ is then

$$\theta(x, t) = \sum_{n=1}^{\infty} b_n e^{-n^2 \nu t} \sin \frac{n \pi x}{L}, \qquad \nu = \left(\frac{\gamma \pi}{L}\right)^2 \qquad (1.6.6)$$

Since $\theta(x, 0) = u(x, 0) - u_{ss}(x)$, the coefficients b_n are

$$b_n = \frac{2}{L} \int_0^L u(x, 0) - u_{ss}(x)] \sin \frac{n \pi x}{L} dx$$

$$= \frac{2}{L} \int_0^L \left(u_1 \sin \frac{\pi x}{L} - u_0 \frac{x}{L}\right) \sin \frac{n \pi x}{L} dx \qquad (1.6.7)$$

Due to the orthogonality of the functions $\sin n\pi x/L$ over the interval $(0, L)$, the term involving $u_1 \sin \pi x/L$ will only contribute to the coefficient b_1. Integration yields

$$b_1 = u_1 - 2u_0/\pi, \qquad b_n = (-1)^n(2u_0/n\pi), \qquad n > 1 \qquad (1.6.8)$$

For the complete solution $u(x, t)$ we now obtain

$$u(x, t) = \frac{u_0 x}{L} + u_1 e^{-\nu t} \sin \frac{\pi x}{L} + \frac{2u_0}{\pi} \sum \frac{(-1)^n}{n} e^{-n^2 \nu t} \sin \frac{n\pi x}{L} \qquad (1.6.9)$$

Some further analysis of this result is worthwhile. Since the total amount of heat on the beam at any time t is given by

$$H(t) = \rho c \int_0^L u(x, t)\, dx \qquad (1.6.10)$$

the initial and final amounts are

$$H(0) = \rho c \int_0^L u_1 \sin \frac{\pi x}{L}\, dx = \frac{2}{\pi} \rho c u_1 L$$

$$\qquad (1.6.11)$$

$$H(\infty) = \rho c \int_0^L u_0 \frac{x}{L}\, dx = \frac{1}{2} \rho c u_0 L$$

Note that if we choose $u_1 = (\pi/4)u_0$, the initial and final values will be equal. We can then specialize the solution given in (1.6.9) to determine whether the total heat on the beam rises above or falls below this limiting value at intermediate times.

From (1.6.9)–(1.6.11) we have

$$\frac{H(t)}{H(\infty)} = \frac{\pi}{2u_0 L} \int_0^L u(x, t)\, dx = 1 + e^{-\nu t} - \frac{8}{\pi^2} \sum_{n=1,3,5\ldots} \frac{1}{n^2} e^{-n^2 \nu t} \qquad (1.6.12)$$

For $t = 0$ the sum equals $\pi^2/8$ (Table F.1, item 5) and we recover the expected results $H(0)/H(\infty) = 1$. At intermediate times the series, which converges quite rapidly, may be evaluated numerically. A graph of the result is shown as the solid line in Figure 1.2a. The dotted line represents an approximate result, good for small values of νt, that is developed in Problem 2.3.4. The rapid increase in temperature is to be expected since imposition of the new boundary condition at $x = L$ introduces a large temperature gradient at that end and produces a flow of heat *onto* the beam. The time of maximum heat on the beam is readily estimated by retaining only the first two terms of the series in (1.6.12) and differentiating $u(x, t)$ with respect to time. When the resulting equation is set equal to zero and solved for νt, one finds $\nu t = \ln(\pi^2/8 - 1) \approx 0.182$. Initial and final temperature profiles, as well as the one at the time of maximum heat content on the beam, $\nu t = 0.182$, are shown in Figure 1.2b.

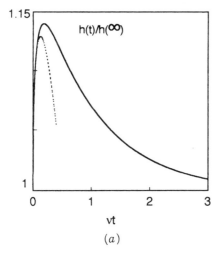
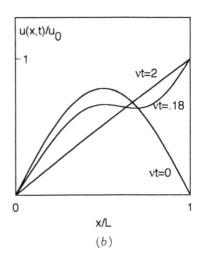

(a)   (b)

Figure 1.2. (a) Time dependence of total heat on beam in Example 1.4. Dotted line refers to an approximate solution valid at early time. (b) Initial and final temperature profiles and profile at time of maximum heat content for Example 1.4.

Problems

1.6.1 A beam of length L initially at zero temperature is maintained at that temperature at $x = 0$ while the end at $x = L$ is placed at temperature u_0. Determine the temperature at any later time.

1.6.2 A beam of length L has an initial temperature distribution $u(x, 0) = -u_0 \sin \pi x/L$. For $t > 0$ the boundary conditions are $u(x, 0) = 0$, $u(x, L) = u_0$. As $t \to \infty$, the temperature on the beam will thus be positive.

  (a) Determine the temperature on the beam at any later time.

  (b) Use the first two terms of the solution to determine the dimensionless time vt at which the temperature at the midpoint, which is initially negative, passes through zero.

1.6.3 Initially a beam of length L has temperature zero except for an amount of heat H_0 at a "hot spot" at its midpoint. The initial temperature may be written $u(x, 0) = (H_0/\rho c)\delta(x - L/2)$. For $t > 0$ boundary conditions $u(0, t) = 0$ and $u(L, t) = u_0$ are imposed. For large values of t, the heat flow at all points on the beam is in the negative x direction.

  (a) Determine H_∞, the total heat on the beam as $t \to \infty$.

  (b) Show that the temperature on the beam at any later time is given by

$$u(x, t) = \frac{u_0 x}{L} + \sum b_n e^{-n^2 vt} \sin \frac{n \pi x}{L}$$

$$b_n = \frac{2u_0}{\rho cL} \sin \frac{n\pi}{2} + \frac{2u_0}{n\pi} \cos n\pi$$

(c) If H_0 is sufficiently large, a point to the right of the source, say $x = 3L/4$, will initially experience a heat flow in the positive x direction (away from the source) before it reverses direction. If $H_0 = 10H_\infty$ determine the time at which the reversal takes place by using the first two terms in the solution.

(d) Show that the second term in the infinite series vanishes for all time and determine the magnitude of the first two nonvanishing terms in the series for $u_x(3L/4, t)$ at the estimated time of reversal.

1.6.4 An inhomogeneous beam is kept at zero temperature at ends located at $x = L$ and $2L$. In this coordinate system the conductivity of the beam is given by $K(x) = K_0(x/L)^2$. The initial temperature is $u(x, 0) = (H_0/\rho c)\delta(x - 3L/2)$.

(a) Show that the diffusion equation for the beam has the form

$$\frac{\partial}{\partial x}\left[\left(\frac{x}{L}\right)^2 \frac{\partial u}{\partial x} \right] - \frac{1}{\gamma^2}\frac{\partial u}{\partial t} = 0, \qquad \gamma^2 = \frac{K_0}{\rho c}$$

(b) Separate variables by writing $u(x, t) = X(x)T(t)$ and obtain

$$x^2 X'' + 2xX' + (kL)^2 X = 0, \qquad T' + (\gamma k)^2 T = 0$$

(c) Since the equation satisfied by $X(x)$ is of Cauchy-Euler type, solutions are of the form $X(x) = x^\alpha$. Show that $\alpha = -\frac{1}{2} \pm i\mu$ where $\mu^2 = (kL)^2 - \frac{1}{4}$ and that the general solution for $X(x)$ may be written

$$X(x) = x^{-1/2}[A \cos(\mu \ln x) + B \sin(\mu \ln x)]$$

(d) Impose boundary conditions $X(L) = X(2L) = 0$ and obtain separated solutions in the form

$$u_n(x, t) = \frac{B_n}{\sqrt{x}} e^{-(\gamma k_n)^2 t} \sin\left(\mu_n \ln \frac{x}{L}\right), \qquad \gamma^2 = \frac{K_0}{\rho c}$$

with

$$\mu_n^2 = (k_n L)^2 - \frac{1}{4} = \left(\frac{n\pi}{\ln 2}\right)^2$$

(e) Show that the total heat flowing out at each end, given by $\int_0^\infty dt\,[-K(x)u_x(x,\,t)]|_{x=L,\,2L}$, is

$$H(L) = -\frac{\ln 2}{\pi}\,\rho c\,\sqrt{L}\,\sum_{n=1}^{\infty}\frac{nB_n}{n^2 + (\ln 2/2\,\pi)^2}$$

$$H(2L) = -\frac{\ln 2}{\pi}\,\rho c\,\sqrt{2L}\,\sum_{n=1}^{\infty}\frac{(-1)^n\,nB_n}{n^2 + (\ln 2/2\,\pi)^2}$$

with

$$B_n = \frac{2}{\ln 2}\int_L^{2L}\frac{dx}{\sqrt{x}}\,u(x,\,0)\sin\left(\mu_n\ln\frac{x}{L}\right)$$

$$= \frac{2H_0}{\rho c\ln 2}\sqrt{\frac{2}{3L}}\,\sin\left(\mu_n\ln\frac{3}{2}\right)$$

(f) Use series 23 and 24 in Table F.1 to show that

$$\frac{H(2L)}{|H(L)|} = \frac{\frac{2}{3}}{\frac{1}{3}} = 2$$

1.7 INHOMOGENEOUS HEAT EQUATION—SOURCE TERMS

The Fourier series method may also be applied to inhomogeneous partial differential equations such as the diffusion equation

$$u_{xx} - \gamma^{-2}u_t = -\frac{1}{K}\,s(x,\,t),\qquad \gamma^2 = \frac{K}{\rho c} \tag{1.7.1}$$

As in the previous solution of the homogeneous diffusion equation, the choice of functions for the Fourier series expansion of the solution is determined by the boundary conditions for the specific problem being considered. For a beam of length L that is maintained at zero temperature at both ends, the temperature is again expanded in the form

$$u(x,\,t) = \sum_{n=1}^{\infty}b_n\sin\frac{n\pi x}{L} \tag{1.7.2}$$

where the coefficients are allowed to have an as yet unknown time dependence, that is, $b_n = b_n(t)$. This time dependence of the coefficients will be determined partly by the initial temperature distribution $u(x, 0)$ and partly by the source term $s(x, t)$. We introduce a similar expansion for the source term and write

$$s(x, t) = \sum_{n=1}^{\infty} s_n(t) \frac{n\pi x}{L} \tag{1.7.3}$$

Substitution of these series expansions into (1.7.1) yields

$$\sum_{n=1}^{\infty} \left[\left(\frac{n\pi}{L} \right)^2 b_n + \gamma^2 \frac{db_n}{dt} \right] \sin \frac{n\pi x}{L} = \frac{1}{K} \sum_{n=1}^{\infty} s_n \sin \frac{n\pi x}{L} \tag{1.7.4}$$

Because of the orthogonality of the functions $\sin n\pi x/L$, we may equate coefficients of each term in the series separately and obtain

$$\frac{db_n}{dt} + n^2 \nu b_n = \frac{\gamma^2}{K} s_n(t), \qquad n = 1, 2, 3, \dots \tag{1.7.5}$$

where $\nu = (\pi\gamma/L)^2$. We thus obtain a set of first order ordinary differential equations for the time dependent coefficients $b_n(t)$. Employing a standard technique for solving such equations, we multiply each equation by the appropriate integrating factor $\exp(n^2 \nu t)$ (cf. Hildebrand, 1976, p. 7) and obtain

$$e^{n^2 \nu t} \left(\frac{db_n}{dt} + n^2 \nu b_n \right) = \frac{d}{dt} (e^{n^2 \nu t} b_n) = \frac{\gamma^2}{K} e^{n^2 \nu t} s_n(t) \tag{1.7.6}$$

Integration from the initial time $t = 0$ to the current time t yields

$$e^{n^2 \nu t} b_n(t) \Big|_0^t = \frac{\gamma^2}{K} \int_0^t e^{n^2 \nu t'} s_n(t') \, dt' \tag{1.7.7}$$

The final expression for the coefficient $b_n(t)$ is

$$b_n(t) = b_n(0) e^{-n^2 \nu t} + \frac{\gamma^2}{K} e^{-n^2 \nu t} \int_0^t e^{n^2 \nu t'} s_n(t') \, dt' \tag{1.7.8}$$

where $b_n(0)$ is the coefficient of $\sin n\pi x/L$ in the Fourier expansion of the *initial* value of $u(x, t)$, that is

$$u(x, 0) = \sum_{n=1}^{\infty} b_n(0) \sin \frac{n\pi x}{L} \tag{1.7.9}$$

Note that if $s(x, t) = 0$, the result in (1.7.8) reduces to $b_n(t) = b_n(0) \exp(-n^2 \nu t)$, which is the solution of the homogeneous equation (1.4.7). We now consider some applications of the general result obtained here.

Example 1.5. A beam of length L initially at zero temperature is insulated at both ends so that the boundary conditions $u_x(0, t) = u_x(L, t) = 0$ are satisfied. A point heat source of strength Q_0 is located at the midpoint. The equation governing the temperature on the beam is thus

$$u_{xx} - \gamma^{-2}u_t = -\frac{Q_0}{K} \delta\left(x - \frac{L}{2}\right) \tag{1.7.10}$$

Determine the temperature profile on the beam as $t \to \infty$.

Since the ends are insulated, the source will produce a continued increase in the temperature of the beam. We expand both the solution $u(x, t)$ and the source $s(x)$, which has no time dependence in the present problem, in terms of a Fourier series composed of functions that satisfy the given boundary conditions. We thus write

$$u(x, t) = a_0(t) + \sum_{n=1}^{\infty} a_n(t)\cos\frac{n\pi x}{L}$$

$$s(x) = s_0 + \sum_{n=1}^{\infty} s_n \cos\frac{n\pi x}{L} \tag{1.7.11}$$

The coefficients s_0 and s_n are available from the results for the expansion of a delta function quoted in Problem 1.5.3, that is, $s_0 = Q_0/L$ and $s_n = (2Q_0/L)\cos n\pi/2$. When these expansions are now substituted into the diffusion equation (1.7.10) and coefficients of each term $\cos n\pi x/L$ are equated, the resulting differential equations are found to be

$$\frac{da_0}{dt} = \frac{\gamma^2 s_0}{K} = \frac{s_0}{\rho c}$$

$$\frac{da_n}{dt} + n^2 \nu a_n = \frac{s_n}{\rho c}, \qquad n \geq 1, \qquad \nu = \left(\frac{\gamma\pi}{L}\right)^2 \tag{1.7.12}$$

Solving these equations by the method used to obtain (1.7.8), we find

$$a_0(t) = a_0(0) + \frac{s_0 t}{\rho c}$$

$$a_0(t) = a_0(0) + \frac{1}{\rho c}\int_0^t e^{-n^2\nu(t-t')}s_n\, dt \tag{1.7.13}$$

In the present problem all $a_n(0)$ equal zero since the beam is initially at zero temperature. Also, using the coefficients for the expansion of the source term noted above, the coefficients in the expansion of $u(x, t)$ are now found from (1.7.13) to be

$$a_0 = \frac{Q_0 t}{\rho c L}$$

$$a_n = \frac{2Q_0}{\rho c L\nu}\cos\frac{n\pi}{2}\frac{1 - e^{-n^2\nu t}}{n^2} \tag{1.7.14}$$

With these coefficients, an expression for the temperature at any time is available from
(1.7.11). As $t \to \infty$ the terms a_n ($n \geq 1$) reduce to $(2Q_0/\rho cLvn^2)\cos n\pi/L$ and we
have

$$u(x, t) \to \frac{Q_0 t}{\rho cL} + \frac{2Q_0}{\rho cLv} \sum_{n=1}^{\infty} \frac{1}{n^2} \cos \frac{n\pi}{2} \cos \frac{n\pi x}{L} \qquad (1.7.15)$$

The sum of the series may be obtained from Table F.1, series 31, by identifying a
with $\pi x/L$ and b with $\pi/2$. The result is

$$u \cong \begin{cases} \dfrac{Q_0 t}{\rho cL} + \dfrac{Q_0}{24KL}(12x^2 - L^2), & 0 \leq x < \dfrac{L}{2} \\[4mm] \dfrac{Q_0 t}{\rho cL} + \dfrac{Q_0}{24KL}[12(L-x)^2 - L^2], & \dfrac{L}{2} < x \leq L \end{cases} \qquad (1.7.16)$$

The linear time dependence is to be expected since heat is being deposited on the beam
at a constant rate.

Problems

1.7.1 During a time T, an amount of heat H_0 with spatial variation proportional to $\sin \pi x/L$ is deposited on a beam of length L. The source term
is thus

$$s(x, t) = \begin{cases} \dfrac{\pi}{2} \dfrac{H_0}{LT} \sin \dfrac{\pi x}{L}, & 0 \leq t < T \\[4mm] 0, & t > T \end{cases}$$

(a) If the ends of the beam are held at zero temperature, determine the
temperature on the beam at any later time.

(b) If $H_<(0)$ and $H_>(0)$ refer to the total amount of heat flowing off
the beam at $x = 0$ in the time intervals $0 \leq t \leq T$ and $t > T$,
respectively, show that

$$|H_<(0)| = \frac{H_0}{2}\left(1 - \frac{1 - e^{-vt}}{vt}\right)$$

$$H_>(0) = \frac{H_0}{2}\frac{1 - e^{-vt}}{vt}$$

(c) Show that for large conductivity, that is, as $vT \to \infty$, one obtains
$|H_<(0)| \to H_0/2$ and $|H_>(0)| \to 0$, while as $vT \to 0$, one obtains
$|H_<(0)| \to 0$ and $|H_>(0)| \to H_0/2$.

1.7.2 Determine the temperature distribution at any later time if the beam in the previous problem has insulated ends.

1.7.3 A beam of length L insulated at both ends is subjected to the oscillatory external heating

$$s(x, t) = q_0 \sin \Omega t \cos \frac{n \pi x}{L} = q_0 \mathrm{Im}\,[e^{i \Omega t}] \cos \frac{n \pi x}{L}$$

where Im refers to the imaginary part of the expression in the brackets. Show that after a long time (i.e., $\nu T \gg 1$), the temperature on the beam is

$$u(x, t) = \frac{q_0}{\rho c \sqrt{\nu^2 + \Omega^2}} \sin (\Omega t - \phi) \cos \frac{n \pi x}{L}$$

where $\phi = \tan^{-1} (\Omega/\nu)$. Note that for high conductivity ($\nu/\Omega \gg 1$), u and s are in phase while for $\nu/\Omega \ll 1$, u lags s by a phase of $\pi/2$.

1.7.4 A beam of length L initially at zero temperature is heated at the midpoint by a source of q calories per second. The ends of the beam are maintained at zero temperature. The equation governing temperature on the beam is thus

$$u_{xx} - \gamma^{-2} u_t = -\frac{q_0}{K} \delta \left(x - \frac{L}{2} \right), \qquad u(0, t) = u(L, t) = u(x, 0) = 0$$

(a) Determine the subsequent temperature on the beam.

(b) Show that as $t \to \infty$

$$u(x, t) = \frac{2 q_0 L}{\pi^2 K} \sum_{n=1}^{\infty} \frac{1}{n^2} \sin \frac{n \pi}{2} \sin \frac{n \pi x}{L}$$

which is the Fourier series expression for the steady state problem considered in Section 1.5.

1.7.5 The following four examples should be considered together since in certain limits their solutions reduce to four closely related series known as theta functions.

(a) (i) Solve $u_{xx} - \gamma^{-2} u_t = 0$, $u(0, t) = u_x(L, t) = 0$, $u(x, 0) = \delta(x - x_0)$.

(ii) Show that as the hot spot x_0 approaches the insulated end $x = L$ the solution reduces to

$$u(x, t) = 2 \sum_{n=0}^{\infty} (-1)^n e^{-(n + 1/2)^2 \nu t} \sin \frac{(2n + 1) \pi x}{2L}$$

(iii) On making the substitution $\tau = iv t/\pi$, $v = x/2L$, this solution becomes

$$u\left(2Lv, \frac{-i\pi\tau}{v}\right) = 2 \sum (-1)^n e^{-(n + 1/2)^2 \pi \tau} \sin \pi(2n + 1)v$$

This series is usually represented $\vartheta_1(v, \tau)$.

(b) Interchange the boundary conditions in the previous problem; that is, set $u_x(0, t) = u(L, t) = 0$. Solve and again let x_0 approach the insulated end. Show that

$$u(x, t) = 2 \sum_{n=0}^{\infty} e^{-(n + 1/2)^2 vt} \cos \frac{(2n + 1)\pi x}{2L}$$

The change of variables used in 1 yields the function $\vartheta_2(v, \tau)$.

(c) If the beam is insulated at both ends, show that

$$u(x, t) = 1 + 2 \sum_{n=1}^{\infty} e^{-n^2 vt} \cos \frac{n\pi x_0}{L} \cos \frac{n\pi x}{L}$$

As $x_0 \rightarrow 0$ the solution is related to the function $\vartheta_3(v, \tau)$ while as $x_0 \rightarrow L$ the solution is related to the function $\vartheta_4(v, \tau)$.

An introductory discussion of theta functions may be found in the literature (Davis, 1960).

1.8   WAVE EQUATION—VIBRATING STRING

The equation often referred to as the wave equation, namely

$$y_{xx} - \frac{1}{c^2} y_{tt} = 0 \tag{1.8.1}$$

arises in many physical contexts. Derivations leading to this equation usually include a number of simplifications and approximations. The final result that yields (1.8.1) is merely a theoretical model (albeit a very good one in many instances) for describing the wave motion that actually takes place.[8] Here we outline a very simple derivation that includes the effects of variable string tension, damping, external sources, and the possibility that the string is embedded in a rubberlike material that provides an elastic restoring force to the string.

---

[8]A derivation of a wave equation for a vibrating string that includes many of the effects that are frequently neglected may be found elsewhere (Weinberger, 1965, Section 1).

The final equation obtained will enable us to use the vibrating string as a prototype for examining many effects that occur in the field of wave propagation.

1.8.1 General Wave Equation

The equation governing the motion of the string is obtained from an application of Newton's second law of motion to an element of the string, as shown in Figure 1.3. The mass of the element of string between x and $x + dx$ is $\rho(x)$ dx. We neglect the stretching of the string that takes place when the string moves out of its equilibrium position along the x axis. An estimate of this stretching is essential for determining the potential energy of the string, however, and is considered below in (1.8.23).

Newton's second law when applied to the vertical motion of the string yields

$$\rho \, dx \, \frac{\partial^2 y}{\partial t^2} = F \qquad (1.8.2)$$

where $y(x, t)$ is the displacement of the string from its equilibrium position along the x axis and F refers to the net effect of the forces mentioned above. We now briefly consider each of these forces separately:

1. *String Tension.* We assume that the vertical motion of the string is due to the vertical component of unbalanced string tension

$$F_{\text{tension}} = T(x + dx)\sin\vartheta\,(x + dx) - T(x)\sin\vartheta\,(x) \qquad (1.8.3)$$

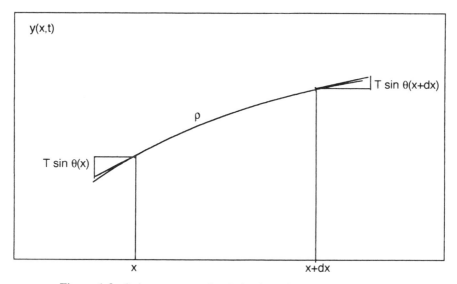

Figure 1.3. String geometry for derivation of wave equation (1.8.1).

For small displacements we can expect $\vartheta \ll 1$ and replace $\sin \vartheta$ by $\tan \vartheta$ since $\cos \vartheta$ will be close to unity in this limit. Recognizing that the tangent of ϑ may be identified with $\partial y / \partial x$, we then have

$$F_{\text{tension}} = \left(T \frac{\partial y}{\partial x} \right)_{x+dx} - \left(T \frac{\partial y}{\partial x} \right)_{x}$$

$$= \frac{\partial}{\partial x} \left[T(x) \frac{\partial y}{\partial x} \right] dx \tag{1.8.4}$$

where we have multiplied the first expression in (1.8.4) by dx/dx and used the definition of the derivative from introductory calculus to obtain the second form in (1.8.4). For the special case of constant tension this result reduces to

$$F_{\text{tension}} = T \frac{\partial^2 y}{\partial x^2}, \qquad T = \text{const} \tag{1.8.5}$$

2. *Friction.* The customary assumption used here is the same as that employed in elementary mechanics; that is, the frictional force acting on a segment of length dx is assumed to be proportional to the vertical velocity of the string. We thus write

$$F_{\text{friction}} = -R(x) \, dx \, \frac{\partial y}{\partial t} \tag{1.8.6}$$

The resistance coefficient $R(x)$ refers to a force *per unit length* of the string. Hence the resistance coefficient for a length dx is equal to $R(x) \, dx$.

3. *Elastic Restoring Force.* Assume that the string is surrounded by a medium that provides a restoring force proportional to $y(x, t)$, the displacement of the string from its equilibrium position. Then, as in the case of the mechanical oscillator, we write

$$F_{\text{elastic}} = -K(x) \, dx \, y(x, t) \tag{1.8.7}$$

The coefficient of the elastic restoring force $K(x)$ also refers to a force per unit length.

4. *External Forces.* This term includes such effects as the force of gravity or forces due to electric and magnetic fields if the string is a wire that carries an electric charge or current. In general such forces are described by a force density function $\mathcal{F}(x, t)$ so that

$$F_{\text{ext}} = \mathcal{F}(x, t) \, dx \tag{1.8.8}$$

When these various forces are combined (we assume that they act independently and can be superposed in a linear fashion), we have

$$\rho \frac{\partial^2 y}{\partial t^2} = \frac{\partial}{\partial x}\left(T \frac{\partial y}{\partial x}\right) - R \frac{\partial y}{\partial t} - Ky + \mathcal{F}(x, t) \tag{1.8.9}$$

As noted above, each of the parameters ρ, T, R, K, and $\mathcal{F}$ may be assumed to have spatial dependence. The simplest wave equation occurs when we assume $R = K = \mathcal{F} = 0$ and T is a constant. We then obtain the equation given in (1.8.1) with $c^2 = T/\rho$. Since T has the dimensions of force and ρ is a linear density, it is readily seen that c has the dimensions of velocity. That c represents the velocity at which waves actually propagate on the string will be shown later in Section 1.9.

1.8.2 Solution by Separation of Variables

Although the physical processes described by the wave equation are fundamentally different from those described by the diffusion equation, some of the mathematical techniques used to solve the diffusion equation may be applied to the wave equation as well. In fact, the elementary product solutions $X_n(x) T_n(t)$ introduced in the separation-of-variables method have a more natural physical interpretation for the wave equation since they correspond to the various modes of vibration of the system and can actually be observed on, say a guitar string when stroboscopic equipment is employed.

Since the equation for the string has been obtained by treating it as a mechanical system, we expect that both an initial displacement $y(x, 0)$ and an initial velocity $y_t(x, 0)$ must be specified in addition to the boundary conditions imposed at the ends of the string. To obtain the subsequent motion described by a solution of the wave equation (1.8.1), we introduce an elementary product solution $y(x, t) = X(x) T(t)$, as was done for the diffusion equation in Section 1.4. Substitution into the wave equation (1.8.1) leads to

$$\frac{X''}{X} = \frac{1}{c^2}\frac{T''}{T} = \pm \alpha^2 \tag{1.8.10}$$

Again, choice of sign for the separation constant α^2 is determined by the requirement that the solution be nonzero and satisfy certain boundary conditions. For a string of length L that is fixed at both ends, we require $X(0) = X(L) = 0$. We thus find ourselves dealing with the same equation and boundary conditions for $X(x)$ that we encountered in Section 1.4 when solving the diffusion equation. As in that previous consideration, the choice of sign $-\alpha^2$ is again found to be appropriate and we obtain

$$X(x) = A\cos\alpha x + B\sin\alpha x$$

$$T(t) = C\cos\alpha ct + D\sin\alpha ct \tag{1.8.11}$$

Application of the fixed end boundary conditions again yields $A = 0$ and $\alpha = n\pi/L$ with $n = \pm1, \pm2, \ldots$ while B is undetermined. Allowing for different values for the integration constants B, C, and D for each value of n, setting $a_n = B_n C_n$, $b_n = B_n D_n$ and then summing the various solutions, we arrive at

$$y(x, t) = \sum_{n=1}^{\infty} \sin \frac{n\pi x}{L} \left(a_n \cos \frac{n\pi ct}{L} + b_n \sin \frac{n\pi ct}{L} \right) \quad (1.8.12)$$

As noted in our consideration of the diffusion equation in Section 1.4, the summation need only be taken over positive values of n. Another common boundary condition for the string is considered in Problem 1.8.4.

Before proceeding to the determination of the constants a_n and b_n in terms of initial conditions for the string, it is useful to contrast the solution obtained here with the solution obtained for the diffusion equation given in (1.4.7). Superficially, there is merely a replacement of an exponentially decaying time dependence in the diffusion equation by an oscillatory time dependence for the wave equation. This is certainly to be expected on the basis of the physical processes being described. However, this difference is quite important when the convergence of these series is examined. For the diffusion equation, all terms in the series beyond $n = 0$ approach zero *exponentially* as n increases. Consequently a rapid convergence of the series solutions for the diffusion equation can be expected at later time. For the wave equation, however, convergence must be provided by the constant coefficients a_n and b_n. These coefficients usually decrease only *algebraically* with increasing n. In fact, term-by-term differentiation of the series solution for the wave equation may lead to a series that does not converge at all! This topic will be treated in more detail after a number of examples have been considered.

Returning to the series representation for $y(x, t)$ given in (1.8.12), we note that the coefficients a_n and b_n may be determined by considering the expansion for the initial conditions $y(x, 0)$ and $y_t(x, 0)$. From (1.8.12) we obtain

$$y(x, 0) = \sum_{n=1}^{\infty} a_n \sin \frac{n\pi x}{L}$$

$$y_t(x, 0) = \sum_{n=1}^{\infty} b_n \frac{n\pi c}{L} \sin \frac{n\pi x}{L} \quad (1.8.13)$$

We shall assume the convergence of these series. On applying the usual Fourier series methods summarized in Appendix A, we have

$$a_n = \frac{2}{L} \int_0^L y(x, 0) \sin \frac{n\pi x}{L} dx$$

$$\frac{n\pi c}{L} b_n = \frac{2}{L} \int_0^L y_t(x, 0) \sin \frac{n\pi x}{L} dx \quad (1.8.14)$$

As in the solution of the heat equation, the subsequent motion of the string due to a specific initial displacement and velocity is given by (1.8.12) with the coefficients a_n and b_n determined by (1.8.14). We now consider some examples.

Example 1.6. A string of length L with fixed ends at $x = 0, L$ has the initial displacement $y(x, 0) = 4 y_0 x(L - x)/L^2$ and $y_t(x, 0) = 0$. The coefficients are readily calculated and are found to be

$$a_n = \frac{2}{L} \frac{4 y_0}{L^2} \int_0^L x(L - x) \sin \frac{n \pi x}{L} dx$$

$$= \begin{cases} \dfrac{32 y_0}{(n \pi)^3}, & n = 1, 3, 5, \ldots \\ 0, & n = 2, 4, 6, \ldots \end{cases} \tag{1.8.15}$$

Since there is no initial velocity, the second of (1.8.14) yields $b_n = 0$. As noted during the solution of the diffusion equation in Section 1.4, the vanishing of the coefficients a_n for even values of n is to be expected on the basis of symmetry considerations. Equation (1.8.12) now provides the subsequent motion of the string in the form

$$y(x, t) = \frac{32 y_0}{\pi^3} \left(\sin \frac{\pi x}{L} \cos \frac{\pi ct}{L} + \frac{1}{3^3} \sin \frac{3 \pi x}{L} \cos \frac{3 \pi ct}{L} + \cdots \right) \tag{1.8.16}$$

Each individual term in this series is customarily referred to as a mode of vibration of the string, the lowest mode, $\sin \pi x/L \cos \pi ct/L$, being called the fundamental mode. The time dependence of the fundamental mode, expressed by $\cos \pi ct/L$, begins to repeat itself when the argument $\pi ct/L$ has increased to $\pi ct/L + 2\pi$, which may be rewritten $\pi c/L(t + 2L/c)$. Thus, the period of the fundamental mode is $P = 2L/c$ and the fundamental frequency is

$$\nu = \frac{1}{P} = \frac{1}{2L} \sqrt{\frac{T}{\rho}} \tag{1.8.17}$$

The subsequent terms in the series have frequencies that are integral multiples of this fundamental frequency.[9]

Example 1.7. The string of the previous example is given no initial displacement but is struck a blow at $t = 0$ in such a way as to impart to it the initial velocity $y_t(x, 0) = v_0 x(L - x)/L^2$. Determine (a) the initial kinetic energy of the string and (b) the subsequent motion of the string and (c) show that at any time the kinetic energy of the string averaged over one cycle of the fundamental period equals one-half of the initial

---

[9] A number of the properties of stringed musical instruments may be inferred from the solution of this problem. As an example, when the length of the E string (vibration frequency $\nu_E = 82.4$ Hz) on a guitar is shortened to three-fourths of its length, it vibrates at the frequency of the A string ($\nu_A = \frac{4}{3}\nu_E = 110$ Hz).

kinetic energy imparted to the string. The energy unaccounted for is in the form of potential energy and is given below in (1.8.24).

(a) Since the string is being treated as an extended mechanical system, the initial kinetic energy is given by $\frac{1}{2}\int_0^L \rho [y_t(x, 0)]^2 \, dx = \frac{1}{2}(\rho v_0^2/L^4)\int_0^L x^2 (L - x)^2 \, dx = \rho L v_0^2/60$.

(b) The string displacement is obtained from (1.8.12) with

$$a_n = 0$$

$$b_n = \frac{2v_0}{n \pi c L^2} \int_0^L x(L - x)\sin \frac{n \pi x}{L} \, dx$$

$$= \begin{cases} \dfrac{8v_0 L}{c(n\pi)^4}, & n = 1, 3, 5, \ldots \\ 0, & n = 2, 4, 6, \ldots \end{cases} \tag{1.8.18}$$

and is thus expressed as

$$y(x, t) = \frac{8v_0 L}{c \pi^4} \sum_{n = 1,3,5,\ldots}^{\infty} \frac{1}{n^4} \sin \frac{n \pi x}{L} \cos \frac{n \pi c t}{L} \tag{1.8.19}$$

(c) The kinetic energy at any later time is given by

$$KE = \frac{\rho}{2} \int_0^L [y_t(x, t)]^2 \, dx$$

$$= \frac{\rho}{2} \left(\frac{8v_0 L}{c \pi^4}\right)^2 \sum_n \sum_m \frac{1}{(nm)^3} \left(\frac{c\pi}{L}\right)^2 \int_0^L \sin k_n x \sin k_m x \, dx \sin \omega_n t \sin \omega_m t$$

$$\tag{1.8.20}$$

where $k_n = n \pi/L$, $\omega_n = n \pi c/L$, and m and n are summed over only odd values. Due to orthogonality, only the terms having $n = m$ will yield a nonzero result when the integration over x is performed. The double sum then reduces to a single sum and we obtain

$$KE = \frac{16 \rho v_0^2 L}{\pi^6} \sum_{n = 1,3,5,\ldots}^{\infty} \frac{1}{n^6} \sin^2 \omega_n t \tag{1.8.21}$$

Averaging this result over the fundamental period, which has duration $P = 2L/c$, we obtain

$$KE_{ave} = \frac{1}{P} \int_0^P KE \, dt = \frac{16 \rho v_0^2 L}{\pi^2} \sum_{n = 1,3,5,\ldots} \frac{1}{n^6} \frac{1}{P} \int_0^P \sin^2 \frac{2 n \pi t}{P} \, dt$$

$$= \frac{8 \rho v_0^2 L}{\pi^6} \sum_{n = 1,3,5,\ldots} \frac{1}{n^6} = \frac{\rho v_0^2 L}{120} \tag{1.8.22}$$

The value of the sum has been obtained from Table F.1, series 12. This result is one-half of the total energy imparted to the string as calculated in part (a).

The half of the initial energy that was not accounted for in the previous example is present in the form of potential energy. This is the energy involved in the stretching of the string at tension T that takes place during vibration. It may be written

$$PE = T \int_0^L ds - T \int_0^L dx = T \int_0^L [\sqrt{(dx)^2 + (dy)^2} - dx]$$

$$= T \int_0^L \left[\sqrt{1 + \left(\frac{dy}{dx}\right)^2} - 1 \right] dx \qquad (1.8.23)$$

On approximating the radical by using $\sqrt{1 + \epsilon} \sim 1 + \frac{1}{2}\epsilon$, we obtain

$$PE = \frac{1}{2}T \int_0^L (y_x)^2 \, dx \qquad (1.8.24)$$

Combining this result with that for the kinetic energy, we find that the total energy on the string is

$$E_{total} = \frac{1}{2}\rho \int_0^L [(y_t)^2 + c^2(y_x)^2] \, dx \qquad (1.8.25)$$

1.8.3 Energy Flow on the String

The result given for the total energy on the entire string in (1.8.25) may also be applied to a subregion $a \le x \le b$ by integrating only over that interval. Indicating the energy in this region by $E_{ab}(t)$, we have

$$E_{ab}(t) = \frac{1}{2}\rho \int_a^b [(y_t)^2 + c^2(y_x)^2] \, dx \qquad (1.8.26)$$

The flow of energy along the string may be obtained by considering the change in $E_{ab}(t)$ with time. Thus, we have

$$\frac{dE_{ab}}{dt} = \rho \int_a^b (y_t y_{tt} + c^2 y_x y_{xt}) \, dx \qquad (1.8.27)$$

and on setting $y_{tt} = c^2 y_{xx}$, we arrive at

$$\frac{dE_{ab}}{dt} = \rho c^2 \int_a^b \frac{\partial}{\partial x} (y_x y_t) \, dx$$

$$= \rho c^2 y_x y_t \big|_a^b \qquad (1.8.28)$$

Introducing an energy flow $J(x, t) = -\rho c^2 y_x y_t$, which has units of energy per unit time, we may write

$$\frac{dE_{ab}}{dt} = J(a, t) - J(b, t) \tag{1.8.29}$$

This result may be interpreted as the net change of energy in the region $a \le x \le b$ due to a flow into the region at $x = a$ minus a flow out at $x = b$.

1.8.4   Source Terms—Inhomogeneous Wave Equation

The treatment of source terms in the diffusion equation that was developed in Section 1.7 may be applied to the wave equation as well. Since the elementary product solutions are observable as modes of vibration of the system, the time dependence associated with both the free and forced motion of the system is of special interest in vibration problems.

Consider a string of length L that is fixed at both ends and driven by an external force at one frequency ω. More specifically, consider the wave equation

$$\frac{\partial^2 y}{\partial x^2} - \frac{1}{c^2}\frac{\partial^2 y}{\partial t^2} = -\frac{1}{T}\mathcal{F}(x)\sin \omega t \tag{1.8.30}$$

Since the ends of the string are fixed, it is appropriate to expand both $y(x, t)$ and $\mathcal{F}(x)$ in a Fourier sine series. We thus write

$$y(x, t) = \sum_{n=1}^{\infty} b_n(t)\sin \frac{n\pi x}{L}$$

$$\mathcal{F}(x, t) = \sum_{n=1}^{\infty} f_n(t)\sin \frac{n\pi x}{L} \tag{1.8.31}$$

Substitution of these forms into (1.8.30) and use of orthogonality yields

$$\frac{d^2 b_n}{dt} + \left(\frac{n\pi c}{L}\right)^2 b_n = \frac{1}{\rho}f_n\sin \omega t \tag{1.8.32}$$

This equation has a particular solution of the form $K_n \sin \omega t$ with $K_n = L^2 f_n/\pi^2 T[n^2 - (\omega L/\pi c)^2]$. When this solution is added to the homogeneous solution, we have

$$b_n(t) = A_n \cos \frac{n\pi ct}{L} + B_n\sin \frac{n\pi ct}{L} + \frac{L^2 f_n}{\pi^2 T}\frac{\sin \omega t}{n^2 - (\omega L/\pi c)^2} \tag{1.8.33}$$

If the system is started from rest, that is, $b_n(0) = db_n(t)/dt|_{t=0} = 0$, we find that the constants are $A_n = 0$ and $B_n = -K_n$. The solution is thus

$$y(x, t) = \frac{L^2}{\pi^2 T} \sum_{n=1}^{\infty} \frac{f_n \sin(n\pi x/L)}{n^2 - (\omega L/\pi c)^2} \left(-\frac{\omega}{\omega_n} \sin \omega_n t + \sin \omega t \right) \quad (1.8.34)$$

It should be noted that although both the free vibration of the string (the homogeneous solution) and the forced vibration (the particular solution) have the same *spatial* modes, the *time* dependence of the free vibration of each mode is that of its own characteristic frequency while the time dependence of the forced vibration of that mode is determined by the time dependence of the external source. It should also be noted that when the driving frequency ω equals one of the natural frequencies of the string, that is, $\omega = n\pi c/L$ for some integer n, the expression in (1.8.34) becomes indeterminate at that frequency. Examination of the expression in the limit as ω approaches ω_n shows that the time dependence takes the form $\omega_n t \cos \omega_n t - \sin \omega_n t$. The amplitude of this mode thus increases linearly with time. The amplitude would be limited if we were to include the resistance term $R \partial y/\partial t$ in the wave equation (1.8.30).

Problems

1.8.1   A string of length L is pulled away from equilibrium at its midpoint by a distance d so that

$$y(x, 0) = \begin{cases} 2\,dx/L, & 0 \le x < L/2 \\ 2d(L - x)/L, & L/2 < x \le L \end{cases}$$

(a) Show that the initial energy on the string (all in the form of potential energy that equals $\frac{1}{2} T \int_0^L [y_x(x, 0)]^2 \, dx$) is $2d^2 T/L$, where T is the tension on the string.

(b) Show that the displacement of the string at any later time is given by

$$y(x, t) = \frac{8d}{\pi^2} \sum_{n=1,3,5,\ldots} \frac{1}{n^2} \sin \frac{n\pi}{2} \sin \frac{n\pi x}{L} \cos \frac{n\pi ct}{L}$$

(c) Evaluate $1/P \int_0^P dt \int_0^L \frac{1}{2} T [y_x(x, t)]^2 \, dx$, the average of the potential energy over one period ($P = 2L/c$) and show that the result is equal to one-half of the initial energy on the string.

1.8.2   A string of length L is pulled aside an amount h at a distance d from the end at $x = 0$. Show that the coefficients in the Fourier series are

$$a_n = \frac{2\,hL^2}{\pi^2 d(L - d)\, n^2} \sin \frac{n\pi d}{L}$$

and that if $L = Nd$, where N is an integer, the frequency $\omega_N = N\pi c/L$ will be absent from the vibration.

1.8.3 A string of length L that is fixed at both ends is struck over the interval $L/2 - a/2 < x < L/2 + a/2$. The initial condition on the string is assumed to be

$$y(x, 0) = 0$$

$$y_t(x, 0) = \begin{cases} v_0, & \dfrac{L}{2} - \dfrac{a}{2} < x < \dfrac{L}{2} + \dfrac{a}{2} \\ 0, & \text{otherwise} \end{cases}$$

where v_0 is the initial velocity imparted to the string.

(a) Determine the displacement of the string at any later time and show that only odd harmonics are excited.

(b) Show that the initial energy on the string is $\frac{1}{2}\rho a v_0^2$ and that the total kinetic energy on the string, averaged over one cycle, is equal to one-half of this initial energy.

1.8.4 The string excitation of the previous problem is applied to a string that satisfies the boundary conditions $y(0) = 0$, $\zeta y(L) + Ly_x(L) = 0$.

(a) Show that the string excitation at any time is of the form

$$y(x, t) = \sum_{n=1}^{\infty} B_n \sin \frac{\alpha_n x}{L} \sin \frac{\alpha_n ct}{L}$$

where the α_n are solutions of $\zeta \tan \alpha_n + \alpha_n = 0$, $n = 1, 2, 3, \ldots$.

(b) Determine the coefficients B_n.

(c) Due to the boundary condition at $x = L$, the vibration of the string is no longer symmetric about the midpoint and modes of vibration for all values of n can occur. Show that for $\zeta \gg 1$, so that the end at $x = L$ is almost fixed, the solution corresponding to $n = 2$ is $\alpha_2 \approx 2\pi(1 + 1/\zeta)$ and determine the amplitude B_2.

1.8.5 An inhomogeneous string has fixed ends located at $x = L$ and $2L$. In this coordinate system the density of the string is given by $\rho(x) = \rho_0(L/x)^2$. Assume that the tension has the constant value T_0.

(a) Show that the wave equation for the string has the form

$$(x/L)^2 y_{xx} - c_0^{-2} y_{tt} = 0, \qquad c_0^2 = T_0/\rho_0$$

(b) Separate variables by writing $y(x, t) = X(x)T(t)$ and obtain

$$x^2 X'' + (kL)^2 X = 0, \qquad T'' + (kc_0)^2 T = 0$$

(c) Since the equation satisfied by $X(x)$ is of Cauchy-Euler type, solutions are of the form $X = x^\alpha$. Show that $\alpha = \frac{1}{2} \pm i\mu$, where $\mu = (kL)^2 - \frac{1}{4}$, and that the general solution for $X(x)$ may be written

$$X(x) = \sqrt{x}[A\cos(\mu\ln x) + B\sin(\mu\ln x)]$$

(d) Show that the requirement $X(L) = X(2L) = 0$ yields the eigenfunctions $X_n(x) = \sqrt{x}\sin[\mu_n\ln(x/L)]$ with $\mu_n = n\pi/\ln 2$ and hence $(k_nL)^2 = (n\pi/\ln 2)^2 + \frac{1}{4}$. The lowest few values of k_nL are 4.56, 9.08, and 13.61.

(e) Show that the average density of the string is $\rho_0/2$ and that for a *uniform* string with this density $k_nL = n\pi\sqrt{2}$. The lowest three values are 4.44, 8.88, and 13.33 and should be compared with the results in part (d).

1.9 D'ALEMBERT'S SOLUTION OF THE WAVE EQUATION

There is an alternate way of approaching the solution of the wave equation. The method may be motivated by examining the solution already obtained in some of the previous examples. In Example 1.6 in the previous section, the motion of a string of length L with fixed ends and initial displacement $y(x, 0) = 4y_0 x(L - x)/L^2$ was found to be

$$y(x, t) = \frac{32y_0}{\pi^3} \sum_{n=1,3,5,\ldots} \frac{1}{n^3} \sin\frac{n\pi x}{L} \cos\frac{n\pi ct}{L} \qquad (1.9.1)$$

This result was interpreted in terms of standing waves on the string.

Using the trigonometric identity $2\sin a\cos b = \sin(a + b) + \sin(a - b)$, we may rewrite the string displacement given above in the form

$$y(x,t) = \frac{16y_0}{\pi^3} \sum_{n=1,3,5,\ldots} \frac{1}{n^3}\left[\sin\frac{n\pi x}{L}(x - ct) + \sin\frac{n\pi}{L}(x + ct)\right] \qquad (1.9.2)$$

The grouping of variables in the form $x \pm ct$ leads to a physical interpretation of the solution that is altogether different from that already given in the previous section in terms of standing waves on the string.

We first digress and note the interpretation of a function with argument $x - ct$, for example, the function $\exp[-(x - ct)^2/L^2]$. At $t = 0$ the function reduces to $\exp(-x^2/L^2)$ and takes on its maximum value of unity at the point $x = 0$. At a later time $t = t_1$ the peak of the curve is located at the point where $x - ct_1 = 0$, that is, at $x = ct_1$. The peak has thus moved in the positive x direction a distance ct_1 in a time t_1; that is, it has moved with the velocity c. A similar consideration applied to any other point on the curve shows that the entire shape moves without distortion at a velocity c in the positive direction.

Similarly, a function of $x + ct$ moves in the negative x direction at the same speed.

The solution for $u(x, t)$ given in (1.9.2) thus corresponds to two displacements that propagate in opposite directions along the string. Since the coefficients in the Fourier series are the same as those in the expansion of the shape of the string at $t = 0$ we may write the solution (1.9.2) as

$$y(x, t) = \tfrac{1}{2}[f(x - ct) + f(x + ct)] \tag{1.9.3}$$

where

$$f(x) = y(x, 0) = 4 y_0 x (L - x)/L^2 \tag{1.9.4}$$

The solution at later times is thus related in a very simple way to the initial displacement. We shall find later that the relation is not as immediate when the string is given an initial velocity.

The Fourier series for $u(x, 0)$ provides a periodic extension of this initial displacement in both directions outside the region $0 \leq x \leq L$ that contains the string. According to the traveling wave solution in (1.9.3), these periodic extensions travel in opposite directions along the x axis so as to pass through the region of the x axis that actually contains the string. The summation of these two traveling waves yields an expression that is mathematically equivalent to the standing wave solution obtained previously by the separation-of-variables method.

Whenever a combination of variables presents itself naturally in the solution of a problem, it is worth recasting the original equations in terms of these new variables. Let us therefore set $\xi = x - ct$ and $\eta = x + ct$ and ask what form is assumed by the wave equation (1.8.1) when it is expressed in terms of ξ and η. To develop the relation between derivatives with respect to x, t and those with respect to ξ, η we may examine the differential expression of a function $f(\xi, \eta)$

$$
\begin{aligned}
df &= f_\xi \, d\xi + f_\eta d\eta \\
&= f_\xi (dx - c \, dt) + f_\eta (dx + c \, dt) \\
&= (f_\xi + f_\xi) \, dx + c(-f_\xi + f_\eta) \, dt \\
&= f_x \, dx + f_t \, dt \tag{1.9.5}
\end{aligned}
$$

Since dx and dt are independent, we may equate corresponding coefficients of the last two expressions in (1.9.5) and obtain the transformation relations

$$\frac{\partial}{\partial x} = \frac{\partial}{\partial \xi} + \frac{\partial}{\partial \eta}, \qquad \frac{\partial}{\partial t} = c\left(-\frac{\partial}{\partial \xi} + \frac{\partial}{\partial \eta}\right) \tag{1.9.6}$$

Hence, for the required second derivatives we have

$$\frac{\partial^2 y}{\partial x^2} = \left(\frac{\partial}{\partial \xi} + \frac{\partial}{\partial \eta}\right)\left(\frac{\partial y}{\partial \xi} + \frac{\partial y}{\partial \eta}\right) = y_{\xi\xi} + 2y_{\xi\eta} + y_{\eta\eta}$$

$$\frac{1}{c^2}\frac{\partial^2 y}{\partial t^2} = \left(-\frac{\partial}{\partial \xi} + \frac{\partial}{\partial \eta}\right)\left(-\frac{\partial y}{\partial \xi} + \frac{\partial y}{\partial \eta}\right) = y_{\xi\xi} - 2y_{\xi\eta} + y_{\eta\eta} \quad (1.9.7)$$

Since the string displacement will be continuous, we may assume $y_{\xi\eta} = y_{\eta\xi}$. The wave equation (1.8.1) now becomes

$$4y_{\xi\eta} = 0 \tag{1.9.8}$$

The most general function of u that yields zero when differentiated successively with respect to both ξ and η is easily recognized to be of the form

$$y(\xi, \eta) = f(\xi) + g(\eta) \tag{1.9.9}$$

In terms of the x, t variables we obtain

$$y(x, t) = f(x - ct) + g(x + ct) \tag{1.9.10}$$

which is of the form given above in (1.9.3).

The functions f and g are, of course, determined by the initial and boundary conditions for a specific problem. We now consider the determination of f and g for the case of an infinite string. Using a prime to denote a derivative with respect to the argument of a particular function, we have[10]

$$\frac{\partial y}{\partial t} = \frac{\partial}{\partial t}f(x - ct) + \frac{\partial}{\partial t}g(x + ct)$$

$$= -cf'(x - ct) + cg'(x + ct) \tag{1.9.11}$$

so that

$$y_t(x, 0) = -cf'(x) + cg'(x) \tag{1.9.12}$$

Introducing the initial conditions $y(x, 0) = \varphi(x)$, $y_t(x, 0) = \psi(x)$, we find that Eqs. (1.9.10) and (1.9.12) reduce to

$$f(x) + g(x) = \varphi(x) \tag{1.9.13a}$$

---

[10]Setting $\xi = x - ct$ so that $f(x - ct) = f(\xi)$ and using the chain rule for differentiation, we have $\partial f/\partial t = df/d\xi \ \partial\xi/\partial t = f'(\xi)(-c) = -cf'(x - ct)$ and similarly for $g(x + ct)$.

$$f'(x) - g'(x) = -\frac{1}{c}\psi(x) \tag{1.9.13b}$$

Integration of the second of these equations yields

$$f(x) - g(x) = -\frac{1}{c}\int_\alpha^x \psi(x')\,dx' \tag{1.9.14}$$

where α is a constant of integration (which will later be found to play no role in the solution). Combining (1.9.13a) and (1.9.14) we obtain

$$f(x) = \tfrac{1}{2}\varphi(x) - \frac{1}{2c}\int_\alpha^x \psi(x')\,dx'$$

$$g(x) = \tfrac{1}{2}\varphi(x) + \frac{1}{2c}\int_\alpha^x \psi(x')\,dx' \tag{1.9.15}$$

Replacing x by $x - ct$ in the function f, and by $x + ct$ in g, and then using (1.9.10), we have the final result known as the d'Alembert solution of the wave equation

$$y(x, t) = \tfrac{1}{2}[\varphi(x - ct) + \varphi(x + ct)] + \frac{1}{2c}\int_{x-ct}^{x+ct} \psi(x')\,dx' \tag{1.9.16}$$

This is a convenient form for displaying the motion of a string *that extends to infinity in both directions* so that effects of boundaries play no role. The displacement of the string is seen to be a certain linear combination of the initial displacement $\varphi(x)$ and the initial velocity $\psi(x)$. An initial displacement $\varphi(x)$ decomposes into two similarly shaped displacements having one-half of the initial amplitude that then propagate in opposite directions. This result has already been obtained in (1.9.3). The shape of the displacement due to an initial velocity $\psi(x)$ is less immediate. However, the reasonableness of the expression involving $\psi(x)$ may be readily understood by considering some simple examples. If $\varphi(x) = 0$ and $\psi(x) = v_0$ so that the entire string is given in initial upward velocity, then the subsequent displacement is merely

$$y(x, t) = \frac{1}{2c}\int_{x-ct}^{x+ct} v_0\,dx' = v_0 t \tag{1.9.17}$$

The entire string thus continues to move upward with the initial velocity v_0.

As another simple example, if a point x_0 is given an initial velocity so that $u_t(x, 0) = \psi(x) = 2cy_0\delta(x - x_0)$ where $\delta(x - x_0)$ is the delta function and the amplitude has been chosen for later convenience, we have

$$y(x, t) = y_0 \int_{x-ct}^{x+ct} \delta(x' - x_0)\, dx' = y_0[H(x + ct) - H(x - ct)] \quad (1.9.18)$$

where H refers to the unit step function introduced in (1.5.13) and (1.5.14).

As indicated in Figure 1.4a, this combination of step functions yields a string displacement at some time t_1 that is equal to y_0 for $|x - x_0| < ct_1$ and is zero outside of this region.

As a function of time, the string displacement or "signal" that arrives at a point $x_1 > x_0$ is shown in Figure 1.4b.

When the initial velocity is applied over an extended region of the string, the range of integration in the d'Alembert solution must be determined with some care. The integration in (1.9.16) extends over the region of x for which there is an overlap between nonzero values of $\psi(x)$ and the interval $x - ct < x' < x + ct$. As an example, consider a string with no initial displacement so that $\varphi(x) = 0$ and initial velocity

$$\psi(x) = \begin{cases} \dfrac{K}{\sqrt{L^2 - x^2}}, & |x| < L \\ 0, & |x| > L \end{cases} \quad (1.9.19)$$

The constant K will be chosen at the end of the calculation in such a way that the final amplitude of the string will have a value y_0.

According to the d'Alembert solution, the displacement of the string is given by the integral of $\psi(x)$ over the region $x - ct < x' < x + ct$ for which $\psi(x)$ is nonzero. This region is readily visualized by considering the diagram in Figure 1.5, where the range of integration at various times is indicated. Until enough time has elapsed for a signal originating at $x = L$ at time $t = 0$ to reach the observation point, there is no displacement. This is the situation at a time such as t_1 shown in Figure 1.5. The time $t_2 = (x - L)/c$ at which the signal first reaches x is shown in the figure at point B. After this time the shape of the string displacement is given by

$$y(x, t) = \frac{K}{2c} \int_{x-ct}^{L} \frac{dx'}{\sqrt{L^2 - x'^2}}, \quad \frac{x - L}{c} < t < \frac{x + L}{c} \quad (1.9.20)$$

The range of integration is the segment PL in Figure 1.5. At time t_3 the range of integration becomes $-L$ to L and remains so for later times. The string displacement then has the constant value obtained by evaluating

$$y(x, t) = \frac{K}{2c} \int_{-L}^{L} \frac{dx'}{\sqrt{L^2 - x'^2}}$$

$$= \frac{K\pi}{2c}, \quad t > \frac{x + L}{c} \quad (1.9.21)$$

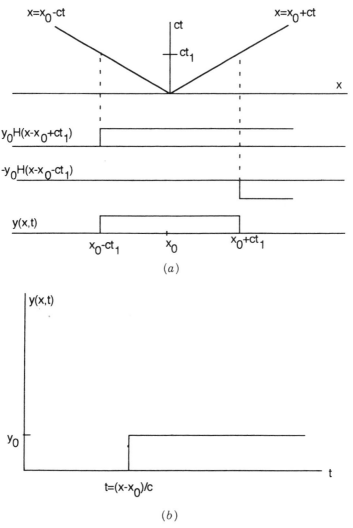

Figure 1.4. (a) Amplitude of string displacement due to initial velocity $y_t(x, 0) = 2cy_0 \delta(x - x_0)$. (b) Temporal response of string at $x = x_1$ due to initial velocity $y_t(x, 0) = 2cy_0 \delta(x - x_0)$.

If we set this result equal to y_0, that is, choose $K = 2cy_0/\pi$ in (1.9.19), then integration of (1.9.20) and (1.9.21) yields a string displacement of the form

$$y(x, t) = \begin{cases} \dfrac{y_0}{\pi} \left(\dfrac{\pi}{2} - \sin^{-1} \dfrac{x - ct}{L} \right), & \dfrac{x - L}{c} < t < \dfrac{x + L}{c} \\ y_0, & t > \dfrac{x + L}{c} \\ 0, & \text{otherwise} \end{cases} \quad (1.9.22)$$

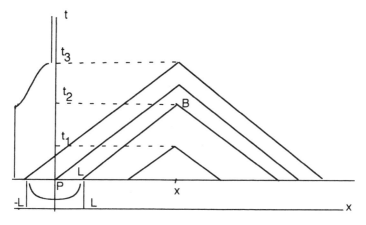

Figure 1.5. Space-time description of string response to initial displacement (1.9.19).

The string displacement as a function of time is shown in Figure 1.6 as well as in Figure 1.5. Note that the transition region or "rise time" of the signal is $2L/c$. As L, the width of the source, approaches zero, we obtain the step function result given previously for the delta function source. This leads us to expect that in the limit $L \to 0$ the initial velocity given in (1.9.19) must become a delta function. This is indeed the case. Hence, another (somewhat cumbersome) representation for a delta function is

$$\delta(x) = \lim_{L \to 0} \begin{cases} \dfrac{1}{\sqrt{L^2 - x^2}}, & |x| < L \\ 0, & |x| > L \end{cases} \tag{1.9.23}$$

and could be used in much the same way as the other representations of the delta function that were given in (1.5.1).

The d'Alembert solution is thus seen to provide a convenient way of describing traveling waves on an infinite string. For a string of finite length,

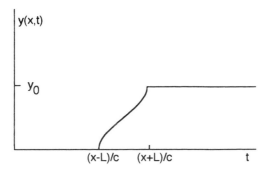

Figure 1.6. Temporal response of string due to initial displacement (1.9.19).

information concerning reflection from the ends must be incorporated into the functions $f(x - ct)$ and $f(x + ct)$. This can be rather cumbersome. In such cases it is frequently more convenient to express the solution in terms of standing waves by using the separation-of-variables method or use the integral transform technique to be introduced in Chapter 6.

As a simple example in which only one boundary is present, consider a semi-infinite string that extends along the positive x axis $(0 < x < \infty)$. The string is initially quiescent so that $y(x, 0) = y_t(x, 0) = 0$. Beginning at $t = 0$ the end at $x = 0$ is moved with a prescribed motion $S(t)$ so that $y(0, t) = S(t)$. The propagation of the disturbance on the string is to be determined.

The most general form for the shape of the string on $x > 0$ is still that given by (1.9.10), namely

$$y(x, t) = f(x - ct) + g(x + ct) \tag{1.9.24}$$

Differentiating with respect to time, we have

$$y_t(x, t) = -cf'(x - ct) + cg'(x + ct), \qquad x > 0 \tag{1.9.25}$$

where the prime indicates a derivative with respect to argument. Imposing the initial conditions we have

$$y(x, 0) = f(x) + g(x) = 0, \qquad x > 0$$
$$y_t(x, 0) = cf'(x) + cg'(x) = 0, \qquad x > 0 \tag{1.9.26}$$

Integration of the second equation yields $f(x) - g(x) = K$ where K is a constant of integration. Solving for f and g we obtain

$$f(x) = -g(x) = \tfrac{1}{2}K \tag{1.9.27}$$

The form of the solution for *positive* values of the argument x is now known. We may thus write

$$y(x, t) = \tfrac{1}{2}K - \tfrac{1}{2}K = 0, \qquad x > ct \tag{1.9.28}$$

This result is merely a statement of causality. The string experiences no displacement until enough time has elapsed for the signal propagating at velocity c to reach the point x. Since x and t are positive, we never need to know the value of $g(x + ct)$ for negative values of the argument. However, for $f(x - ct)$ we will be interested in values of x and t for which $x - ct < 0$. In order to find the form of f for negative values of the argument we consider the boundary conditions $y(0, t) = S(t)$. From (1.9.3) we obtain

$$y(0, t) = f(-ct) + g(ct) = S(t), \qquad t > 0 \tag{1.9.29}$$

Setting $ct = \xi$, we have

$$f(-\xi) + g(\xi) = S(\xi/c), \qquad \xi > 0 \tag{1.9.30}$$

Since $g(\xi) = \tfrac{1}{2}K$ for $\xi > 0$, we have

$$f(-\xi) = S(\xi/c) - \tfrac{1}{2}K, \qquad \xi > 0 \tag{1.9.31}$$

which gives the form of f for negative values of the argument. Setting $\xi = ct - x$ we have $\xi > 0$ for $x < ct$ and may write $f(x - ct)$ in the form

$$f(x - ct) = S(t - x/c) - \tfrac{1}{2}K, \qquad x < ct \tag{1.9.32}$$

We thus have

$$\begin{aligned} f(x - ct) + g(x + ct) &= S(t - x/c) - \tfrac{1}{2}K + \tfrac{1}{2}K \\ &= S(t - x/c), \qquad x < ct \end{aligned} \tag{1.9.33}$$

This result in conjunction with (1.9.28) yields the solution

$$y(x, t) = \begin{cases} 0, & x > ct \\ S(t - x/c), & x < ct \end{cases} \tag{1.9.34}$$

The movement of the end at $x = 0$ thus generates a disturbance that propagates without changing shape. We shall consider this problem again in Section 2.1 and show that this result is more immediately obtained by a transform method. The "method of characteristics" presented here is especially useful for inhomogeneous systems and is considered at greater length in Sections 8.1 and 9.8.

Problems

1.9.1 An infinite string has initial conditions

$$y(x, 0) = y_0 \operatorname{sech}^2 \frac{x}{L}$$

$$y_t(x, 0) = \frac{y_0 c}{L} \operatorname{sech}^2 \frac{x}{L} \tanh \frac{x}{L}$$

(a) Use the d'Alembert solution to determine the subsequent motion of the string.

(b) Show that the kinetic and potential energy are always equal and have the value $\frac{8}{15}\rho c^2 y_0^2/L$. (Note: $\int_{-\infty}^{\infty} \text{sech}^{2n} x \, dx = 2^{2n-1}[(n - 1)!]^2/(2n - 1)!$.)

1.9.2 Determine the motion of an infinite string for which

$$y(x, 0) = y_0(\text{sech } \xi^- - \text{sech } \xi^+)$$

$$y_t(x, 0) = -\frac{y_0 c}{L} (\text{sech } \xi^- \tanh \xi^+ + \text{sech } \xi^- \tanh \xi^+)$$

where $\xi^+ = (x + x_0)/L$ and $\xi^- = (x - x_0)/L$. Evaluate $u(x, x_0/c)$ and $u(0, t)$. Sketch the solution for various times when $x_0 \gg L$.

1.9.3 A string of length L is pulled aside at the midpoint a distance d and then released at $t = 0$.

(a) Draw a sketch of the initial displacement as well as the appropriate extension on the x axis.

(b) Use the d'Alembert solution to sketch the shape of the string at $t = L/4c, L/2c, 3L/4c, L/c$.

1.9.4 An infinite string has the initial conditions

$$y(x, 0) = 0, \qquad y_t(x, 0) = v_0 \text{sech } x/L$$

(a) Determine the subsequent motion.

(b) Sketch the wave motion and express v_0 in terms of the total energy on the string. [*Hint:* Use $\int \text{sech } x \, dx = 2 \tan^{-1} (e^x)$.]

(c) If the total energy E is held fixed as $L \to 0$, can the limit of $u_t(x, 0)$ be interpreted as a delta function? Consider also $[u_t(x, 0)]^2$.

1.9.5 A mass M is located at $x = 0$ on an infinite string. The mass may be described as an inhomogeneity of the string density by writing $\rho(x) = \rho_0 + M \delta(x)$. Equation (1.8.9) may now be used to write the wave equation for the string as

$$y_{xx} - \frac{1}{c^2} y_{tt} = \frac{M}{T_s} \delta(x) y_{tt}$$

where T_s is the string tension.

Now consider the reflection and transmission of a wave $F(t - x/c)$ that is incident upon the mass from $x = -\infty$. The string displacement is

$$y(x, t) = \begin{cases} F(t - x/c) + R(t + x/c), & x < 0 \\ T(t - x/c), & x > 0 \end{cases}$$

(a) Use the continuity condition $F(t) + R(t) = T(t)$ as well as integration of the wave equation across $x = 0$ to show that $T(t)$, which equals the mass displacement $y(0, t)$, satisfies

$$T'' + \gamma T' = \gamma F', \qquad \gamma = sT_s/Mc$$

(b) Integrate to obtain

$$T(t) = \gamma e^{-\gamma t} \int_0^t e^{\gamma t'} F(t') \, dt'$$

(c) Determine the transmitted pulse shape if the incident pulse is the delta function $F(t) = \delta(t - t_0)$.

1.9.6 The scattering of a wave by a mechanical oscillator attached to a string may be analyzed by setting $\rho(x) = \rho_0 + M\,\delta(x)$ and $K(x) = k_0\,\delta(x)$ in (1.8.9).

(a) Obtain the wave equation

$$y_{xx} - \frac{1}{c^2} y_{tt} = \frac{M}{T_s} \delta(x)(\, y_{tt} - \omega_0^2 y_{xx})$$

where $\omega_0^2 = k/M$.

(b) Follow the procedure used in the previous problem to show that the transmitted wave now satisfies

$$T'' + \gamma T' + \omega_0^2 T = \gamma F', \qquad \gamma = 2T_s/Mc$$

Since $y(0, t) = T(t)$, the oscillator has a damping term due to radiation on the string.

(c) Show that if $F(t) = f_0 \sin \Omega t$, then when $\Omega = \omega_0$, there is perfect transmission ($R = 0, T = F$).

2 Laplace Transform Method

In the previous chapter the separation-of-variables method was used to reduce the solution of a partial differential equation to the solution of a set of ordinary differential equations. The solution of the partial differential equation was then expressed, in general, as an infinite sum of the solutions to these ordinary differential equations. One of the main limitations on the usefulness of this method is the possibility of slow convergence of the series. We now take up a method, based upon the Laplace transform, that has much greater flexibility than the series method. It can be used to recover the solution in series form if one desires, but it can also be used to display the solution in other forms as well. It may also be used to develop various approximate expressions for the solution. However, many of these approximation techniques rely upon the use of complex variable methods and will not be considered in this volume.

The Laplace transform procedure will be introduced by considering an example of a vibrating string since application of the method to the wave equation is somewhat simpler than to the diffusion equation. The reader is assumed to have some experience with the use of Laplace transforms to solve ordinary differential equations. A summary of this background material is contained in Appendix B.

2.1 VIBRATING STRING

From (1.1.1), the simplest equation governing the motion of a vibrating string is

$$u_{xx} - \frac{1}{c^2} u_{tt} = 0 \tag{2.1.1}$$

To solve this equation by the Laplace transform method, we first introduce the transform of the solution in the form[1]

$$U(x, s) = \int_0^\infty e^{-st} u(x, t)\, dt \tag{2.1.2}$$

---

[1] We shall use the corresponding uppercase letter to indicate the transform of a function of t that is designated by a lowercase letter.

In taking the transform of the wave equation (2.1.1) we assume that the order of integration and differentiation may be interchanged. Validity of this procedure follows from the uniform convergence of the integral that defines the Laplace transform (cf. Section B.1). We then obtain

$$\int_0^\infty e^{-st} u_{xx}(x, t)\, dt = \frac{\partial^2}{\partial x^2} \int_0^\infty e^{-st} u(x, t)\, dt = \frac{\partial^2 U(x, s)}{\partial x^2} \qquad (2.1.3)$$

Using the standard expression for the transform of a second time derivative as obtainable from Eq. (B.4), we find that the transform of the wave equation is

$$U_{xx}(x, s) - \frac{1}{c^2} [s^2 U(x, s) - su(x, 0) - u_t(x, 0)] = 0 \qquad (2.1.4)$$

The Laplace transform of the string displacement $U(x, s)$ is now found by solving this second order ordinary differential equation. It should be noted that success with the determination of $u(x, t)$ from $U(x, s)$ will depend upon the complexity of the s dependence that ultimately occurs in the solution of this differential equation for $U(x, s)$. When solving partial differential equations by the Laplace transform method, this s dependence can be expected to be more complicated than the simple algebraic dependence encountered when solving ordinary differential equations.

The boundary conditions imposed upon $U(x, s)$ are the Laplace transform of the boundary conditions imposed upon $u(x, t)$. The solution of (2.1.4), the differential equation for $U(x, s)$, may be obtained for arbitrary initial conditions $u(x, 0)$ and $u_t(x, 0)$ by using the method of variation of parameters. We shall at first, however, consider only a simple example in which the solution may be obtained rather immediately.

Consider a semi-infinite string ($0 \le x < \infty$) that is initially at rest. For $t > 0$ the end at $x = 0$ is given a displacement $u(0, t) = f(t)$. We determine the shape of the wave that propagates on the string. This example was considered previously in Section 1.9.

Since the initial conditions are $u(x, 0) = u_t(x, 0) = 0$, Eq. (2.1.4) for $U(x, s)$ reduces to the homogeneous equation

$$U_{xx} - \left(\frac{s}{c}\right)^2 U = 0 \qquad (2.1.5)$$

which has the solution

$$U(x, s) = A(s) e^{-sx/c} + B(s) e^{sx/c} \qquad (2.1.6)$$

It should be noted that the "constants" of integration are functions of the parameter s. Since x is positive, the second solution $e^{sx/c}$ is not bounded as

$s \to \infty$. It is thus not an admissible Laplace transform function (cf. Appendix B). This term is eliminated by choosing $B(s) = 0$. We shall see later that the solution that we have eliminated involving $e^{sx/c}$ also corresponds to a wave moving *toward* the source at $x = 0$ and is thus not a physically acceptable solution in the present problem on these grounds as well. Writing the Laplace transform of the boundary condition as $F(s)$, we have $U(0, s) = F(s)$. Evaluating the solution of the ordinary differential equation (2.1.6) at $x = 0$, we obtain $U(0, s) = A(s)$ and thus the integration constant is given as $A(s) = F(s)$. The transform of the solution is then $U(x, s) = F(s)e^{-sx/c}$. The solution $u(x, t)$ for this problem may now be obtained very simply by using a shift theorem for Laplace transforms (B.20). We have finally

$$u(x, t) = f(t - x/c)H(t - x/c) \tag{2.1.7}$$

which is the solution previously obtained in (1.9.34). Note that a corresponding use of the shift theorem with the discarded solutions $e^{sx/c}$ would yield a solution containing the argument $t + x/c$. Such an expression would represent a wave traveling in the negative x direction, that is, toward the source.

Another way of recovering the result obtained here is to use the convolution theorem (B.27). Since $e^{-sx/c}$ is the Laplace transform of the delta function $\delta(t - x/c)$, we may write[2]

$$u(x, t) = \int_0^t f(t - \tau)\delta(\tau - x/c)\, d\tau \tag{2.1.8}$$

For $t > x/c$ the range of integration will include the peak of the delta function. The integral is then evaluated by setting $\tau = x/c$ in $f(t - \tau)$ and we again obtain

$$u(x, t) = f(t - x/c), \qquad t > x/c \tag{2.1.9}$$

For $t < x/c$, not enough time has elapsed for the disturbance to reach the point x/c and the signal should be zero. The peak of the delta function is now outside the range of integration and the value of the integral is zero, as expected (cf. Fig. 2.1).

Note that if the time dependence of the displacement is modeled by a step function $u(0, t) = u_0 f(t) = H(t)$, then $F(s) = u_0/s$ and $u(x, t) = u_0 H(t - x/c)$. A graph of this result is shown in Figure 2.2 as a function of both x and t.

When the end at $x = 0$ oscillates at a single frequency ω_0, for example, $u(x, 0) = u_0 \cos \omega_0 t H(t)$, the motion of the string is given by

$$u(x, t) = u_0 \cos \omega_0(t - x/c)H(t - x/c) \tag{2.1.10}$$

---

[2]We could also write the solution in the form $u(x, t) = \int_0^t f(\tau)\delta(t - \tau - x/c)\, d\tau$.

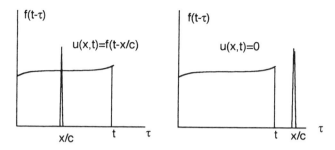

Figure 2.1. Location of delta function for evaluation of (2.1.8).

In more complicated situations it is convenient to write $u(0, t) =$ $\text{Re}[u_0\exp(\pm i\omega_0 t)]H(t)$ and then the solution is the real part of

$$u(x, t) = \text{Re}[u_0 e^{\pm i\omega_0(t - x/c)} H(t - x/c)] \qquad (2.1.11)$$

We will choose the lower (minus) sign in our subsequent usage of this convention.

After passage of the leading edge of the signal, the string executes the steady oscillation $u(x, t) = u_0 \cos \omega_0(t - x/c)$. Frequently one is only interested in this so-called steady state response. It can also be obtained from the separation-of-variables procedure developed in the previous chapter and is most readily accomplished by using the complex quantities introduced here. We write the solution of the separated wave equation (1.8.10) in the form

$$X(x) = Ae^{i\alpha x} + Be^{-i\alpha x}, \qquad T(t) = Ce^{i\alpha ct} + De^{-i\alpha ct} \qquad (2.1.12)$$

The choice of sign for the complex time dependence referred to above is now made by choosing only the term $De^{-i\alpha ct}$ in $T(t)$. Then $u(x, t) = ADe^{i\alpha(x - ct)} +$ $BDe^{-i\alpha(x + ct)}$. Since the wave moves away from the source, that is, in the positive x direction, it is only a function of $x - ct$. We therefore set $B = 0$ and obtain $u(x, t) = ADe^{i\alpha(x - ct)}$. At $x = 0$ we impose the boundary condition

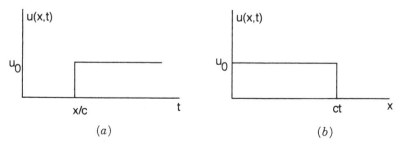

Figure 2.2. Response of string to a step function boundary condition as a function of time (a) and space (b).

$u(0, t) = u_0 \exp(-i\omega_0 t)$ and obtain $AD = u_0$ and $\alpha c = \omega_0$. We now have $u(x, t) = u_0 \exp[-i\omega_0(t - x/c)]$. The real part of this result is again the steady state response of the string.

As an application of the Laplace transform method to a string of finite length, consider a string of length L fixed at both ends so that $u(0, t) = u(L, t) = 0$ that has an initial displacement $u(x, 0) = \sin \pi x/L$ and no initial velocity, that is, $u_t(x, 0) = 0$. In this example the transformed equation (2.1.4) takes the form

$$U_{xx} - \left(\frac{s}{c}\right)^2 U = -\frac{u_0 s}{c^2} \sin \frac{\pi x}{L} \qquad (2.1.13)$$

which must satisfy the transformed boundary conditions $U(0, s) = U(L, s) = 0$. The solution of the ordinary differential equation (2.1.10) is composed of a solution of the homogeneous equation U_h plus a particular solution U_p. For a region of finite length it is usually convenient to express the homogeneous solution of (2.1.13) in terms of hyperbolic functions although the present example is sufficiently simple that no real advantage is gained. A particular solution may be seen by inspection to be of the form $U_p = K \sin \pi x/L$. The amplitude K, which depends upon s, is found by substituting U_p into (2.1.13) and equating coefficients of $\sin \pi x/L$. One finds $K = u_0 s/[s^2 + (\pi c/L)^2]$. The solution U is thus of the form

$$U(x, s) = A(s)\cosh(sx/L) + B(s)\sinh(sx/L) + K(s)\sin(\pi x/L) \quad (2.1.14)$$

with $K(s)$ given above. The constants $A(s)$ and $B(s)$ in the homogeneous solution are determined by requiring that the entire solution satisfy the homogeneous boundary conditions $U(0, s) = U(L, s) = 0$. One readily finds that $A(s) = B(s) = 0$ and therefore

$$U(x, s) = \frac{u_0 s}{s^2 + (\pi c/L)^2} \sin \frac{\pi x}{L} \qquad (2.1.15)$$

From entry 11 in Table F.2 the s dependence is seen to be that associated with $\cos(\pi ct/L)$. The inverse transform is thus found to be

$$U(x, t) = u_0 \cos \frac{\pi ct}{L} \sin \frac{\pi x}{L} \qquad (2.1.16)$$

The string is seen to oscillate in its fundamental mode of vibration.

Problems

Use Laplace transform methods to solve the following problems.

2.1.1 $y_{xx} - \frac{1}{c^2} y_{tt} = -\frac{\mathcal{F}_0}{T} \sin \frac{\pi x}{L}$

with initial conditions $y(x, 0) = y_t(x, 0) = 0$ and boundary conditions $y(0, t) = y(L, t) = 0$.

2.1.2 $y_{xx} - c^{-2}y_{tt} = 0$, $y(x, 0) = 0$, $y_t(x, 0) = v_0 \sin \pi x/L$, $y(0, t) = y(L, t) = 0$.

2.1.3 $y_{xx} - c^{-2}y_{tt} = \mathcal{F}_0/Te^{-t/T}\sin \pi x/L$, $y(x, 0) = y_t(x, 0) = 0$, $y(0, t) = y(L, t) = 0$.

(a) Show that for $t > 0$ the string displacement is

$$y(x, t) = \frac{\mathcal{F}_0 L^2}{T(\pi c)^2} \frac{\sin \pi x/L}{1 + (L/\pi c\tau)^2} \left(e^{-t/\tau} - \cos \frac{\pi ct}{L} + \frac{L}{\pi ct} \sin \frac{\pi ct}{L} \right)$$

Note that as $\tau \to \infty$ the solution of Problem 2.1.1 is obtained.

(b) Show that as $\tau \to 0$ while $\mathcal{F}_0 \to \infty$ such that the impulse density $\mathcal{F}_0 \int_0^\infty e^{-t/\tau} dt = \mathcal{F}_0\tau$ remains constant, one recovers the solution of Problem 2.1.1 when $\mathcal{F}_0\tau$ is set equal to ρv_0, the momentum density of the string. (Recall that $\rho = T/c^2$.)

2.2 DIFFUSION EQUATION

When the Laplace transform method is applied to the diffusion equation, it is found that the inversion of the transform $U(x, s)$ to obtain the solution $u(x, t)$ requires a somewhat more elaborate calculation than that for the wave equation. The transform of the heat equation $u_{xx} - \gamma^{-2}u_t = 0$ is

$$U_{xx} - (s\gamma^{-2})U = -\gamma^{-2}u(x, 0) \tag{2.2.1}$$

and for the homogeneous initial condition $u(x, 0) = 0$, the solution is

$$U(x, s) = A(s)e^{-x\sqrt{s}/\gamma} + B(s)e^{x\sqrt{s}/\gamma} \tag{2.2.2}$$

If we again consider a semi-infinite system $(0 < x < \infty)$, then the solution that diverges for large x must again be discarded. Assuming a positive sign for $\sqrt{s}$ we must set $B(s) = 0$ and obtain

$$U(x, s) = A(s)e^{-x\sqrt{s}/\gamma} \tag{2.2.3}$$

The presence of the term $\sqrt{s}$ in the exponent is the source of the complication mentioned above. There is no simple shift theorem associated with $e^{-x\sqrt{s}/\gamma}$.

In order to compare solutions for the wave equation and the diffusion equation, let us now solve the diffusion equation for the same step function boundary condition and homogeneous initial condition as considered above for the wave equation, that is, $u(0, t) = H(t)$, $u(x, 0) = 0$. Then $U(0, s) = u_0/s$ and from

(2.2.3) the transform of the solution is

$$U(x, s) = \frac{u_0}{s} e^{-x\sqrt{s}/\gamma} \qquad (2.2.4)$$

The function s that arises in this problem is one of a number of expressions that frequently occur when heat conduction problems are solved by the Laplace transform method. A brief table listing the simplest of these expressions,[3] along with the corresponding function $u(x, t)$, are shown in Figure 2.3. A graph of each function $u(x, t)$ and the boundary condition at $x = 0$ that gives rise to this solution are also included. It should be noted that these transforms arise in many other problems besides those indicated in the figure.

The third entry, with $k = x/\gamma$, is the one appropriate for the present problem. It contains the complimentary error function defined as $\text{erfc}(x) = 1 - \text{erf}(x)$ where $\text{erf}(x)$ is merely an abbreviation for the integral

$$\text{erf}(x) = \frac{2}{\sqrt{\pi}} \int_0^x e^{-y^2} \, dy \qquad (2.2.5)$$

This latter integral is known as the error function. A graph of the error function is shown in Figure 2.4. The general shape of this curve is as should be expected since it is proportional to the area under the bell-shaped curve $\exp(-x^2)$ from $x = 0$ to some specified value of x. The numerical factor $2/\sqrt{\pi}$ is introduced so that $\text{erf}(x)$ will approach unity for large values of x.

According to the third entry in the table in Figure 2.3, the temperature distribution in the problem considered here is

$$u(x, t) = u_0 \text{erfc}(x/2\gamma\sqrt{t}) \qquad (2.2.6)$$

The graph associated with this entry thus corresponds to a rise in temperature from zero to u_0, an expected result. On comparing this solution of the diffusion equation with the previous solution of the wave equation, it is seen that the net effect of the $e^{-x\sqrt{s}/\gamma}$ dependence has been to introduce a "smearing out" of the abrupt step function solution that resulted from the $e^{-sx/c}$ dependence in the transform for the wave equation. It should also be noted that for the diffusion process a nonzero (albeit small) value of $u(x, t)$ occurs for all values of x as soon as t is greater than zero. The diffusion equation thus predicts a signal that moves with an infinite velocity. This nonphysical result is not of practical concern, however. Temperature variations large enough to be observable can be expected to evolve in a manner that conforms to experiment.

The solution of the diffusion problem given in Eq. (2.2.6) contains the *two* independent variables x and t in the *single* ratio $x/\sqrt{t}$. If this ratio or equivalently

---

[3]For a more extensive table see Erdélyi et al. (1954).

$$F(s) = e^{-k\sqrt{s}}$$

$$f(t) = \frac{k}{2\sqrt{\pi}t^{3/2}} e^{-k^2/4t}$$

$$u(0, t) = \delta(t)$$

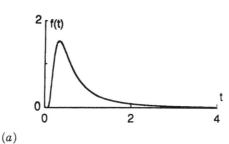

(a)

$$F(s) = \frac{e^{-k\sqrt{s}}}{\sqrt{s}}$$

$$f(t) = \frac{1}{\sqrt{\pi}t^{1/2}} e^{-k^2/4t}$$

$$-u_x(x) = \delta(t)$$

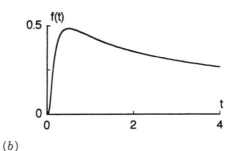

(b)

$$F(s) = \frac{e^{-k\sqrt{s}}}{s}$$

$$f(t) = Erfc\left(\frac{k}{2\sqrt{t}}\right)$$

$$u(0, t) = H(t)$$

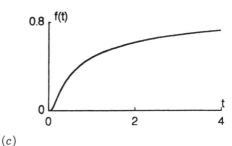

(c)

$$F(s) = \frac{e^{-k\sqrt{s}}}{s^{3/2}}$$

$$f(t) = 2\sqrt{\frac{t}{\pi}} e^{-k^2/4t} - kErfc\left(\frac{k}{2\sqrt{t}}\right)$$

$$-u_x(0, t) = H(t)$$

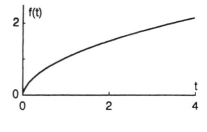

(d)

Figure 2.3. Various boundary conditions for a semi-infinite system, the resulting temperature variation (for $k = x/\gamma = 1$) and the Laplace transform of that temperature variation $F(s)$.

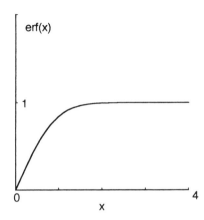

Figure 2.4. The error function $\mathrm{erf}(x) = (2/\sqrt{\pi})\int_0^x d\zeta \exp(-\zeta^2)$.

x^2/t were used as a new independent variable, and the original partial differential equation expressed in terms of it, we would be able to reduce the original partial differential equation to an ordinary differential equation. A solution in which two or more independent variables appear in such an interrelated form is usually referred to as a similarity solution. An example is considered in Problem 2.2.6.

The error function that occurred in the previous problem can be used in other problems to obtain an extremely useful and convenient approximate solution to diffusion problems for small values of t. As an example, recall from Section 1.4 that at early times the infinite-series solution provided by the separation-of-variables method may be very slowly convergent and thus require many terms. To show how the error function provides a rapidly convergent expansion at early times we now reexamine the problem considered in Section 1.4 in which the ends of a beam of length L that is initially at temperature u_0 are placed at zero temperature. We thus have the initial and boundary conditions $u(x, 0) = u_0$ and $u(0, t) = u(L, t) = 0$.

The Laplace transform of the diffusion equation in this case is

$$U_{xx} - (s/\gamma^2)U = -u_0/\gamma^2 \tag{2.2.7}$$

Since this is an inhomogeneous differential equation, the solution will be a sum of a particular solution and a solution of the homogeneous equation. In the present example the particular solution is sufficiently simple that it may be constructed by inspection and is merely the constant $U_p(x, s) = u_0/s$. The solution of the homogeneous equation has already been given in (2.2.2). However, for systems of finite extent it is usually advantageous to express this solution in terms of hyperbolic functions. To satisfy the boundary conditions with a minimum of algebraic manipulation, in the present instance we use the expression[4]

$$U(x, s) = A(s)\sinh x\sqrt{s}/\gamma + B(s)\sinh(L - x)\sqrt{s}/\gamma + u_0/s \quad (2.2.8)$$

Imposition of the boundary conditions $U(0, s) = U(L, s) = 0$ yields $A(s) = B(s) = -(u_0/s \sin L\sqrt{s}/\gamma)$. The solution is thus

$$U(x, s) = \frac{u_0}{s}\left[1 - \frac{\sinh x\sqrt{s}/\gamma + \sinh\sqrt{s}(L - x)/\gamma}{\sinh L\sqrt{s}/\gamma}\right]$$

$$= \frac{u_0}{s}\left[1 - \frac{\cosh(L - 2x)\sqrt{s}/2\gamma}{\cosh L\sqrt{s}/2\gamma}\right] \quad (2.2.9)$$

where the identities $\sinh a + \sinh b = 2 \sinh[(a + b)/2]\cosh[(a - b)/2]$ and $\sinh 2a = 2 \sinh a \cosh a$ have been used to obtain the second form of the solution.

An extensive table of Laplace transforms would include this second expression for $U(x, s)$ and give the inverse transform as precisely the infinite-series solution obtained by the separation-of-variables method in (1.4.17). The Fourier series form of the solution obtained in Chapter 1 may be recovered from the Laplace transform result developed here by expanding $U(x, s)$ in a Fourier sine series. The expansion is available from entry 25 in Table F.1 by replacing x by $\pi x/L$ and a by $L\sqrt{s}/\gamma$. The Fourier series for unity is also available from entry 25 by setting $a = 0$. Employing the definition $\nu = (\gamma\pi/L)^2$ from (1.4.8), we obtain

$$U(x, s) = \frac{u_0}{\pi}\sum_{n\,\text{odd}}\frac{1}{n(s + \nu n^2)}\sin\frac{n\pi x}{L} \quad (2.2.10)$$

The inverse transform of this series, obtained through term-by-term application of the relation $L^{-1}[1/(s + n^2)] = \exp(-n^2\nu t)$, is the result given in (1.4.17).

As noted above, such an infinite series is slowly convergent at early times ($\nu t \ll 1$). As a guide toward developing an approximate expression for the inverse transform that is useful (i.e., rapidly convergent) at early times, we recall the complementarity between s and t in Laplace transform theory (Section B.2). An expression that approximates the above result for $U(x, s)$ at *large* values of s may be expected to approximate the solution $u(x, t)$ for *small* values of t. We thus develop an approximate expression for $U(x, s)$ that is valid as s

---

[4]The reader unfamiliar with the minor computational advantages associated with proper choice of form of solution is urged to obtain (2.2.8) by using the other two forms for the homogeneous solution, namely $U_h = c_1(s)e^{-x\sqrt{s}/\gamma} + c_2(s)e^{x\sqrt{s}/\gamma}$ and $U_h = c_3(s)\sinh x\sqrt{s}/\gamma + c_4(s)\cosh x\sqrt{s}/\gamma$.

$\rightarrow \infty$. To this end let us first simplify the algebra by introducing the dimensionless variables $\xi = x\sqrt{s}/\gamma$ and $\lambda = L\sqrt{s}/\gamma$. Note that as $s \rightarrow \infty$, λ will always be large while ξ may be large or small since x may approach zero. When the hyperbolic functions are written as exponentials, the solution (2.2.9) takes the form

$$U(x, s) = \frac{u_0}{s} \left(1 - \frac{e^{-1/2\lambda - \xi} + e^{-1/2\lambda + \xi}}{e^{1/2\lambda} + e^{-1/2\lambda}} \right)$$

$$= \frac{u_0}{s} \left(1 - \frac{e^{-\xi} + e^{-\lambda + \xi}}{1 + e^{-\lambda}} \right) \tag{2.2.11}$$

For $\lambda \gg 1$ we may expand the denominator using $(1 + x)^{-1} = 1 - x + x^2 \cdots$ and obtain

$$U(x, s) = \frac{u_0}{s} - \frac{u_0}{s} [e^{-\xi} - e^{-(\xi + \lambda)} + e^{-(\xi + 2\lambda)} - \cdots$$

$$+ e^{-(\lambda - \xi)} - e^{-(2\lambda - \xi)} + e^{-(3\lambda - \xi)} - \cdots] \tag{2.2.12}$$

In terms of s dependence, each exponential term has the form given in entry 3 of the brief table of Laplace transforms given in Figure 2.3. When the inverse transform of this series is taken term by term, we obtain

$$u(x, t) = u_0 - u_0 \left(\text{erfc} \frac{x}{2\gamma\sqrt{t}} - \text{erfc} \frac{x + L}{2\gamma\sqrt{t}} + \cdots \right)$$

$$- u_0 \left(\text{erfc} \frac{L - x}{2\gamma\sqrt{t}} - \text{erfc} \frac{2L - x}{2\gamma\sqrt{t}} + \cdots \right) \tag{2.2.13}$$

For small values of t, all terms involving erfc approach zero unless x approaches one of the values nL where $n = 0, \pm 1, \pm 2, \ldots$. For x not near one of these values only the first term in the solution is significantly different from zero. We then have $u(x, t) \approx u_0$, which is the expected temperature on the beam at early times. Since the beam is located in the region $0 < x < L$, only the values $x = nL$ where $n = 0, 1$ correspond to values of x that are on the beam. As x approaches zero, the function $\text{erfc}(x/2\gamma\sqrt{t})$ approaches unity while $\text{erfc}[x(L - x)/2\gamma\sqrt{t}]$ is nearly zero. The boundary condition $u(0, t) = 0$ is thus satisfied to a high degree of approximation. Similar considerations apply at $x = L$. As t increases, the "tail" from each of these error functions lengthens and gives a contribution to the opposite end of the beam. The boundary conditions are now no longer approximately satisfied by just these first terms in the series and it becomes necessary to include additional terms. Eventually it is preferable to return to the series obtained previously in (1.4.17) by the separation-of-variables method.

Problems

2.2.1 Solve the following equations by Laplace transform methods:
(a) $u_{xx} - \gamma^{-2} u_t = 0$, $u_x(0, t) = u_x(L, t) = 0$, $u(x, 0) = u_0$. Why is the answer to this problem obvious from physical considerations?
(b) $u_{xx} - \gamma^{-2} u_t = 0$, $u(0, t) = u(L, t) = 0$, $u(x, 0) = u_0 \sin^3 \pi x/L$.
(c) $u_{xx} - \gamma^{-2} u_t = 0$, $u_x(0, t) = u_x(L, t) = 0$, $u(x, 0) = u_0 \sin^2 \pi/L$.

2.2.2 A beam of length L at temperature zero has insulated ends. An amount of heat H_0 that varies linearly along the beam is supplied to the beam at time $t = t_0$ so that the governing equation is

$$u_{xx} - \gamma^{-2} u_t = -\frac{2H_0 x}{KL^2} \delta(t - t_0)$$

Justify the form of the source term and evaluate the Laplace transform in a suitable limit to determine the temperature distribution as $t \to \infty$. The result should be obvious from physical considerations.

2.2.3 A semi-infinite beam is heated at a constant rate at the end $x = 0$. There is therefore a boundary condition $u_x(0, t) = -J_0/K$ where J_0 is a constant and K is the conductivity of the beam. If the beam is initially at zero temperature, determine the temperature of the beam at $x = 0$ as a function of time.

2.2.4 A beam of length L has boundary conditions $u(0, t) = u(L, t) = 0$ and initial temperature distribution $u(x, 0) = u_0 \sin \pi x/L$. It receives an instantaneous cooling at $t = t_0$ so that the equation governing temperature on the beam is

$$\frac{\partial^2 u}{\partial x^2} - \frac{1}{\gamma^2} \frac{\partial u}{\partial t} = q_0 \delta(t - t_0) \sin \frac{\pi x}{L}$$

(a) Determine the temperature on the beam at later time.
(b) What is the relation between u_0, q_0, and t_0 so that $u(x, t) = 0$ for $t > t_0$?

2.2.5 A beam of length L kept at zero temperature at both ends has the initial temperature distribution $u(x, 0) = (\pi H_0/2\rho cL) \sin \pi x/L$. The amount of heat on the beam initially is thus H_0. Using methods developed previously, the temperature distribution at any later time is obtained by solving a homogeneous heat equation with the inhomogeneous initial condition given above to yield the result

$$u(x, t) = \frac{\pi H_0}{2\rho cL} e^{-(\pi\gamma/L)^2 t} \sin \frac{\pi x}{L}, \qquad \gamma^2 = \frac{K_0}{\rho c}$$

Now consider a situation that reduces to the above result in a certain limit. Assume that the beam is initially at zero temperature so that $u(x, 0) = 0$, but a source term

$$s(x, t) = \frac{\pi H_0}{2L} \frac{1}{T} e^{-t/T} \sin \frac{\pi x}{L}$$

supplies an amount of heat H_0 to the beam in a time roughly equal to T.

(a) Solve

$$u_{xx} - \gamma^{-2} u_t = \frac{1}{K_0} s(x, t), \qquad u(x, 0) = 0$$

by Laplace transform techniques.

(b) Show that as $T \to 0$ the solution obtained in part (a) reduces to the solution quoted above.

2.2.6 Similarity solution of the heat equation $u_{xx} - \gamma^{-2} u_t = 0$.

(a) Set $\xi = x^\alpha t$ where α is to be determined and show that

$$\frac{\partial u}{\partial x} = \alpha x^{\alpha - 1} \frac{du}{d\xi}, \qquad \frac{\partial u}{\partial t} = x^\alpha \frac{du}{d\xi}$$

(b) Substitute into the diffusion equation and show that the result depends only on ξ if $\alpha = -2$.

(c) Obtain the heat equation in the form

$$4\xi^2 \frac{d^2 u}{d\xi^2} + (6\xi - \gamma^{-2}) \frac{du}{d\xi} = 0$$

(d) Show that the solution of this equation that satisfies $u(0) = u_0$, $u(+\infty) = 0$ is $u_0 = u \operatorname{erf}(x/2\gamma \sqrt{t})$.

(e) What is the initial temperature distribution that evolves according to this solution?

2.2.7 A semi-infinite beam has a temperature variation $u(0, t) = f(t)$ imposed at $x = 0$ for $t \geq 0$.

(a) If the initial temperature on the beam is zero, show that the heat flux $J(t) = -Ku_x(0, t)$ onto the beam is given by

$$J(t) = \sqrt{\frac{K\rho c}{\pi}} \int_0^t (t - \tau)^{-1/2} \frac{df}{d\tau} d\tau$$

(b) Note that when an integration by parts is attempted, the integrated term will diverge at the upper limit. To examine this situation, integrate from 0 to $t - \epsilon$ and obtain

$$J(t) = \epsilon^{-1/2}f(t) - \frac{1}{2}\int_0^{t-\epsilon}(t - \tau)^{-3/2}f(\tau)$$

Now add and subtract the term $f(t)/\sqrt{t}$ and show that one obtains the result

$$J(t) = \sqrt{\frac{K\rho c}{\pi}}\left[\frac{f(t)}{\sqrt{t}} + \frac{1}{2}\int_0^t \frac{f(\tau) - f(t)}{(t - \tau)^{3/2}}d\tau\right]$$

which converge at the upper limit provided $|f(\tau) - f(t)| \sim |t - \tau|^\mu$ with $\mu > \frac{1}{2}$.

2.2.8 Solve $u_{xx} - \gamma^{-2}u_t = 0$, $u(x, 0) = u_0$, $u(0, t) = 0$, $t > 0$, $x \geq 0$, by using a Laplace transform on *space*, that is,

$$L\{u(x, t)\} = \int_0^\infty e^{-px}u(x, t)\,dx = U(p, t)$$

(a) Obtain the solution to the resulting first order differential equation in the form

$$U(p, t) = e^{(\gamma p)^2 t}\left[\frac{u_0}{p} - \gamma^2\int_0^t u_x(0, t')e^{-(\gamma p)^2 t'}\,dt'\right]$$

where $u_x(0, t')$ is unknown.

(b) The unknown $u_x(0, t)$ is determined from the requirement

$$\frac{u_0}{p} = \gamma^2\int_0^\infty u_x(0, t')e^{-(\gamma p)^2 t'}\,dt'$$

Why?

(c) Use entry 3 in Table F.2 to determine $u_x(0, t)$ from the equation given in (b).

(d) With $u_x(0, t)$ thus determined, show that

$$U(p, t) = \frac{u_0}{p}e^{(\gamma p)^2 t}\text{erfc}(\gamma p \sqrt{t})$$

and by using the transform pair 18 in Table F.2, show that

$$u(x, t) = u_0\text{erf}(x/2\gamma\sqrt{t})$$

2.2.9   Write $\text{erfc}(x)$ in the form

$$\text{erfc}(x) = \frac{2}{\sqrt{\pi}} \int_x^\infty y e^{-y^2} \frac{dy}{y}$$

and integrate by parts to show that

$$\text{erfc}(x) \approx \frac{e^{-x^2}}{\sqrt{\pi}x} \left(1 - \frac{1}{2x^2} + \frac{1 \cdot 3}{(2x^2)^2} - \frac{1 \cdot 3 \cdot 5}{(2x^2)^3} + \cdots \right)$$

a result that is useful when $x \gg 1$.

2.3 MISCELLANEOUS EXAMPLES

Advantages of the Laplace transform method include the ease with which step function and delta function sources may be treated as well as that the *transformed quantity itself* frequently contains useful physical information. Thus there is often no need to even consider the question of obtaining the inverse transform $u(x, t)$ from $U(x, s)$. We now consider three very simple examples that emphasize these aspects of the Laplace transform method.

2.3.1   Impulse Acting on a String

Consider a string of length L with fixed ends and the initial displacement $u(x, 0) = u_0 \sin \pi x / L$. The string is acted on by an impulsive driving force density $(\pi u_0 T / cL) \sin \pi x (t - L/2c)L$ where T is the tension in the string and c is the velocity of waves on the string. To determine the motion of the string both before and after the time $t = L/2c$ at which the force acts, we solve

$$u_{xx} - \frac{1}{c^2} u_{tt} = -\frac{\pi u_0}{cL} \sin \frac{\pi x}{L} \delta\left(t - \frac{L}{2c} \right), \qquad u(0, t) = u(L, t) = 0$$

$$(2.3.1)$$

The Laplace transform of this equation is

$$U_{xx} - \left(\frac{s}{c} \right)^2 U = -u_0 \left(\frac{s}{c^2} + \frac{\pi}{cL} e^{-sL/2c} \right) \sin \frac{\pi x}{L} \qquad (2.3.2)$$

A particular solution of this ordinary differential equation may be obtained in the form $U_p = D \sin \pi x / L$ where the constant D is determined by substitution of U_p into the differential equation (2.3.2). One readily finds

$$D = \frac{u_0}{s^2 + (\pi c/L)^2} \left(s + \frac{\pi c}{L} e^{-sL/2c} \right) \tag{2.3.3}$$

The general solution of the equation for $U(x, s)$ can be written

$$u(x, s) = A(s)\sinh sx/L + B(s)\sinh s(L - x)/L + D \sin \pi x/L \tag{2.3.4}$$

with D given in (2.3.3). The boundary conditions $U(0, s) = U(L, s) = 0$ now yield $A(s) = B(s) = 0$. The solution is thus

$$U(x, s) = D \sin \frac{\pi x}{L}$$

$$= u_0 \left[\frac{s}{s^2 + (\pi c/L)^2} + \frac{\pi c}{L} \frac{e^{-sL/2c}}{s^2 + (\pi c/L)} \right] \sin \frac{\pi x}{L} \tag{2.3.5}$$

The first term in the square brackets is the transform of $u_0 \cos \pi ct/L$. The inverse transform of the second term is obtained by use of a shift theorem [Eq. (B.20)]. The string displacement is found to be

$$u(x, t) = u_0 \left[\cos \frac{\pi ct}{L} + \sin \frac{\pi c}{L} \left(t - \frac{L}{2c} \right) H \left(t - \frac{L}{2c} \right) \right] \sin \frac{\pi x}{L} \tag{2.3.6}$$

Rewriting the term by multiplying the step function as $\sin[(\pi c/L)(t - L/2c)] = \sin[(\pi ct/L) - \pi/2] = \cos(\pi ct/L)$, we see that the final result is

$$u(x, t) = \begin{cases} u_0 \sin \dfrac{\pi x}{L} \cos \dfrac{\pi ct}{L}, & 0 \leq t < \dfrac{L}{2c} \\[3mm] 0, & t > \dfrac{L}{2c} \end{cases} \tag{2.3.7}$$

The driving force imposed on the string at $t = L/2c$ is thus exactly that required to stop the string at that time.

2.3.2 Long-Time Behavior

The relation obtainable from Eq. (B.15), namely,

$$\lim_{s \to 0} [sU(x, s)] = u(x, \infty) \tag{2.3.8}$$

is frequently useful in obtaining information on the long-time behavior of a system. This limit is a simple example of how physically interesting information may be obtained from the transform itself.

As an illustration, consider a beam of length L that is kept at temperature u_0 at $x = L$ and has a constant flux of heat J_0 supplied to it at $x = 0$. If the beam is initially at zero temperature, the conditions imposed upon the temperature $u(x, t)$ are the initial condition $u(x, 0) = 0$ and the boundary conditions $u(L, t) = u_0$ and $u_x(0, t) = -J_0/K$ where K is the conductivity. We expect that as $t \to \infty$, the system will "forget" about its initial conditions and reduce to the steady state situation governed by the boundary conditions. We now consider how this long-time behavior may be extracted from the Laplace transform solution without actually performing the inversion from $U(x, s)$ back to $u(x, t)$. The Laplace transform of the diffusion equation is

$$U_{xx} - s\gamma^{-2}U = 0 \tag{2.3.9}$$

While any linear combination of solutions $e^{\pm x\sqrt{s}/\gamma}$, $\sinh x\sqrt{s}/\gamma$ or $\cosh x\sqrt{s}/\gamma$, could be used as a solution, as noted in conjunction with (2.2.8), it is convenient to take a combination that will simplify the algebraic procedure involved in satisfying the boundary conditions. In this example the convenient combination is

$$U(x, s) = A(s)\cosh x\sqrt{s}/\gamma + B(s)\sinh \sqrt{s}(L - x)/\gamma \tag{2.3.10}$$

as is evident when the transformed boundary conditions $U_x(0, s) = -J_0/Ks$ and $U(L, s) = u_0/s$ are imposed. The constants A and B are immediately found to be $A = u_0/(s \cosh L\sqrt{s}/\gamma)$ and $B = J_0/(Ks^{3/2} \cosh L\sqrt{s}/\gamma)$. Forming the expression $sU(x, s)$, we obtain

$$sU(x, s) = \frac{\cosh x\sqrt{s}/\gamma}{\cosh L\sqrt{s}/\gamma} + \frac{J_0\gamma}{K\sqrt{s}} \frac{\sinh (L - x)\sqrt{s}/\gamma}{\cosh L\sqrt{s}/\gamma} \tag{2.3.11}$$

Taking the limit $s \to 0$ we find[5]

$$u(x, \infty) = \lim_{s \to 0} [sU(K, s)] = u_0 + \frac{J_0}{K}(L - x) \tag{2.3.12}$$

This result could, of course, be obtained more directly since as $t \to \infty$ the problem reduces to the steady state situation that is governed by $u_{xx} = 0$. The solution of the ordinary differential equation that satisfies the boundary conditions of the problem is the limiting solution obtained here.

---

[5]A simple way to do this is to replace the hyperbolic functions by the first few terms in their Taylor expansions and then let $s \to 0$.

2.3.3 Total Heat Flow through an End

We now consider another instance in which physically interesting information may be obtained directly from the transform. We reconsider the problem solved in Section 1.4 where the series method was used to determine the temperature distribution on a beam of length L for which $u(x, 0) = u_0$ and $u(0, t) = u(L, t) = 0$. The series solution was then used to determine the total amount of heat flowing off the beam at one end. We now show how this information is contained in the Laplace transform solution of the problem. The appropriate Laplace transform has already been obtained in Section 2.2 where it was used to determine the early-time behavior of the solution to this problem. The transform was given in (2.2.9) as

$$U(x, s) = \frac{u_0}{s}\left[1 - \frac{\cosh(L - 2x)\sqrt{s/2\gamma}}{\cosh L\sqrt{s/2\gamma}}\right] \tag{2.3.13}$$

The total heat flow out the end at $x = 0$ is given by

$$h(0) = \int_0^\infty J(0, t)\, dt = -K\int_0^\infty u_x(0, t)\, dt \tag{2.3.14}$$

This integral over u_x may be expressed as a Laplace transform by noting that

$$\int_0^\infty u_x(0, t)\, dt = \lim_{s \to 0}\int_0^\infty e^{-st}u_x(0, t)\, dt = U_x(0, 0) \tag{2.3.15}$$

The total heat flow is thus

$$h(0) = -KU_x(0, 0) \tag{2.3.16}$$

In the example considered here

$$U_x(x, s) = \frac{u_0}{\gamma\sqrt{s}}\frac{\sinh(L - 2x)\sqrt{s/2\gamma}}{\cosh L\sqrt{s/2\gamma}} \tag{2.3.17}$$

As s approaches zero we have

$$U_x(x, 0) = \frac{u_0}{2\gamma^2}(L - 2x) \tag{2.3.18}$$

and thus $U_x(0, 0) = u_0/2\gamma^2$. Since $K = \gamma^2\rho c$, we obtain $h(0) = -\rho c u_0 L/2$ as given previously in Section 1.4. Similarly, at the other end we obtain $h(L) = \rho c u_0 L/2$.

Problems

2.3.1 **(a)** Use Laplace transform techniques to obtain $U(x, s)$ for

$$u_{xx} - \gamma^{-2}u_t = 0, \qquad u(0, t) = u_x(L, t) = 0,$$

$$u(x, 0) = u_0 \sin \pi x/L$$

(b) Use the transform $U(x, s)$ to show that all the heat flows out the end at $x = 0$.

2.3.2 Solve $u_{xx} - \gamma^{-2}u_t = 0$, $u(x, 0) = u_x(L, t) = 0$, $u_x(0, t) = J_0 e^{-\alpha t}$ by Laplace transform methods.

(a) Evaluate $H = \int_0^\infty - K u_x(0, t) \, dt$.

(b) Determine $u(\infty)$, the final (constant) temperature on the beam, by evaluating $H = \int_0^L \rho c u(\infty) \, dx$ where H is as obtained in part (a).

(c) Evaluate the Laplace transform $U(x, s)$ in an appropriate limit to recover the value of $u(\infty)$ obtained in part (b).

2.3.3 Take the Laplace transform of $u_{xx} - \gamma^{-2}u_t = 0$, $u(0, t) = u_x(L, t) = 0$, $u(x, 0) = u_0 \sin \pi x/L$ and obtain

$$U(x, s) = \frac{u_0}{s + \nu} \left(\frac{\pi \gamma}{L} \frac{\sinh \sqrt{s} x/\gamma}{\sqrt{s} \cosh \sqrt{s} L/\gamma} + \sin \frac{\pi x}{L} \right)$$

Do not attempt to obtain $u(x, t)$ but show that all the heat on the beam flows out the end at $x = 0$ by evaluating

$$h(0) = -K \int_0^\infty u_x(0, t) \, dt = -LU_x(0, 0)$$

2.3.4 A beam of length L has the initial temperature distribution $u(x, 0) = u_1 \sin \pi x/L$. The boundary conditions are $u(0, t) = 0$ and $u(L, t) = u_0$.

(a) Show that the Laplace transform of the temperature is

$$U(x, s) = \frac{u_1 \sin \pi x/L}{s + \nu} + \frac{u_0 \sin \sqrt{s} x/\gamma}{s \sinh \sqrt{s} L/\gamma}$$

(b) Show that the Laplace transform of the total heat on the beam at any time is

$$H(s) = \rho c \left[\frac{2u_1 L}{\pi (s + \nu)} + \frac{\gamma u_0}{s^{3/2} \sinh (\sqrt{s}/L/\gamma)} \left(\cosh \frac{\sqrt{s} L}{\gamma} - 1 \right) \right]$$

(c) Use the relation between $f(t)$ and its transform $F(s)$ at early time to show that $h(t)$, the total heat on the beam, is

$$h(t) \cong \frac{2}{\pi} \rho c u_1 L \left(1 + \frac{u_0}{u_1} \sqrt{\frac{\nu t}{\pi}} - \nu t \right)$$

2.3.5 A semi-infinite beam initially at zero temperature is given a pulse of heat flux at the end $(x = 0)$ at some time t_0 so that the boundary condition there is

$$J(0, t) = -K u_x(0, t) = q\delta(t - t_0)$$

(a) Calculate the temperature distribution on the beam at any later time and relate q to the total heat on the beam.

(b) Determine the total heat flux past any point x on the beam by evaluating

$$\int_0^\infty J(x, t)\, dt$$

2.4 POINT SOURCE PROBLEM—PREVIEW OF GREEN'S FUNCTION

We now consider the diffusion of an amount of heat H_0 that is deposited within a small region of a beam in a brief interval of time. If we are not interested in the exact size of the source region or in the exact duration of time that the source acts (i.e., if we are only interested in lengths and times much longer than those associated with the source), then a source of this type is conveniently modeled by employing delta functions. For a source of heat H_0 acting at x' at time t', we may write

$$s(x, t) = H_0 \delta(x - x')\delta(t - t') \tag{2.4.1}$$

Note that the source term $s(x, t)$ has the appropriate dimensions of calories per centimeter per second for a heat source in one dimension.

To avoid the complications of boundary conditions, we consider a beam of infinite extent. We also assume the initial condition $u(x, 0) = 0$. With the source term given above, and recalling that $L[\delta(t - t')] = e^{-st'}$, we find that the Laplace transform of the diffusion equation (1.7.1) with the source term used here is

$$U_{xx} - s\gamma^{-2} U = -\frac{H_0}{K} \delta(x - x') e^{-st'} \tag{2.4.2}$$

Since $\delta(x - x')$ is zero for $x \neq x'$, U satisfies a homogeneous equation for all values of x except at the source point $x = x'$. As in Section 1.5, we may thus solve the homogeneous equation on both sides of x' and then determine how these solutions must be joined at $x = x'$. Denoting the solutions for $x > x'$ and $x < x'$ by $U_>(x, s)$ and $U_<(x, s)$, respectively, we have

$$U_>(x, s) = A(s)e^{x\sqrt{s}/\gamma} + B(s)e^{-x\sqrt{s}/\gamma}, \qquad x > x' \tag{2.4.3}$$
$$U_<(x, s) = C(s)e^{x\sqrt{s}/\gamma} + D(s)e^{-x\sqrt{s}/\gamma}, \qquad x < x'$$

We again choose $\sqrt{s}$ to be positive, and to avoid a divergent result as $x \to +\infty$, we must require $A(s) = 0$. Similarly, for convergence as $x \to -\infty$ we require $D(s) = 0$. The solutions are then $U_> = B(s)e^{-x\sqrt{s}/\gamma}$ and $U_< = C(s)e^{x\sqrt{s}/\gamma}$. To have a continuous variation in $u(x, t)$ and thus[6] also in $U(x, s)$ at $x = x'$, we require $U_>(x', s) = U_<(x', s)$, which yields

$$B(s)e^{-x'\sqrt{s}/\gamma} = C(s)e^{x'\sqrt{s}/\gamma} \tag{2.4.4}$$

A second condition on the constants $B(s)$ and $C(s)$ is obtained by integrating the differential equation (2.4.2) across the point x'. We obtain

$$\int_{x'-\epsilon}^{x'+\epsilon} (U_{xx} - \gamma^{-2}U) \, dx = -\frac{H_0}{K} \int_{x'-\epsilon}^{x'+\epsilon} \delta(x' - x) \, dx = -\frac{H_0}{K} e^{-st'} \tag{2.4.5}$$

Since U_{xx} is a perfect derivative, the first term in the integral is immediately evaluated and we have

$$U_x(x, s)\Big|_{x'-\epsilon}^{x'+\epsilon} - \gamma^{-2} \int_{x'-\epsilon}^{x'+\epsilon} U(x, s) \, dx = -\frac{H_0}{K} e^{-st'} \tag{2.4.6}$$

With U designated by $U_>$ for $x > x'$ and by $U_<$ for $x < x'$, the integrated term may be written $U_{>x}(x', s) - U_{<x}(x', s)$. The integral proportional to γ^{-2} represents the area under the curve $U(x, s)$ in the region between $x' - \epsilon$ and $x' + \epsilon$. Since U is continuous, this area goes to zero as $\epsilon \to 0$. We then have

$$U_{>x}(x', s) - U_{<x}(x', s) = -\frac{H_0}{K} e^{-st'} \tag{2.4.7}$$

Introducing the expressions $U_> = Be^{-x\sqrt{s}/\gamma}$ and $U_< = Ce^{x\sqrt{s}/\gamma}$, we finally obtain

$$B(s)e^{-x'\sqrt{s}/\gamma} + C(s)e^{x'\sqrt{s}/\gamma} = \frac{H_0\gamma}{K\sqrt{s}} e^{-st'} \tag{2.4.8}$$

---

[6]Since the derivative of a step function is a delta function, (2.4.2) for $U(x, s)$ suggests that U_x must have a step function behavior at $x = x'$. This in turn implies that U has a change in slope at $x = x'$.

When this relation is solved simultaneously with (2.4.4) we find $B = (\gamma H_0/(2K\sqrt{s})e^{x'\sqrt{s}/\gamma - st'}$ and $C = \gamma H_0/(2K\sqrt{s})e^{-x'\sqrt{s}/\gamma - st'}$. The final result for both $U_>$ and $U_<$ may now be combined into the single expression

$$U(s, x) = \frac{H_0\gamma}{2K\sqrt{s}} e^{-|x - x'|\sqrt{s}/\gamma - st'} \tag{2.4.9}$$

As is to be expected, the result depends only upon the magnitude of the distance between the observation point x and the source point x'. The temperature itself may now be obtained by noting that the s dependence of $U(x, s)$ is precisely that given by the second entry in Figure 2.3. Recalling that $\gamma^2 = K/\rho c$, we may write the temperature on the beam as

$$\frac{\gamma\rho c u(x, t)}{H_0} = \frac{1}{2\sqrt{\pi(t - t')}} e^{(x - x')^2/4\gamma^2(t - t')} H(t - t') \tag{2.4.10}$$

where $H(t - t_0)$ is the unit step function. If we set the temperature at the source equal to u_0 at some time $t = t_0$, that is, $u(0, t_0) = u_0$, then we may introduce the dimensionless variables $\tau = t/t_0$ and $\xi = x/(2\gamma\sqrt{t_0})$ and write (2.4.10) in the dimensionless form

$$\frac{u(\xi, \tau)}{u_0} = \frac{e^{-\xi^2/\tau}}{\sqrt{\tau}} \tag{2.4.11}$$

A graph of this result for $t' = x' = 0$ is shown in Figure 2.5.

The solution that we have obtained gives the temperature at any space-time point (x, t) when a source is applied at a single (earlier) space-time point (x', t'). The result obtained here is an example of a Green's function. In Chapter 4 it will be shown that a solution such as the one just developed can be used to obtain the solutions to more complicated heat conduction problems.

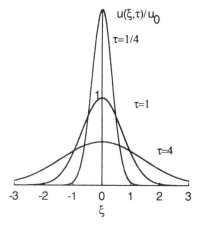

Figure 2.5. Temperature diffusion on an infinite beam due to an instantaneous deposition of heat at $\xi = 0$.

Problems

2.4.1 An infinite beam initially at zero temperature is heated at a constant rate Q_0 calories per second at a point $x = 0$.

(a) Determine the temperature distribution on the beam at later times by solving

$$u_{xx} - \gamma^{-2}u_t = -\frac{1}{K}s(x, t) = -\frac{Q_0}{K}\delta(x), \qquad u(x, 0) = 0$$

(b) Show that the total heat on the beam increases linearly with time, that is,

$$\int_0^\infty \rho c u(x, t)\, dx = Q_0 t$$

2.4.2 A mass M is attached to an infinite string of uniform density ρ_0 that is stretched to a tension T. Interpret the mass as an inhomogeneity of the string by writing the string density as $\rho(x) = \rho_0 + M\delta(x)$. The equation governing the motion of the string displacement $y(x, t)$ is thus

$$y_{xx} - \frac{1}{c^2}y_{tt} = -\frac{M}{T}\delta(x)y_{tt}(0, t)$$

At $t = 0$ the mass is struck a blow that gives it an initial velocity v_0.

(a) Obtain the Laplace transform of the wave equation in the form

$$Y_{xx} - \left(\frac{s}{c}\right)^2 Y = \frac{M}{T}\delta(x)[s^2 Y(0, s) - v_0]$$

(b) By matching solutions of the homogeneous equation across the origin, show that the transform of the displacement is

$$Y(x, s) = \frac{v_0 e^{-s|x|/c}}{s(s + 2\rho c/M)}$$

and that the string displacement is

$$y(x, t) = \frac{1}{2}\frac{Mv_0}{\rho c}\left(1 - \exp\left(-\frac{2\rho}{M}\right)(ct - |x|)\right)H(ct - |x|)$$

Note that the limit $M \to 0$, $v_0 \to \infty$ such that $Mv_0 = $ const yields the result given in (1.9.18) with $y_0 = Mv_0/(2\rho c)$.

(c) Show that the kinetic energy of the mass is $\frac{1}{2}Mv_0^2 e^{-4\rho ct/M}$ and the total energy of the string on either side of the mass is $\frac{1}{4}Mv_0^2(1 - e^{-4\rho ct/M})$.

2.4.3 Solve the first order equation

$$\frac{\partial z}{\partial x} + \frac{1}{v}\frac{\partial z}{\partial t} + \gamma z = 0, \qquad z(x, 0) = f(x),$$

$$-\infty < x < \infty, \qquad 0 \le t < \infty$$

by using the Laplace transform method.

(a) Show that $Z(x, s)$, the transform of $z(x, t)$, satisfies a first order equation that can be put in the form

$$\frac{d}{dx}[Z(x, s)e^{(\gamma + s/v)x}] = v^{-1}z(x, 0)e^{(\gamma + s/v)x}$$

and integrate to obtain

$$Z(x, s) = \frac{1}{v}\int_0^\infty d\xi\, z(x - \xi, 0)e^{-(\gamma + s/v)\xi}$$

(b) If the value of $z(x, t)$ is given along the line $x = Vt$, that is, $z(Vt, t) = F(t)$, where $F(t)$ is known, use the result in (a) to show that $z(x, t)$ may be expressed in terms of $F(t)$ in the form

$$z(x, t) = e^{\gamma v[(x - Vt)/(V - v)]}F\left(\frac{x - vt}{V - v}\right), \qquad v \ne V$$

(c) Reconsider the development leading to the result in (b) for the case in which $v = V$, that is, $z(vt, t) = F(t)$ and show that there is no solution unless $F(t) = Ae^{-\gamma vt}$ where A is an undetermined constant. This problem will be reconsidered by a different method in Section 8.3.

2.4.4 A beam of length L is insulated at both ends. Initially the temperature is zero. For $t > 0$ a constant heat source is placed at the center. Temperature on the beam is thus governed by

$$u_{xx} - \gamma^{-2}u_t = -\frac{Q_0}{K}\delta\left(x - \frac{L}{2}\right),$$

$$u_x(0, t) = u_x(L, t) = 0, \qquad u(x, 0) = 0$$

(a) Show that the Laplace transform of the temperature is

$$U(x, s) = \begin{cases} A(s)\cosh(\sqrt{s}x/\gamma), & 0 \leq x < L/2 \\ A(s)\cosh[\sqrt{s}(L - x)/\gamma], & L/2 < x \leq L \end{cases}$$

where

$$A(s) = \frac{Q_0\gamma}{2Ks^{3/2}} \frac{1}{\sinh(L\sqrt{s}/2\gamma)}$$

(b) Determine the temperature at large time by first expanding $U(x, s)$ for small s and then inverting the transform term by term to obtain the result given in (1.7.16).

3 Two and Three Dimensions

The one-dimensional problems analyzed previously are, of course, idealizations. They are useful because they display in a simple way many of the physical concepts that are described more realistically but in a more cumbersome mathematical framework by problems in two and three dimensions. For instance, consideration of temperature variation over the beam cross section in any of the previous one-dimensional diffusion problems introduces one of the main limitations in treating partial differential equations by analytical methods. It is found that only certain shapes for the cross-sectional area are amenable to exact solution. In this chapter we introduce some of the extensions of the previous developments that are associated with this continuation into additional space dimensions.

3.1 INTRODUCTION

Although the *solutions* of partial differential equations in more than one dimension may be more difficult to obtain, the *equations themselves* are readily extended to more space dimensions by merely introducing vector concepts. For diffusion problems, the heat flow $J = -Ku_x$ is replaced by the heat flow vector $\mathbf{J} = -K(\mathbf{i}u_x + \mathbf{j}u_y + \mathbf{k}u_z) = -K\nabla u$ where $\mathbf{i}$, $\mathbf{j}$, and $\mathbf{k}$ are unit vectors in the x, y, and z directions, respectively, and $\nabla = \mathbf{i}\,\partial/\partial x + \mathbf{j}\,\partial/\partial y + \mathbf{k}\,\partial/\partial z$. In place of the conservation law $\rho c\,\partial u/\partial t + \partial J/\partial x = s(x, t)$ there is the corresponding vector extension $\rho c u_t + \nabla \cdot \mathbf{J} = s(\mathbf{r}, t)$ where $s(\mathbf{r}, t)$ is an abbreviation for $s(x, y, z, t)$. The diffusion equation now becomes $\rho c u_t - \nabla \cdot (K \nabla u) = s(\mathbf{r}, t)$. For constant K, we obtain

$$\frac{1}{\gamma^2} u_t - \nabla^2 u = \frac{1}{K} s(\mathbf{r}, t), \qquad \gamma^2 = \frac{K}{\rho c} \tag{3.1.1}$$

where $\nabla^2 u = \nabla \cdot \nabla u = u_{xx} + u_{yy} + u_{zz}$. For problems involving only two space dimensions, $\nabla^2 u = u_{xx} + u_{yy}$ and $s(\mathbf{r}, t)$ then refers to $s(x, y, t)$. For steady state problems $u_t = 0$ and we have

$$\nabla^2 u = -\frac{1}{K} s(\mathbf{r}) \tag{3.1.2}$$

an equation sometimes referred to as Poisson's equation.

Similar extensions can be applied to the wave equation. For two dimensions, in place of waves on a string we may consider waves on a membrane. The expression for the force $\mathcal{F}(\mathbf{r}, t)$ acting on a square element of a membrane is a direct extension of that used for an element of a string, and as will be shown in Section 3.4, we obtain the two-dimensional wave equation in the form

$$\nabla^2 w(\mathbf{r}, t) - \frac{1}{c^2} \frac{\partial^2 w}{\partial t^2} = \frac{1}{T} \mathcal{F}(\mathbf{r}, t) \tag{3.1.3}$$

Waves in three dimensions may be considered in many contexts, one of which is sound propagation. Again, the only mathematical extension required is that of the three-dimensional Laplacian plus a consideration of the boundary conditions that model various physical situations.

3.2 STEADY STATE TEMPERATURE DISTRIBUTION IN RECTANGULAR COORDINATES—LAPLACE'S EQUATION

As a simple example of the extension to two dimensions of a one-dimensional problem treated previously, consider a plate in an XY plane as shown in Figure 3.1. The two surfaces of the plate that face in the Z direction (i.e., perpendicular to the page) are assumed to be insulated, as was the case for the length of the beam in the one-dimensional problems, and can thus be ignored. The two rims of the plate that face in the Y direction will also be assumed to be perfectly insulated, but since we will now consider temperature variations in the Y direction, we must introduce the insulation-type boundary condition explicitly and write $u_y(x, 0) = u_y(x, W) = 0$. In addition, the rim at $x = 0$ is kept at zero temperature while at $x = L$ we specify the temperature variation in the

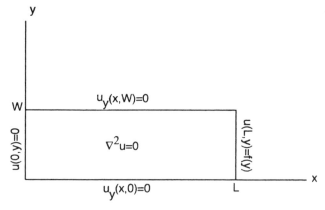

Figure 3.1. Boundary conditions determining two-dimensional steady state temperature distribution.

form $u(L, y) = f(y)$ where $f(y)$ is presumed given. Note that if $f(y) = $ const, there would be no variation in the y direction at all and we would actually have a simple one-dimensional problem of the type considered in Chapter 1.

Since there is no source of heat within the plate, $s(\mathbf{r})$ in (3.1.2) is zero and the steady state temperature distribution on the plate is given by the solution of

$$\nabla^2 u = \frac{\partial^2 u}{\partial x^2} + \frac{\partial^2 u}{\partial y^2} = 0 \tag{3.2.1}$$

that satisfies the boundary conditions of the problem. We shall again construct the solution from elementary product solutions and set $u(x, y) = X(x)Y(y)$. By the same technique used in Chapter 1, this form of $u(x, y)$ enables us to write (3.2.1) in the form

$$\frac{X''}{X} = \frac{Y''}{Y} = \text{const} \tag{3.2.2}$$

The choice of sign for the constant is determined by the location of the rim on which the spatially varying temperature is specified. Extension of the method to situations in which more than one rim has an inhomogeneous boundary condition is considered at the end of this section. In the problem being considered here, the temperature is given as a function of y, so ultimately we will find it necessary to be able to construct a Fourier series in the y coordinate. To prepare for the required sinusoidal solutions in the y coordinate, we therefore set the constant in (3.2.2) equal to k^2 and obtain the two equations

$$X'' - k^2 X = 0, \qquad Y'' + k^2 Y = 0 \tag{3.2.3}$$

Note that for $k \neq 0$ it is the function $Y(y)$ that will have solutions $\sin ky$ and $\cos ky$ and thus allow for the construction of the expected Fourier series in the y coordinate. The solutions are

$$k = 0: \quad X = a + bx$$
$$Y = c + dy$$
$$k \neq 0: \quad X = A \cosh kx + B \sinh ky$$
$$Y = C \cos ky + D \sin ky \tag{3.2.4}$$

To apply the boundary conditions to the solution $Y(y)$, namely $Y'(0) = Y'(W) = 0$, we must first determine $Y'(y)$. For $k = 0$ we obtain $Y'(y) = d$ while for $k \neq 0$ we have $Y'(y) = k(-C \sin ky + D \cos ky)$. Then, since $Y'(0) = Y'(W) = 0$, we find $d = 0$, $D = 0$, and $k = n\pi/W$, where $n = 1, 2, 3, \ldots$. For the function $X(x)$, the condition $X(0) = 0$ yields $a = A = 0$ and we finally have the product solutions

$$u_0(x, y) = bcx$$

$$u_n(x, y) = B_n C_n \sinh \frac{n\pi x}{W} \cos \frac{n\pi y}{W}, \qquad n = 1, 2, 3, \ldots \qquad (3.2.5)$$

The solution associated with $k = 0$ is seen to describe the linear spatial variation in temperature along a one-dimensional system, as already encountered in Section 1.3. The solutions for $k \neq 0$ describe the structure of the temperature variation across the width of the plate. Summing these solutions we have

$$u(x, y) = a_0 x + \sum_{n=1}^{\infty} a_n \sinh \frac{n\pi x}{W} \cos \frac{n\pi y}{W} \qquad (3.2.6)$$

where $a_0 = bc$ and $a_n = B_n C_n$. Note that the first term in (3.2.6) is not obtained by using $n = 0$ in the infinite series. The coefficients in the series are now determined by the unusual Fourier series considerations. At $x = L$ we impose the boundary condition $u(L, y) = f(y)$ and write

$$u(L, y) = f(y) = a_0 L + \sum_{n=1}^{\infty} a_n \sinh \frac{n\pi x}{W} \cos \frac{n\pi y}{W} \qquad (3.2.7)$$

Since $\int_0^W \cos n\pi y/W \, dy = 0$, integration of the series over y yields

$$\int_0^W f(y) \, dy = a_0 L W \qquad (3.2.8)$$

Multiplying the series by $\cos m\pi y/W$, again integrating over y, and using the orthogonality of the cosine terms, we obtain

$$\int_0^W f(y) \cos \frac{m\pi y}{W} \, dy = a_m \sinh \frac{m\pi L}{W} \frac{W}{2} \qquad (3.2.9)$$

The constants are thus given as

$$a_0 = \frac{1}{LW} \int_0^W f(y) \, dy, \qquad a_n = \frac{2}{W \sinh(n\pi L/W)} \int_0^W f(y) \cos \frac{n\pi y}{W} \, dy$$

$$(3.2.10)$$

As an example of the method, assume that the plate is heated by a hot spot at the midpoint of the rim at $x = L$. It is then appropriate to set $u(L, y) = f(y) = \Gamma \delta(y - W/2)$. The constant Γ can eventually be related to various physical quantities in the problem such as the total heat on the plate, the average temperature of the plate, or the average temperature of the edge at $x = L$. Since the delta function has dimensions L^{-1}, the constant Γ must have dimensions (temperature) xL. From (3.2.10) we obtain

$$a_0 = \frac{\Gamma}{LW}, \qquad a_n = \frac{2\Gamma}{W} \frac{\cos(n\pi/2)}{\sinh(n\pi L/W)} \qquad (3.2.11)$$

and from (3.2.6), the temperature distribution is

$$u(x, y) = \frac{\Gamma}{W} \frac{x}{L} + \frac{2\Gamma}{W} \sum_{n=1}^{\infty} \cos \frac{n\pi}{2} \frac{\sinh(n\pi x/W)}{\sinh(n\pi L/W)} \cos \frac{n\pi y}{W} \qquad (3.2.12)$$

The constant Γ will now be expressed in terms of the total heat on the plate, H_0. Writing

$$H_0 = \int_0^L dx \int_0^W dy \, \sigma c u(x, y) = \tfrac{1}{2}\Gamma \, \sigma c L \qquad (3.2.13)$$

we have $\Gamma = 2H_0/\sigma c L$. Note that the total contribution to H_0 comes from the first term in $u(x, y)$. The remaining terms in the series for $u(x, y)$ all vanish on integration over y due to the factor $\cos(n\pi y/W)$.

If we were only able to determine an average temperature across the plate at any position x along the plate, this average would be given by

$$\langle u(x) \rangle = \frac{1}{W} \int_0^W u(x, y) \, dy = \frac{\Gamma}{W} \frac{x}{L} \qquad (3.2.14)$$

and the evaluation of this average temperature at $x = L$ would yield $\Gamma = \langle u(L) \rangle W$. The average temperature at any point along the beam, given by (3.2.14), would be

$$\langle (x) \rangle = \langle u(L) \rangle \frac{x}{L} \qquad (3.2.15)$$

that is, the one-dimensional result used in Section 1.6.

The average temperature of the entire plate is given by

$$\langle u \rangle = \frac{1}{LW} \int_0^L dx \int_0^W dy \, u(x, y) = \frac{\Gamma}{2W} = \frac{H_0}{\sigma c L W} \qquad (3.2.16)$$

Using this result to eliminate Γ in (3.2.12), we find that the plate temperature is

$$\frac{u(x, y)}{\langle u \rangle} = 2 \frac{x}{L} + 4 \sum_{n=1}^{\infty} \cos \frac{n\pi}{2} \frac{\sinh(n\pi x/W)}{\sinh(n\pi L/W)} \cos \frac{n\pi y}{W} \qquad (3.2.17)$$

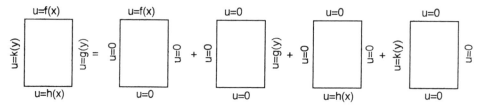

Figure 3.2. Decomposition into subproblems when inhomogeneous boundary conditions are imposed on more than one rim.

For a square plate, the heat flow out the edge at $x = 0$ turns out to be quite uniform in the y direction since

$$J_x(0, y)$$

$$= -K \left. \frac{\partial u}{\partial x} \right|_{x=0}$$

$$= -\frac{2K\langle u \rangle}{L} \left[1 - 4\pi \left(\operatorname{csch} 2\pi \cos \frac{2\pi y}{W} - 2 \operatorname{csch} 4\pi \cos \frac{4\pi y}{W} + \cdots \right) \right]$$

$$= -\frac{2K\langle u \rangle}{L} \left[1 - 0.047 \cos \frac{2\pi y}{W} + 0.00018 \cos \frac{4\pi y}{W} \cdots \right] \qquad (3.2.18)$$

The second term in the series only makes a contribution of less than 5%.

When more than one rim has an inhomogeneous boundary condition, we may exploit the linearity of the problem and decompose it into a number of problems each of which has an inhomogeneous boundary condition on only one rim. The procedure is perhaps best summarized in terms of a diagram such as that shown in Figure 3.2. Some examples are considered in the problems.

Problems

3.2.1 Show that the solution of Laplace's equation in the rectangular region $0 \le x \le L, 0 \le y \le H$ that satisfies $u(x, 0) = u_0 \sin \pi x/L$, $u(x, H) = u(0, y) = u(L, y) = 0$ is $u(x, y) = u_0 \sin \pi x/L \sinh \pi (H - y)/L$ csch $\pi H/L$.

3.2.2 Determine the solution of Laplace's equation that satisfies the boundary conditions in each of the following three examples:

(a) $u(0, y) = f(y) = u(L, y) = 0$, $u_y(x, 0) = 0$, $u_y(x, H) = 0$.

(b) $u_x(0, y) = 0$, $u_x(L, y) = f(y)$, $u_y(x, 0) = 0$, $u_y(x, H) = 0$. Discuss the condition that must be imposed upon $f(y)$ in this example.

(c) $u(0, y) = 0$, $u(L, y) = 0$, $u(x, 0) = -u_0 \sin 2\pi x/L$, $u(x, L) = u_0 \sin \pi x/L$. Obtain $u(L/2, y)$. Use physical intuition to sketch the contour on the plate (isotherm) for which $u(x, y) = 0$.

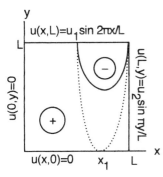

Figure 3.3.

3.2.3 The temperature at the rim of a square plate is maintained as shown in Figure 3.3. Assume u_1, $u_2 > 0$.

(a) Show that the steady state temperature is given by

$$u = u_1 \frac{\sin 2\pi x/L \, \sinh 2\pi y/L}{\sinh 2\pi} + u_2 \frac{\sin \pi y/L \, \sinh \pi x/L}{\sinh \pi}$$

Intuitively, one expects that the plate will be divided into regions of positive and negative temperatures as shown in the figure. If the temperature u_1 is increased, it should be possible to bring the $u = 0$ contour down to the line $y = 0$ at some point $x = x_1$ as shown by the dotted line in the figure.

(b) Discuss why x may be determined by solving

$$u_y(x, 0) = 0$$

and

$$u_{yx}(x, 0) = 0$$

simultaneously. Show that these two equations lead to the transcendental equation

$$\tan \frac{2\pi x}{L} = 2 \tanh \frac{\pi x}{L}$$

(c) Sketch $y = \tan 2\pi x/L$ and $y = 2 \tanh \pi x/L$ on the same graph, and from inspection obtain the bounds $\frac{5}{8} < x_1/L < \frac{3}{4}$.

(d) Obtain an approximate solution and determine the corresponding value of u_1/u_2.

3.2.4 A rectangular plate satisfies the boundary conditions $u(0, y) = 0$, $u(L, y) = 0$, $u_y(x, 0) = 0$, $u_y(x, H) = 0$. A constant heat source Q_0 is located at the center of the plate. The steady state temperature on the plate is thus governed by

$$\nabla^2 u = -\frac{Q_0}{K}\,\delta\left(x - \frac{L}{2}\right)\delta\left(y - \frac{H}{2}\right)$$

Carry out the calculation that yields the obvious result that half of the heat flows off the edge at each end $(x = 0, L)$.

3.2.5 A square plate kept at zero temperature at the rim has the steady source $q(x, y) = H_0\delta(x - L/2)\delta(y - L/2)$, that is, a point source located at its midpoint. The steady state temperature on the plate is governed by

$$\nabla^2 u = -\frac{H_0}{K}\,\delta\left(x - \frac{L}{2}\right)\delta\left(y - \frac{L}{2}\right)$$

(a) Introduce the expansions

$$u(x, y) = \sum_{n=1}^{\infty} f_n(x)\sin\frac{n\pi y}{L}$$

$$\delta\left(y - \frac{L}{2}\right) = \frac{2}{L}\sum_{n=1}^{\infty}\sin\frac{n\pi}{2}\sin\frac{n\pi y}{L}$$

and show that

$$f_n'' - \left(\frac{n\pi}{L}\right)^2 f_n = -\frac{2}{L}\frac{H_0}{K}\sin\frac{n\pi}{2}\,\delta\left(x - \frac{L}{2}\right)$$

(b) Use solutions

$$f_{n<}(x) = A_n\sinh n\pi x/L, \qquad\qquad 0 \le x < L/2$$
$$f_{n>}(x) = A_n\sinh n\pi(L - x)/L, \qquad L/2 < x \le L$$

and show that

$$A_n = \frac{1}{n\pi}\frac{H_0}{K}\sin\frac{n\pi}{2}\,\text{sech}\,\frac{n\pi}{2}$$

(c) The heat out through the edge at $x = 0$ is given by

$$H\big|_{x=0} = \int_0^L dy(-K)\frac{\partial u}{\partial x}\bigg|_{x=0}$$

With the help of the series used in the last problem, show that $H|_{x=0} = H_0/4$.

3.2.6 The steady state temperature on a rectangular plate $(0 \le x \le L, 0 \le y \le H)$ satisfies the boundary conditions $u(0, y) = u_y(x, 0) = u_y(x, H) = 0$ and $u_x(L, y) = (Q/K)\sin^2(2\pi y/H)$. On the edge at $x = L$ there are thus maxima in the heat flux at $y = H/4$ and $y = 3H/4$.

(a) Determine the steady state temperature on the plate and the heat flux off the plate, $J(0, y)$.

(b) One expects that the larger the ratio L/H, the less resolved the two peaks in the heat flow will be at $x = 0$. If $R(L)$, the resolution of the peaks at $x = 0$, is defined as

$$R(L) = \frac{J_{max}(L) - J_{min}(L)}{J_{max}(L) + J_{min}(L)}$$

show that $J_{max}(L) = (Q/2)(1 + \text{sech } 4\pi L/H)$, $J_{min}(L) = (Q/2)(1 - \text{sech } 4\pi L/H)$, and thus that $R(L) = \text{sech}(4\pi L/H)$.

3.3 TIME DEPENDENT DIFFUSION IN RECTANGULAR COORDINATES

Time dependent problems in two-space dimensions may be solved by a minor extension of the separation-of-variables procedure used previously. We will find that there are now two separation constants, and summation of the various elementary solution leads to an expression for the solution in terms of a double Fourier series.

As an example, consider the diffusion of an initial temperature distribution $u(x, y, 0)$ on a rectangular plate of length L and width W that has zero temperature at the ends $x = 0, L$ and is insulated on the sides $y = 0$ and W. The boundary conditions are thus $u(0, y) = u(L, y) = 0$ and $u_y(x, 0) = u_y(x, W) = 0$. The diffusion equation is

$$\rho c \frac{\partial u}{\partial t} - K\left(\frac{\partial^2 u}{\partial x^2} + \frac{\partial^2 u}{\partial y^2}\right) = 0 \qquad (3.3.1)$$

Considering the elementary product solutions $u(x, y, t) = X(x)Y(y)T(t)$, we obtain

$$\frac{1}{\gamma^2}\frac{T'}{T} = \frac{X''}{X} + \frac{Y''}{Y}, \qquad \gamma^2 = \frac{K}{\rho c} \qquad (3.3.2)$$

After setting each side of this equation equal to a separation constant $-\alpha^2$, we can also write the x and y dependence in the separated form

$$\frac{X''}{X} = -\alpha^2 - \frac{Y''}{Y} = -\kappa^2 \tag{3.3.3}$$

where $-\kappa^2$ is another separation constant. We thus obtain $X'' + \kappa^2 X = 0$ and $Y'' + (\alpha^2 - \kappa^2)Y = 0$. If we set $\alpha^2 - \kappa^2 = \mu^2$, then we can write the three separated equations as

$$T' + (\alpha\gamma)^2 T = 0$$

$$X'' + \kappa^2 X = 0$$

$$Y'' + \mu^2 Y = 0 \tag{3.3.4}$$

with $\alpha^2 = \lambda^2 + \mu^2$. The solutions of these equations are

$$
\begin{array}{ll}
T = \text{const}, & \alpha = 0 \\
T = e^{-(\alpha\gamma)^2 t}, & \alpha^2 > 0 \\
X = a + bx, & \kappa = 0 \\
X = A \cos \kappa x + B \sin \kappa x, & \kappa^2 > 0 \\
Y = c + dy, & \mu = 0 \\
Y = C \cos \mu y + D \sin \mu y, & \mu^2 > 0
\end{array}
\tag{3.3.5}
$$

As in the case of one dimension, no useful (i.e., oscillatory) solutions are obtainable for negative values of κ^2 or μ^2.

To satisfy the boundary conditions for the problem being considered, we require that $Y(y)$ vanish at $y = 0$, W. These requirements give $D = d = 0$ and $\mu W = m\pi$, where $m = 1, 2, 3, \ldots$. Similarly, to satisfy $X(0) = X(L) = 0$, we require $a = b = A = 0$ and $\kappa L = k\pi$, $k = 1, 2, 3, \ldots$. The elementary product solutions are thus constant multiples of

$$X_k = \sin \frac{k\pi x}{L}, \qquad k = 1, 2, 3, \ldots$$

$$Y_m = \cos \frac{m\pi y}{W}, \qquad m = 0, 1, 2, \ldots \tag{3.3.6}$$

Note that the special cases $\kappa, \mu = 0$, which are $X_0(x) = 0$ and $Y_0(y) = \text{const}$, are taken into account in this listing of solutions. For each k, m combination we have $\alpha^2_{km} = (k\pi/L)^2 + (m\pi/W)^2$ and the complete solution is thus

$$u(x, y, t) = \sum_{k=1}^{\infty} \sum_{m=0}^{\infty} a_{km} e^{-(\alpha_{km}\gamma)^2 t} \sin \frac{k\pi x}{L} \cos \frac{m\pi y}{W} \qquad (3.3.7)$$

which is a Fourier series in both x and y.

The coefficients a_{km} are obtained by applying orthogonality considerations at $t = 0$, since the series expansion must reduce to the initial temperature distribution on the plate at this time. Setting $t = 0$ in (3.3.7), multiplying by $\sin(p\pi x/L)$ and $\cos(q\pi y/W)$, and then integrating over the area of the plate, we have

$$\int_0^L dx \int_0^W dy \, u(x, y, 0) \sin \frac{p\pi x}{L} \cos \frac{q\pi y}{W} = a_{pq} \frac{LW}{4}, \qquad p, q > 0 \quad (3.3.8)$$

For $q = 0$ the result is

$$\int_0^L dx \int_0^W dy \, u(x, y, 0) \sin \frac{p\pi x}{L} = a_{p0} \frac{LW}{2} \qquad (3.3.9)$$

Note that if we average the solution (3.3.7) over y by determining

$$\langle u(x, t) \rangle = \frac{1}{W} \int_0^W dy \, u(x, y, t) = \sum_{k=1}^{\infty} a_{k0} e^{-(\alpha_{k0}\gamma)^2 t} \sin \frac{k\pi x}{L} \qquad (3.3.10)$$

we obtain the result for the corresponding one-dimensional problem for a beam with insulated sides that was considered in Section 1.4. The one-dimensional solution also results if the initial temperature distribution $u(x, y, 0)$ is independent of y.

Problems

3.3.1 A rectangular plate $0 \le x \le L$, $0 \le y \le W$ is initially at uniform temperature u_0. The edges are then placed at zero temperature.

(a) Show that the temperature on the plate at any later time, which is governed by

$$\nabla^2 u - \frac{1}{\gamma^2} \frac{\partial w}{\partial t} = 0, \qquad u(x, y, 0) = u_0$$

is given by

$$u(x, y, t) = \frac{16 u_0}{\pi^2} \sum_{\substack{n, m \\ \text{odd}}}^{\infty} \frac{e^{-\gamma^2 k_{nm}^2 t}}{nm} \sin \frac{n\pi x}{L} \sin \frac{m\pi y}{W}$$

in which $k_{nm}^2 = (n\pi/L)^2 + (m\pi/W)^2$.

(b) Show that the total heat flowing out through the edge at $x = 0$ is

$$|H|_{x=0} = \frac{32\rho cu_0 LW}{\pi^4} \sum_{m=1,3,5\ldots} \frac{1}{m^2} \sum_{n=1,3,5\ldots} \frac{1}{n^2 + (L/W)^2 m^2}$$

Use the result (Table F.1, series 22)

$$\sum_{m=\text{odd}} \frac{1}{n^2 + a^2} = \frac{\pi}{4a} \tanh \frac{\pi a}{L}$$

to show that

$$|H|_{x=0} = \frac{8}{\pi^3} H_0 \frac{W}{L} \sum_{m=1,3,5\ldots} \frac{1}{m^3} \tanh \left(\frac{\pi m}{2} \frac{L}{W} \right)$$

where $H_0 = \rho cu_0 LW$.

(c) When $W = L$, we may expect to have $|H|_{x=0} = \frac{1}{4}H_0$ on the basis of symmetry. Hence infer that the value of the series in this case is

$$\sum_{\text{odd}} \frac{1}{m^3} \tanh \frac{\pi m}{2} = \frac{\pi^3}{32} = 0.9689 \ldots$$

Note that the first three terms yield

$$0.91715 + 0.03703 + 0.0080 = 0.9622$$

3.3.2 A rectangular plate with boundary conditions $u(0, y) = u(L, y) = 0$, $u_y(x, 0) = u_y(x, W) = 0$ has a "hot spot" at the center that contains heat H_0.

(a) Use separation of variables to solve $\gamma^{-2} u_t - \nabla^2 u = 0$ with the initial condition

$$u(x, y, 0) = \frac{H_0}{\sigma c} \delta \left(x - \frac{L}{2} \right) \delta \left(y - \frac{w}{2} \right)$$

(b) Obtain an expression for the total amount of heat flowing out the edge at $x = 0$. Since this result is clearly $-H_0/2$, with the minus sign associated with the direction of heat flow, obtain the evaluation of a "well-known" sum.

3.3.3 The edges of a rectangular plate are kept at zero temperature. An amount of heat H_0, in the form of a hot spot, is initially deposited at the center of the plate. The initial temperature distribution on the plate may thus be written

$$u(x, y, 0) = \frac{H_0}{\sigma c} \delta \left(x - \frac{L}{2} \right) \delta \left(y - \frac{w}{2} \right)$$

(a) Show that at any later time

$$u(x, y, t) = \sum_{n=1}^{\infty} \sum_{m=1}^{\infty} \dot{a}_{nm} e^{-\nu_{nm} t} \sin \frac{n \pi x}{L} \sin \frac{m \pi y}{W}$$

where

$$a_{nm} = \frac{4 H_0}{\sigma c L W} \sin \frac{n \pi}{2} \sin \frac{m \pi}{2}, \qquad \nu_{nm} = \frac{\pi^2 K}{\sigma c} \left(\frac{n^2}{L^2} + \frac{m^2}{W^2} \right)$$

(b) The total amount of heat that flows out through the edges at $x = 0, L$ is

$$H \big|_{x=0,L} = -K \int_0^\infty dt \int_0^\infty dy \, \frac{\partial u}{\partial x} \bigg|_{x=0,L}$$

Show that

$$H \big|_{x=0} = -\frac{8 H_0}{\pi^2} \sum_{n=1}^{\infty} \sum_{m=1}^{\infty} \frac{n}{m} \frac{\sin (n \pi/2) \sin (m \pi/2)}{n^2 + (L/W)^2 m^2}$$

(c) Use the result (cf. Table F.1, series 23)

$$\sum_{n=1}^{\infty} \frac{n \sin n\theta}{n^2 + a^2} = \frac{\pi}{2} \frac{\sinh a(\pi - \theta)}{\sinh \pi a}, \qquad 0 < \theta < 2\pi$$

to sum the series on n to obtain

$$\frac{H \big|_{x=0}}{H_0} = -\frac{2}{\pi} \sum_{n=1}^{\infty} \frac{1}{m} \sin \frac{m\pi}{2} \, \text{sech} \left(\frac{\pi}{2} \frac{L}{W} m \right)$$

(d) Use this series to show that you would expect the following limiting results:
For $L/w \ll 1$, $|H_{x=0}|/H_0 \rightarrow \frac{1}{2}$.
For $L/w \gg 1$, $|H_{x=0}|/H_0 \approx (4/\pi) e^{-\pi L/2W}$.

For $L/w = 1$, use the result that $|H_{x=0}|/H_0 = \frac{1}{4}$, obvious on the basis of symmetry, to conclude that

$$\sum_{n=1}^{\infty} \frac{1}{m} \sin \frac{m\pi}{2} \operatorname{sech} \frac{m\pi}{2} = \frac{\pi}{8}$$

Note that the first term alone is $0.39854 = 1.015 \times (\pi/8)$.

3.4 WAVES ON A MEMBRANE—RECTANGULAR COORDINATES

Previous considerations of waves on a string are readily extended to the case of waves on a rectangular membrane. A rectangular segment of the membrane $\Delta x\,\Delta y$ is assumed to have a surface density σ and thus a mass $\sigma\,\Delta x\,\Delta y$. As shown in Figure 3.4, each side of the rectangular element is considered to be stretched by a tension T_m that is spread along an entire edge of the element so that the force acting on a section Δx is $T_m\,\Delta x$. The membrane tension T_m thus has dimensions of force per length. The force along Δy is $T_m\,\Delta y$. The derivation of the wave equation for the vertical displacement $w(x, y, t)$, which is perpendicular to the page in Figure 3.4, is an immediate extension of that for the string. Determining the forces on the membrane in a manner analogous to that used for the string, we have

$$\sigma\,\Delta x\,\Delta y\,\frac{\partial^2 w}{\partial t^2} = \left|\left(T_m\,\Delta y\,\frac{\partial w}{\partial x}\right)\right|_{x+\Delta x} - \left(T_m\,\Delta y\,\frac{\partial w}{\partial x}\right)\Big|_{x}$$

$$+ \left(T_m\,\Delta x\,\frac{\partial w}{\partial y}\right)\Big|_{y+\Delta y} - \left(T_m\,\Delta x\,\frac{\partial w}{\partial y}\right)\Big|_{y}$$

$$+ \mathcal{F}(x, y, t)\,\Delta x\,\Delta y \qquad (3.4.1)$$

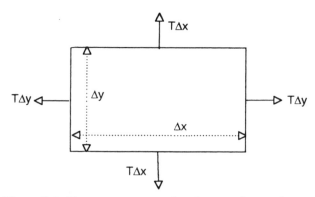

Figure 3.4. Forces on a rectangular element of a membrane.

where $\mathcal{F}(x, y, t)$ is an external force per unit *area* acting on the membrane. We may carry out an immediate extension of the development used for the string in Section 1.8 to obtain

$$\sigma \frac{\partial^2 w}{\partial t^2} = \frac{\partial}{\partial x}\left(T_m \frac{\partial w}{\partial x}\right) + \frac{\partial}{\partial y}\left(T_m \frac{\partial w}{\partial y}\right) + \mathcal{F}(x, y, t) \qquad (3.4.2)$$

For the case of constant tension this expression reduces to

$$\nabla^2 w - \frac{1}{c^2}\frac{\partial^2 w}{\partial t^2} = -\frac{1}{T_m}\mathcal{F}(x, y, t) \qquad (3.4.3)$$

where $c^2 = T_m/\sigma$. A resistive term $R\,\partial w/\partial t$ and an elastic restoring force $K(x,y)w$ could also be included, as was done for the string in (1.8.9). Note that the time independent situation in which $\partial^2 w/\partial t^2 = 0$ reduces to the equation given in (1.1.3) when we set $\mathcal{F} = \rho g$.

3.4.1 Normal Modes

As was the case for the string, use of the separation-of-variables procedure for the membrane equation leads to elementary solutions that are physically realizable and are referred to as normal modes. Setting $\mathcal{F} = 0$ and writing $w(x, y, t) = X(x)Y(y)T(t)$ we can rewrite the wave equation (3.4.3) in the form

$$\frac{1}{c^2}\frac{T''}{T} = \frac{X''}{X} + \frac{Y''}{Y} \qquad (3.4.4)$$

Following the procedure used in the previous section for the diffusion equation, we can associate separation constants α, κ, and μ with T, X, and Y, respectively. We then obtain the three equations

$$T'' + (\alpha c)^2 T = 0$$

$$X'' + \kappa^2 X = 0$$

$$Y'' + \mu^2 Y = 0 \qquad (3.4.5)$$

For a rectangular membrane that covers an XY plane inside the region $0 \leq x \leq L$, $0 \leq y \leq W$ and is fixed at all edges, the functions $X(x)$ and $Y(y)$ must satisfy the boundary conditions $X(0) = X(L) = 0$ and $Y(0) = Y(W) = 0$. By following the same procedure as that used for the diffusion equation in the previous section, we find that these boundary conditions lead to the solutions

$$X(x) = \sin \frac{k\pi x}{L}, \qquad k = 1, 2, 3, \ldots$$

$$Y(y) = \sin \frac{m\pi y}{W}, \qquad m = 1, 2, 3, \ldots \qquad (3.4.6)$$

Thus α is determined to have the double infinity of values

$$\alpha_{km}^2 = \frac{\pi^2}{c^2}\left[\left(\frac{k}{L}\right)^2 + \left(\frac{m}{W}\right)^2\right], \qquad k, m = 1, 2, 3, \ldots \qquad (3.4.7)$$

The time factor $T(t)$ is composed of the terms $\sin \omega_{km} t$ and $\cos \omega_{km} t$ where $\omega_{km} = \alpha_{km} c$. The motion of the membrane will be a superposition of the form

$$w(x, y, t) = \sum_{k=1}^{\infty} \sum_{n=1}^{\infty} \sin \frac{k\pi x}{L} \sin \frac{m\pi y}{W} (a_{km}\cos \omega_{km} t + b_{km}\sin \omega_{km} t)$$

$$= \sum_{k,m} w_{km}(x, y, t) \qquad (3.4.8)$$

When an initial displacement $w(x, y, 0)$ and an initial velocity $w_t(x, y, 0)$ are specified, the coefficients a_{km} and b_{km} are determined by using the orthogonality in both the x and y coordinates as was done for the heat equation in Section 3.3. One obtains

$$a_{km} = \frac{4}{LW} \int_0^L dx \int_0^W dy\, w(x, y, 0)\sin \frac{k\pi x}{L} \sin \frac{m\pi y}{W} \qquad (3.4.9)$$

and from the time derivative of (3.4.8).

$$b_{km} = \frac{4}{\omega_{km} LW} \int_0^L dx \int_0^W dy\, w_t(x, y, 0)\sin \frac{k\pi x}{L} \sin \frac{m\pi y}{W} \qquad (3.4.10)$$

The individual modes of vibration for the membrane $w_{km}(x, y, t)$ are of interest in themselves. For k and/or m greater than unity the factors $\sin k\pi x/L$ and $\sin m\pi y/W$ will be zero for all time at x or y values for which the sine functions vanish, that is, along straight lines parallel to the sides of the membrane. On these lines the membrane displacement $w(x, y, t)$ will vanish. Such nodal lines are readily displayed by sprinkling sand over the membrane surface when it is made to vibrate at the frequency corresponding to the specified values of k and m as given by (3.4.7). When the membrane is vibrating at one of the characteristic frequencies ω_{kn}, the sand moves about until it is on a line of zero vibration amplitude. The sand patterns formed in this manner are usually referred to as Chladni patterns (after Ernst Chladni, a nineteenth-century acoustician).

3.4.2 Guided Waves

Wave motion may be transmitted over large distances by confining it within some sort of channel. In optics, electromagnetic theory, and acoustics, such guided waves have many uses. Many of the features of such guided wave propagation are displayed by waves on the surface of a membrane. We now consider some introductory aspects of this topic.

Consider a membrane in the form of a long ribbon located in the region $0 \leq x \leq \infty$ and $0 \leq y \leq W$. For the boundary conditions imposed on the membrane displacement $w(x, y, t)$ at $y = 0$ and W, we shall consider two distinct cases. The condition $w = 0$; which corresponds to that imposed on an electric field in an electromagnetic waveguide, and the case $\partial w / \partial y = 0$, which corresponds to a situation frequently encountered in an acoustic wave guide, will be examined in two separate problems. In both of these cases the boundary conditions that we will impose on x are similar to those already considered for waves on a semi-infinite string (cf. Section 2.1). We assume that there is some specified displacement at $x = 0$ so that $w(0, y, t) = f(y, t)$ while as x goes to $+\infty$ we impose the requirement that there be propagation in only the positive x direction, that is, away from the source.

If the separation-of-variables solution is carried out by first writing $w(x, y, t) = f(x, y)T(t)$, then the wave equation (3.4.3), with $\mathcal{F} = 0$, becomes

$$\frac{\nabla^2 f}{f} = \frac{1}{c^2} \frac{T''}{T} = -k_0^2 \qquad (3.4.11)$$

where k_0^2 is a separation constant. The minus sign has been introduced to provide oscillatory solutions for $T(t)$ when k_0 is real. Following the procedure used in Section 1.8 for the semi-infinite string, we then have $T(t) = a \cos k_0 ct + b \sin k_0 ct$. In the following development it will be convenient to write this solution in the form

$$T(t) = \text{Re}(Ae^{-ik_0ct}) \qquad (3.4.12)$$

where A is a complex constant and Re refers to "real part of."[1] As in the case of the semi-infinite string considered in Section 2.1, the separation constant k_0 is not determined by the boundary conditions but is free to be used to express the frequency of vibration imposed by some external source at the boundary. If we now set $f(x, y) = X(x)Y(y)$ and separate variables, we may write

$$\frac{X''}{X} + k_0^2 = -\frac{Y''}{Y} = k_y^2 \qquad (3.4.13)$$

where k_y^2 is the second separation constant.

---

[1]Note that if we assume A to be complex and set $A = \alpha e^{i\varphi}$, then $\text{Re}[Ae^{-ik_0ct}] = \text{Re}[\alpha e^{i(\varphi - k_0ct)}] = \alpha \cos(\varphi - k_0ct) = \alpha \cos \varphi \cos k_0 ct + \alpha \sin \varphi \sin k_0 ct$ and the constants a and b in $T(t)$ are given by $a = \alpha \cos \varphi$, $b = \alpha \sin \varphi$.

At this stage we distinguish the two cases associated with the boundary conditions at $y = 0$ and W. For fixed edges we have $w(x, 0, t) = w(x, W, t) = 0$ and thus set $Y(0) = Y(W) = 0$. The solution $Y(y)$ is thus $\sin k_y y$ with $k_y = n\pi/W$, $n = 1, 2, 3, \ldots$. The solution for $X(x)$ is then

$$X(x) = A e^{ix\sqrt{k_0^2 - k_y^2}} + B e^{-ix\sqrt{k_0^2 - k_y^2}} \tag{3.4.14}$$

To obtain waves propagating in the positive x direction, that is, functions of only $x - ct$, in conjunction with e^{-ik_0ct} time dependence, we keep only the first solution in (3.4.14) and set $B = 0$ in that equation. The general solution, labeled w_1, is now a sum of these solutions for all values of n and we have

$$w_1(x, y, t) = \sum_{n=1}^{\infty} b_n \sin \frac{n\pi y}{W} e^{ix\sqrt{k_0^2 - (n\pi/W)^2} - i\omega_0 t} \tag{3.4.15}$$

where $\omega_0 = k_0 c$.

For the second boundary condition, namely $w_y(x, 0, t) = w_y(0, W, t) = 0$, a similar calculation yields

$$w_2(x, y, t) = \sum_{n=0}^{\infty} a_n \cos \frac{n\pi y}{W} e^{ix\sqrt{k_0^2 - (n\pi/W)^2} - i\omega_0 t} \tag{3.4.16}$$

Both of these solutions refer to waves at the single frequency ω_0 that, as mentioned above, is at our disposal.

That the series for w_2 begins at $n = 0$ while the series for w_1 begins at $n = 1$ provides a fundamental difference between these two problems. Note first that for values of n such that $n\pi/W > k_0$, the imaginary exponent in each case becomes real so that the factor $X_n(x)$ does not correspond to a propagating disturbance but rather to one that decreases exponentially with increasing values of x. For the solution w_1, which does not contain the case $n = 0$, the lowest value of k_0 that can be associated with a propagating disturbance is for $n = 1$, that is, $k_0 = \pi/W$. The corresponding value of $\omega_0 = k_0 c$ is known as the cutoff frequency for waves on the membrane. For solutions w_2, on the other hand, the summation includes the term $n = 0$. In this case the solution $a_0 e^{ik_0 x - i\omega_0 t}$ occurs. It represents a propagating disturbance for arbitrarily low frequencies. It has no spatial variation across the width of the membrane and is usually referred to as a plane wave.

Problems

3.4.1 The energy flow on a string was obtained in Section 1.8.3 by multiplying the equation for a string by y_t and integrating over an arbitrary section of the string. Apply a similar procedure to the membrane equation $\rho w_{tt} - T \nabla^2 w = 0$ and integrate over an arbitrary rectangular surface element to obtain

$$\frac{\partial u}{\partial t} + \nabla \cdot \mathbf{J} = 0$$

where

$$u = \tfrac{1}{2}\rho(w_t)^2 + \tfrac{1}{2}T(\nabla w \cdot \nabla w), \qquad \mathbf{J} = -Tw_t \nabla w$$

3.4.2 A rectangular membrane ($0 \le x \le L$, $0 \le y \le H$) is given the initial displacement

$$w(x, y, 0) = \begin{cases} \dfrac{2y_0 x}{L} \sin \dfrac{\pi y}{H}, & 0 \le x < \dfrac{L}{2} \\[3mm] \dfrac{2y_0(L - x)}{L} \sin \dfrac{\pi y}{H}, & \dfrac{L}{2} < x \le L \end{cases}$$

and no initial velocity

(a) Show that the subsequent displacement of the membrane is given by

$$w(x, y, t) = \sum_{m,n} A_{mn} \sin \frac{m\pi x}{L} \sin \frac{n\pi y}{H} \cos\omega_{mn} t$$

where

$$\omega_{mn} = \frac{\pi c}{L} \sqrt{m^2 + \left(\frac{nL}{H}\right)^2}$$

and

$$A_{mn} = \frac{8y_0}{\pi^2 m^2} \delta_{n1} \sin \frac{m\pi}{2}$$

(b) Use the result from Problem 3.4.1 to show that the initial energy (potential) on the membrane is

$$u(0) = \int dS\, u(x, y, 0) = T \int_0^{L/2} dx \int_0^H dy\, |\nabla w|^2$$

$$= Ty_0^2 \frac{H}{L} \left[1 + \frac{\pi^2}{12} \left(\frac{L}{H}\right)^2 \right]$$

(c) Use the solution obtained in part (a) to show that at any later time

$$u_k(t) = \frac{\rho}{2} \int dS(w_t)^2 = \frac{\pi^2}{8} T \frac{H}{L} \sum A_{m1}^2 \left[m^2 + \left(\frac{L}{H} \right)^2 \right] \sin^2 \omega_{m1} t$$

$$u_p(t) = \frac{T}{2} \int dS |\nabla w|^2 = \frac{\pi^2}{8} T \frac{H}{L} \sum A_{m1}^2 \left[m^2 + \left(\frac{L}{H} \right)^2 \right] \cos^2 \mu_{m1} t$$

(d) Use sums given in Table F.1 to show that energy is conserved on the membrane.

3.4.3 A two-dimensional duct composed of a membrane is contained within the region $0 \le x < \infty$, $0 \le y \le H$. The edges at $y = 0, H$ are fixed while the edge at $x = 0$ is given the displacement $w(0, y, t) = f(y)\cos \omega_0 t$ where $\omega_0 = 4\pi c/H$ and

$$f(y) = \begin{cases} w_0 \sin \dfrac{n\pi y}{H}, & 0 \le y < \dfrac{H}{2} \\[3mm] 0, & \dfrac{H}{2} < y \le H \end{cases}$$

(a) Show that the disturbance that propagates on the membrane is given by

$$w(\mathbf{r}, t) = \sum_{n=1}^{3} a_n \sin \frac{n\pi y}{H} \cos(k_n x - \omega_0 t)$$

where

$$k_n = \sqrt{(\omega_0/c)^2 - (n\pi/H)^2}$$

and

$$a_n = \begin{cases} \dfrac{w_0}{2}, & n = 2 \\[3mm] \dfrac{4w_0}{\pi} \dfrac{\sin n\pi/2}{4 - n^2}, & n \ne 2 \end{cases}$$

(b) As obtained in problem 3.4.1, energy flow on the membrane is given by $\mathbf{J} = -Tw_t \nabla w$. Show that the time average of the energy flow in the present example is

$$\langle \mathbf{J} \rangle = \frac{1}{P} \int_0^P dt \, \frac{1}{H} \int_0^H dy \, \mathbf{J}(x, y, t)$$

$$= \frac{\omega_0 T}{4} \mathbf{i} \sum_{n=1}^{3} a_n^2 k_n, \qquad P = \frac{2\pi}{\omega_0}$$

and show that

$$a_n^2 k_n \left(\frac{\pi}{4w_0}\right)^2 \frac{H}{\pi} = \frac{\sqrt{15}}{9}, \left(\frac{\pi}{8}\right)^2 \sqrt{12}, \frac{\sqrt{7}}{25}$$

for $n = 1, 2, 3$, respectively. The energy in the three modes is thus in the ratio $1:1.24:0.25$.

3.5 ORTHOGONAL CURVILINEAR COORDINATES

Thus far, problems involving more than one space dimension have been confined to those having rectangular boundaries. To a considerable extent, our previous success with the separation-of-variables method was due to the circumstance that the boundary conditions could be specified in terms of each rectangular coordinate separately. The constraint imposed at a boundary could then be incorporated into the individual functions $X(x)$ and/or $Y(y)$. A condition imposed at a circular boundary, on the other hand, involves the simultaneous variation of both x and y. This situation precludes the application of the separation-of-variables method as it has been developed thus far. To effect a separation procedure in other than rectangular coordinates, it is helpful to first recast the governing partial differential equation in terms of a coordinate system that does specify the boundary through the variation of a single coordinate. For the circular geometry mentioned above, this extension of the method would obviously imply the use of polar coordinates. While the transformation of the second partial derivatives in the Laplacian from rectangular to polar coordinates is quite straightforward, such calculations can, especially for three-dimensional coordinate systems, become rather lengthy. In addition, the details of such calculations usually add little to one's understanding of a problem. For these reasons we develop here a standard prescription of writing out this transformation of derivatives rather immediately. We shall first summarize the results and then apply them to some standard examples. As will be noted in the following development, the main constraint on the use of these results is that the coordinate lines of the curvilinear coordinate system must intersect at right angles (i.e., be orthogonal).

For a three-dimensional orthogonal coordinate system u_1, u_2, u_3 specified by the transformation equations

$$x = x(u_1, u_2, u_3)$$

$$y = y(u_1, u_2, u_3)$$

$$z = z(u_1, u_2, u_3) \tag{3.5.1}$$

we first consider the position vector $\mathbf{r} = \mathbf{i}x + \mathbf{j}y + \mathbf{k}z$ with x, y, and z specified in terms of the three coordinates u_1, u_2, u_3. A tangent vector $\mathbf{U}_i$ along the curve generated by allowing one of the coordinates u_i to vary will be

$$\mathbf{U}_i = \frac{\partial \mathbf{r}}{\partial u_i}$$

$$= \mathbf{i}\,\frac{\partial x}{\partial u_i} + \mathbf{j}\,\frac{\partial y}{\partial u_i} + \mathbf{k}\,\frac{\partial z}{\partial u_i}, \qquad i = 1, 2, 3 \qquad (3.5.2)$$

The square of the length of $\mathbf{U}_i$ will be

$$h_i^2 = \mathbf{U}_i \cdot \mathbf{U}_i = \left(\frac{\partial x}{\partial u_i}\right)^2 + \left(\frac{\partial y}{\partial u_i}\right)^2 + \left(\frac{\partial z}{\partial u_i}\right) \qquad (3.5.3)$$

The quantities h_i are known as the scale factors for the corresponding coordinates u_i. If $\mathbf{a}_i$ is a *unit* vector along the curve generated by allowing one u_i to vary, then

$$\mathbf{a}_i = \frac{1}{h_i}\frac{\partial \mathbf{r}}{\partial u_i} \qquad (3.5.4)$$

In addition, the differential length increment along an arbitrary curve will be

$$ds^2 = d\mathbf{r} \cdot d\mathbf{r} = dx^2 + dy^2 + dz^2$$

$$= \mathbf{U}_1 \cdot \mathbf{U}_1\, du_1^2 + \mathbf{U}_2 \cdot \mathbf{U}_2\, du_2^2 + \mathbf{U}_3 \cdot \mathbf{U}_3\, du_3^2$$

$$= h_1^2\, du_1^2 + h_2^2\, du_2 + h_3^2\, du_3 \qquad (3.5.5)$$

From this result it is evident that when only one coordinate u_i varies, the distance ds_i along that coordinate is given by $ds_i = h_i\, du_i$. The gradient of some scalar function $f(u_1, u_2, u_3)$ is thus

$$\nabla f = \mathbf{a}_1\,\frac{\partial f}{\partial s_1} + \mathbf{a}_2\,\frac{\partial f}{\partial s_2} + \mathbf{a}_3\,\frac{\partial f}{\partial s_3}$$

$$= \frac{\mathbf{a}_1}{h_1}\,\frac{\partial f}{\partial u_1} + \frac{\mathbf{a}_2}{h_2}\,\frac{\partial f}{\partial u_2} + \frac{\mathbf{a}_3}{h_3}\,\frac{\partial f}{\partial u_3} \qquad (3.5.6)$$

The divergence of a vector $\mathbf{V} = \mathbf{a}_1 V_1 + \mathbf{a}_2 V_2 + \mathbf{a}_3 V_3$ is obtained quite readily by applying the divergence theorem to a small volume element with sides of length ds_1, ds_2, ds_3 along the coordinate lines u_1, u_2, u_3, respectively, as shown in Figure 3.5. When higher order quantities are neglected, the volume element may be written $dv = ds_1\, ds_2\, ds_3 = h_1 h_2 h_3\, du_1\, du_2\, du_3$. A net flow of $\mathbf{V}$ through the surface perpendicular to, say, the direction u_2 is given by the

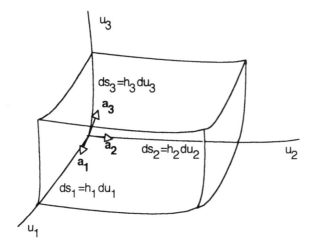

Figure 3.5. Coordinate lines for a general orthogonal curvilinear coordinate system.

expression

$$(V_2 \, ds_1 \, ds_3)|_{u_2 + du_2} - (V_2 \, ds_1 \, ds_3)|_{u_2} = \frac{\partial}{\partial u_2} (V_2 \, ds_1 \, ds_3) \, du_2$$

$$= \frac{\partial}{\partial u_2} (h_1 h_3 V_2) \, du_1 \, du_2 \, du_3$$

$$= \frac{1}{h_1 h_2 h_3} \frac{\partial}{\partial u_2} (h_1 h_3 V_2) \, dv \qquad (3.5.7)$$

When the corresponding contributions associated with directions u_1 and u_3 are added to this expression, the result is equivalent to $\mathbf{V} \cdot \mathbf{n} \, dS$ integrated over all faces of the volume element dv. By the divergence theorem this expression is equivalent to $\nabla \cdot \mathbf{V} \, dv$. We thus obtain

$$\int \mathbf{V} \cdot \mathbf{n} \, dS = \int \frac{1}{h_1 h_2 h_3} \left[\frac{\partial}{\partial u_1} (h_3 h_2 V_1) + \frac{\partial}{\partial u_2} (h_1 h_3 V_2) + \frac{\partial}{\partial u_3} (h_2 h_1 V_3) \right] dv$$

$$= \int \nabla \cdot \mathbf{V} \, dv \qquad (3.5.8)$$

From this result we can infer the form of $\nabla \cdot \mathbf{V}$ in curvilinear coordinates, namely,

$$\nabla \cdot \mathbf{V} = \frac{1}{h} \left[\frac{\partial}{\partial u_1} \left(\frac{h}{h_1} V_1 \right) + \frac{\partial}{\partial u_2} \left(\frac{h}{h_2} V_2 \right) + \frac{\partial}{\partial u_3} \left(\frac{h}{h_3} V_3 \right) \right] \qquad (3.5.9)$$

where $h = h_1 h_2 h_3$.

Finally since $\nabla^2 f = \nabla \cdot \nabla f$, combination of (3.5.6) and (3.5.9) yields the Laplacian in the form

$$\nabla^2 f = \frac{1}{h} \left[\frac{\partial}{\partial u_1} \left(\frac{h}{h_1} \frac{\partial f}{\partial u_1} \right) + \frac{\partial}{\partial u_2} \left(\frac{h}{h_2} \frac{\partial f}{\partial u_2} \right) + \frac{\partial}{\partial u_3} \left(\frac{h}{h_3} \frac{\partial f}{\partial u_3} \right) \right] \quad (3.5.10)$$

We now consider some examples of the application of these results.

3.5.1 Cylindrical Coordinates

The coordinates u_1, u_2, u_3 are taken to be ρ, φ, z, respectively, with transformation equations

$$x = \rho \cos \varphi, \qquad y = \rho \sin \varphi, \qquad z = z \qquad (3.5.11)$$

The scale factors are obtained from (3.5.3). We find

$$h_\rho^2 = \left(\frac{\partial x}{\partial \rho} \right)^2 + \left(\frac{\partial y}{\partial \rho} \right)^2 + \left(\frac{\partial z}{\partial \rho} \right)^2 = \cos^2 \varphi + \sin^2 \varphi = 1 \quad (3.5.12)$$

Similarly, we calculate $h_\varphi^2 = \rho^2$ and $h_z^2 = 1$. Writing the position vector as $\mathbf{r} = \mathbf{i} \rho \cos \varphi + \mathbf{j} \rho \sin\varphi + \mathbf{k} z$, and using (3.5.4), we determine that the unit vectors in the ρ, φ, z coordinate system are

$$\mathbf{a}_\rho = \frac{\partial \mathbf{r}}{\partial \rho} = \mathbf{i} \cos \varphi + \mathbf{j} \sin \varphi$$

$$\mathbf{a}_\varphi = \frac{1}{\rho} \frac{\partial \mathbf{r}}{\partial \varphi} = -\mathbf{i} \sin \varphi + \mathbf{j} \cos \varphi$$

$$\mathbf{a}_z = \frac{\partial \mathbf{r}}{\partial z} = \mathbf{k} \qquad (3.5.13)$$

The cylindrical volume element is $dv = h_\rho h_\varphi h_z \, d\rho \, d\varphi \, dz = \rho \, d\rho \, d\varphi \, dz$. A surface element on the curved wall of the cylinder is $ds_\varphi \, ds_z = h_\varphi h_z \, d\varphi \, dz = \rho \, d\varphi \, dz$ while on the top or base of the cylinder an element of area is $ds_\rho \, ds_\varphi = \rho \, d\rho \, d\varphi$. Finally the Laplacian is given by

$$\nabla^2 f = \frac{1}{\rho} \frac{\partial}{\partial \rho} \left(\rho \frac{\partial f}{\partial \rho} \right) + \frac{1}{\rho^2} \frac{\partial^2 f}{\partial \varphi^2} + \frac{\partial^2 f}{\partial z^2} \qquad (3.5.14)$$

3.5.2 Spherical Coordinates

In this case the transformation equations are

$$x = r \sin \vartheta \cos \varphi, \qquad y = r \sin \vartheta \sin \varphi, \qquad z = r \cos \vartheta \quad (3.5.15)$$

The angle ϑ is measured away from the positive z axis, that is, $\vartheta = 0$ and π correspond to the north and south poles, respectively. The spherical coordinate system can be thought of as being generated by rotating the polar coordinates $y = r \sin \vartheta$, $z = r \cos \vartheta$ about the z axis through angles measured by φ. It should be noted that many writers use a notation in which the role played by the angles ϑ and φ is interchanged.

For spherical coordinates the scale factors are readily found to be $h_r = 1$, $h_\vartheta = r$, $h_\varphi = r \sin \vartheta$. The Laplacian is

$$\nabla^2 f = \frac{1}{r^2} \frac{\partial}{\partial r} \left(r^2 \frac{\partial f}{\partial r} \right) + \frac{1}{r^2 \sin \vartheta} \frac{\partial}{\partial \vartheta} \left(\sin \vartheta \frac{\partial f}{\partial \vartheta} \right) + \frac{1}{r^2 \sin^2 \vartheta} \frac{\partial^2 f}{\partial \varphi^2}$$

(3.5.16)

In Section 3.6 it will be found that the radial portion of this result is more conveniently expressed by using the identity

$$\frac{1}{r^2} \frac{\partial}{\partial r} \left(r^2 \frac{\partial f}{\partial r} \right) = \frac{1}{r^2} \frac{\partial^2}{\partial r^2} (rf)$$

(3.5.17)

Chapter 5 will be devoted to the solution of problems with spherical symmetry. As a preliminary example of the separation-of-variables technique in other than rectangular coordinates, we now obtain a simple solution of Laplace's equation in spherical coordinates. For simplicity we only consider solutions that are independent of the azimuthal angle φ. In this case the Laplacian (3.5.16) reduces to

$$\frac{1}{r^2} \frac{\partial}{\partial r} \left(r^2 \frac{\partial f}{\partial r} \right) + \frac{1}{r^2 \sin \vartheta} \frac{\partial}{\partial \vartheta} \left(\sin \vartheta \frac{\partial f}{\partial \vartheta} \right) = 0$$

(3.5.18)

On introducing the product form $f(r, \vartheta) = R(r) \Theta(\vartheta)$ and multiplying (3.5.18) by $r^2/R\Theta$, we can obtain the separated expression

$$\frac{(r^2 R')'}{R} = -\frac{(\Theta' \sin \vartheta)'}{\Theta' \sin \vartheta} = \text{const}$$

(3.5.19)

A prime on a function of r signifies an r derivative and similarly for ϑ. As was the case in rectangular coordinates, each side of the separated equation must be a constant, as indicated in (3.5.19). For the present we shall only consider the simplest solutions of these separated equations, namely, the ones obtained by setting the separation constant equal to zero.

On canceling off factors of R and $\Theta \sin \vartheta$ in (3.5.19) and integrating over r and ϑ, we obtain

$$r^2 R' = c_1, \qquad \Theta' \sin \vartheta = d_1$$

(3.5.20)

A second integration of each of these equations yields

$$R = -\frac{c_1}{r} + c_2$$

$$\Theta = d_1 \int \cos \vartheta \, d\vartheta = \frac{d_1}{2} \ln \frac{1 - \cos \vartheta}{1 + \cos \vartheta} + d_2 \qquad (3.5.21)$$

The product solution $f = R\Theta$ will thus diverge logarithmically at $\vartheta = 0$ and π unless we set $d_1 = 0$. Also, to have a solution that vanishes as $r \to \infty$ (a commonly occurring requirement when dealing with solutions of Laplace's equation), we must also set $c_2 = 0$. We then obtain the solution

$$f = A/r \qquad (3.5.22)$$

where $A = c_1 d_2$. On the other hand, if the solution must remain finite as R approaches 0, we must set $c_1 = 0$ and obtain $f = B$ where $B = c_2 d_2$.

As an example of the use of this result, we consider f to refer to a temperature field and ask for the temperature distribution in a solid from which a spherical hole has been excised when the temperature on the surface of the hole is u_0 and the temperature of the solid is maintained at zero at large distances from the hole. The boundary conditions $f(a) = u_0$ and $f(\infty) = 0$ yield $A = u_0 a$ and we obtain

$$f(r) = u_0(a/r) \qquad (3.5.23)$$

Since distance away from the surface of the sphere is measured in terms of an increase in the coordinate r, the component of the heat flow vector $\mathbf{J}$ that is perpendicular to the surface $R = a$ is given by $J_r = \mathbf{a}_r \cdot \mathbf{J}(a)$ where

$$\mathbf{J}(a) = -K\nabla f|_{r=a} = -K\mathbf{a}_r \left.\frac{\partial f}{\partial r}\right|_{r=a} = K\mathbf{a}_r u_0 \left.\frac{a}{r^2}\right|_{r=a}$$

$$= K\frac{u_0}{a}\mathbf{a}_r \qquad (3.5.24)$$

The unit vector $\mathbf{a}_r$ follows from (3.5.4). We have

$$\mathbf{a}_r = \frac{\partial \mathbf{r}}{\partial r} = \mathbf{i} \sin \vartheta \cos \varphi + \mathbf{j} \sin \vartheta \sin \varphi + \mathbf{k} \cos \vartheta \qquad (3.5.25)$$

The solution corresponding to zero separation constant can thus be used to obtain the temperature in a region that is heated by a spherically symmetric source, a result of the sort considered in introductory physics courses. We now proceed to a somewhat less familiar coordinate system and show that when the

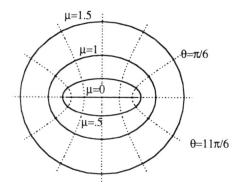

Figure 3.6. Elliptical coordinates used to generate oblate spheroidal coordinates.

solution corresponding to zero separation constant is considered, we obtain a more interesting result.

3.5.3 Oblate Spheroidal Coordinates

Just as the spherical coordinates given in (3.5.15) could be thought of as having been generated by the rotation of circles in the YZ plane about the Z axis, so also a coordinate system can be generated by rotating an ellipse in the YZ plane about the Z axis. When the ellipses have their major axes along the Y axis, the surfaces generated are referred to as oblate spheroids. Note that in the limit that the minor axis of the ellipse approaches zero, the corresponding oblate spheroid reduces to a disk. We now show that when Laplace's equation is written in oblate spheroidal coordinates, the separation-of-variables procedure for the simplest case of zero separation constant can be used to determine the temperature distribution about a disklike crack in a solid.

The ellipse used to generate the oblate spheroids is shown in Figure 3.6. The hyperbolas that form the mutually orthogonal coordinate lines for this ellipse are also indicated. The transformation equations are conveniently written

$$y = c \cosh \mu \sin \vartheta, \qquad z = c \sinh \mu \cos \vartheta \qquad (3.5.26)$$

Note that $\mu = 0$ corresponds to the line segment $|y| < c, z = 0$. On eliminating ϑ by substituting these relations into the identity $\sin^2 \vartheta + \cos^2 \vartheta = 1$, we obtain the ellipses

$$\left(\frac{y}{c \cosh \mu} \right)^2 + \left(\frac{z}{c \sinh \mu} \right)^2 = 1 \qquad (3.5.27)$$

while a corresponding elimination of μ yields the hyperbolas

$$\left(\frac{y}{c \sin \vartheta} \right)^2 - \left(\frac{z}{c \cos \vartheta} \right)^2 = 1 \qquad (3.5.28)$$

Note that for large μ both $\cosh\mu$ and $\sinh\mu$ approach the same value $(\frac{1}{2})e^{\mu}$. Thus, for large values of μ the ellipses become circles with radii $(\frac{1}{2})ce^{\mu}$. Also, the hyperbolas approach the asymptotes $y = z\tan\vartheta$. These curves are the radial lines of polar coordinates. Thus we see that for large μ the elliptical coordinate system approaches polar coordinates.

If rotation about the z axis is again described by an angle φ, the equation for the ellipse in (3.5.26) leads to

$$x = c \cosh \mu \sin \vartheta \cos \varphi$$

$$y = c \cosh \mu \sin \vartheta \sin \varphi$$

$$z = c \sinh \mu \cos \varphi \tag{3.5.29}$$

Note that $\mu = 0$ corresponds to the disk $x^2 + y^2 \le c^2$, $z = 0$. Using (3.5.3), we find that the scale factors are $h_{\mu}^2 = h_{\vartheta}^2 = c^2(\cosh^2\mu - \cos^2\vartheta)$ and $h_{\varphi}^2 = c^2\cosh^2\mu \sin^2\vartheta$. By (3.5.10), the Laplacian is

$$\nabla^2 f = \frac{c}{h}\left[\sin\vartheta\frac{\partial}{\partial\mu}\left(\cosh\mu\frac{\partial f}{\partial\mu}\right) + \cosh\mu\frac{\partial}{\partial\vartheta}\left(\sin\vartheta\frac{\partial f}{\partial\vartheta}\right)\right.$$

$$\left. + \frac{\cosh^2\mu - \sin^2\vartheta}{\cosh\mu\sin\vartheta}\frac{\partial^2 f}{\partial\varphi^2}\right] \tag{3.5.30}$$

When solutions are specialized to be independent of the angle φ and a separated solution $f(\mu, \vartheta) = M(\mu)\Theta(\vartheta)$ is introduced, Laplace's equation is readily separated to give

$$\frac{(M'\cosh\mu)'}{M\cosh\mu} = -\frac{(\Theta'\sin\vartheta)'}{\Theta\sin\vartheta} = \text{const} \tag{3.5.31}$$

If we again investigate only the simplest solution in which the separation constant is zero, we obtain

$$\Theta(\vartheta) = d_1\ln\frac{1 - \cos\vartheta}{1 + \cos\vartheta} + d_2 \tag{3.5.32}$$

the same result as that obtained for spherical coordinates in (3.5.21). To avoid singularities at $\vartheta = 0$ and π, we again set $d_1 = 0$.

The integration of the equation for the coordinate μ yields

$$M(\mu) = c_1\int \text{sech}\,\mu\,d\mu + c_2 = -2c_1\tan^{-1}(e^{-\mu}) + c_2 \tag{3.5.33}$$

In order to have $M(\mu)$ approach zero for large μ, we must set $c_2 = 0$. A solution of Laplace's equation is thus

$$f(\mu, \vartheta) = M(\mu)\Theta(\vartheta) = A\tan^{-1}(e^{-\mu}) \qquad (3.5.34)$$

where $A = -2c_1 d_2$. If $f = u_0$ on the disk $\mu = 0$, we have $u_0 = A\tan^{-1}(1) = \pi A/4$, or

$$f(\mu) = \frac{4u_0}{\pi}\tan^{-1}(e^{-\mu}) \qquad (3.5.35)$$

Note that distance away from the surface of the disk is measured in terms of an increase in the coordinate μ. Heat flow normal to the disk is thus given by determining $\mathbf{a}_\mu \cdot \mathbf{J}$ for $\mu = 0$ where

$$\mathbf{a}_\mu \cdot \mathbf{J} = -K\mathbf{a}_\mu \cdot \nabla f = -\frac{K}{h_\mu}\frac{\partial f}{\partial \mu} \qquad (3.5.36)$$

The unit vector $\mathbf{a}_\mu$ is

$$\mathbf{a}_\mu = \frac{1}{h_\mu}\frac{\partial \mathbf{r}}{\partial \mu}$$

$$= \frac{c}{h_\mu}[\sinh \mu \sin \vartheta(\mathbf{i}\cos \varphi + \mathbf{j}\sin \varphi) + \mathbf{k}\cosh \mu \cos \vartheta] \qquad (3.5.37)$$

On the surface of the disk we have $\mu = 0$ and thus $h_\mu = c\sqrt{1 - \sin^2\vartheta} = c\cos \vartheta$ and $\mathbf{a}_\mu = \mathbf{k}$. The heat flow is therefore

$$\mathbf{J}_\mu = -\frac{K}{a\cos \vartheta}\frac{4u_0}{\pi}\frac{\partial}{\partial \mu}(\tan^{-1}e^{-\mu})|_{\mu=0}$$

$$= \frac{2u_0}{\pi a}\frac{\mathbf{k}}{\cos \vartheta} \qquad (3.5.38)$$

From (3.5.29) we obtain $\rho = \sqrt{x^2 + y^2} = c\cosh \mu \sin \vartheta$. For $\mu = 0$ we have $\rho = c\sin \vartheta$ and thus $c\cos \vartheta = \sqrt{c^2 - \rho^2}$. The heat flow normal to the disc is thus

$$J_\mu = \frac{2K}{\pi}\frac{u_0}{\sqrt{a^2 - \rho^2}} \qquad (3.5.39)$$

The use of oblate spheroidal coordinates is one of the simpler ways of obtaining this result.

Problems

3.5.1   A coordinate transformation that corresponds to the rotation of a two-dimensional coordinate system through an angle φ is

$$x = x'\cos \varphi + y'\sin \varphi$$
$$y = -x'\sin \varphi + x'\cos \varphi$$

Calculate the scale factors for the transformation and give the Laplacian in the primed coordinate system.

3.5.2   It may be shown that if u and v are related to x and y by $x + iy = f(u + iv)$ where $i = \sqrt{-1}$, then the real and imaginary parts of the equation constitute a two-dimensional orthogonal transformation. As an example, consider $x + iy = (u + iv)^2$. Separate real and imaginary parts to obtain

$$x = u^2 - v^2, \qquad y = 2uv$$

Calculate the scale factors h_u and h_v and obtain the Laplacian in the u, v coordinate system. Calculate the vectors $\mathbf{U}_u$ and $\mathbf{U}_v$ and show that $\mathbf{U}_u \cdot \mathbf{U}_v = 0$. Show that if v is eliminated between the transformation equations, a parabola opening out along the negative x axis is obtained, while if u is eliminated, a parabola opening out along the positive x axis is obtained.

3.5.3   Separate real and imaginary parts in $x + iy = a \cosh(u + iv)$ and obtain the transformation for elliptic coordinates in the form

$$x = a \cosh u \cos v, \qquad y = a \sinh u \sin v$$

show that

$$\left(\frac{x}{a \cosh u}\right)^2 + \left(\frac{y}{a \sinh u}\right)^2 = 1$$
$$\left(\frac{x}{a \cos v}\right)^2 - \left(\frac{x}{a \sin v}\right)^2 = 1$$

Calculate the scale factors and obtain the Laplacian.

3.6 SPHERICAL SYMMETRY

The alternate form for the radial part of the Laplacian in spherical coordinates that was noted in (3.5.17) frequently permits an immediate simplification of a calculation when it is used in the solution of problems having spherical symmetry. As will be seen in the next section, no such simplification is available for two-dimensional problems in polar coordinates or for three-dimensional problems having cylindrical symmetry. Since the solutions required for polar and cylindrical symmetry are better appreciated when contrasted with the simplicity of those occurring for spherical symmetry, we briefly consider this latter situation in the present section. Spherical coordinates will be considered in more detail in Chapter 5.

For many types of wave motion that take place in three dimensions the quantity representing the propagated disturbance (e.g., air pressure in a small-amplitude sound wave) may be shown to satisfy the three-dimensional generalization of the wave equation (1.1.1), namely,

$$\nabla^2 p - \frac{1}{c^2} \frac{\partial^2 p}{\partial t^2} = 0 \qquad (3.6.1)$$

If there is no angular dependence to the wave field, as would be the case for waves emanating from a point source, then only the radial part of the Laplacian obtained in (3.5.16) is needed. When this radial part is rewritten according to the identity given in (3.5.17), the wave equation becomes

$$\frac{1}{r} \frac{\partial^2}{\partial r^2} (rp) - \frac{1}{c^2} \frac{\partial^2 p}{\partial t^2} = 0 \qquad (3.6.2)$$

On multiplying this equation by r and setting $f(r, t) = rp(r, t)$, we obtain

$$\frac{\partial^2 f}{\partial r^2} - \frac{1}{c^2} \frac{\partial^2 f}{\partial t^2} = 0 \qquad (3.6.3)$$

which is the same equation as that considered previously for waves in *one* space dimension. Hence, the solution of problems in three dimensions that have complete spherical symmetry will require essentially no additional mathematical considerations beyond those already introduced to solve problems in one dimension.

3.6.1 Spherical Standing Waves

The solution of the wave equation (3.6.3) by the separation-of-variables method proceeds in exactly the same manner as for the one-dimensional problems considered previously. The only difference lies in the application of the boundary conditions. The product solution of (3.6.3) is $f(r, t) = (A \cos kr + B \sin kr)(E \cos kct + F \sin kct)$ where k is the separation constant. The original dependent variable $p(r, t)$ is thus

$$p(r, t) = \frac{1}{r} (A \cos kr + B \sin kr)(E \cos kct + F \sin kct) \qquad (3.6.4)$$

We can use this solution to discuss the types of spherically symmetric standing waves that can occur within a spherical enclosure. In order for $p(r, t)$ to remain finite at the center of the region, however, we must set $A = 0$ in the product solution given above. If the boundary condition at $r = a$ is that $p(a, t) = 0$, then we again obtain the condition $\sin ka = 0$ or $ka = n\pi$ with $n = 1, 2, 3, \ldots$. The general solution is then

$$p(r, t) = \frac{1}{r} \sum_{n=1}^{\infty} \sin \frac{n\pi r}{a} \left(a_n \cos \frac{n\pi ct}{a} + b_n \sin \frac{n\pi ct}{a} \right) \quad (3.6.5)$$

If the initial conditions $p(r, 0)$ and $p_t(r, 0)$ are specified, then application of the usual orthogonality relations to the quantities $rp(r, 0)$ and $rp_t(r, 0)$ yield

$$a_n = \frac{2}{a} \int_0^a r \, dr \, p(r, 0) \sin \frac{n\pi xr}{a}, \qquad b_n = \frac{2}{n\pi c} \int_0^a r \, dr \, p_t(r, 0) \sin \frac{n\pi r}{a}$$

$$(3.6.6)$$

3.6.2 Spherical Traveling Waves

When a wave with spherical symmetry radiates to large distances, the solutions of (3.6.3) are most conveniently expressed in complex exponential notation as was done for the membrane in Section 3.4. Hence we write

$$p(r, t) = \frac{1}{r} (Ae^{ikr} + Be^{-ikr}) e^{-ikct} \quad (3.6.7)$$

instead of the standing wave expression used in (3.6.4). As was the case for the membrane, A and B are allowed to be complex as well. Then, outgoing spherical waves (i.e., those traveling in the positive r direction) will be those with r and t dependence in the form $r - ct$. Thus we set $B = 0$ in (3.6.7) and obtain

$$p(r, t) = \frac{1}{r} Ae^{ik(r - ct)} \quad (3.6.8)$$

For an incoming wave we would set $A = 0$ instead of B and obtain $p(r, t) = r^{-1} Be^{-ik(r + ct)}$. A similar simplification of the radial part of the Laplacian applies to diffusion problems having spherical symmetry. An example is considered in the problems.

Problems

3.6.1 The function $u(r, t)$ satisfies

$$\nabla^2 u - \frac{1}{2} \frac{\partial^2 u}{\partial t^2} = 0$$

within a spherical region of radius a and satisfies the boundary condition $u(a, t) = 0$.

(a) Write the general solution that remains finite at the origin.

(b) Write, and sketch as a function of r, the function that describes the spatial variation of the lowest mode of vibration.

3.6.2 *Cooling of a uniformly heated sphere.* A sphere of radius a initially has the uniform temperature u_0. At $t = 0$ the surface $r = a$ is set to zero temperature.

Use separation of variables to solve

$$\frac{1}{\gamma^2}\frac{\partial u}{\partial t} - \frac{1}{r}\frac{\partial^2}{\partial r^2}(ru) = 0, \qquad \gamma^2 = \frac{K}{\rho c}$$

with $u(r, 0) = u_0$, $u(a, t) = 0$, and $u(0, t)$ bounded to obtain

$$u(r, t) = \frac{2u_0}{\pi}\frac{a}{r}\sum_1^\infty \frac{(-1)^n}{n}e^{-n^2 vt}\sin\frac{n\pi r}{a}, \qquad v = \left(\frac{\gamma\pi}{a}\right)^2$$

To check that the initial condition is satisfied, consider the result just obtained in the limit t approaches 0. If we interchange limit and summation and use

$$\sin x - \tfrac{1}{2}\sin 2x + \tfrac{1}{3}\sin 3x - \ldots = \tfrac{1}{2}x, \qquad -\pi < x < \pi$$

we obtain, with $x = \pi r/a$, the expected result $u(r, 0) = u_0$. Validity of this interchange of limiting process and summation is considered in Appendix A.

3.6.3 Use the Laplace transform method to recover the result obtained in Problem 3.6.2.

(a) Show that the transform of the solution is

$$U(r, s) = \frac{u_0}{s}\left(1 - \frac{a}{r}\frac{\sinh\sqrt{x}\,r/\gamma}{\sinh\sqrt{s}a/\gamma}\right)$$

(b) To carry out the inversion, first rewrite $\sinh r\sqrt{s}/\gamma$ as a Fourier series, that is, set

$$\sinh\frac{\sqrt{s}r}{\gamma} = \sum_{n=1}^\infty A_n\sin\frac{n\pi r}{a}$$

and find

$$\frac{\sinh\sqrt{s}r/\gamma}{\sinh\sqrt{s}a/\gamma} = \frac{2v}{\pi}\sum_1^\infty \frac{(-1)^{n-1}n}{s + n^2 v}\sin\frac{n\pi r}{a}$$

(c) Use this result in part (a) to recover the result obtained by the separation-of-variables method.

3.7 CIRCULAR AND CYLINDRICAL SYMMETRY

A transformation of the type used in the last section to simplify the radial part of the Laplacian in *spherical* coordinates is not available for the Laplacian in two-dimensional polar coordinates or three-dimensional cylindrical coordinates. As a result, the separation-of-variables procedure leads to new types of ordinary differential equations for the radial coordinate. For the two-dimensional *Laplace* equation we obtain a Cauchy-Euler equation while for the *wave* equation we will encounter Bessel's equation. An introduction to the solutions of this latter equation, known as Bessel functions, is contained in Appendix D. In the following development it will be assumed that the reader is familiar with the material summarized there. We now take up four examples that show how the concepts already developed in the two previous chapters are applied in polar and cylindrical coordinates.

3.7.1 Steady State Temperature in a Pie-Shaped Region

If the temperature and its normal derivative are specified as shown in Figure 3.7, then the temperature within the region is determined by solving

$$\nabla^2 u = \frac{1}{\rho}\frac{\partial}{\partial \rho}\left(\rho \frac{\partial u}{\partial \rho}\right) + \frac{1}{\rho^2}\frac{\partial^2 u}{\partial \vartheta^2} = 0 \qquad (3.7.1)$$

and then applying the given boundary conditions. This form of the two-dimensional Laplacian is obtainable from the expression given in (3.5.14) for cylindrical coordinates by ignoring the z dependence. Separation of variables in the form $u(\rho, \vartheta) = R(\rho)\Theta(\vartheta)$ leads to

$$\frac{\rho}{R}(\rho R')' = -\frac{\Theta''}{\Theta} = \pm k^2 \qquad (3.7.2)$$

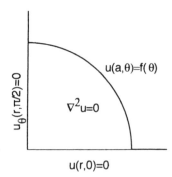

Figure 3.7. Boundary conditions for solving Laplace's equation in a quarter circle.

where primes on R or Θ refer to derivatives with respect to ρ or ϑ, respectively. To satisfy the boundary conditions at $\vartheta = 0$ and $\pi/2$, we choose the upper sign[2] for k^2 and obtain

$$\Theta = \begin{cases} A \cos k\vartheta + B \sin k\vartheta, & k^2 > 0 \\ A + B\vartheta, & k^2 = 0 \end{cases} \tag{3.7.3}$$

For $k = 0$, the only solution satisfying the required boundary conditions $\Theta(0) = 0$, $\Theta'(\pi/2) = 0$ is $\Theta = 0$. For $k^2 > 0$, the requirement $\Theta(0) = 0$ yields $A = 0$ while the requirement $\Theta'(\pi/2) = 0$ imposes $\cos(k\pi/2) = 0$, that is, $k = 1, 3, 5, \ldots$. We thus obtain $\Theta = B \sin k\vartheta$ with $k = 2n + 1$, where $n = 0, 1, 2, \ldots$.

The radial equation obtained in (3.7.2) may be written

$$\rho^2 R'' + \rho R' - k^2 R = 0 \tag{3.7.4}$$

which is of the Cauchy-Euler form (Hildebrand, 1976, Section 1.6). Such equations have solutions $R = \rho^\alpha$ where the constant α is determined by substitution into Eq. (3.7.4). In the present instance, the result is $\alpha^2 = k^2$. Hence we have $\alpha = \pm k$ and the radial equation has the solution

$$R(\rho) = C\rho^k + D\rho^{-k} \tag{3.7.5}$$

Although the possibility $k = 0$ has already been ruled out for this problem, we note in passing that, for $k = 0$, Eq. (3.7.2) yields $(\rho R')' = 0$. On integrating twice we have

$$R = C + D \ln \rho \tag{3.7.6}$$

For those problems in which the $k = 0$ term *is* present, the constant D must be set equal to zero if the solution is to remain finite at the origin. In the present problem the solutions involving ρ^{-k} must also be eliminated since they would lead to R being infinite at the origin. We thus set $D = 0$ in both solutions. The elementary product solutions $u(\rho, \vartheta) = R(\rho)\Theta(\vartheta)$ are accordingly of the form

$$u_n(\rho, \theta) = \rho^{2n+1} \sin(2n + 1)\theta, \qquad n = 0, 1, 2, \ldots \tag{3.7.7}$$

and the general solution is given by a sum of such solutions with arbitrary amplitude. We thus obtain

$$u(\rho, \vartheta) = \sum_{n=0}^{\infty} B_n \rho^{2n+1} \sin(2n + 1)\vartheta \tag{3.7.8}$$

---

[2]The lower sign would lead to the solution $\Theta = A \cosh k\theta + B \sinh k\theta$. The only values of the constants A and B that satisfy the given boundary condition are $A = B = 0$.

The expansion of a function in terms of a Fourier series containing only odd multiples of ϑ is considered in Appendix A. The results obtained there may be applied to the present situation by choosing $L = \pi/2$. Then, on setting $\rho = a$ in (3.7.8), we obtain

$$u(a, \vartheta) = \sum_{n=0}^{\infty} B_n a^{2n+1} \sin(2n + 1)\vartheta \tag{3.7.9}$$

with the coefficients B_n given by

$$B_n = \frac{4}{\pi a^{2n+1}} \int_0^{\pi/2} u(a, \vartheta) \sin(2n + 1)\vartheta \, d\vartheta \tag{3.7.10}$$

In order to display some of the details associated with calculations in polar coordinates, we now show that, for the boundary conditions being considered here, the total heat flowing out through the horizontal edge at $\vartheta = 0$ equals that flowing in through the curved edge at $r = a$. Recall that the heat flow through a surface equals the integral over that surface of the normal component of the heat flow vector. The heat flow through the edge at $r = a$ is therefore given by

$$H\big|_{\rho=a} = \int_0^{\pi/2} J_\rho a \, d\vartheta = -K \int_0^{\pi/2} \frac{\partial u}{\partial \rho}\bigg|_{\rho=a} d\vartheta$$

$$= -K \sum_0^{\infty} (2n + 1) B_n a^{2n+1} \int_0^{\pi/2} \sin(2n + 1)\vartheta \, d\vartheta$$

$$= K \sum_0^{\infty} B_n a^{2n+1} \tag{3.7.11}$$

On the edge at $\vartheta = 0$, the heat flow is

$$H\big|_{\vartheta=0} = \int_0^a J_\vartheta \, d\rho = -K \int_0^a \frac{1}{\rho} \frac{\partial u}{\partial \vartheta}\bigg|_{\vartheta=0} d\rho$$

$$= -K \sum_0^{\infty} (2n + 1) B_n \int_0^a \rho^{2n} \, d\rho$$

$$= -K \sum_0^{\infty} B_n a^{2n+1} \tag{3.7.12}$$

The difference in sign merely distinguishes an inflow from an outflow.

The calculation considered here can be carried out for a pie-shaped region with any central angle α such that $0 < \alpha < 2\pi$. Determination of the solution to Laplace's equation in a pie-shaped region when the temperature is specified along one of the straight edges is a somewhat more complicated problem. An example is considered in Section 9.2.

Example 3.1. The temperature distribution on the rim of the quarter circle shown in Figure 3.7 is given by

$$f(\vartheta) = \begin{cases} 0, & 0 \le \vartheta < \pi/4 \\ u_0, & \pi/4 < \vartheta \le \pi/2 \end{cases} \tag{3.7.13}$$

Determine the temperature profile along the insulated edge $\vartheta = \pi/2$.
    From (3.7.10) we obtain

$$B_n = \frac{4u_0}{\pi a^{n+1}} \int_{\pi/4}^{\pi/2} \sin(2n+1)\vartheta \, d\vartheta$$

$$= \frac{4u_0}{\pi a^{n+1}} \frac{\cos(2n+1)\pi/4}{2n+1} \tag{3.7.14}$$

The temperature at any point on the disk is thus given by

$$u(\rho, \vartheta) = \frac{4u_0}{\pi} \sum_{n=0}^{\infty} (2n+1)^{-1} \left(\frac{\rho}{a}\right)^{2n+1} \cos(2n+1)\frac{\pi}{4} \sin(2n+1)\vartheta \tag{3.7.15}$$

This series may be summed by first using the identity $2 \sin a \cos b = \sin(a+b) + \sin(a-b)$ and then employing series 26 in Table F.1. We obtain

$$u(\rho, \vartheta) = \frac{u_0}{\pi} \left(\tan^{-1} \frac{2(\rho/a)\sin(\vartheta - \pi/4)}{1 - (\rho/a)^2} + \tan^{-1} \frac{2(\rho/a)\sin(\vartheta + \pi/4)}{1 - (\rho/a)^2} \right)$$

$$\tag{3.7.16}$$

Along the edge $\vartheta = \pi/2$ this general result becomes

$$u\left(\rho, \frac{\pi}{2}\right) = \frac{2u_0}{\pi} \tan^{-1} \frac{\sqrt{2}(\rho/a)}{1 - (\rho/a)^2} \tag{3.7.17}$$

Despite the somewhat cumbersome form of this result, the radial dependence is very nearly linear. A graph of $u(\rho, \pi/2)$ is shown in Figure 3.8.

3.7.2 Laplace's Equation in an Annular Circle

For a *complete* circle (i.e., when the angle α in the previous section becomes 2π) the boundary conditions imposed on the function $\Theta(\vartheta)$ given in (3.7.3) are provided by the requirement that both Θ and Θ' must be continuous throughout the circular region, so that for any value of ϑ we require $\Theta(\vartheta) = \Theta(\vartheta + 2\pi)$ and $\Theta'(\vartheta) = \Theta'(\vartheta + 2\pi)$. The condition that continuity imposes on the solution is the same for any choice of ϑ and we merely set $\vartheta = 0$. The

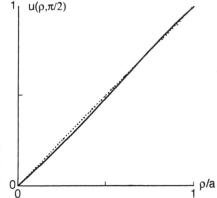

Figure 3.8. Steady state temperature on the rim $\theta = \pi/2$ for boundary conditions shown in Fig. 3.7 for $f(\theta)$ specified by equation (3.7.13).

requirement $\Theta(0) = \Theta(2\pi)$ yields

$$A = A \cos 2\pi k + B \sin 2\pi k \qquad (3.7.18)$$

while continuity of Θ' gives

$$kB = -kA \sin 2\pi k + kB \cos 2\pi k \qquad (3.7.19)$$

The continuity conditions thus provide a homogeneous system of equations for the coefficients A and B. One way to express the condition that such a homogeneous system of equations should yield nonzero values of A and B is that Δ, the determinant of the system of equations, be equal to zero. This requirement yields

$$\Delta = \begin{vmatrix} 1 - \cos 2\pi k & -\sin 2\pi k \\ \sin 2\pi k & 1 - \cos 2\pi k \end{vmatrix} = 0 \qquad (3.7.20)$$

Evaluation of the determinant leads to $\Delta = (1 - \cos 2\pi k)^2 - \sin^2 2\pi k = 4 \sin^2 \pi k$ and the condition $\Delta = 0$ is satisfied by choosing $k = 1, 2, \ldots$ For $k = 0$ we have $\Theta = A + B\vartheta$ and the continuity conditions imply $B = 0$. The solution is then $\Theta = A$, a constant. The solutions are thus of the form

$$\Theta(\vartheta) = A_n \cos n\vartheta + B_n \sin n\vartheta, \qquad n = 0, 1, 2, \ldots \qquad (3.7.21)$$

Turning now to the radial solution, we note that if the solution of Laplace's equation is sought in the annular region $\rho_1 \leq \rho \leq \rho_2$, then the terms in the solutions (3.7.5) and (3.7.6) that are singular at the origin must be retained since the point $\rho = 0$ is now excluded from the region in which the solution is to be obtained. The solution will thus be the sum

$$u(\rho, \vartheta) = A_0(C_0 + D_0 \ln \rho)$$

$$+ \sum_{1}^{\infty} (C_n \rho^n + D_n \rho^{-n})(A_n \cos n\vartheta + B_n \sin n\vartheta) \quad (3.7.22)$$

The integration constants are determined by Fourier series methods once the boundary conditions are specified at the inner and outer radii ρ_1 and ρ_2, respectively. Setting $u(\rho_1, \vartheta) = f_1(\vartheta)$ and $u(\rho_2, \vartheta) = f_2(\vartheta)$ we obtain

$$f_i(\theta) = a_0 + b_0 \ln \rho_i + \sum_{n=1}^{\infty} (a_n \rho_i^n + b_n \rho_i^{-n}) \cos n\vartheta$$

$$+ \sum_{n=1}^{\infty} (c_n \rho_i^n + d_n \rho_i^{-n}) \sin n\vartheta, \quad i = 1, 2 \quad (3.7.23)$$

where $a_n = A_n C_n$, $b_n = A_n D_n$, $c_n = B_n C_n$, and $d_n = B_n D_n$. Employing the orthogonality of $\sin n\vartheta$ and $\cos n\vartheta$ over the interval $0 \le \vartheta \le 2\pi$, we arrive at the following three pairs of equations:

$$a_0 + b_0 \ln \rho_i = \frac{1}{2\pi} \int_0^{2\pi} f_i(\vartheta) \, d\vartheta, \quad i = 1, 2$$

$$a_n \rho_i^n + b_n \rho_i^{-n} = \frac{1}{\pi} \int_0^{2\pi} f_i(\vartheta) \cos n\vartheta \, d\vartheta, \quad i = 1, 2$$

$$c_n \rho_i^n + d_n \rho_i^{-n} = \frac{1}{\pi} \int_0^{2\pi} f_i(\vartheta) \sin n\vartheta \, d\vartheta, \quad i = 1, 2 \quad (3.7.24)$$

These results must, of course, be appropriately modified when other boundary conditions are imposed at either of the rims.

The two special cases $\rho_1 \to 0$, a solid disk, and $\rho_2 \to \infty$, an infinite two-dimensional region from which a circular hole has been excised, yield the following results:

Case 1: $\rho_1 \to 0$. To avoid divergence at $\rho = 0$, we must set $b_0 = b_n = d_n = 0$. Also, $f_1(\vartheta)$ is now the temperature at the origin, $u(0, \vartheta)$, so that $a_0 = u(0, \vartheta)$. For $i = 2$ in (3.7.24) we have

$$a_0 = \frac{1}{2\pi} \int_0^{2\pi} f_2(\vartheta) \, d\vartheta \quad (3.7.25)$$

On equating these two values of a_0 we have an example of the mean value theorem, which will be considered in Section 8.2, namely

$$u(0, \vartheta) = \frac{1}{2\pi} \int_0^{2\pi} f_2(\vartheta) \, d\vartheta \quad (3.7.26)$$

that is, the value of the solution at the center equals the average value on the rim. In addition, for $i = 2$ we obtain

$$a_n = \frac{1}{2\pi\rho_2^n} \int_0^{2\pi} f_2(\vartheta)\cos n\vartheta \; d\vartheta \qquad (3.7.27)$$

$$c_n = \frac{1}{2\pi\rho_2^n} \int_0^{2\pi} f_2(\vartheta)\sin n\vartheta \; d\vartheta \qquad (3.7.28)$$

The results for $i = 1$ merely yield the identity $0 = 0$. On introducing the dimensionless coefficients $A_n = a_n \rho_2^n$ and $B_n = c_n \rho_2^n$ we obtain

$$u(\rho, \vartheta) = A_0 + \sum_1^\infty \left(\frac{\rho}{\rho_2}\right)^n (A_n\cos n\vartheta + B_n\sin n\vartheta) \qquad (3.7.29)$$

with

$$A_0 = \frac{1}{2\pi} \int_0^{2\pi} f_2(\vartheta) \; d\vartheta$$

$$A_n = \frac{1}{\pi} \int_0^{2\pi} f_2(\vartheta)\cos n\vartheta \; d\vartheta, \qquad n \geq 1$$

$$B_n = \frac{1}{\pi} \int_0^{2\pi} f_2(\vartheta)\sin n\vartheta \; d\vartheta, \qquad n \geq 1 \qquad (3.7.30)$$

as the solution to Laplace's equation in the circular region $\rho \leq \rho_2$ with $u(\rho_2, \vartheta) = f_2(\vartheta)$.

Case 2: $\rho_2 \to \infty$. The logarithmic divergence at large values of ρ is avoided by again setting $b_0 = 0$. We must also set $a_n = c_n = 0$ to avoid divergences for large ρ so that (3.7.23) reduces to

$$u(\rho, \vartheta) = a_0 + \sum_1^\infty \rho^{-n}(b_n\cos n\vartheta + d_n\sin n\vartheta) \qquad (3.7.31)$$

As ρ increases, all terms in the series approach zero and the temperature approaches a_0. From (3.7.25), the constants are seen to be

$$a_0 = \frac{1}{2\pi} \int_0^{2\pi} f_1(\vartheta) \; d\vartheta = \frac{1}{2\pi} f_2(\vartheta) \; d\vartheta = u(\infty, \vartheta)$$

$$b_n = \frac{\rho_1^n}{\pi} \int_0^{2\pi} f_1(\vartheta)\cos n\vartheta \; d\vartheta$$

$$d_n = \frac{\rho_1^n}{\pi} \int_0^{2\pi} f_1(\vartheta)\sin n\vartheta \; d\vartheta \qquad (3.7.32)$$

Hence the solution of $\nabla^2 u = 0$, $\rho > \rho_1$ with $u(\rho_1, \vartheta) = f_1(\vartheta)$ and u finite as $\rho \to \infty$ is

$$u(\rho, \vartheta) = A_0 + \sum_1^\infty \left(\frac{\rho_1}{\rho}\right)^n (B_n \cos n\vartheta + D_n \sin n\vartheta) \quad (3.7.33)$$

with

$$A_0 = \frac{1}{2\pi} \int_0^{2\pi} f_1(\vartheta) \, d\vartheta$$

$$B_n = \frac{1}{\pi} \int_0^{2\pi} f_1(\vartheta) \cos n\vartheta \, d\vartheta, \quad n \geq 1$$

$$D_n = \frac{1}{\pi} \int_0^{2\pi} f_1(\vartheta) \sin n\vartheta \, d\vartheta, \quad n \geq 1 \quad (3.7.34)$$

These results must, of course, be appropriately modified when other boundary conditions are specified at the rim.

Example 3.2. We consider a situation in which the outer rim ρ_2 is maintained at zero temperature while the inner rim ρ_1 is insulated except at $\varphi = 0$ where there is an influx of heat H_0 (in calories per centimeter per second). The boundary condition at the inner rim may be written $u_\rho(\rho_1, \vartheta) = -(H_0/K\rho_1)\delta(\vartheta)$ where K is the conductivity of the medium, for then $\int_0^{2\pi} \rho_1(-Ku_\rho(\rho_1, \vartheta)) \, d\vartheta = H_0$, as required.

Since the source is symmetric about $\vartheta = 0$, only the $\cos n\vartheta$ term in the general solution will be present. The temperature is thus given by

$$u(\rho, \vartheta) = a_0 + b_0 \ln \rho + \sum_1^\infty (a_n \rho^n + b_n \rho^{-n}) \cos n\vartheta \quad (3.7.35)$$

This expression must vanish at the outer rim $\rho = \rho_2$. Since the terms $\cos n\vartheta$ are linearly independent, each term in the series must vanish separately, and we have

$$a_0 + b_0 \ln \rho_2 = 0$$

$$a_n \rho_2^n + b_n \rho_2^{-n} = 0, \quad n = 1, 2, 3, \ldots \quad (3.7.36)$$

At the inner rim $\rho = \rho_1$ the radial derivative $u_\rho(\rho_1, \vartheta)$ must satisfy the boundary condition $u_\rho(\rho_1, \vartheta) = -(H_0/K\rho_1)\delta(\vartheta)$, that is,

$$\frac{-H_0}{K\rho_1} \delta(\vartheta) = \frac{b_0}{\rho_1} + \sum_1^\infty n(a_n \rho_1^{n-1} - b_n \rho_1^{-n-1}) \cos n\vartheta \quad (3.7.37)$$

On multiplying by $\cos p\vartheta$ and integrating over ϑ from 0 to 2π, we obtain

$$b_0 = -H_0/2\pi K$$

$$n(a_n \rho_1^n - b_n \rho_1^{-n}) = \frac{-H_0}{\pi K}, \quad n = 1, 2, 3, \ldots \quad (3.7.38)$$

When Eqs. (3.7.36) and (3.7.38) are solved simultaneously, we arrive at

$$a_0 = \frac{H_0}{2\pi K} \ln \rho_2, \qquad b_0 = -\frac{H_0}{2\pi K}$$

$$a_n = -\frac{H_0}{\pi n K D_n} \rho_2^{-n}, \qquad b_n = \frac{h_0}{\pi n K D_n} \rho_2^{n}, \qquad n = 1, 2, 3, \ldots \qquad (3.7.39)$$

where $D_n = (\rho_1/\rho_2)^n + (\rho_2/\rho_1)^n$. On setting $\rho_2/\rho_1 = e^\gamma$, we have $D_n = 2 \cosh n\gamma$ and the temperature distribution in the disk takes the form

$$u(\rho, \vartheta) = \frac{H_0}{2\pi K} \ln \frac{\rho_2}{\rho} - \frac{H_0}{2\pi K} \sum_1^\infty \left[\left(\frac{\rho}{\rho_2}\right)^n - \left(\frac{\rho}{\rho_2}\right)^{-n} \right] \operatorname{sech} n\gamma \cos n\vartheta \qquad (3.7.40)$$

As a specific application of this result, the total heat flowing out through the outer rim in the semicircular region $-\pi/2 \leq \vartheta \leq \pi/2$ is

$$H = \int_{-\pi/2}^{\pi/2} -K u_\rho(\rho_2, \vartheta) \rho_2 \, d\vartheta$$

$$= \frac{H_0}{2} \left(1 + \frac{4}{\pi} \sum_1^\infty \frac{1}{n} \operatorname{sech} n\gamma \sin \frac{n\pi}{2} \right) \qquad (3.7.41)$$

If we choose a ratio $\rho_2/\rho_1 = e^{\pi/2} = 4.81$ so that $\gamma = \pi/2$, the series reduces to that considered in Problem 3.3.3, where it was noted (on the basis of symmetry consider- ations appropriate for that problem) that the sum equals $\pi/8$. We thus find $H = H_0(1 + \frac{1}{2})/2 = 3H_0/4$, that is, 75% of the heat from the source flows out through this half of the circumference.

3.7.3 Vibrating Membrane

We now proceed to problems that give rise to Bessel functions, the special functions that are customarily identified with circular and cylindrical geometry. A standard example is the description of waves on the surface of a circular membrane. We thus return to consideration of the type of wave motion first encountered in Section 3.4, where two-dimensional waves in rectangular co- ordinates were introduced. For waves bounded by a circular rim, we again solve the two-dimensional wave equation, but in order to satisfy conditions at the rim, we express the Laplacian in polar coordinates and write

$$\frac{1}{\rho} \frac{\partial}{\partial \rho} \left(\rho \frac{\partial u}{\partial \rho} \right) + \frac{1}{\rho^2} \frac{\partial^2 u}{\partial \vartheta^2} - \frac{1}{c^2} \frac{\partial^2 u}{\partial t^2} = 0 \qquad (3.7.42)$$

If we confine attention to waves at a single frequency ω and set[3] $u(\rho, \vartheta, t) = f(\rho, \vartheta)e^{-i\omega t}$, then we obtain the Helmholtz equation

---

[3] As noted in Section 3.4, $\operatorname{Re}[(a + ib)e^{-i\omega t}] = a \cos \omega t + b \sin \omega t$, which is the complete time dependent solution obtained by separation of variables.

$$\frac{1}{\rho} \frac{\partial}{\partial \rho} \left(\rho \frac{\partial f}{\partial \rho} \right) + \frac{1}{\rho^2} \frac{\partial^2 f}{\partial \vartheta^2} + k^2 f = 0 \qquad (3.7.43)$$

where $k = \omega/c$. The separation-of-variables procedure follows as developed previously in this section. With $f(\rho, \vartheta) = R(\rho)\Theta(\vartheta)$, the Helmholtz equation (3.7.43) becomes

$$\frac{\rho}{R} (\rho R')' + k^2 \rho^2 = -\frac{\Theta''}{\Theta} = \text{const} \qquad (3.7.44)$$

For the full circular membrane, the requirement of periodicity introduced in (3.7.20) is again appropriate with the separation constant of the form m^2 where m is an integer. We thus obtain

$$\Theta'' + m^2 \Theta = 0$$
$$\rho^2 R'' + \rho R' + (k^2 \rho^2 - m^2)R = 0 \qquad (3.7.45)$$

The equation for Θ again yields the solutions $\sin m\vartheta$ and $\cos m\vartheta$. However, for the radial function $R(\rho)$, the term $k^2 \rho^2$ gives us an equation that is not of Cauchy-Euler type. Rather, $R(\rho)$ satisfies the equation that defines Bessel functions. It will be assumed that the reader is familiar with this subject to the extent that it is developed in Appendix D. The solution of the second equation in (3.7.45) is given by

$$R_m(\rho) = A_m J_m(k\rho) + B_m Y_m(k\rho), \qquad m = 0, 1, 2, \ldots \qquad (3.7.46)$$

If the membrane extends over the entire surface within a circular region so that the origin ($\rho = 0$) is included, then the solutions $Y_m(k\rho)$ must be discarded since these functions are singular at $\rho = 0$ while the membrane displacement presumably remains finite at that point. Setting $B_m = 0$ in (3.7.46) we obtain the elementary solutions

$$f_m(\rho, \theta) = J_m(k\rho)(a_m \cos m\vartheta + b_m \sin m\vartheta) \qquad (3.7.47)$$

If the membrane is attached at the rim $\rho = a$ so that the displacement vanishes there, then this boundary condition is incorporated into these elementary solutions by requiring $J_m(ka) = 0$. From the discussion of the function $J_m(x)$ given in Appendix D and the graphs shown in Figure D.2, it is clear that for each value of m there is an infinite sequence of values of ka for which the condition $J_m(ka) = 0$ is satisfied. The first few values, labeled α_{mn}, are given in Table D.1. The index m refers to the order of the Bessel function and the index n refers to the nth zero in the sequence of zeros of $J_m(x)$. Since the frequency ω is related to k through $k = \omega/c$, each selection of m, n values has a specific frequency ω_{mn} associated with it. A product solution

$$u_{mn}(\rho, \theta, t) = J_m(\alpha_{mn}\rho/a)\cos m\vartheta\, e^{-i\omega_{mn}t} \qquad (3.7.48)$$

thus specifies a mode of vibration of the circular membrane in the same manner as was described for the rectangular membrane in Section 3.4. The Chladi patterns referred to in that section may be observed on the circular membrane as well. A few examples are indicated in Figure 3.9. The numbers above each pattern refer to the mode numbers (m, n), while the number below each pattern refers to the ratio of the frequency of that mode to the fundamental frequency ω_{01}.

It should be noted that the higher frequencies of vibration, being proportional to α_{mn}, are not integral multiples of the fundamental (lowest) frequency as was the case for the vibrating string.

The complete description of the membrane vibration is provided by the sum

$$u(\rho, \vartheta, t) = \sum_{m=0}^{\infty}\sum_{n=1}^{\infty} J_m\left(\alpha_{mn}\frac{\rho}{a}\right)(a_{mn}\cos m\vartheta + b_{mn}\sin m\vartheta)$$

$$\times\, (c_{mn}\cos \omega_{mn}t + d_{mn}\sin \omega_{mn}t) \qquad (3.7.49)$$

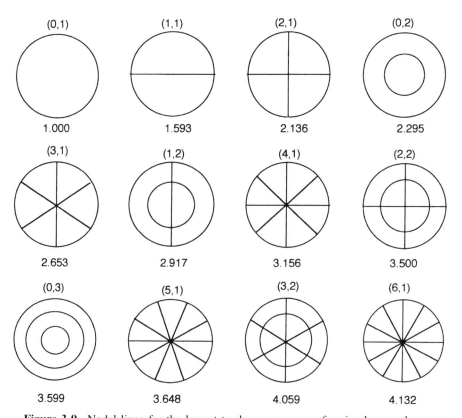

Figure 3.9. Nodal lines for the lowest twelve resonances of a circular membrane.

If the only displacements considered are those without any angular dependence, then only the $m = 0$ terms can occur and we obtain

$$u(\rho, t) = \sum_{n=1}^{\infty} J_0\left(\alpha_{0n} \frac{\rho}{a}\right) (A_n \cos \omega_n t + B_n \sin \omega_n t) \qquad (3.7.50)$$

where $A_n = a_{0n} C_{0n}$, $B_n = a_{0n} d_{0n}$, and $\omega_n = \omega_{0n}$. The coefficients A_n and B_n are determined by an application of the orthogonality considerations for Bessel functions that are developed in Appendix D. There it is shown that

$$\int_0^a J_0\left(\alpha_{0n} \frac{\rho}{a}\right) J_0\left(\alpha_{0m} \frac{\rho}{a}\right) \rho \, d\rho = \begin{cases} \dfrac{a^2}{2} [J_1(\alpha_{0n})]^2, & n = m \\ 0, & n \neq m \end{cases} \qquad (3.7.51)$$

Setting $t = 0$ in (3.7.50), multiplying by $\rho J_0(a_{0m} \rho/a)$, and integrating over ρ from 0 to a, we obtain

$$A_n = \frac{2}{a^2 [J_1(\alpha_{0n})]^2} \int_0^a u(\rho, 0) J_0\left(\alpha_{0n} \frac{\rho}{a}\right) \rho \, d\rho \qquad (3.7.52)$$

Also, differentiating (3.7.50) with respect to time, setting $t = 0$, and again employing the orthogonality considerations, we obtain

$$B_n = \frac{2}{\omega_n a^2 [J_1(\alpha_{0n})]^2} \int_0^a u_t(\rho, 0) J_0\left(\alpha_{0n} \frac{\rho}{a}\right) \rho \, d\rho \qquad (3.7.53)$$

As might be expected, there are only a limited number of expressions for $u(\rho, 0)$ and $u_t(\rho, 0)$ for which the integrals defining A_n and B_n in (3.7.52) and (3.7.53) may be performed analytically. One very useful example is $(a^2 - \rho^2)^\nu$ with $\nu > -1$. Here we shall only consider ν to be an integer or zero. The required integrals are performed by using the relation (Watson, 1944, p. 373)

$$\int_0^{\pi/2} \sin\theta \cos^{2\nu+1}\theta J_0(z \sin\theta) \, d\theta = \frac{2^\nu \Gamma(\nu + 1)}{z^{\nu+1}} J_{\nu+1}(z) \qquad (3.7.54)$$

where $\Gamma(\nu + 1)$ is the gamma function considered in Appendix B. On setting $\sin\vartheta = \rho/a$ and $z = \alpha_{0n}$ this integral becomes

$$\int_0^a (a^2 - \rho^2)^\nu J_0\left(\frac{\alpha_{0n} \rho}{a}\right) \rho \, d\rho = \frac{2^\nu \Gamma(\nu + 1) a^{2\nu+2}}{\alpha_{0n}^{\nu+1}} J_{\nu+1}(\alpha_{0n}) \qquad (3.7.55)$$

Example 3.3. A circular membrane with surface density σ and no initial displacement is given an initial velocity $u_t(\rho, 0) = v_0(1 - \rho^2/a^2)$. Determine the kinetic energy of the membrane at any later time and show that, when averaged over many cycles of the lowest frequency, this energy is equal to one-half of the kinetic energy initially imparted to the membrane. (For a similar example involving the vibrating string, see Section 1.8.)

The initial kinetic energy of the membrane is

$$KE_{initial} = \tfrac{1}{2}\sigma \int_0^a 2\pi [u_t(\rho, 0)]^2 \rho \, d\rho$$

$$= \pi\sigma v_0^2 \int_0^a (1 - \rho^2/a^2)^2 \rho \, d\rho$$

$$= \tfrac{1}{6}\pi\sigma v_0^2 a^2 \tag{3.7.56}$$

Since the initial displacement is assumed to be zero, the coefficients A_n in (3.7.52) are zero. From (3.7.53) the coefficients B_n are

$$B_n = \frac{2v_0}{\omega_n a^4 [J_1(\alpha_{0n})]^2} \int_0^a (a^2 - \rho^2) J_0(\alpha_{0n}\rho/a) \rho \, d\rho \tag{3.7.57}$$

The integral is evaluated by using (3.7.55) with $\nu = 1$. We obtain

$$B_n = \frac{4v_0 J_2(\alpha_{0n})}{\omega_n a_{0n}^2 [J_0(\alpha_{0n})]^2} \tag{3.7.58}$$

This result may be simplified by using the identity (D.41)

$$J_{n+1}(kx) = \frac{2n}{kx} J_n(kx) - J_{n-1}(kx) \tag{3.7.59}$$

from Appendix D with $n = 1$. Since $J_0(\alpha_{0n}) = 0$, the general result in the previous equation reduces to $J_2(\alpha_{0n}) = (2/\alpha_{0n})J_1(\alpha_{0n})$ and we obtain

$$B_n = \frac{8v_0}{\omega_n a_{0n}^3 J_1(\alpha_{0n})} \tag{3.7.60}$$

The displacement at any later time is thus

$$u(\rho, t) = \Sigma B_n J_0 \frac{\alpha_{0n}\rho}{a} \sin \omega_n t \tag{3.7.61}$$

with B_n given by (3.7.60).

The kinetic energy at any time is

$$KE(t) = \tfrac{1}{2}\sigma \int_0^a 2\pi [u_t(\rho, t)]^2 \rho \, d\rho \tag{3.7.62}$$

On calculating $u_t(\rho, t)$ from (3.7.50) we obtain

$$\text{KE}(t) = \pi\sigma \int_0^a \rho \, d\rho \sum_n \sum_m B_n B_m \omega_n \omega_m J_0(k_{0n}\,\rho) J_0(k_{0m}\,\rho) \cos \omega_n t \cos \omega_m t \quad (3.7.63)$$

Integrating over ρ and using the orthogonality of the Bessel functions (3.7.51), we have

$$\text{KE}(t) = \tfrac{1}{2}\pi\sigma a^2 \sum \omega_n^2 B_n^2 [J_1(\alpha_{0n})]^2 \cos^2 \omega_n t \quad (3.7.64)$$

At this point the calculation differs slightly from that for the string. Since the higher frequencies of the membrane are not integral multiples of the lowest (fundamental) frequency, we cannot conclude that the average of each term $\cos^2 \omega_n t$ in (3.7.64), when averaged over the fundamental period, will be equal to $\tfrac{1}{2}$. However, when averaged over a time T that is long compared to the fundamental period, we obtain

$$\frac{1}{T} \int_0^T \cos^2 \omega_n t \, dt = \frac{1}{2} - \frac{\sin 2\omega_n T}{4\omega_n T} \quad (3.7.65)$$

which does approach $\tfrac{1}{2}$ for $\omega_n T \gg 1$.

The average kinetic energy $(\text{KE})_{\text{ave}}$ is thus

$$\text{KE}_{\text{ave}} = \tfrac{1}{4}\pi\sigma a^2 \sum \omega_n^2 B_n^2 [J_1(\alpha_{0n})]^2 \quad (3.7.66)$$

and when the expression for $\omega_n B_n$ is introduced from (3.7.60), we have

$$\text{KE}_{\text{ave}} = 16\pi\sigma a^2 v_0^2 \sum \frac{1}{\alpha_{0n}^6} \quad (3.7.67)$$

The sum is given in series 18 of Table F.1 as being equal to 1/192. (The first three terms, with $\alpha_{01} = 2.405$, $\alpha_{02} = 5.320$, and $\alpha_{03} = 8.654$, yield 1/191.8.) The average value of the kinetic energy is seen to be $\tfrac{1}{12}\pi\sigma a^2 v_0$, which is one-half of the initial kinetic energy. (The other half, of course, resides in potential energy.)

3.7.4 Steady State Temperature in a Cylinder

Bessel functions arise in contexts other than those related to wave motion. We now consider solutions of Laplace's equation for cylindrical geometry and show how these functions can arise in steady state situations as well. We shall encounter not only the functions $J_n(kr)$ and $Y_n(kr)$ but also nonoscillatory Bessel functions labeled $I_n(kr)$ and $K_n(kr)$. These functions bear a relation to J_n and Y_n that is somewhat analogous to the relation between hyperbolic functions and trigonometric functions. For the sake of algebraic simplicity we confine attention to problems that do not contain angular dependence.

As was the case for the rectangular region considered in Section 3.2, a problem involving inhomogeneous boundary conditions on more than one boundary may be decomposed into a number of separate problems involving homogeneous boundary conditions on all but one of the bounding surfaces. In rectangular coordinates the function specified by the inhomogeneous boundary condition was then expressed as a Fourier series. We could be guided in choosing the sign of the separation constant by the requirement that trigonometric rather than hyperbolic functions should be available in order to carry out the Fourier expansion. In cylindrical coordinates a similar choice must be made between the oscillatory Bessel functions J_0 and Y_0 and the monotonically varying Bessel functions I_0 and K_0. Properties of these latter two functions are also summarized in Appendix D. As examples of the two situations that arise, we consider the two problems labeled A and B in Figure 3.10.

In problem A, where the temperature is specified along the vertical surface as a function of z, we expect to expand $f(z)$ in an infinite series. It will turn out to be the standard Fourier series. In problem B, where the temperature is specified over the base as a function of ρ, we expect to expand $f(\rho)$ in an appropriate infinite series. Here the expansion will turn out to be in terms of the oscillatory Bessel functions J_0. The details of this so-called Fourier-Bessel expansion will be very similar to those used for the membrane vibration problem considered in the previous section.

For both problems A and B we will use the expression for Laplace's equation in cylindrical coordinates that was given in (3.5.14). Neglecting the angular dependence, we have

$$\frac{1}{\rho}\frac{\partial}{\partial\rho}\left(\rho\frac{\partial u}{\partial\rho}\right) + \frac{\partial^2 u}{\partial z^2} = 0 \tag{3.7.68}$$

We now introduce separation of variables by writing $u(\rho, z) = R(\rho)Z(z)$ and obtain

$$\frac{1}{\rho R}(\rho R')' = -\frac{Z''}{Z} = \text{const} \tag{3.7.69}$$

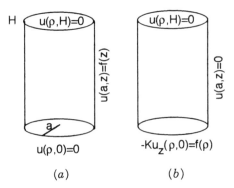

Figure 3.10. Two examples of boundary conditions for: (a) heated wall, (b) heated base.

H   u(ρ,H)=0   u(ρ,H)=0

u(a,z)=f(z)   u(a,z)=0

a

u(ρ,0)=0   -Ku$_z$(ρ,0)=f(ρ)

(a)   (b)

The choice of signs for the separation constant is determined by the series expansion that must eventually be developed.

Problem A. We first consider the case in which the temperature is specified along the vertical surface. In order to express this temperature in terms of a Fourier series in the variable z, we choose the separation constant to be $+k^2$ and obtain

$$(\rho R')' - k^2 \rho R = 0, \qquad Z'' + k^2 z = 0 \qquad (3.7.70)$$

The solution for $Z(z)$ is then

$$Z(z) = A \cos kz + B \sin kz \qquad (3.7.71)$$

while the solution for $R(\rho)$ is expressed in terms of $I_0(k\rho)$ and $K_0(k\rho)$. Since the function $K_0(k\rho)$ is singular at $\rho = 0$, it cannot contribute to the solution of this problem. We thus use only $I_0(k\rho)$. Imposing the boundary conditions $Z(0) = Z(H) = 0$ we obtain $B = 0$ and $k_n = n\pi/H$, $n = 1, 2, 3, \ldots$. The general solution is thus a series of the form

$$u(\rho, z) = \sum_{n=1}^{\infty} B_n I_0(k_n \rho) \sin k_n z, \qquad k_n = \frac{n\pi}{H} \qquad (3.7.72)$$

The constants B_n are determined by imposing the boundary condition at $\rho = a$. We therefore write

$$f(z) = u(a, z) = \sum_{n=1}^{\infty} B_n I_0(k_n a) \sin k_n z \qquad (3.7.73)$$

Employing the orthogonality of the functions $\sin k_n z$ over the interval $0, \ldots, H$, we obtain

$$B_n = \frac{2}{H I_0(k_n a)} \int_0^H f(z) \sin k_n z \, dz \qquad (3.7.74)$$

Upon substituting this result for the coefficients B_n into the solution (3.7.72), we obtain the solution in the form

$$u(\rho, z) = \frac{2}{\pi} \int_0^H dz' \, f(z') \sum_{n=1}^{\infty} \frac{I_0(k_n \rho)}{I_0(k_n a)} \sin k_n z' \sin k_n z \qquad (3.7.75)$$

This result will be encountered again in Section 4.12 when we reconsider this problem with Green's function techniques.

If, instead of the temperature, the heat flux $J(z) = -K u_\rho(a, z)$ were specified on the vertical wall, we would have the boundary condition

$$J(z) = -Ku_\rho(a, z) = -K \sum B_n k_n I_1(k_n a) \sin k, z \qquad (3.7.76)$$

where the identity [cf. (D.37) with $n = 0$] $I_0'(x) = I_1(x)$ has been employed. The coefficients B_n in this case would be

$$B_n = \frac{-2}{KHk_n I_1(k_n a)} \int_0^H J(z) \sin k_n z \, dz \qquad (3.7.77)$$

We now use the method developed here to determine the temperature within a cylinder due to a ring source around the midpoint of the cylinder.

Example 3.4. A heat influx Q_0 calories per square centimeter per second into a cylinder wall is due to a heated ring source around the midpoint $z = H/2$. The top and base are maintained at zero temperature and the cylinder has radius a. We also consider the limiting cases $a/H \gg 1$ and $a/H \ll 1$.

The boundary condition at the wall may be written

$$J(z) = -\frac{Q_0}{2\pi a} \delta \left(z - \frac{H}{2} \right) \qquad (3.7.78)$$

From (3.7.77) we then obtain

$$B_n = \frac{Q_0}{\pi KaHk_n I_1(k_n a)} \qquad (3.7.79)$$

and the temperature within the cylinder is

$$u(\rho, z) = \frac{Q_0}{\pi KaH} \sum_1^\infty \frac{\sin n\pi/2}{k_n I_1(k_n a)} I_0(k_n \rho) \sin k_n z \qquad (3.7.80)$$

Due to the symmetry of the problem, we expect that half of the heat input will flow out through the base of the cylinder. This result is readily verified by evaluating

$$H = \int_0^a 2\pi\rho \, d\rho \, [-Ku_z(\rho, 0)]$$

$$= \frac{2Q}{aH} \sum_1^\infty \frac{\sin n\pi/2}{k_n I_1(k_n a)} \int_0^a \rho \, d\rho \, I_0(k_n \rho) \qquad (3.7.81)$$

Using Eq. (D.34) we see that the integral over ρ is equal to $aI_1(k_n a)$, and since $k_n = 2\pi/H$, we finally obtain

$$H = \frac{2Q_0}{\pi} \sum_1^\infty \frac{1}{n} \sin \frac{n\pi}{2} = \frac{Q_0}{2} \qquad (3.7.82)$$

since the value of the series is $\pi/4$.

Further insight into the solution may be obtained by determining the temperature on the axis of the cylinder. Noting that $I_0(0) = 1$ and setting $k_n = n\pi/H$, we obtain

$$u(0, z) = \frac{Q_0}{\pi^2 Ka} \sum_1^\infty \frac{\sin(n\pi/2)\sin n\pi z/H}{nI_1(n\pi a/H)} \tag{3.7.83}$$

For $a/H \gg 1$, so that the cylinder becomes a thin disk, we may use the large-argument approximation $I_1(x) \sim e^x/(2\pi x)^{1/2}$ given in Eq. (D.32). Successive terms in the series for $u(0, z)$ thus contain factors $e^{-\pi a/H}$, $e^{-2\pi a/H}$, $e^{-3\pi a/H}$, etc. For sufficiently large values of a/H (i.e., thin disks) successive terms diminish very rapidly. If we retain only the first term, we obtain

$$u(0, z) = \frac{Q_0}{\pi K}\sqrt{\frac{2}{aH}}\, e^{-\pi a/H}\sin\frac{\pi z}{H} \tag{3.7.84}$$

As is to be expected in this limit of a thin disk, the temperature on axis is very small since it is governed by the more proximate value of the temperature at the top and base than by the input of heat at the wall.

For $a/H \ll 1$, so that the cylinder becomes a wire, we should expect that the temperature would not vary much over the cross section of the cylinder. The problem should then become equivalent to the one-dimensional problem considered at the end of Section 1.5. To see that this is the case, replace $I_1(x)$ by its small-argument approximation, namely, $x/2$ [cf. (D.33)], and write (3.7.83) as

$$u(0, z) \approx \frac{2Q_0 H}{\pi^2 Ka^2}\sum_1^\infty \frac{1}{n^2}\sin\frac{n\pi}{2}\sin\frac{n\pi z}{H} \tag{3.7.85}$$

According to the result listed in Table F.1, series 30, for this sum we have

$$u(0, z) = \begin{cases} \dfrac{1}{2}\dfrac{Q_0}{\pi a^2 K}z, & z < \dfrac{H}{2} \\[2ex] \dfrac{1}{2}\dfrac{Q_0}{\pi a^2 K}(H - z), & z > \dfrac{H}{2} \end{cases} \tag{3.7.86}$$

which agrees with the result given in Eq. (1.5.21). In the present problem the heat is introduced around the *surface* of the cylinder at the midpoint. When the source amplitude Q_0 is divided by the cross-sectional area of the cylinder, πa^2, we obtain the strength of the equivalent *volume* source term employed in the example in Section 1.5.

Problem B. When the temperature or the heat flux is specified on the *base* of the cylinder, we expect to employ a Fourier-Bessel expansion. This will occur if we reverse the sign of the separation constant k^2 in Eq. (3.7.70). We then obtain solutions $J_0(k\rho)$ and $Y_0(k\rho)$ as well as hyperbolic functions for the solution $Z(z)$. The solution $Y_0(k\rho)$ is absent by virtue of its singularity at $\rho = 0$. We thus obtain the solutions

$$R(\rho) = J_0(k\rho), \qquad Z(z) = A\cosh kz + B\sinh kz \tag{3.7.87}$$

As boundary conditions on the other surfaces (top and wall) let us choose $u_z(\rho, H) = 0$ and $u(a, z) = 0$, respectively. The separation constant k is then determined by setting $R(a) = J_0(ka) = 0$. As in Section 3.7.3 we set $k_n a = \alpha_{0n}$. Also, the condition $Z'(H) = 0$ is satisfied by $Z(z) = \cosh k_n (H - z)$ and we have the solution

$$u(\rho, z) = \Sigma A_n J_0(k_n \rho)\cosh k_n (H - z), \qquad k_n a = \alpha_{0n} \quad (3.7.88)$$

If the heat influx is specified on the base as

$$J(\rho) = -K u_z(\rho, 0) = f(\rho) \tag{3.7.89}$$

then the z derivative of (3.7.88) yields

$$f(\rho) = -K \Sigma A_n k_n J_0(k_n \rho)\sinh k_n H \tag{3.7.90}$$

After multiplying this equation by $\rho J_0(k_p \rho)$, integrating over ρ from 0 to a, and using the orthogonality relation for Bessel functions (C.27) and (D.29), we obtain

$$A_n = -\frac{2 \displaystyle\int_0^a f(\rho)J_0(k_n \rho)\,\rho\, d\rho}{K k_n a^2 [J_1(k_n a)]^2 \sinh k_n H} \tag{3.7.91}$$

The integrals are readily evaluated if, as in Section 3.7.3, the function $f(\rho)$ is expressed in terms of the functions $(a^2 - \rho^2)^\nu$. Examples are considered in the problems.

Problems

3.7.1   Determine the steady state temperature distribution in the quarter circle $0 \le r \le a, 0 \le \theta \le \pi/2$ for the following boundary conditions:
  (a) $u(r, 0) = u(r, \pi/2) = 0, u(a, \theta) = u_0 \sin 4\theta$.
  (b) $u_\theta(r, 0) = u(r, \pi/2) = 0, u(a, \theta) = u_0 \sin 4\theta$.

3.7.2   The temperature at the rim of the annular quarter circle $a \le r \le b$, $0 \le \theta \le \pi/2$, is given by $u_\theta(r, 0) = u(a, \theta) = u(b, \theta) = 0, u(r, \pi/2) = f(r)$. Show that

$$u(r, \theta) = \sum_1^\infty A_n \cosh(n\alpha\theta) \sin[n\alpha \ln(r/\alpha)]$$

where $\alpha = \pi/\ln(b/a)$ and

$$A_n = \frac{2}{\ln(b/a)} \int_a^b \frac{dr}{r} f(r)\sin[n\alpha \ln(r/a)]$$

Show that if $f(r) = u_0 \sin[\alpha \ln(r/a)]$, the heat flow values out through the rim at $r = a$ and $r = b$ are equal and have the value $K u_0 \tanh(\alpha \pi / 2)$ where K is the conductivity of the plate.

3.7.3 A membrane in the form of a quarter circle of radius a is fixed at the rim so that $w(a, \theta) = 0$, $w(r, 0) = w(r, \pi/2) = 0$. Show that the membrane displacement is of the form

$$w(r, \theta) = \sum_{m=1}^{\infty} \sum_{n=1}^{\infty} A_{2m,n} J_{2m}(k_{2m,n} r) \sin 2m\theta \, \cos(k_{2m,n} t - \varphi_{2m,n})$$

where the $k_{2m,n}$ are solutions of $J_{2m}(k_{2m,n} a) = 0$.

3.7.4 A circular membrane of radius a fixed at the rim has the initial displacement $w(r, 0) = w_0[J_0(\alpha_{01} r/a) - J_0(\alpha_{02} r/a)]$ and zero initial velocity. Determine the kinetic energy of the membrane at any later time $[\mathrm{KE} = (\rho/2) \int dA [w_t(r, t)]^2]$.

3.7.5 A circular region of radius a has a temperature at the rim given by $u(a, \theta) = u_0 \sin \theta/2$, $0 \le \theta \le 2\pi$.

(a) Show that the steady temperature inside the circle is

$$\frac{u(r, \theta)}{u_0} = \frac{2}{\pi} - \frac{1}{\pi} \sum_{1}^{\infty} \left(\frac{r}{a}\right)^n \frac{\cos n\theta}{n^2 - \frac{1}{4}}$$

(b) Show that the temperature on the diameter $\theta = 0$, π is

$$\frac{u(r, 0)}{u_0} = \frac{2}{\pi}\left(1 - \frac{r}{a}\right) \frac{\tanh^{-1}\sqrt{r/a}}{\sqrt{r/a}}$$

$$\frac{u(r, \pi)}{u_0} = \frac{2}{\pi}\left(1 + \frac{r}{a}\right) \frac{\tan^{-1}\sqrt{r/a}}{\sqrt{r/a}}$$

The series may be summed by writing

$$\frac{1}{n^2 - \frac{1}{4}} = \frac{2}{2n - 1} - \frac{2}{2n + 1}$$

and using series 20 and 21 in Table F.1.

3.7.6 A circular cylinder of height H and radius a is insulated at the top and kept at zero temperature on the vertical wall. The base is kept at the temperature $u(r, 0) = u_0 J_0(\alpha_{0n} r/a)$.

(a) Determine the steady state temperature within the cylinder.

    (b) Determine the total heat flow $H = \int dS \cdot (-K \nabla u)$ out through the vertical wall.

    (c) How does the result in (b) vary as $n \rightarrow \infty$?

3.7.7   A solid cylinder of radius a and height H has surface temperatures maintained as follows: $u(r, 0) = 0$, $u(r, H) = u_0 J_0(\alpha_{02} r/a)$, $u(a, z) = 0$.

    (a) Sketch $u(r, H)$.

    (b) Determine $u(r, z)$.

    (c) Is there any locus on the base for which the heat flow out of the cylinder $J_z(r, 0)$ $[= -Ku_z(r, 0)]$ equals zero? If so, where?

3.7.8   An annular circular membrane has an inner radius a and outer radius b. The amplitude of vibration $u(r, t)$ is independent of angle θ.

    (a) Using separation of variables $u(r, t) = R(r)T(t)$, write the equation governing $R(r)$.

    (b) Write the solution of the equation in terms of the functions that describe incoming and outgoing cylindrical waves.

    (c) If the membrane is fixed at both rims, write the equation that must be satisfied by the functions introduced in part (b). (The roots of this equation determine the frequencies of vibration.)

3.7.9   The base of a circular cylinder of height H and radius a is maintained at temperature $u(r, 0) = f(r)$. The temperature is zero on the other surfaces, that is, $u(r, H) = u(a, z) = 0$.

    (a) Show that

$$u(r, z) = \sum_1^\infty A_n J_0(k_n r) \sinh k_n (H - z)$$

where $k_n = \alpha_{0n}/a$ with $J_0(\alpha_{0n}) = 0$ and

$$A_n = \frac{2}{a^2 J_1^2(k_n a)} \sinh kH \int_0^a r\, dr\, f(r) J_0(k_n r)$$

    (b) Show that the heat flow through the various surfaces is given by

$$H_{\text{base}} = 2\pi \int_0^a r\, dr[-Ku_z(r, 0)] = 2\pi aK \sum_1^\infty A_n J_1(k_n a) \cosh k_n H$$

$$H_{\text{top}} = 2\pi \int_0^a r\, dr[-Ku_z(r, H)] = 2\pi aK \sum_1^\infty A_n J_1(k_n a)$$

$$H_{\text{wall}} = 2\pi \int_0^H dz[-Ku_r(a, z)] = 2\pi aK \sum_1^\infty A_n J_1(k_n a)(\cosh k_n H - 1)$$

What general property of the Laplace equation leads one to expect the conservation property exhibited by these results?

3.7.10 A circular cylinder of height H and radius a has an insulated base while the top is maintained at zero temperature. The temperature on the wall depends only on the height and is given by $u(a, z) = u_0\cos\pi z/2H$.

(a) Determine the steady state temperature inside the cylinder.

(b) Show explicitly (i.e., without merely invoking conservation considerations) that the total heat flowing in through the walls equals that flowing out through the top.

3.7.11 The rim of a circular disk is held at temperature zero. Initially the disk has the temperature distribution $u(r, 0) = u_0(a^2 - r^2)/a^2$.

(a) Show that the amount of heat initially on the disk is $h(0) = Mu_0 c$ where $M = \pi a^2\sigma$ is the mass of the disk and c is a specific heat.

(b) Show that the appropriate solution of the two-dimensional diffusion equation

$$K\nabla^2 u - \sigma c\,\frac{\partial u}{\partial t} = 0$$

has the form

$$u(r, t) = \sum_1^\infty A_n J_0(k_n r)e^{-\gamma^2 k_n^2 t}, \qquad \gamma^2 = \frac{K}{\sigma c}$$

where the k_n satisfy $J(k_n a) = 0$ (i.e., $k_n a = \alpha_{0n} = 2.40483$, 5.52008, 8.65373, 11.79153, ...) and that the amplitudes A_n are obtained by evaluating

$$A_n = \frac{2u_0}{a^4[J_1(k_n a)]^2}\int_0^a r\,dr(a^2 - r^2)J_0(k_n r)$$

(c) Evaluate the integral by setting $r = a\sin\varphi$ and using the general result (Watson, 1944, p. 373)

$$\int_0^{\pi/2} d\varphi\,\sin^{\mu+1}\varphi\,\cos^{2\nu+1}\varphi J_\mu(z\sin\varphi) = \frac{2^\nu\Gamma(\nu + 1)}{z^{\nu+1}}J_{\mu+\nu+1}(z)$$

(which is valid for $\mu, \nu > -1$) to obtain [Note: $\Gamma(n + 1) = n!$]

$$A_n = \frac{4u_0}{(k_n a)^2}\frac{J_2(k_n a)}{[J_1(k_n a)]^2}$$

(d) Use the general relation $J_{n+1}(x) = (2n/x)J_n(x) - J_{n+1}(x)$ for $x = k_n a = \alpha_{0n}$ to obtain

$$A_n = \frac{8u_0}{(k_n a)^3} \frac{1}{J_1(k_n a)}$$

(e) Calculate the total heat flow off the disk by evaluating

$$h(\infty) = 2\pi a \int_0^\infty dt \, (-K) u_r(a, t)$$

and show that the result equals the initial amount of heat obtained in part (a) provided (Table F.1, series 17)

$$\sum_1^\infty \frac{1}{\alpha_{0n}^4} = \frac{1}{32} = 0.03125$$

(The first four terms yield 0.03121.)

3.7.12 A solid cylinder of radius a and height H has surface temperatures maintained as follows:

$$u(\rho, 0) = 0, \qquad u(\rho, H) = \Gamma\delta(\rho - a/2), \qquad u(a, z) = 0$$

that is, there is a hot ring of radius $a/2$ at the top of the cylinder and Γ is a constant.
(a) Determine $u(\rho, z)$ within the cylinder.
(b) Discuss how you would relate the constant Γ to physically significant quantities.

3.7.13 A chain of length L and density ρ swings freely under its own weight from a rigid support.
(a) Use the coordinate system shown in Figure 3.11 to show that the equation governing the motion of the string is

$$\frac{\partial^2 y}{\partial t^2} = g \frac{\partial}{\partial x}\left(x \frac{\partial y}{\partial x}\right)$$

(b) Set $y(x, t) = f(x)e^{-i\omega t}$ and introduce the dimensionless length variable $w^2 = x/L$. Show that, as a function of w, f satisfies

$$\frac{d}{dw}\left(w \frac{df}{dw}\right) + (2\alpha)^2 wf = 0, \qquad \alpha^2 = \frac{L\omega^2}{g}$$

which has the solution $J_0(2\alpha w)$.

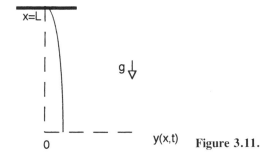

x=L

g

0   y(x,t)   **Figure 3.11.**

(c) Show that the lowest frequency of vibration, $\nu_1 = \omega_1/2\pi$, is approximately equal to $1.203\nu_p$ where ν_p is the frequency of a pendulum of length L $(= \frac{1}{2}\pi\sqrt{g/L})$. Note that for a rigid bar the oscillation frequency is $\frac{1}{2}\sqrt{6}\,\nu_p = 1.225\nu_p$.

3.7.14 The rim of a circular plate of radius a is held at zero temperature. An amount of heat H_0, in the form of a hot spot, is initially deposited at the center of the plate. The initial temperature distribution on the plate may then be written

$$u(r, 0) = \frac{H_0}{\rho c}\,\delta^2(\mathbf{r}) = \frac{H_0}{\rho c}\,\frac{\delta(r)}{2\pi r}$$

(a) Determine the temperature on the plate at any later time.

(b) Determine the total amount of heat that flows out through the rim and from the result infer the value of a certain infinite sum involving Bessel functions.

4 Green's Functions

We now introduce a technique for solving partial differential equations (as well as ordinary differential equations) that only achieves its full usefulness when rather elaborate problems are considered, problems that, for the most part, are more complicated than those encountered in this book. One of the main advantages of the method is that it enables one to develop approximate solutions to problems for which no exact solution may be obtained. When applied to the simpler problems of the sort to be considered here initially, the method will merely provide an alternate procedure for recovering results obtainable by previously developed techniques.

The method to be developed uses the notion of superposition of solutions in a somewhat different way than it has been used up until now. In essence, we obtain the solution to a problem due to a localized source (a delta function) and then show how the solution of problems involving more complicated sources can be expressed in terms of a superposition of solutions to the point source problem. In three-dimensional problems the superposition process employs Green's theorem, and thus the function used in the superposition process, the solution of the point source problem, has come to be known as Green's function. The concept is perhaps best introduced by considering a simple example.

4.1 INTRODUCTORY EXAMPLE

In order to display the Green's function approach in a very simple way, we first consider the one-dimensional problem of the static deflection of a string of length L due to a steady external force. We will thus be solving an ordinary differential equation. The string will be assumed to be fixed at both ends. Setting the time derivative equal to zero in the equation for the vibrating string (1.8.9), since we are considering a static problem, and also setting $K = 0$, we obtain the ordinary differential equation

$$T\frac{d^2y}{dx^2} = -\mathcal{F}(x) \tag{4.1.1}$$

Rather than solve this equation directly, we first obtain the solution of a simpler problem in which the external force is $\mathcal{F}_1(x) = T\delta(x - x_1)$, an external force located at the single point x_1. The amplitude T has been chosen to conveniently eliminate the constant T in (4.1.1). Representing the displacement due to this driving force by z, we have

$$\frac{d^2z}{dx^2} = -\delta(x - x_1), \qquad 0 \leq x \leq L \tag{4.1.2}$$

Differential equations with delta function source terms have been encountered previously (Section 1.5). The solution may be obtained by solving the homogeneous equation in the regions on either side of x_1 and then matching the solutions appropriately at $x = x_1$. Distinguishing these two solutions by the subscript notation introduced in Section 1.5, we have

$$z_<(x) = ax, \qquad\qquad 0 \leq x < x_1$$

$$z_>(x) = b(L - x), \qquad x_1 < x \leq L \tag{4.1.3}$$

Note that the required boundary conditions $z_<(0) = 0$, $z_>(L) = 0$ have already been incorporated into these solutions. Since the string does not break at $x = x_1$, we also require $z_<(x_1) = z_>(x_1)$. As in the development introduced in Section 1.5, an additional condition is obtained by integrating (4.1.2) across the singularity at $x = x_1$. This procedure yields

$$\int_{x_1-\epsilon}^{x_1+\epsilon} \frac{d^2z}{dx^2} = \left.\frac{dz_>}{dx}\right|_{x_1} - \left.\frac{dz_<}{dx}\right|_{x_1} = -1 \tag{4.1.4}$$

Continuity of the solution at $x = x_1$ and the discontinuity in the slope obtained in (4.1.4) provide two conditions for the determination of the constants a and b in (4.1.3). Following the procedure used in Section 1.5, we find that $ax_1 = b(L - x_1)$ and $b + a = 1$. Solving for a and b we obtain $a = (1 - x_1/L)$ and $b = x_1/L$. Thus the final solution to this problem is

$$z_<(x) = x(L - x_1)/L$$

$$z_>(x) = x_1(L - x)/L \tag{4.1.5}$$

The symmetry of this result with respect to interchange of the source point x_1 and the observation point x will be considered in more detail later in this section.

The solution of (4.1.5) obtained here, which is the deflection (a more general term is the response) of the string at x due to a unit source located at x_1, is known as the Green's function for this problem and is written

$$G(x|x_1) = \begin{cases} x(L - x_1)/L, & 0 \leq x < x_1 \\ x_1(L - x)/L, & x_1 < x \leq L \end{cases} \tag{4.1.6}$$

There are two commonly used ways of abbreviating this result. If we use $G_<$ to refer to G for $x < x_1$ and $G_>$ to refer to G for $x > x_1$, then $G_<(x|x_1) =$

$x(L - x_1)/L$ and $G_>(x|x_1) = x_1(L - x)/L$. An alternate way of writing the solution is to use $x_<$ for the lesser of the two points and use $x_>$ for the larger of the two points. We may then abbreviate the solution as

$$G(x|x_1) = x_<(L - x_>)/L \qquad (4.1.7)$$

where $x_<$ and $x_>$ are to be appropriately identified with x and x_1. We have now obtained the solution of $z'' = -\delta(x - x_1)$ in the form $z = G(x|x_1)$.

It should be noted that the boundary conditions satisfied by $G(x|x_1)$ are satisfied in both the x and x_1 coordinates. From (4.1.6) it is clear that

$$G(x|0) = 0, \qquad G(x|L) = 0 \qquad (4.1.8)$$

as well as

$$G(0|x_1) = 0, \qquad G(L|x_1) = 0 \qquad (4.1.9)$$

It is a property of linear differential equations that if the inhomogeneous term (the source term) is multiplied by a constant, then the solution is multiplied by the same constant. Thus the solution of $z'' = -(F_1/T)\delta(x - x_1)$ is $z = (F_1/T)G(x|x_1)$. Another property of linear differential equations is that the solution corresponding to the sum of two such source terms is just the sum of the solutions due to each source term separately. Thus the solution of $z'' = -(F_1/T)\delta(x - x_1) - (F_2/T)\delta(x - x_2)$ is $z = (F_1/T)G(x|x_1) + (F_2/T)G(x|x_2)$.

In general, the solution of $z'' = -(1/T)\sum_i F_i \delta(x - x_i)$ is $z = (1/T)\sum_i F_i G(x|x_i)$. If we now introduce a source density $\mathcal{F}(x)$ by writing $F_i = \mathcal{F}(x_i) \Delta x_i$, then the solution of $z'' = -(1/T)\sum_i \mathcal{F}_i(x_i) \Delta x_i \delta(x - x_i)$ is $z = (1/T) \sum_i \mathcal{F}(x_i) \Delta x_i G(x|x_i)$. Introducing a standard limiting procedure, we may replace the sums by integrals and obtain the solution of the differential equation

$$z'' = -\frac{1}{T} \int_0^L dx' \, \mathcal{F}(x')\delta(x - x') = -\frac{1}{T}\mathcal{F}(x) \qquad (4.1.10)$$

in the form $z = (1/T) \int_0^L dx' \, \mathcal{F}(x')G(x|x')$. However, the function $z(x)$ in this last equation satisfies the same equation and boundary conditions as $y(x)$ in the original equation (4.1.1). We have therefore expressed the solution to (4.1.1) in the form

$$y = \frac{1}{T} \int_0^L dx' \, \mathcal{F}(x')G(x|x') \qquad (4.1.11)$$

that is, the solution to the problem involving the arbitrary force $\mathcal{F}(x)$ has been expressed in terms of $G(x|x')$, which is the solution to the problem involving the elementary force $\delta(x - x')$.

Example 4.1. Use the Green's function method to determine the deflection of a string due to gravitational force.

For this problem we set $\mathcal{F}(x) = -\rho g$ in (4.1.1) and, from (4.1.11), immediately write the solution in the form

$$y(x) = -\frac{\rho g}{T} \int_0^L dx' \, G(x|x') \tag{4.1.12}$$

Using the appropriate expressions for $G(x|x')$ from (4.1.5) in the two regions $x > x'$ and $x < x'$, we have

$$
\begin{aligned}
y(x) &= -\frac{\rho g}{T} \left[\int_0^x dx' \, G_>(x|x') + \int_x^l dx' \, G_<(x|x') \right] \\
&= -\frac{\rho g}{T} \left[\int_0^x dx' \, \frac{(L-x)x'}{L} + \int_x^L dx' \, \frac{(L-x')x}{L} \right] \\
&= -\frac{\rho g}{T} \left[\frac{(L-x)x^2}{2L} + \frac{x(L-x)^2}{2L} \right] \tag{4.1.13}
\end{aligned}
$$

After algebraic simplification this solution reduces to

$$y(x) = -\frac{\rho g}{2T} x(L-x) \tag{4.1.14}$$

It should be noted that for this simple problem the final result is obtained more immediately by direct integration of the original differential equation $y'' = -(\rho g/T)$. We would then have $y = -(\rho g/2T)x^2 + ax + b$ where a and b are constants of integration. Application of the boundary conditions $y(0) = y(L) = 0$ leads directly to the result given in (4.1.14). As alluded to earlier, the advantage of the Green's function approach only becomes apparent in more elaborate problems.

4.1.1 Reciprocity

As has been noted, $G(x|x')$ given in (4.1.6) displays a symmetry with respect to the interchange of x and x'. In physical terms this symmetry represents the statement that the deflection of the string at x due to a unit point source at x' has the same value as the deflection of the string at x' due to a unit point source at x. It should be noted that the shape of the string deflection will, in general, be completely different for these two different source locations (cf. Problem 4.1.1).

The reciprocity relation $G(x|x') = G(x'|x)$ may be derived in general by noting that for sources located at x_1 and x_2 we have

$$\frac{d^2 G(x|x_1)}{dx^2} = -\delta(x - x_1)$$

$$\frac{d^2 G(x|x_2)}{dx^2} = -\delta(x - x_2) \tag{4.1.15}$$

Multiplying the first of these equations by $G(x|x_2)$, the second by $G(x|x_1)$, subtracting, and integrating over the length of the string, we obtain

$$\int_0^L dx \left[G(x|x_2) \frac{d^2 G(x|x_1)}{dx^2} - G(x|x_1) \frac{d^2 G(x|x_2)}{dx^2} \right]$$

$$= -G(x_1|x_2) + G(x_2|x_1) \qquad (4.1.16)$$

Upon adding and subtracting the product $dG(x|x_1)/dx \, dG(x|x_2)/dx$ under the integral sign we find that the integrand is a perfect derivative. Integration then yields

$$\left[G(x|x_2) \frac{dG(x|x_1)}{dx} - G(x|x_1) \frac{dG(x|x_2)}{dx} \right]\Bigg|_{x=0}^{x=L}$$

$$= -G(x_1|x_2) + G(x_2|x_1) \qquad (4.1.17)$$

Since the Green's function vanishes at $x = 0, L$, the left-hand side of this result is zero. The right-hand side then yields the reciprocity relation

$$G(x_1|x_2) = G(x_2|x_1) \qquad (4.1.18)$$

Note that the reciprocity condition is also satisfied for any combination of boundary conditions for which the integrated term vanishes.

The limiting procedure used above to express the solution of $y'' = -\mathcal{F}(x)/T$ in terms of the Green's function $G(x|x')$ is too lengthy to be employed for more than pedagogic purposes. As a more direct way of obtaining the result in (4.1.11) we adopt the "multiply-and-subtract" procedure just employed in obtaining the reciprocity relation. We combine

$$\frac{d^2 y}{dx^2} = -\frac{1}{T} \mathcal{F}(x) \qquad (4.1.19)$$

with

$$\frac{d^2 G(x|x')}{dx^2} = -\delta(x - x') \qquad (4.1.20)$$

by multiplying these equations by $G(x|x')$ and $y(x)$, respectively. Then, subtracting and integrating over the length of the string, we obtain

$$\int_0^L \left[G(x|x') \frac{d^2 y}{dx^2} - y(x) \frac{d^2 G(x|x')}{dx^2} \right] dx$$

$$= -\frac{1}{T} \int_0^L \mathcal{F}(x) G(x|x') \, dx + \int_0^L y(x) \delta(x - x') \, dx \qquad (4.1.21)$$

The left-hand side may again be written as a perfect derivative and integrated. We obtain

$$\left[G(x\,|\,x') \frac{dy}{dx} - y(x) \frac{dG(x\,|\,x')}{dx} \right]\Bigg|_0^L + \frac{1}{T} \int_0^L dx\; \mathcal{F}(x) G(x\,|\,x')$$

$$= \begin{cases} y(x'), & 0 \le x' \le L \\ 0, & \text{otherwise} \end{cases} \tag{4.1.22}$$

While there are instances in which it is useful to consider the implications of having x' outside the range of integration, such examples will only be briefly considered in this text (cf. Problems 4.1.9, 4.3.4, and 4.12.6). Since both $y(x)$ and $G(x\,|\,x')$ vanish at $x = 0, L$, the integrated term vanishes and we have $y(x') = (1/T) \int_0^L \mathcal{F}(x') G(x'\,|\,x)\, dx'$. This result is not yet the same as that obtained previously since the coordinates on G are interchanged. However, due to the reciprocity relation we may write $G(x'\,|\,x) = G(x\,|\,x')$ and finally arrive at

$$y(x) = \frac{1}{T} \int_0^L \mathcal{F}(x') G(x\,|\,x')\, dx' \tag{4.1.23}$$

in agreement with (4.1.11).

A more direct procedure for carrying out this interchange of primed and unprimed coordinates is to introduce the derivatives with respect to the primed coordinates initially. Since $d^2 G/dx^2 = d^2 G/dx'^2$, as may be seen by carrying out the differentiations in a manner similar to that outlined in Problem 4.1.8, we may replace Eq. (4.1.19) and (4.1.20) by

$$\frac{d^2 y}{dx'^2} = -\frac{1}{T} \mathcal{F}(x')$$

$$\frac{d^2 G(x\,|\,x')}{dx'^2} = -\delta(x - x') \tag{4.1.24}$$

and integrate over x'. The delta function is an even function of its argument so that $\delta(x' - x) = \delta(x - x')$. Repetition of the multiply-and-subtract procedure now leads directly to (4.1.23). The fact that the equation for G is the same in both the primed and unprimed coordinates is a result of the special form of the equation satisfied by G in this example. In Section 4.6 the equation being considered here will be shown to be an example of a so-called self-adjoint equation. In the non-self-adjoint situations to be developed in Sections 4.5 and 4.6, either the equation or the boundary conditions (or both) satisfied by G may differ from those for the equation being solved.

Note that the integrated terms in (4.1.22) contain both y and y' evaluated at each boundary. In any given problem only two of these four quantities can be specified by the boundary conditions of the problem. The remaining two quantities, say the value of $y'(0)$ and $y'(L)$ if $y(0)$ and $y(L)$ are specified, are only available after the problem has been solved. The choice of boundary conditions to be satisfied by the Green's function is motivated by the requirement that we eliminate the unknown quantities at the boundaries. In general this is accomplished by having G satisfy the homogeneous counterpart of the boundary conditions satisfied by y; for example, if $y'(0) = c_1$ and $y(L) = c_2$, where c_1 and c_2 are assumed to be known, then we require $dG(x|0)/dx = 0$ and $G(x|L) = 0$. If y satisfies some set of homogeneous boundary conditions, then both G and y satisfy the same homogeneous boundary conditions. For the inhomogeneous boundary conditions quoted above, (4.1.22), with x and x' interchanged, would reduce to

$$y(x) = \frac{1}{T} \int_0^L \mathcal{F}(x') G(x|x') \, dx' - c_1 G(x|0) - c_2 \frac{dG(x|L)}{dx} \quad (4.1.25)$$

In this case the Green's function is constructed so as to satisfy the corresponding homogeneous boundary conditions $G(x|L) = G'(x|0) = 0$. Repetition of the procedure leading to (4.1.6) yields the appropriate Green's function in the form

$$G(x|x') = \begin{cases} L - x', & x < x' \\ L - x, & x > x' \end{cases} \quad (4.1.26)$$

When this Green's function is used, Eq. (4.1.25) will express the solution solely in terms of known quantities. A general procedure that will facilitate the construction of Green's functions for various boundary conditions will be developed in the next section.

In order to display in a simple way some of the uses of Green's functions as they arise in more complicated problems, let us consider the association of Green's function developed above for the *static* deflection of the string with the differential equation for *steady vibrations* of that string, namely,

$$\frac{d^2 y}{dx^2} + k^2 y = -\frac{1}{T} \mathcal{F}(x), \quad y(0) = y(L) = 0 \quad (4.1.27)$$

The term $k^2 y$ may be moved to the right-hand side of (4.1.27) and treated as an additional source term. Then Eq. (4.1.23) giving y in terms of $G(x|x')$ becomes

$$y(x) = -k^2 \int_0^L G(x|x') y(x') \, dx' + \frac{1}{T} \int_0^L \mathcal{F}(x') G(x|x') \, dx' \quad (4.1.28)$$

Since $y(x)$ appears under the integral sign, the use of the Green's function for the static deflection of the string has merely enabled us to transform our original *differential* equation plus boundary conditions into an *integral* equation. In more complicated problems this may be a useful procedure or even the only possible one.

In the present case, however, we can obtain a solution, rather than merely an integral equation, for the differential equation for the vibrating string (4.1.27) by introducing a *new* Green's function. We consider the Green's function $G_k(x|x')$ satisfying

$$\frac{d^2 G_k}{dx^2} + k^2 G_k = -\delta(x - x'), \qquad G_k(0|x') = G_k(L|x') = 0 \quad (4.1.29)$$

We may solve this equation by the same technique as that used previously. We solve the homogeneous counterpart of (4.1.29) for $x \neq x'$ and obtain

$$G_k(x|x') = \begin{cases} a \sin kx, & x < x' \\ b \sin k(L - x), & x > x' \end{cases} \quad (4.1.30)$$

The two forms of the solution will be labeled $G_{k<}$ and $G_{k>}$, respectively. Since the string does not break at $x = x'$, we again require equality of the two forms at $x = x'$ and write $G_{k>}(x'|x') = G_{k<}(x'|x')$. The second condition imposed upon G_k, obtained by integrating (4.1.29) across the singularity at $x = x'$, yields

$$\left.\frac{dG_k}{dx}\right|_{x'-\epsilon}^{x'+\epsilon} + k^2 \int_{x'-\epsilon}^{x'+\epsilon} G_k(x|x')\, dx = -1 \quad (4.1.31)$$

The integral in the second term represents the area under the curve for $G_k(x|x')$ from $x = x' - \epsilon$ to $x = x' + \epsilon$. Since, as noted above, the string does not break at the source point $x = x'$, $G_k(x|x')$ is continuous at $x = x'$. The area under the curve between $x - \epsilon$ and $x + \epsilon$, and hence the value of the integral, will vanish as ϵ goes to zero. There is thus no contribution from the integral to this limit and we obtain

$$\left.\frac{dG_{k>}}{dx}\right|_{x=x'} - \left.\frac{dG_{k<}}{dx}\right|_{x=x'} = -1 \quad (4.1.32)$$

This relation and the continuity relation at $x = x'$ provide two conditions for determining the constants a and b in (4.1.30). The results are $a = \sin k(L - x')/k \sin kL$ and $b = \sin kx'/k \sin kL$. We therefore obtain

$$G(x|x') = \begin{cases} \sin k(L - x')\sin kx/k \sin kL, & x < x' \\ \sin k(L - x)\sin kx'/k \sin kL, & x > x' \end{cases} \quad (4.1.33)$$

This result again exhibits the reciprocity relation $G_k(x|x') = G_k(x'|x)$. Note also that in the limit $kL \to 0$ the result reduces to the form for the static deflection given previously in (4.1.6). It is frequently, although not always, the case that a static solution can be obtained from a zero frequency limit of a wave problem.

When the multiply-and-subtract procedure is applied to (4.1.27) and (4.1.29), we obtain the solution (rather than merely an integral equation) in the form

$$y(x) = \frac{1}{T} \int_0^L dx' \; \mathcal{F}(x') G_k(x|x') \tag{4.1.34}$$

with $G_k(x|x')$ given by (4.1.33).

In this second example it has been a relatively simple matter to obtain a Green's function that would lead to an expression of the solution in terms of an integral over the source term. In more complicated situations (such as wave propagation in an inhomogeneous medium) it may be too difficult to construct the appropriate Green's function. In such cases one must settle for a simpler Green's function and attempt to solve the resulting integral equation.

Problems

4.1.1   (a) Use the Green's function method to obtain solutions in the region $L/2 < x \le L$ for

$$u_{xx} = -\frac{1}{K} s(x), \qquad u_x(0) = 0, \qquad u(L) = 0$$

where

$$s(x) = \begin{cases} s_0, \text{ a constant,} & 0 \le \quad x < L/2 \\ 0, & L/2 < y \le L \end{cases}$$

(b) Use Green's function methods to determine $u(x)$ in the region $0 \le x < L/2$ for

$$u_{xx}(x) = -\frac{1}{K} s(x), \qquad u(0) = 0, \qquad u_x(L) = 0$$

where

$$s(x) = \begin{cases} 0, & 0 \le \quad x < L/2 \\ s_0, \text{ a constant,} & L/2 < x \le L \end{cases}$$

4.1.2   Derive Eq. (4.1.26).

4.1.3   It was shown in the text that

$$y'' = -\frac{1}{T}\mathcal{F}(x), \qquad y(0) = y(L) = 0$$

has the solution

$$y(x) = \frac{1}{T}\int_0^L dx' \, \mathcal{F}(x') G(x|x')$$

where

$$G(x|x') = x_< (L - x_>)/L$$

In Example 4.1 the integral over $\mathcal{F}$ and G was carried out for the case $\mathcal{F}(x) = -\rho g$. Now obtain that same solution by using the Green's function satisfying the "wrong" boundary conditions

$$G(0|x') = \left.\frac{dG(x|x')}{dx}\right|_{x=L} = 0$$

(a) Show that now

$$G_> (x|x') = x', \qquad G_< (x|x') = x$$

Note that this result is independent of L.

(b) Show that

$$y(x) = G_< (x|L)y'(L) + \frac{1}{T}\int_0^L dx' \, G(x|x')\mathcal{F}(x')$$

where $y'(L)$ is unknown.

(c) Use the boundary condition $y(L) = 0$ to determine $y'(L)$ and finally recover the same result obtained in Example 4.1 of the text.

4.1.4   Determine the solution of

$$\frac{d^2 G}{dx^2} = -\delta(x - x'), \qquad 0 \le x \le L$$

satisfying $G(0)|x') = 0$ as well as an impedance boundary condition of the form $LG'(L|x') + i\zeta G(L|x') = 0$, where ζ is a constant.

4.1.5   (a) Express the solution of $y'' - k^2 y = 0$, $y(0) = y_0$, $y(\infty)$ bounded, in the form

$$y(x) = y(0) \left. \frac{dG_>(x|x')}{dx'} \right|_{x'=0} - G_>(x|0) y'(0)$$

where

$$\frac{d^2 G}{dx^2} - k^2 G = -\delta(x - x'), \qquad G(0|x') = 0, \qquad G(\infty|x') \to 0$$

(b) Show that

$$G(x|x') = -\frac{1}{k} e^{-kx_>} \sinh kx_<$$

(c) Obtain the solution for y in the form $y_0(x) = ye^{-kx}$.

4.1.6   Reconsider Problem 4.1.5 with the Green's function used in Problem 4.1.3. (As noted there, the result is independent of L and we now consider $L \to \infty$.)

(a) Show that now

$$y(x) = y(0) \left. \frac{dG_>(x|x')}{dx'} \right|_{x'=0} - k^2 \int_0^\infty dx' \, G(x|x') y'(x')$$

which, since the unknown $y(x)$ occurs under the integral sign, is an integral equation for $y(x)$.

(b) Since

$$G(x|x') = \begin{cases} x, & x < x' \\ x', & x > x' \end{cases}$$

show that

$$y(x) = y(0) - k^2 \left[\int_0^x dx' \, x' \, y(x') + \int_x^\infty dx' \, x y(x') \right]$$

$$= y(0) - k^2 \left[\int_0^x dx' \, (x' - x + x) y(x') \right.$$

$$\left. + x \int_x^\infty dx' \, y(x') \right]$$

$$= y(0) - k^2 \left[-\int_0^x dx' \, (x - x') y(x') + x\Gamma \right]$$

where $\Gamma = \int_0^\infty dx' \, y(x')$. Solve this integral equation for y by using Laplace transforms (note that it is in a form for which the convolution theorem is appropriate) and show that, for $y(0)$ to remain bounded as $x \to \infty$, one must require $\Gamma k = y(0)$.

(c) Show that $y(x) = y_0 e^{-kx}$, which agrees with the result obtained in Problem 4.1.5.

4.1.7 (a) Use the Green's function method to solve

$$u_{xx} - k^2 u = \mathcal{F}(x), \qquad u(0) = 0, \qquad u \text{ (and } u_x) \to 0 \text{ as } x \to \infty$$

(b) If $\mathcal{F}(x) = \mathcal{F}_0 \operatorname{sech}(x - x_0)$, write the solution for $x \gg x_0$.

4.1.8 Write the equation for the static deflection of the string given in Eq. (4.1.5) in the form

$$z(x) = (x/L)(L - x_1)H(x_1 - x) + (x_1/L)(L - x)H(x - x_1)$$

where $H(x)$ refers to the Heaviside step function, and show by differentiation of this expression that the defining differential equation for $z(x)$ given in Eq. (4.1.2) is satisfied.

4.1.9 When performing calculations with Green's functions it is frequently useful to be able to represent them as a Fourier series. Write the Green's function for static deflection of a string as $G(x|x') = \Sigma_1^\infty g_n(x') \sin(n\pi x/L)$. To determine the coefficients $g_n(x')$, substitute this series, along with the series for the delta function obtained in Problem 1.4.3, into Eq. (4.1.20) for $G(x|x')$ and obtain

$$G(x|x') = \frac{2L}{\pi^2} \sum_{n=1}^{\infty} \frac{1}{n^2} \sin \frac{n\pi x'}{L} \sin \frac{n\pi x}{L}$$

4.1.10 The Green's function

$$G_0(x|x') = -\tfrac{1}{2}|x - x'| = \begin{cases} \tfrac{1}{2}(x' - x) = G_{0>}(x|x'), & x > x' \\ \tfrac{1}{2}(x - x') = G_{0<}(x|x'), & x < x' \end{cases}$$

satisfies the inhomogeneous equation (4.1.24) but neither of the two boundary conditions appropriate for a fixed string. We now develop a method for recovering the result in (4.1.23) by using the Green's function $G_0(x|x')$. This is accomplished by using the vanishing of the right-hand side of (4.1.22) for field points outside the range of integration.

(a) Combine the equation satisfied by $G_0(x|x')$ with that for the string (4.1.24), and show that for a string with fixed ends, one obtains

$$f(x) = G_{0<}(x|L)f_{x'}(L) - G_{0>}(x|0)f_{x'}(0)$$

$$+ \frac{1}{T}\int_0^L dx' \, G_0(x|x')\mathcal{F}(x'), \qquad 0 \le x \le L$$

(b) Show that for $x < 0$, (4.1.22) yields

$$0 = G_{0<}(x|L)f_{x'}(L) - G_{0<}(x|0)f_{x'}(0)$$

$$+ \frac{1}{T}\int_0^L dx' \, G_{0<}(x|x')\mathcal{F}(x')$$

(A result containing equivalent information is obtained by requiring $x > L$.)

(c) Introduce the above expression for $G_0(x|x')$ and separate terms according to their x dependence. Since the equation in (b) holds for *all* $x < 0$, conclude that

$$f_{x'}(L) - f_{x'}(0) + \frac{1}{T}\int_0^L dx' \, \mathcal{F}(x') = 0$$

$$f_{x'}(L) + \frac{1}{LT}\int_0^L dx' \, x' \, \mathcal{F}(x') = 0$$

(d) Finally, introduce these expressions into the result in (b) to obtain

$$f(x) = \frac{1}{T}\int_0^L dx' \, \mathcal{F}(x')G(x|x')$$

where

$$G(x|x') = \frac{1}{2L}[-L|x - x'| - 2xx' + L(x + x')]$$

$$= \begin{cases} x(L - x')/L, & 0 \le x < x' \\ x'(L - x)/L, & x' < x \le L \end{cases}$$

which is the result given in (4.1.23) with $G(x|x')$ given by (4.1.6).

4.2 GENERAL PROCEDURE FOR CONSTRUCTING GREEN'S FUNCTIONS IN ONE DIMENSION

The two examples considered in the previous section should make it clear that the success of the Green's function method depends upon our ability to construct Green's functions that incorporate various boundary conditions, that is, to impose various boundary conditions on the solution of an ordinary differential equations with a delta function inhomogeneity. In order to expedite this process and circumvent the detailed calculations used in the previous section, we now develop a general procedure for obtaining Green's functions for specific equations and boundary conditions. As might be expected, in dealing with coordinate systems other than rectangular or for inhomogeneous systems having variable density or conductivity, etc., we require a method that is applicable to differential equations having variable coefficients. The most commonly occurring equation is the inhomogeneous Sturm-Liouville equation

$$\frac{d}{dx}\left[p(x)\frac{dy}{dx} \right] + [q(x) + \lambda s(x)]\, y = f(x), \qquad a \le x \le b \quad (4.2.1)$$

with specified boundary conditions at $x = a, b$. Some of the elementary properties of Sturm-Liouville equations are summarized in Appendix C. We begin by assuming that the *homogeneous* equation has linearly independent solutions $u_1(x)$ and $u_2(x)$ that are *known*. Each of these solutions is chosen to satisfy *one* of the boundary conditions to be satisfied by the solution $y(x)$, e.g., if $y(a) = y'(b) = 0$, then we can take $u_1(a) = 0$, $u_2'(b) = 0$. Then, by using the method of variation of parameters, we can write the solution for $y(x)$ in the form

$$y(x) = A(x)u_1(x) + B(x)u_2(x) \quad (4.2.2)$$

The method of variation of parameters is usually applied to an equation for which the coefficient of y'' has been set equal to unity. We thus rewrite the Sturm-Liouville equation (4.2.1) as

$$y'' + (p'/p)\, y' + (q + \lambda s)y/p = f(x)/p \quad (4.2.3)$$

A standard usage of the method of variation of parameters (Hildebrand, 1976, p. 26) then yields

$$A'u_1 + B'u_2 = 0$$

$$A'u_1' + B'u_2' = f(x)/p \quad (4.2.4)$$

where the prime refers to a derivative with respect to x.

Before proceeding with the solution of this system of equations, we incorporate the boundary conditions for our particular example. Since $u_1(x)$ and

$u_2(x)$ already satisfy boundary conditions, (4.2.2) yields $y(a) = B(a)u_2(a) = 0$ so that we have $B(a) = 0$. Similarly, differentiating (4.2.2) and using the first of (4.2.4) yield $A(b) = 0$.

Solving the system (4.2.4) for A' and B' and then integrating we obtain

$$\int_{A(x)}^{A(b)} dA = -A(x) = -\int_x^b \frac{f(x')u_2}{pW} \, dx'$$

$$\int_{B(a)}^{B(x)} dB = B(x) = \int_0^x \frac{f(x')u_1}{pW} \, dx' \qquad (4.2.5)$$

where $W(x)$ is the Wronskian

$$W(x) = u_1(x)u_2'(x) - u_2(x)u_1'(x) \qquad (4.2.6)$$

As noted in Appendix C, the Wronskian for a Sturm-Liouville equation is $W(x) = c/p(x)$ where c is a constant. Since the integrals for A and B contain the product $W(x)p(x) = c$, they are thus simplified and (4.2.2), the solution for $y(x)$, is

$$y(x) = \frac{u_1}{c} \int_x^b f(x')u_2 \, dx' + \frac{u_2}{c} \int_a^x f(x')u_1 \, dx' \qquad (4.2.7)$$

where $c = W(x)p(x)$.

We now specialize this general result to an inhomogeneity of the form

$$f(x) = -J\delta(x - x_0) \qquad (4.2.8)$$

In later applications the quantity J may be either a constant or a function of x. If it is a function of x, it is evaluated at $x = x_0$ because of the presence of the delta function. When this form of $f(x)$ is used in (4.2.7), only one of the two terms will be integrated across the delta function and thus be nonzero. We obtain

$$y_<(x) = -Ju_1(x)u_2(x_0)/c, \qquad x < x_0$$

$$y_>(x) = -Ju_2(x)u_1(x_0)/c, \qquad x > x_0 \qquad (4.2.9)$$

It is convenient to express the constant c as $W(x_0)p(x_0)$ and obtain the final result

$$y(x) = G(x|x_0) = \begin{cases} y_<(x) = -Ju_1(x)u_2(x_0)/p(x_0)W(x_0), & x < x_0 \\ y_>(x) = -Ju_2(x)u_1(x_0)/p(x_0)W(x_0), & x > x_0 \end{cases}$$

$$= -\frac{J}{pW} u_1(x_<)u_2(x_>) \qquad (4.2.10)$$

where $x_>$ is the greater of x and x_0 while $x_<$ is the lesser of the two. The same general form of the solution is obtained when other boundary conditions are applied to $y(x)$. The reciprocity relation $G(x|x_0) = G(x_0|x)$ should be noted.

Example 4.2. As a first usage of this general expression, we recover one of the results of the previous section, namely the solution of

$$y'' + k^2 y = -\delta(x - x'), \qquad y(0) = y(L) = 0 \qquad (4.2.11)$$

Here we have $J = 1$ and $p = 1$. Also, solutions to the homogeneous equation that satisfy the boundary conditions are $u_1(x) = \sin kx$, which satisfies the boundary condition at $x = 0$, and $u_2(x) = \sin k(L - x)$, which is appropriate at $x = L$. Then

$$W = u_1 u_2' - u_2 u_1' = -k[\sin kx \cos k(L - x) + \cos kx \sin k(L - x)]$$

$$= -k \sin kL \qquad (4.2.12)$$

Since $p = 1$ we have $c = pW = -k \sin kL$. The solution, which is the Green's function for this problem, is

$$G(x|x') = \begin{cases} \sin kx \sin k(L - x')/k \sin kL, & x < x' \\ \sin kx' \sin k(L - x)/k \sin kL, & x > x' \end{cases} \qquad (4.2.13)$$

as obtained in (4.1.33).

On the other hand, for the Neumann boundary conditions $y'(0) = y'(L) = 0$ we would use $u_1(x) = \cos kx$ and $u_2(x) = \cos k(L - x)$. Then, $W = k \sin kL$ and

$$G(x|x') = \begin{cases} -\cos kx \cos k(L - x')/k \sin kL, & x < x' \\ -\cos kx' \cos k(L - x)/k \sin kL, & x > x' \end{cases} \qquad (4.2.14)$$

Many other examples of the use of the general expression for a Green's function given in Eq. (4.2.10) will occur later.

Problems

4.2.1 Recover the Green's functions in Eq. (4.1.26) and (4.2.14) as well as in Problems 4.1.3–4.1.5 by using the general result (4.2.10).

4.2.2 Use the general method developed in the text (Wronskian, etc.) to determine the Green's function for harmonic waves on an inhomogeneous string $\rho(x) = \rho_0(L/x)^2$ stretched between a and b, that is, solve

$$\frac{d^2 G}{dx^2} + k_0^2 (L/x)^2 G = -\delta(x - x'), \qquad G(a|x') = G(b|x') = 0$$

4.2.3 Use the general method (Wronskian, etc.) to obtain the Green's function associated with

$$2\sqrt{x}(\sqrt{x}y')' + \lambda y = 0, \qquad y(0) = y(L) = 0$$

if you are given that the solution of this equation is

$$y(x) = A\cos(\sqrt{2\lambda x}) + B\sin(\sqrt{2\lambda x})$$

4.3 ONE-DIMENSIONAL STEADY WAVES

For waves on a string of infinite length, the appropriate form of the Green's function is one that represents waves radiating outward from the source point at x'. When the time dependence is that of a single frequency ω and is expressed as $e^{-i\omega t}$ as in Section 3.4, the outgoing wave is of the form e^{ikx} as $x \to +\infty$ and e^{-ikx} as $x \to -\infty$ (cf. Section 3.4). To see how these results are incorporated into the construction of the Green's function for one-dimensional waves, consider a localized source $\mathcal{F}(x)e^{-i\omega t}$ applied to an infinite string. The meaning of "localized" will become clear when we examine the solution of the problem. We now use the Green's function procedure to express the waves on the string in terms of an integral over the source. To do this we combine

$$y'' + k^2 y = -\frac{1}{T}\mathcal{F}(x), \qquad k = \frac{\omega}{c} \tag{4.3.1}$$

with

$$G_k'' + k^2 G_k = -\delta(x - x') \tag{4.3.2}$$

The appropriate form of G_k is immediately obtained from the results of the previous section by noting that as a, the left-hand boundary, approaches $-\infty$ we require an outgoing wave e^{-ikx}, while as b approaches $+\infty$ we must again have an outgoing wave, which this time is of the form e^{ikx}. Thus $u_1(x) = e^{-ikx}$, $u_2(x) = e^{ikx}$, and the Wronskian $W = u_1 u_2' - u_2 u_1' = 2ik$. Since $p = 1$, we have $c = W = 2ik$ as well as $J = 1$. Thus, from (4.2.10)

$$G_k(x|x') = \begin{cases} \dfrac{i}{2k}e^{ik(x-x')}, & x < x' \\[3mm] \dfrac{i}{2k}e^{ik(x-x')}, & x < x' \end{cases} \tag{4.3.3}$$

Note that these two expressions are conveniently combined into the form

$$G_k(x|x') = \frac{i}{2k}e^{ik|x-x'|} \tag{4.3.4}$$

When we interchange x and x' in (4.3.1) and (4.3.2) and then combine them by the multiply-and-subtract procedure, we obtain

$$
y(x) = \frac{1}{T} \int_{-\infty}^{\infty} dx' \, \mathcal{F}(x') G_k(x|x')
$$

$$
+ \left[G_k(x|x') \frac{dy}{dx'} - y(x') \frac{dG_k(x|x')}{dx'} \right]\Bigg|_{x'=-\infty}^{x'=+\infty}
\tag{4.3.5}
$$

We now show that the integrated terms vanish. This result may be anticipated on physical grounds since we expect all contributions to the wave motion $y(x)$ to be due solely to the source term $\mathcal{F}(x)$. Of course, if there were boundaries to the string (as in Problem 4.3.3), then there would also be a reflected wave that would be associated with the boundary term.

We now examine in detail the evaluation at the upper limit $x' = +\infty$ in (4.3.5). Consider at first the upper limit to be at $x' = L$ where L is much larger than any point x at which the field is to be evaluated. Then $x < x'$ and we use $G_{k<}(x|x')$ in the evaluation at this upper limit $x' = L$. We must also evaluate $y(x')$ at the upper limit but $y(x')$ is unknown until the problem has been solved. We may proceed, however, by noting that far from a localized source we require $\mathcal{F}(x) = 0$ and in that region the wave equation (4.3.1) reduces to the homogeneous equation

$$
y'' + k^2 y = 0
\tag{4.3.6}
$$

for which solutions are known to within an unknown amplitude factor. As x' approaches $+\infty$ we use the outgoing wave solution $y(x') = Ce^{ikx'}$ where C is the unknown amplitude referred to previously. At the upper limit $x' = L$ we thus evaluate

$$
\left[G_{k<}(x|x') \frac{dy}{dx'} - \frac{dG_{k<}(x|x')}{dx'} y(x') \right]_{x'=L}
$$

$$
= \frac{Ci}{2k} [e^{ik(x-L)} ike^{ikL} - e^{ikL} ike^{ik(x-L)}]
$$

$$
= 0
\tag{4.3.7}
$$

A similar result is obtained at the lower limit. There is thus no contribution from the integrated terms, as conjectured, and we obtain the solution

$$
y(x) = \frac{i}{2kT} \int_{-\infty}^{\infty} \mathcal{F}(x') e^{ik|x-x'|} \, dx'
\tag{4.3.8}
$$

We now examine this result by breaking up the range of integration into regions for which $x' > x$ and $x' < x$. This decomposition yields

$$y(x) = \frac{i}{2kT}\left[e^{ikx} \int_{-\infty}^{x} \mathcal{F}(x')e^{-ikx'}\,dx' + e^{-ikx}\int_{x}^{\infty}\mathcal{F}(x')e^{ikx'}\,dx'\right] \quad (4.3.9)$$

As x goes to $\pm\infty$ we obtain

$$y(x) = \begin{cases} \dfrac{i}{2kT}e^{ikx}\displaystyle\int_{-\infty}^{\infty}\mathcal{F}(x')e^{-ikx'}\,dx', & x \to +\infty \\[3mm] \dfrac{i}{2kT}e^{-ikx}\displaystyle\int_{-\infty}^{\infty}\mathcal{F}(x')e^{ikx'}\,dx', & x \to -\infty \end{cases} \quad (4.3.10)$$

As anticipated above, in these limits the form of $y(x)$ is proportional to $e^{\pm ikx}$. The amplitude factor C has now been determined in terms of an integral over the source. We have implicitly assumed that the function $\mathcal{F}(x)$ is such that these integrals converge. The integrals involved are referred to as a Fourier transform, and it is known (cf. Section 6.4) that a sufficient condition to guarantee convergence of this integral is that $\int_{-\infty}^{\infty}|\mathcal{F}(x')|\,dx'$ converges. A localized source is one for which this integral does converge.

Example 4.3. For the case $\mathcal{F}(x) = \mathcal{F}_0 e^{-|x|/L}$, (4.3.8) yields

$$y(x) = \frac{i\mathcal{F}_0}{2kT}e^{ikx}\int_{-\infty}^{\infty}e^{-|x'|/L}e^{-ikx'}\,dx', \quad x \to +\infty \quad (4.3.11)$$

Since only the even part of the integrand will give a nonvanishing contribution, we have

$$y(x) = \frac{i\mathcal{F}_0}{kT}e^{ikx}\int_{0}^{\infty}e^{-x'/L}\cos kx'\,dx'$$

$$= \frac{i\mathcal{F}_0}{kT}\frac{e^{ikx}}{1 + (kL)^2} \quad (4.3.12)$$

The integral is in the form of a Laplace transform and has been evaluated by using 11 in Table F.2. To understand this result physically, it should be recalled that $k = \omega/c = 2\pi\nu/c = 2\pi/\lambda$, where λ is the wavelength of the wave on the string. For $kL \gg 1$ we have $L/\lambda \gg 1$ and the source is many wavelengths wide. There is thus considerable interference in the waves generated by different parts of the source and the radiated signal is small. This is seen to be the case since $y \sim (kL)^{-3}$ in this limit. On the other hand, for $kL \ll 1$ there is no such interference, we find $y \sim (kL)^{-1}$, in the amplitude $y(x)$ is correspondingly larger.

4.3.1 Scattering of Waves on a String

A problem for which there *is* a contribution from the integrated term is the scattering problem. In this situation, a wave is assumed to be incident from, say, $x = -\infty$ upon some sort of inhomogeneity on the string. This incident wave is partly reflected back to $-\infty$ and partly transmitted to $+\infty$. As a specific example, consider a string for which the density has the constant value ρ_0 except for a localized region where it becomes $\rho_0 + \delta\rho(x)$. The wave equation for the string (1.8.9) with $T = \text{const}$, $R = K = \mathcal{F} = 0$, and $\rho = \rho_0 + \delta\rho(x)$ becomes

$$\rho_0 \frac{\partial^2 y}{\partial t^2} - T\frac{\partial^2 y}{\partial x^2} = -\delta\rho(x)\frac{\partial^2 y}{\partial t^2} \tag{4.3.13}$$

For wave motion at one frequency ω we may write $y(x, t) = f(x)e^{-i\omega t}$, as in Section 3.4, to obtain

$$f'' + k_0^2 f = k^2(\delta\rho/\rho_0)f \tag{4.3.14}$$

where $k = \omega/c$. When we combine this equation (written in primed coordinates) with the equation for $G_k(x|x')$ given in (4.3.2), we obtain

$$f(x) = \left[G_k(x|x')\frac{df}{dx'} - f(x')\frac{dG_k(x|x')}{dx'} \right]\Bigg|_{x'=-\infty}^{x'=\infty}$$

$$+ k^2 \int_{-\infty}^{\infty} dx'\, \mu(x')f(x')G_k(x|x') \tag{4.3.15}$$

where $\mu(x) = \delta\rho(x)/\rho$. To evaluate the integrated term we again note that as $x\to\infty$, $f(x')$ has the form $Ce^{ikx'}$, as in our previous radiation problem. Hence we again obtain no contribution from the upper limit. As $x\to-\infty$, however, the solution must represent both an incident wave of the form Ae^{ikx} and a reflected wave Be^{-ikx}. The term Be^{-ikx} again yields no contribution as $x' \to -\infty$, as was the case for the radiation problem. When we repeat the calculation outlined in (4.3.7) for the term Ae^{ikx}, however, we obtain

$$\left[G_{k>}(x|x')\frac{df}{dx'} - \frac{dG_{k>}(x|x')}{dx'}f(x') \right]_{x'=-L} = Ae^{ikx'} \tag{4.3.16}$$

that is, the evaluation at $x = -L$ yields just the incident wave and (4.3.15) becomes

$$f(x) = Ae^{ikx} + k^2 \int_{-\infty}^{\infty} dx'\, \mu(x')G_k(x|x') \tag{4.3.17}$$

Since the unknown wave amplitude $f(x)$ appears under the integral sign in (4.3.17), we have not solved our problem at all but have merely recast it in the form of an integral equation [recall the discussion of (4.1.28)]. Since it is usually too difficult to obtain a Green's function that incorporates the inhomogeneity $\mu(x)$, we must consider the solution of the integral equation in (4.3.17). While any full discussion of this issue is beyond the framework of our development here, we can obtain a standard approximate solution that may be expected to be valid as long as the density variation is small in some sense. In this case we expect that the exact solution for $f(x)$ will differ only slightly from Ae^{ikx} and we may therefore replace $f(x)$ under the integral sign in (4.3.17) with the incident wave Ae^{ikx}. This approximation is frequently referred to as the Born approximation. We now obtain

$$f(x) = Ae^{ikx} + k_0^2 \int_{-\infty}^{\infty} dx' \, \mu(x') Ae^{ikx'} G_k(x|x') \qquad (4.3.18)$$

When we follow the procedure just employed in treating (4.3.8) for the radiation problem, we obtain

$$f(x) = \begin{cases} Ae^{ikx} + \dfrac{ik}{2} Ae^{ikx} \displaystyle\int_{-\infty}^{\infty} dx' \, \mu(x'), & x \to +\infty \\[4mm] Ae^{ikx} + \dfrac{ik}{2} Ae^{-ikx} \displaystyle\int_{-\infty}^{\infty} dx' \, \mu(x') e^{2ikx'}, & x \to -\infty \end{cases} \qquad (4.3.19)$$

The amplitude of the term Ae^{-ikx} as $x \to \infty$ is referred to as the reflection coefficient for the scatterer. Similarly, the total amplitude of the transmitted wave is the coefficient of e^{ikx} in the first of (4.3.19).

4.3.2 Significance of Boundary Terms

For a *semi-infinite* string that is driven by a localized force $\mathcal{F}(x)$, the multiply-and-subtract procedure applied to (4.3.1) and (4.3.2) over the region $0 \le x < \infty$ yields

$$y(x) = \frac{1}{T} \int_0^{\infty} dx' \, \mathcal{F}(x') G_k(x|x') - y'(0) G_k(x|0) + y(0) G_{kx'}(x|0)$$

$$(4.3.20)$$

in place of (4.3.8). The two additional terms associated with the boundary can be given a simple physical interpretation. The term $y'(0)G_k(x|0)$ is the response of the string to a point source of strength $Ty'(0)$ located at the boundary $x' = 0$. To interpret the term containing the spatial derivative of the Green's function $G_{kx'}(x|0)$, we first recall the definition of a derivative and write

$$\left. \frac{\partial G(x|x')}{\partial x'} \right|_{x'=0} = \lim_{\epsilon \to 0} \left. \frac{G_k(x|x'+\epsilon) - G_k(x|x'-\epsilon)}{2\epsilon} \right|_{x'=0} \qquad (4.3.21)$$

and then note that this expression is the response of the system to two equal and opposite sources, that is,

$$s(x) = \lim_{\epsilon \to \infty} \frac{\delta(x' + \epsilon) - \delta(x' - \epsilon)}{2\epsilon}$$

$$= \delta'(x') \tag{4.3.22}$$

Here, $\delta'(x)$ refers to the derivative of the delta function. Such a pair of sources is frequently referred to as a *dipole* source. The last term in (4.3.20) can thus be interpreted as the response of the string to a dipole source of strength $Ty(0)$ located at the boundary. The total string displacement can now be written

$$y(x) = \frac{1}{T} \int_0^\infty dx' [\mathcal{F}(x') - Ty'(0)\delta(x') + Ty(0)\delta'(x')]G(x|x') \tag{4.3.23}$$

The net effect of the boundary can be described as being that due to a point source (usually referred to as a monopole) of strength $Ty'(0)$ and a dipole of strength $Ty(0)$. This interpretation of the boundary terms can be carried over to two- and three-dimensional problems. The construction of a Green's function that has appropriate boundary conditions for the semi-infinite string will be taken up in the next section.

Problems

4.3.1   A dipolelike source for harmonic waves on a string is composed of two point sources separated by a distance $2x_0$ and radiating $180°$ out of phase with each other. This situation may be represented by the source term

$$\mathcal{F}(x, t) = F_0[\delta(x - x_0) - \delta(x + x_0)]e^{-i\omega t}$$

in the wave equation.
   (a) Use Green's function methods to determine the wave motion on the string as $x \to +\infty$.
   (b) What happens when $\omega x_0/c = \pi$? Discuss this situation.

4.3.2   A uniform string of length L has boundary conditions $y(0, t) = 0$, $\zeta y(L, t) + Ly'(L, t) = 0$.
   (a) Determine the steady state Green's function for the string, that is, solve

$$g'' + k^2 g = -\delta(x - x')$$

   when g satisfies the same boundary conditions as the string.
   (b) Discuss the limits $\zeta \to 0$ and ∞.

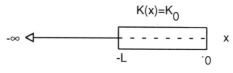

$$K(x) = K_0$$

Figure 4.1

4.3.3 A semi-infinite string extends from $x = -\infty$ to $x = 0$. The string is fixed at $x = 0$. A region of constant elasticity $K(x) = K_0$ surrounds the string from $x = -L$ to $x = 0$. (Cf. Fig. 4.1.)

(a) Show that the integral equation for the reflection problem is

$$f(x) = 2i \sin kx - \frac{1}{T} \int_{-L}^{0} dx' \, g(x|x') K(x) f(x)$$

where g vanishes at $x = 0$.

(b) Solve the integral equation in Born approximation and obtain

$$f(x) = 2i \sin kx - i \frac{K_0 L^2}{T} e^{-ikx} \cdot \frac{1}{kL} \left(1 - \frac{\sin 2kL}{2kL} \right)$$

4.3.4 A semi-infinite string is excited by a sinusoidal point force at $x = x_0$. The string displacement is thus governed by

$$\frac{d^2 f}{dx^2} + k^2 f = -Q_0 \delta(x - x_0), \qquad 0 \le x < \infty$$

(a) Combine this equation, written in primed coordinates, with that for the infinite string Green's function to obtain

$$Q_0 g(x|x_0) + [f(0) g_{x'}(x|0) - f'(0) g(x|0)] = \begin{cases} f(x), & x > 0 \\ 0, & x < 0 \end{cases}$$

(b) Show that

$$g_{x'}(x|0) = \begin{cases} \frac{1}{2} e^{ikx}, & x > 0 \\ -\frac{1}{2} e^{-ikx}, & x < 0 \end{cases}$$

and therefore that

$$f(x) = Q_0 g(x|x_0) - \frac{i}{2k} e^{ikx} [ikf(0) + f'(0)], \qquad x > 0$$

$$0 = Q_0 e^{ikx_0} + [ikf(0) - f'(0)], \qquad x < 0$$

(c) Assume that at $x = 0$ the impedance boundary condition $f'(0) + ik\zeta f(0) = 0$ is obeyed and show that the condition for $x < 0$ obtained in (b) yields

$$Q_0 e^{ikx_0} + ikf(0)(1 - \zeta) = 0$$

(d) Obtain the string displacement in the form

$$f(x) = Q_0 g(x|x_0) + \frac{1 - \zeta}{1 + \zeta} Q_0 g(x|-x_0)$$

by using the relation for $x > 0$ given in (b).

4.4 METHOD OF IMAGES

The form of the Green's function that is required in order to satisfy a boundary condition in a semi-infinite region can sometimes be given a simple geometric interpretation. As an example, let us consider the radiation problem of the previous section for the case of the semi-infinite region $x > 0$ with the string satisfying the fixed end boundary condition $y(0) = 0$. If we impose the same boundary condition on G at $x = 0$ and assume outgoing waves for large positive values of x, then we employ the functions

$$u_1(x) = \sin kx, \qquad u_2(x) = e^{ikx} \tag{4.4.1}$$

in the general form for $G_k(x|x')$ given in (4.2.10). The Wronskian is now found to be $W = u_1 u_2' - u_2 u_1' = -k$ and, since $p(x) = 1$, we obtain

$$G_D(x|x') = \frac{e^{ikx_>} \sin kx_<}{k} \tag{4.4.2}$$

where the subscript D is used to signify that $G_k = 0$ at the boundary, a condition frequently referred to as a homogeneous Dirichlet boundary condition. This result has a simple physical interpretation when it is written entirely in terms of exponentials. We have

$$G_D(x|x') = \frac{i}{2k} (e^{ikx_> - ikx_<} - e^{ikx_> + ikx_<})$$

$$= \frac{i}{2k} (e^{ik|x - x'|} - e^{ik(x + x')}) \tag{4.4.3}$$

Note that $x_> - x_<$ and $|x - x'|$ are merely two equivalent ways of expressing a difference such that the result is always positive. Since x and x' are both positive, the exponent in the second exponent is always positive.

The result for $G_D(x|x')$ in (4.4.3) is the expression for the wave field that would occur at any point $x > 0$ on an *infinite* string with the source located at x' plus an *additional* source of equal magnitude but opposite phase located at $-x'$. The net result of these two sources is to maintain a vanishing displacement at $x = 0$ and thus satisfy the boundary condition at that point, that is, $G_D(0|x') = 0$.

If, on the other hand, a semi-infinite string satisfies the condition $y'(0) = 0$, then the corresponding Green's function is found to be $(i/k)\exp(ikx_>)\cos kx_<$. This result may be written

$$G_N(x|x') = \frac{i}{2k}(e^{ik|x-x'|} + e^{ik(x+x')}) \qquad (4.4.4)$$

where the subscript N signifies that Neumann boundary conditions are satisfied. The interpretation is that the image source is now in phase with the real source at $x = x'$.

Since the image term is due to a source at $x = x'$ that is *outside the region of the physical problem* $(x > 0)$, it is a solution to the *homogeneous* equation for $x > 0$. We can thus give the image term the interpretation that it is the appropriate solution of the homogeneous equation that must be added to the particular solution in order for the whole solution for $(G(x|x')$ to satisfy a specified boundary condition at $x = 0$.

Returning to the problem of waves on a semi-infinite string with a fixed end at $x = 0$, the solution is

$$y(x) = \frac{1}{T}\int_0^\infty dx'\, \mathcal{F}(x')G_D(x|x')$$

$$= \frac{1}{T}\frac{i}{2k}\int_0^\infty dx'\, \mathcal{F}(x')(e^{ik|x-x'|} - e^{ik(x+x')}) \qquad (4.4.5)$$

When the system is of finite length, the source will have a mirror image in each boundary and these images must, in turn, be imaged in the opposite boundary. Continuation of this procedure produces an infinite series of images. An example of this procedure is developed in Problem 4.4.3.

Problems

4.4.1 (a) Solve $G'' - k^2G = -\delta(x - x')$, $x > 0$, subject to the boundary condition $G = 0$ at $x = 0$, ∞.

(b) Write the result in terms of an image source.

4.4.2 The displacement of a semi-infinite string $(0 \le x < \infty)$ satisfies the boundary condition

$$\frac{dy}{dx} + ik\zeta y = 0$$

at $x = 0$.

(a) Obtain the Green's function for this situation by *assuming* that the boundary conditions may be satisfied by using an appropriate image, that is, determine R in the expression

$$g(x|x') = \frac{i}{2k}(e^{ik(x-x')} + Re^{ik(x+x')})$$

(b) Show that you obtain "expected" results as $\zeta \to 0, \infty$.

4.4.3 It has been shown that the Green's function satisfying

$$\frac{d^2G}{dx^2} + k^2G = -\delta(x - x'), \qquad G(0|x') = G(L|x') = 0$$

is

$$G(x|x') = \frac{\sin kx_< \sin k(L - x_>)}{k \sin kL}$$

Interpret this result in terms of an infinite set of image sources. [*Hint*: Consider the development of Eq. (2.2.11) et seq.]

4.5 A NON-SELF-ADJOINT GREEN'S FUNCTION

When we consider time dependent problems in the next section, we will encounter a complication associated with the reciprocity principle that has not appeared thus far. The symmetry associated with interchanging the spatial location of source and receiver does not carry over immediately to the temporal coordinate since in the standard physical situation a signal is always received *after* it has originated. In fact, time dependent problems are usually formulated in such a way that the field is zero before the source begins to act. This temporal asymmetry is usually referred to as causality. Also, if a signal decays in time, it would have to increase in time if the origination time and observation time were merely interchanged.

The differential equations and boundary conditions considered up until now, with which reciprocity symmetries are associated, are referred to as being self-adjoint problems. Problems not possessing these symmetries are referred to as being non-self-adjoint. This topic will be developed further at the end of Section 4.6. Before introducing this topic in the time dependent setting, we first consider a reformulation of one of our previous developments so as to display some of these issues in a space dependent context.

First recall our previous solution of the one-dimensional steady state heat equation

$$\frac{d^2u}{dx^2} = -\frac{1}{K}s(x), \qquad u(0) = u(L) = 0 \tag{4.5.1}$$

We have seen in Section 4.1 that the solution can be expressed as

$$u(x) = \frac{1}{K}\int_0^L dx'\, G(x|x')s(x') \tag{4.5.2}$$

where $G(x|x')$ satisfies

$$\frac{d^2G}{dx^2} = -\delta(x - x'), \qquad G(0|x') = g(L|x') = 0 \tag{4.5.3}$$

and was found to have the form

$$G(x|x') = \begin{cases} x(L - x')/L, & x < x' \\ x'(L - x)/L, & x > x' \end{cases} \tag{4.5.4}$$

On interchanging x and x' we obtain

$$G(x'|x) = \begin{cases} x'(L - x)/L, & x' < x \\ x(L - x')/L, & x' > x \end{cases} \tag{4.5.5}$$

and after merely rearranging these inequalities we have

$$G(x'|x) = \begin{cases} x(L - x')/L, & x < x' \\ x'(L - x)/L, & x > x' \end{cases}$$
$$= G(x|x') \tag{4.5.6}$$

an equality that is usually referred to as the reciprocity relation. This symmetry was used previously to derive the solution given in (4.5.2) by combining Eq. (4.5.1) and (4.5.3) to yield

$$u(x') = \frac{1}{K}\int_0^L dx\, s(x)\, G(x|x') \tag{4.5.7}$$

and then interchange x and x' to obtain

$$u(x) = \frac{1}{K} \int_0^L dx' \, s(x') G(x'|x) = \frac{1}{K} \int_0^L dx' \, s(x') G(x|x') \quad (4.5.8)$$

The integrated term $Gu' - uG'$ was found to vanish at $x = 0, L$ by virtue of the boundary conditions imposed on G in the primed coordinate system. Also, the validity of the interchange of x and x' under the integral sign follows from the reciprocity relation for $G(x|x')$.

As a preview to some of the concepts that we will encounter later, we now express the solution of this equation for $u(x)$ in terms of a different Green's function. To this end, note that we will still have a solution of the differential equation for G if we add any homogeneous solution of the equation for G. This new Green's function will not satisfy the same boundary conditions as the previous Green's function, of course, but as we shall see, it can still be used to provide the solution to the problem. The Green's function that we will use, call it $\underline{G}(x|x')$, is

$$\underline{G}(x|x') = G(x|x') - x(L - x')/L$$

$$= \begin{cases} x(L - x') - x(L - x') = 0, & x < x' \\ x'(L - x) - x(L - x') = x' - x, & x > x' \end{cases} \quad (4.5.9)$$

The difference $\underline{G} - G = -x(L - x')/L$ is clearly a homogeneous solution of the equation for G over the *entire* region $0 < x < L$. Green's functions that vanish for one choice of the inequality between primed and unprimed coordinates, such as $\underline{G}(x|x')$ being introduced here, will be found appropriate for the time dependent problems to be considered later. Note that on interchanging x and x' we now have

$$\underline{G}(x'|x) = \begin{cases} 0, & x' < x \\ x - x', & x' > x \end{cases}$$

$$= \begin{cases} x - x', & x < x' \\ 0, & x > x' \end{cases} \quad (4.5.10)$$

which is *not* the same as $\underline{G}(x|x')$ given in (4.5.9). Graphs of $\underline{G}(x|x')$ and $\underline{G}(x'|x)$ are shown in Figure 4.2. Note, however, that if we replace x by $-x'$ and x' by $-x$ in (4.5.9), we have

$$\underline{G}(-x'|-x) = \begin{cases} 0, & -x' < -x \\ -x + x', & -x' > -x \end{cases}$$

$$= \begin{cases} 0, & x < x' \\ x' - x, & x > x' \end{cases}$$

$$= \underline{G}(x|x') \quad (4.5.11)$$

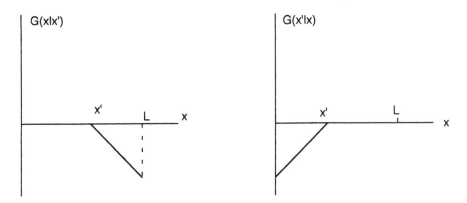

Figure 4.2. Example of non-self-adjoint Green's function [Eq. (4.5.10)].

as a reciprocity relation that is satisfied by $\underline{G}(x|x')$. Hence, if we want to express the solution for $u(x)$ in terms of $\underline{G}(x|x')$, we cannot merely combine the equation for $u(x)$ with that for $\underline{G}(x|x')$ and then interchange coordinates at the end of the calculation as we did for the symmetric Green's function $G(x|x')$.

To obtain a Green's function $\underline{G}(x|x')$ that is zero for $x < x'$ *after* we interchange x and x', we should begin with a Green's function that vanishes for $x > x'$. On the basis of the previous considerations, such an expression, call it $\underline{G}^a(x|x')$, is clearly

$$\underline{G}^a(x|x') = G(x|x') - x'(L - x)/L$$

$$= x(L - x')/L - x'(L - x)/L = x - x', \quad x < x'$$

$$= x'(L - x)/L - x'(L - x)/L = 0, \quad x > x' \quad (4.5.12)$$

For, on interchanging x and x' we obtain

$$\underline{G}^a(x'|x) = \begin{cases} x' - x, & x' < x \\ 0, & x' > x \end{cases}$$

$$= \begin{cases} 0, & x < x' \\ x' - x, & x > x' \end{cases}$$

$$= \underline{G}(x|x') \quad (4.5.13)$$

as we desire. A comparison of (4.5.11) and (4.5.13) yields $\underline{G}^a(x'|x) = \underline{G}(-x'|-x)$.

Since $\underline{G}^a(x|x')$ only differs from $G(x|x')$ by a solution of the homogeneous equation associated with (4.5.3), $\underline{G}^a(x|x')$ also satisfies the inhomogeneous differential equation (4.5.3), that is,

$$\frac{d^2 \underline{G}^a(x|x')}{dx^2} = -\delta(x - x') \tag{4.5.14}$$

Combining the equation for $u(x)$ in (4.5.1) with that for $\underline{G}^a(x|x')$ in (4.5.14), by the usual procedure, we arrive at

$$\left[\underline{G}^a(x|x') \frac{du(x)}{dx} - u(x) \frac{d\underline{G}^a(x|x')}{dx} \right]_{x=0}^{x=L} + \int_0^L dx \, s(x) \, \underline{G}^a(x|x') = u(x') \tag{4.5.15}$$

Since $\underline{G}^a(x|x') = 0$ for $x > x'$, we have $\underline{G}^a(L|x')$ and $d\underline{G}^a(L|x')/dx = 0$. Thus there is no contribution from the upper limit because of $\underline{G}^a(x'|x)$ and *not* because of a homogeneous boundary condition being imposed upon $u(x)$. We obtain

$$u(x') = -\underline{G}^a(0|x') \frac{du(x)}{dx}\bigg|_{x'=0} + \frac{1}{K} \int_0^L dx \, s(x) \underline{G}^a(x|x') \tag{4.5.16}$$

In obtaining this result, the boundary condition $u(0) = 0$ given in (4.5.1) has also been used. Introducing $\underline{G}^a(0|x') = -x'$ from (4.5.12), we have

$$u(x') = x' u'(0) + \frac{1}{K} \int_0^L dx \, s(x) \underline{G}^a(x|x') \tag{4.5.17}$$

On interchanging x and x' and also using $\underline{G}^a(x'|x) = \underline{G}(x|x')$ from (4.5.13), we now obtain

$$u(x) = xu'(0) + \frac{1}{K} \int_0^L dx' \, s(x') \underline{G}(x|x') \tag{4.5.18}$$

Since $\underline{G}(x|x') = 0$ for $x < x'$, we could replace the upper limit by x in (4.5.18).

In the expression for $u(x)$ obtained in (4.5.18), $u'(0)$ is still unknown. [Note that in an initial-value problem where the independent variable has the significance of time and both $u(0)$ and $u'(0)$ are specified, it *would* be known.] We can determine $u'(0)$ here by first setting $x = L$ to obtain

$$u(L) = 0 = Lu'(0) + \frac{1}{K} \int_0^L dx' \, s(x') \underline{G}(L|x') \tag{4.5.19}$$

When this result is solved for $u'(0)$ and introduced into (4.5.18), we obtain

$$u(x) = \frac{1}{K} \int_0^L dx' \, s(x')[\underline{G}(x|x') - (x/L)\underline{G}(L|x')] \qquad (4.5.20)$$

When the expression for $\underline{G}(x|x')$ is used from (4.5.9), we find that $\underline{G}(x|x')$ $- (x/L)\underline{G}(L|x') = G(x|x')$ and the original form of the solution given in (4.5.8) is recovered. While this calculation may have followed a circuitous route in arriving at this result, the procedure has been used to introduce concepts that will prove useful in time dependent problems.

Finally, note that the expression of the solution in terms of $\underline{G}(x|x')$ would have been obtained somewhat more directly if we had begun with our equations in the primed coordinates. Since $d^2\underline{G}(x|x')/dx^2 = d^2\underline{G}(x|x')/dx'^2 = -\delta(x - x')$, we could combine

$$\frac{d^2u}{dx'^2} = -\frac{1}{K}s(x'), \qquad \frac{d^2\underline{G}}{dx'^2} = -\delta(x - x') \qquad (4.5.21)$$

and obtain

$$\left(\underline{G}\frac{du}{dx'} - u\frac{d\underline{G}}{dx'}\right)\Bigg|_{x'=0}^{x'=L} = u(x) - \frac{1}{K}\int_0^L dx' \, s(x')\underline{G}(x|x') \qquad (4.5.22)$$

Since $\underline{G}(x|x')$ is zero for $x < x'$ by (4.5.9) and $\underline{G}(x|0) = -x$, we obtain the result given in (4.5.18).

Problems

4.5.1   Since the solution given in (4.5.17) depends only upon boundary information about the solution at $x = 0$, it is independent of both u and u' at $x = L$, and it may be used to solve problems with boundary conditions other than $u(L) = 0$. Show that if $u_x(L) = 0$, then (4.5.18) yields

$$u_x(0) = \frac{1}{K} \int_0^L dx' \, s(x')$$

which is to be expected on physical grounds (Why?) and then that

$$u(x) = \frac{1}{K} \int_0^L dx' \, s(x')[x + \underline{G}(x|x')]$$

Also, show that $x + \underline{G}(x|x')$ reduces to the customary Green's function for this problem, that is, it satisfies $G(0|x') = G_x(L|x') = 0$, the boundary conditions for this problem. It is given in Problem 4.1.6.

4.5.2 Show that if the methods of this section are applied to the Green's function for the infinite string (4.3.4), one obtains

$$\underline{G}_k(x|x') = \begin{cases} 0, & x < x' \\ -\dfrac{1}{k}\sin k(x - x'), & x > x' \end{cases}$$

$$\underline{G}_k^a(x|x') = \begin{cases} \dfrac{1}{k}\sin k(x - x'), & x < x' \\ 0, & x > x' \end{cases}$$

and therefore

$$\underline{G}_k^a(x'|x) = \underline{G}_k(x|x')$$

Note also that $\underline{G}_k(-x'|-x) = \underline{G}_k(x|x')$ as in (4.5.11).

4.5.3 **(a)** Combine the wave equation (4.3.1) with the differential equation for $\underline{G}^a(x|x')$ and integrate over x from $-L$ to $+L$ where $|L| > x'$ to obtain

$$f(x') = -\left[\underline{G}^a \frac{df}{dx} - f\frac{d\underline{G}^a}{dx} \right]\Bigg|_{x=-L} + \frac{1}{T}\int_{-L}^{x'} dx\ \underline{G}^a(x|x')\mathfrak{F}(x)$$

(b) For values of L outside the source region, (4.3.1) reduces to the homogeneous equation $f'' + k^2 f = 0$, which has solutions $f = Ae^{ikx}$ and Be^{-ikx}. Show that the integrated term in part (a) reduces to Be^{-ikx} when the expression for $\underline{G}^a(x|x')$ is introduced. One then obtains

$$f(x') = Be^{-ikx} + \frac{1}{T}\int_{-\infty}^{x'} dx\ \underline{G}^a(x|x')\mathfrak{F}(x)$$

(c) Use the results of Problem 4.5.2 to interchange x and x' and obtain

$$f(x') = Be^{-ikx} + \frac{1}{T}\int_{-\infty}^{x'} dx'\ \underline{G}(x|x')\mathfrak{F}(x')$$

(d) Introduce the expression for $\underline{G}(x|x')$ from Problem 4.5.2 and examine the result in part (c) in the limit $x \to \pm\infty$ to obtain

$$B = \frac{i}{2kT} \int_{-\infty}^{\infty} dx' \, \mathcal{F}(x') e^{ikx}$$

$$f(x) = \frac{i}{2kT} e^{ikx} \int_{-\infty}^{\infty} dx' \, \mathcal{F}(x') e^{-ikx'}, \qquad x \to +\infty$$

These results agree with those of (4.3.10).

4.6 GREEN'S FUNCTION FOR A DAMPED OSCILLATOR

As alluded to in the previous section, there are certain complexities in the Green's function method that occur when the method is applied to time dependent problems. To introduce this subject we now determine the Green's function for a damped mechanical oscillator. The equation for the oscillator will be taken to be $my'' + Ry' + ky = F(t)$, which will be rewritten

$$y'' + 2\gamma y' + \omega^2 y = f(t) \tag{4.6.1}$$

with $2\gamma = R/m$, $\omega^2 = k/m$, and $f(t) = F(t)/m$. In the spirit of our previous developments, we again consider the solution to the simpler problem

$$\frac{d^2 G}{dt^2} + 2\gamma \frac{dG}{dt} + \omega^2 G = \delta(t - t') \tag{4.6.2}$$

with the homogeneous initial conditions $G_<(0|t') = dG_<(0|t')/dt = 0$, that is, until $t = t'$ the solution $G(t|t')$ will remain equal to zero. After $t = t'$, $G_>(t|t')$ will be a solution to the homogeneous equation, namely,

$$G_>(t|t') = Ae^{-\gamma t}\sin \Omega(t - \varphi) \tag{4.6.3}$$

where $\Omega = \sqrt{\omega^2 - \gamma^2}$ while A and φ are integration constants. Since G is continuous at $t = t'$, we require $G_>(t'|t') = Ae^{-\gamma t'}\sin \Omega(t' - \varphi) = 0$, which is satisfied by $\varphi = t'$. Integrating (4.6.2) across the singularity at $t = t'$, we have

$$\left. \frac{dG_>}{dt} \right|_{t=t'} - \left. \frac{dG_<}{dt} \right|_{t=t'} = 1 \tag{4.6.4}$$

and since $dG_<(0|t')/dt = 0$, the form of $G_>$ in (4.6.3) yields $A = \Omega^{-1}e^{\gamma t'}$. Finally, we obtain

$$G(t|t') = \frac{1}{\Omega} e^{-\gamma(t-t')} \sin \Omega(t-t') H(t-t') \quad (4.6.5)$$

where $H(t-t')$ is the unit step function.

In changing to the primed coordinate for combining the equation for $G(t|t')$ with that for the oscillator we can introduce derivatives with respect to the primed coordinates of G by noting that $dG/dt' = -dG/dt$. We thus have

$$\frac{d^2y}{dt'^2} + 2\gamma \frac{dy}{dt'} + \omega^2 y = f(t')$$

$$\frac{d^2G}{dt'^2} - 2\gamma \frac{dG}{dt'} + \omega^2 G = \delta(t-t') \quad (4.6.6)$$

The change in sign of the term with the first derivative of $G(t|t')$ is to be noted. It will be considered in a more general way later in this section. This change of sign is precisely what is needed to have the multiply-and-subtract procedure yield a perfect t' derivative. We obtain

$$\frac{d}{dt'}\left(G\frac{dy}{dt'} - y\frac{dG}{dt'}\right) + 2\gamma \frac{d}{dt'}(Gy) = Gf(t') - y\delta(t-t') \quad (4.6.7)$$

When this equation is integrated over t' from $t' = 0$ to $t' = t + \epsilon$ (to integrate beyond the peak in the delta function), we obtain

$$\left(G\frac{dy}{dt'} - y\frac{dG}{dt'} + 2\gamma Gy\right)\Bigg|_{t'=0}^{t'=t+\epsilon} = \int_0^{t+\epsilon} G(t|t')f(t')\,dt' - y(t) \quad (4.6.8)$$

At the upper limit both G and dG/dt' are zero because of the step function in (4.6.5), that is, $G(t|t') = 0$ for $t' > t$. Hence only the evaluation at the lower limit provides a contribution, and we have

$$y(t) = \int_0^{t+\epsilon} \mathcal{F}(t')G(t|t')\,dt' + G(t|0)y'(0) \frac{dG(t|0)}{dt'} y(0)$$

$$+ 2\gamma G(t|0)y(0) \quad (4.6.9)$$

This result is the general solution for the motion of the oscillator in terms of initial conditions and driving force.

The first difference between the present problem and those considered previously has already been noted, namely G satisfies a different equation than that satisfied by the oscillator itself. A second difference is in the reciprocity relation. As is evident by inspection, $G(t|t') \neq G(t'|t)$. Rather, we have

$$G(t|t') = G(-t'|-t) \quad (4.6.10)$$

which may be compared with the relation satisfied by $G(x|x')$ in (4.5.11). A relation of this sort is actually to be expected since $G(t|t')$ is the response at a *later* time t to an impulsive source acting at the *earlier* time t'. Note that if $t > t'$, then so also is $-t' > -t$, and the equality expressed by (4.6.10) is not unexpected.

An alternate way of introducing a function that can be combined with the oscillator equation in the *unprimed* coordinate system is to consider a function $G^a(t|t_1)$ that satisfies

$$\frac{d^2G^a}{dt^2} - 2\gamma \frac{dG^a}{dt} + \omega^2 G^a = \delta(t - t_1) \tag{4.6.11}$$

that is, the equation for an oscillator with negative damping. We can understand the significance of $G^a(t|t_1)$ if we combine this equation with (4.6.2), the equation defining $G(t|t')$. Writing t_2 for t', we have

$$\frac{d^2G}{dt^2} + 2\gamma \frac{dG}{dt} + \omega^2 G = \delta(t - t_2) \tag{4.6.12}$$

Combining the equations for G and G^a by the multiply-and-subtract procedure and then integrating over all time, we have

$$\left[G(t|t_2) \frac{dG^a(t|t_1)}{dt} - G(t|t_1) \frac{dG^a(t|t_2)}{dt} - 2\gamma G(t|t_1)G^a(t|t_2) \right]\Bigg|_{t=-\infty}^{t=+\infty}$$

$$= G(t_1|t_2) - G^a(t_2|t_1) \tag{4.6.13}$$

Because of the step function in $G(t|t_2)$ [cf. (4.6.5)], we must set $G(-\infty|t_2) = 0$ in (4.6.13). Note that we will obtain the simple reciprocity relation $G(t_1|t_2) = G^a(t_2|t_1)$ from (4.6.13) if we now impose a *final* condition on $G^a(t_2|t_1)$ by requiring $G^a(\infty|t_1) = 0$. This condition implies that G^a is zero for times *later* than the source time. With this requirement imposed upon G^a, the left-hand side of (4.6.13) vanishes and we obtain (setting $t = t_1$, $t' = t_2$)

$$G(t|t') = G^a(t'|t) \tag{4.6.14}$$

which may be compared with (4.5.13). Thus G^a represents the motion of the original system as it evolves *backward in time*.

Since we have already seen in (4.6.10) that $G(t|t') = G(-t'|-t)$, we also have, from (4.6.14),

$$G^a(t'|t) = G(-t'| - t) \tag{4.6.15}$$

The function G^a is known as the adjoint of the function G. If G and G^a satisfy the same equation and have the same boundary (initial/final) conditions, the problem is said to be *self-adjoint*. Note that if there were no damping in the original equation for the oscillator ($\gamma = 0$), then G and G^a would satisfy the same equation but the problem would still be non-self-adjoint because of the boundary (initial) conditions.

4.6.1 A Generalization

The multiply-and-subtract procedure has been found to be effective when the resulting expression leads to a perfect derivative and an example of the need for an adjoint equation in obtaining this perfect derivative has just been encountered. We now consider the form of the adjoint operator that is associated with a more general second order differential operator. In particular, consider the operator L defined by

$$Lu(x) = \left[f(x) \frac{d^2}{dx^2} + g(x) \frac{d}{dx} + h(x) \right] u(x) \qquad (4.6.16)$$

If we multiply this expression by another function $\underline{u}(x)$ and integrate over x, we obtain an integral that can be recast through integration by parts into the form

$$\int dx\, \underline{u}\, Lu = \int dx\, u \left[\frac{d^2}{dx^2}\, (f\underline{u}) - \frac{d}{dx}\, (g\underline{u}) + h\underline{u} \right] \qquad (4.6.17)$$

$$+ \underline{u} f \frac{du}{dx} - u \frac{d}{dx}\, (f\underline{u}) + u g \underline{u}$$

On defining an operator L^a according to

$$L^a \underline{u} = \frac{d^2}{dx^2}\, (f\underline{u}) - \frac{d}{dx}\, (g\underline{u}) + h\underline{u} \qquad (4.6.18)$$

we can rewrite (4.6.17) in the form

$$\int dx\, (\underline{u}\, Lu - u L^a \underline{u}) = P(x) \qquad (4.6.19)$$

where

$$P(x) = f\underline{u}\, \frac{du}{dx} - u \frac{d}{dx}\, (f\underline{u}) + \underline{u}\, g u \qquad (4.6.20)$$

The quantity P is referred to either as the bilinear concomitant or the conjunct of the functions u and $\underline{u}$. The derivative of (4.6.19) now yields a relation between L and L^a in the form of a perfect derivative, since we have

$$\underline{u} L u - u L^a \underline{u} = \frac{dP}{dx} \qquad (4.6.21)$$

If L and L^a are the same operator, then L is referred to as being formally self-adjoint. The subsequent use of L and L^a entails a choice of boundary conditions such that P vanishes at the boundaries. If u and $\underline{u}$ both satisfy the same boundary conditions, then L is said to be self-adjoint.

With x replaced by t and the choices $f = 1$, $g = 2\gamma$, and $h = \omega^2$, we obtain the adjoint operator G^a that is used in (4.6.11).

Problems

4.6.1 (a) Express the solution of

$$\frac{dy}{dt} + \alpha y = f(t), \qquad t \geq 0, \qquad y(0) = a$$

where α is a constant, in terms of an appropriate Green's function.

(b) Determine the Green's function.

4.6.2 Solve the equation for a forced oscillator, namely

$$\frac{d^2 y}{dt^2} + \omega^2 y = f(t), \qquad y(0) = y_0, \qquad y'(0) = v_0$$

by using the Green's function that satisfies

$$\frac{d^2 G}{dt^2} = \delta(t - t'), \qquad G(0|t') = G'(0|t') = 0$$

(a) Show that $G(t|t') = (t - t')H(t - t')$ where H refers to the unit step function.

(b) Obtain the integral equation

$$y(t) = y_0 + v_0 t - \omega^2 \int_0^t dt'(t - t')y(t') + \int_0^t dt' f(t')(t - t')$$

(c) Solve this integral equation by Laplace transform methods and obtain the expected result

$$y(t) = y_0 \cos \omega_0 t + \frac{v_0}{\omega_0} \sin \omega_0 t + \frac{1}{\omega_0} \int_0^t dt' f(t')\sin \omega_0 (t - t')$$

4.7 ONE-DIMENSIONAL DIFFUSION AND WAVE MOTION

The techniques introduced thus far in this chapter for solving ordinary differential equations in terms of Green's functions are readily applied to partial differential equations. We now examine both the diffusion equation and the wave equation in one space dimension.

4.7.1 Diffusion Equation

Consider the one-dimensional diffusion equation

$$\frac{\partial^2 u}{\partial x^2} - \frac{1}{\gamma^2}\frac{\partial u}{\partial t} = -\frac{1}{K}s(x,\, t), \qquad a \le x \le b, \qquad t \ge 0 \qquad (4.7.1)$$

with the initial condition $u(x,\, 0)$ and boundary conditions $u(a,\, t)$ and $u(b,\, t)$ specified. The solution will now be expressed in terms of a Green's function $g(x,\, t\,|\,x',\, t')$ that satisfies

$$\frac{\partial^2 g}{\partial x^2} - \frac{1}{\gamma^2}\frac{\partial g}{\partial t} = -\delta(x - x')\delta(t - t') \qquad (4.7.2)$$

As seen in the previous section, we will ultimately use the adjoint of this equation. The solution of (4.7.2) represents the temperature distribution on an infinite beam due to a unit heat source applied at $x = x'$ and at time $t = t'$. We impose no boundary conditions on g at this stage of the solution (other than the obvious one that the temperature approaches zero as $x \to +\infty$). We shall thus obtain only the "free-space" Green's function for the one-dimensional diffusion equation. Later we consider modifications due to boundary conditions.

The solution of the equation for g is readily obtained by the Laplace transform method. Indeed, the relevant calculation has already been carried out in Section 2.4. Defining

$$G(x,\, s\,|\,x',\, t') = \int_0^\infty e^{-st}g(x,\, t\,|\,x',\, t')\, dt \qquad (4.7.3)$$

and introducing the initial conditions $g(x,\, 0\,|\,x',\, t') = g_t(x,\, 0\,|\,x',\, t') = 0$, we obtain the transformed equation

$$\frac{\partial^2 G}{\partial x^2} - \frac{s}{\gamma^2}\, G = -e^{-st'}\delta(x - x') \qquad (4.7.4)$$

For $x \ne x'$ the equation has solutions proportional to $e^{\pm x\sqrt{s}/\gamma}$. We can now obtain the appropriate solution for G by using the general method for constructing Green's functions that was developed in Section 4.2. If we assume that $\sqrt{s} > 0$, we obtain convergent solutions as $|x| \to \infty$ by choosing $u_1(x)$ $= e^{x\sqrt{s}/\gamma}$ and $u_2(x) = e^{-x\sqrt{s}/\gamma}$, which converge as x approaches minus and plus infinity, respectively. The Wronskian is $W = -2\sqrt{s}/\gamma$, and since $p = 1$ and $J = e^{-st'}$, the general expression for the Green's function given by (4.2.10) leads to

$$G = \frac{\gamma}{2\sqrt{s}} e^{-\sqrt{s}|x-x'|/\gamma - st'} \tag{4.7.5}$$

The inverse Laplace transform of this result was shown in (2.4.10) to be

$$g(x, t|x', t') = \frac{\gamma e^{-|x-x'|^2/4\gamma^2(t-t')}}{2\sqrt{\pi(t-t')}} H(t - t') \tag{4.7.6}$$

We now consider the solution of the diffusion equation (4.7.1) by combining it with the Green's function just obtained. We first transform to the primed coordinates. Since $\partial g/\partial t' = -\partial g/\partial t$, we are actually combining the diffusion equation with the adjoint equation for g. We have

$$\frac{\partial^2 u}{\partial x'^2} - \frac{1}{\gamma^2}\frac{\partial u}{\partial t'} = -\frac{1}{K}s(x', t')$$

$$\frac{\partial^2 g}{\partial x'^2} + \frac{1}{\gamma^2}\frac{\partial g}{\partial t'} = -\delta(x - x')\delta(t - t') \tag{4.7.7}$$

Multiplying and subtracting as usual, but now integrating over both the length of the system ($a \leq x \leq b$) and over all time from $t' = 0$ to $t' = t + \epsilon$, as was done in obtaining (4.6.8) when we considered the damped oscillator, we have

$$\int_0^{t^+} dt' \int_a^b dx' \left[\frac{\partial}{\partial x'}\left(g\frac{\partial u}{\partial x'} - u\frac{\partial g}{\partial x'}\right) - \frac{1}{\gamma^2}\frac{\partial}{\partial t'}(gu) \right]$$

$$= -\frac{1}{K}\int_0^{t^+} dt' \int_a^b dx'\, s(x', t)g(x, t|x', t') + u(x, t) \tag{4.7.8}$$

where the term $\partial g/\partial x' \cdot \partial u/\partial x'$ has been added and subtracted to obtain the perfect spatial derivative in the first integral. Carrying out the integrations for which there is either a perfect space or perfect time derivative, we obtain

$$u(x, t) = \int_0^{t^+} dt' \left(g\frac{\partial u}{\partial x'} - u\frac{\partial g}{\partial x'}\right)\Bigg|_{x'=a}^{x'=b} - \frac{1}{\gamma^2}\int_a^b dx'\,(gu)\Bigg|_{t'=0}^{t'=t+\epsilon}$$

$$+ \frac{1}{K}\int_0^{t^+} dt' \int_a^b dx'\, s(x', t')g(x, t|x', t') \tag{4.7.9}$$

Note that the integral over t' contains the boundary conditions while the integral over x' contains the initial conditions. Because of the step function in g in (4.7.6), the value of g at the upper limit $t' = t + \epsilon$ is actually zero, that is, $g(x, t|x', t + \epsilon) = 0$.

The function g determined earlier in the section was the free-space Green's function. As in the case for ordinary differential equations, the solution of a problem may be simplified if g is also required to satisfy boundary conditions. Two examples of the use of the general result (4.7.9) will now be considered.

Example 4.4. A semi-infinite beam is initially at zero temperature. Beginning at some time $t = t_0$ a heat flux $J(t) = -Ku_x(0, t)$ is specified at the end located at $x = 0$. Obtain the temperature on the beam at any later time by using the solution (4.7.9).

We set $a = 0$ and $b = \infty$ in (4.7.9). Since $u(x, 0)$ is zero, there is no contribution from the single integral over space. Since the derivative $u_x(0, t)$ is specified $[= -K^{-1}J(t)]$, we shall eliminate the unknown boundary condition $u(0, t)$ by imposing the boundary condition on the Green's function that $\partial g/\partial x = 0$. This is readily accomplished by using the method of images. We write

$$g(x, t|x', t') = \frac{\gamma H(t - t')}{2\sqrt{\pi(t - t')}} (e^{-|x - x'|^2/4\gamma^2(t - t')} + e^{-(x + x')^2/4\gamma^2(t - t')}) \quad (4.7.10)$$

Then, since the source term $s(x, t)$ is also zero, the solution for $u(x, t)$ in (4.7.9) reduces to

$$u(x, t) = -\int_0^{t^+} u_x(0, t')g(x, t|0, t') \, dt' \quad (4.7.11)$$

Writing $J(t) = f(t)H(t - t_0)$, we have $u_x(0, t) = -K^{-1}f(t)H(t - t_0)$. Thus we finally obtain

$$u(x, t) = \frac{1}{K} \int_{t_0}^{t^+} f(t')g(x, t|0, t') \, dt'$$

$$= \frac{\gamma}{K\sqrt{\pi}} \int_{t_0}^{t} f(t') \frac{e^{-x^2/4\gamma^2(t - t')}}{\sqrt{t - t'}} \, dt' \quad (4.7.12)$$

As noted in Appendix B [cf. Eq. (B.36)] the long-time behavior of such an expression can be obtained by letting t_0 approach $-\infty$. In particular, consider $f(t) = \cos \Omega t = \text{Re}(e^{-i\Omega t})$. For later convenience (i.e., to provide convergence for an integral that arises subsequently) we shall consider this oscillation to be turned on slowly from $t = -\infty$ up to the present time and write $f(t) = e^{\epsilon t} \text{Re}(e^{-i\Omega t})$. Then, replacing t_0 by $-\infty$ and setting $t - t' = \tau$, we obtain

$$u(x, t) = \frac{\gamma e^{\epsilon t - i\Omega t}}{\sqrt{\pi K}} \int_0^{\infty} \tau^{-1/2} e^{-(\epsilon - i\Omega)\tau - x^2/4\gamma^2\tau} \, d\tau \quad (4.7.13)$$

The integral may be interpreted as a Laplace transform (with $s = \epsilon - i\Omega$) and evaluated by using entry 8 in Table F.2. We obtain

$$u(x, t) = \frac{\gamma e^{-i\Omega t}}{\sqrt{\pi K}} \sqrt{\frac{i\pi}{\Omega}} e^{-x/\gamma\sqrt{-i\Omega}} \tag{4.7.14}$$

where ϵ has been set equal to zero after the calculation has been performed. Taking the real part of this result we have

$$u(x, t) = \frac{\gamma}{\sqrt{\Omega K}} e^{-x\sqrt{\Omega/2}/\gamma} \cos\left(\Omega t - \frac{x\sqrt{\Omega/2}}{\gamma} - \frac{\pi}{4}\right) \tag{4.7.15}$$

Note that the temperature profile is that of a highly damped wave that propagates into the beam. This result was obtained in a simpler way in Problem 1.4.9.

Example 4.5. Determine the steady state heating of an infinite beam due to a heat source moving with a constant velocity.

The equation to be solved is

$$\frac{\partial^2 u}{\partial x^2} - \frac{1}{\gamma^2} \frac{\partial u}{\partial t} = -\frac{Q_0}{K} \delta(x - Vt), \qquad \infty < x < \infty \tag{4.7.16}$$

where Q_0 (in calories per second) is the strength of the source that moves with the constant velocity V. We now specialize the general solution (4.7.9) to apply to this problem. We set $a = -\infty$, $b = +\infty$ and assume that u and u_x both vanish as $|x| \to \infty$. Also, if $u(x, 0) = 0$, the spatial integral over the initial condition is zero. Finally, setting the lower limit of the time integral over the source equal to $-\infty$ in order to obtain the steady state solution, we have

$$u(x, t) = \frac{Q_0}{K} \int_{-\infty}^{t} dt' \int_{-\infty}^{\infty} dx' \, \delta(x' - Vt') g(x, t|x', t') \tag{4.7.17}$$

with g given by (4.7.6). Carrying out the x' integration and setting $t - t' = \tau$, we obtain

$$u(x, t) = \frac{\gamma Q_0}{2\sqrt{\pi K}} \int_0^{\infty} d\tau \, \tau^{-1/2} e^{-\alpha(\tau - \underline{t})^2/\tau} \tag{4.7.18}$$

where $\alpha = (V/2\gamma)^2$ and $\underline{t} = t - x/V$. This result may be evaluated by writing it as

$$u(x, t) = \frac{\gamma Q_0}{2\sqrt{\pi K}} \int_0^{\infty} \tau^{-1/2} e^{-\alpha(\tau^{1/2} - \underline{t}\tau^{-1/2})^2} d\tau$$

$$= \frac{\gamma Q_0}{\sqrt{\pi K}} e^{2\alpha\underline{t}} \int_0^{\infty} e^{-(\alpha\theta^2 + \alpha\underline{t}^2/\theta^2)} d\theta \tag{4.7.19}$$

where we have set $\tau = \vartheta^2$ and expanded the perfect square in the exponent. Using the result

$$\int_0^{\infty} e^{-(\alpha x^2 + b/x^2)} dx = \frac{1}{2} \sqrt{\frac{\pi}{a}} e^{-2\sqrt{ab}}, \qquad a, b > 0 \tag{4.7.20}$$

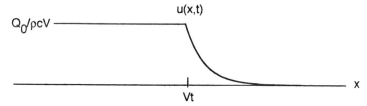

Figure 4.3. Temperature distribution due to moving heat source.

obtained in Example 7 of Appendix B we find

$$u(x, t) = \begin{cases} Q_0/\rho cV, & x < Vt \\ (Q_0/\rho cV)e^{-V(x - Vt)/\gamma^2}, & x > Vt \end{cases}$$ (4.7.21)

which is shown in Figure 4.3. Note that at $x = Vt + \epsilon$ the heat flow in the positive direction is $-Ku_x(Vt, t) = Q_0$, which is just the heat flow into the system.

4.7.2 Wave Equation

The Green's function for the one-dimensional time dependent wave equation may be obtained by a similar procedure. We determine the solution of

$$\frac{\partial^2 g}{\partial x^2} - \frac{1}{c^2}\frac{\partial^2 g}{\partial t^2} = -\delta(x - x')\delta(t - t')$$ (4.7.22)

Taking the Laplace transform, we find

$$\frac{d^2 G}{dx^2} - \left(\frac{s}{c}\right)^2 G = -e^{-st'}\delta(x - x')$$ (4.7.23)

We now solve this equation by the general method for constructing a Green's function that was developed in Section 4.2. The solutions of the homogeneous equation that converge as $|x| \to \infty$ are $u_1(x) = e^{sx/c}$ for $x < 0$ and $u_2(x) = e^{-sx/c}$ for $x > 0$. The corresponding Wronskian is $W(u_1, u_2) = -2s/c$ and since $p = 1$ and $J = e^{-st'}$, (4.2.10) yields

$$G = \frac{c}{2s} e^{-st'} e^{-s|x - x'|/c}$$ (4.7.24)

The inverse transform of this result is

$$g(x, t|x', t') = \frac{c}{2} H\left(t - t' - \frac{|x - x'|}{c}\right)$$ (4.7.25)

where H is the unit step function. The solution g corresponds to a step displacement of amplitude $c/2$ that propagates with velocity c in opposite direction away from the source point x'. Note that $\partial g/\partial t = -\partial g/\partial t'$ and $\partial g/\partial x = -\partial g/\partial x'$. Since the equation for g has only second derivatives, g satisfies the same equation in both primed and unprimed coordinates.

When the wave equation

$$\frac{\partial^2 u}{\partial x'^2} - \frac{1}{c^2}\frac{\partial^2 u}{\partial t'^2} = -\frac{1}{T}\mathfrak{F}(x', t'), \qquad a \leq x \leq b \qquad (4.7.26)$$

is combined with the equation for g (both equations in the primed coordinates) by the multiply-and-subtract procedure, we obtain

$$u(x, t) = \frac{1}{T}\int_0^{t^+} dt' \int_a^b dx' \, \mathfrak{F}(x', t')g(x, t|x', t')$$

$$+ \int_0^{t^+} dt' \, (gu_{x'} - ug_{x'}) \Big|_{x'=a}^{x'=b} + \frac{1}{c^2}\int_a^b dx' \, (gu_{t'} - ug_{t'}) \Big|_{t'=0}$$

$$(4.7.27)$$

As discussed in regard to the corresponding result for the diffusion equation in (4.7.9), we have used the fact that $g(x, t|x', t + \epsilon) = 0$.

Boundary conditions may be imposed upon g to simplify the result in (4.7.27) as was done for the diffusion equation. We now consider an example.

Example 4.6. The end of a semi-infinite string ($x > 0$) is given a prescribed displacement $u(0, t) = f(t)$. Determine the motion of the string.

We assume that the string is initially quiescent so that $u(x, 0) = u_t(x, 0) = 0$ and that there is no external force so that $\mathfrak{F}(x, t) = 0$. Then the solution (4.7.27) reduces to

$$u(x, t) = \int_0^{t^+} dt' \, [f(t')g_{x'} - u_{x'}g] \Big|_{x'=0} \qquad (4.7.28)$$

Since the disturbance propagates at the velocity c, we assume that no disturbance has reached $x = \infty$. Thus we have set u and $u_{x'}$, equal to zero at the upper limit for x'. Since $u_{x'}(0, t)$ is unknown, we require that g vanish at $x' = 0$ in order to eliminate this unknown under the integral sign in (4.7.28). The vanishing of g at $x' = 0$ is readily accomplished by employing the method of images. We use

$$g(x, t|x', t') = \frac{c}{2}\left[H\left(t - t' - \frac{|x - x'|}{c}\right) - H\left(t - t' - \frac{x + x'}{c}\right)\right] \qquad (4.7.29)$$

In calculating $g_{x'}(x, t|0, t')$ for use in (4.7.28), the magnitude signs are removed in g by noting that at the boundary, x' equals zero so that $x > x'$ and thus $|x - x'| = x - x'$. The derivative of a step function is a delta function and so we obtain

$$g_{x'}(x, t|0, t') = \delta(t - t' - x/c) \qquad (4.7.30)$$

The solution $u(x, t)$ is now seen, from (4.7.28), to be

$$u(x, t) = f(t - x/c)H(t - x/c) \tag{4.7.31}$$

as obtained in (2.1.7).

Problems

4.7.1 (a) An infinite string has initial displacement $y(x, 0) = \varphi(x)$ and initial velocity $y(x, 0) = \psi(x)$. Show that the subsequent motion of the string (subject to no external forces) is given by

$$y(x, t) = \frac{1}{c^2} \int_{-\infty}^{\infty} dx' \, [g(x, t|x', 0)\psi(x') - \varphi(x')g_{t'}(x, t|x', 0)]$$

(b) Since the string is of infinite extent, the appropriate Green's function is the free-space form obtained in Eq. (4.7.25). Use this expression for g to obtain the d'Alembert solution (1.9.16).

4.7.2 (a) A semi-infinite string is initially at rest. The end at $x = 0$ is given a time dependent displacement $y(0, t) = f(t)$. Combine the wave equation with the equation for g given in (4.7.25) to show that

$$y(x, t) = \int_0^{t^+} dt' \, y(0, t') \left. \frac{\partial g(x, t|x', t')}{dx'} \right|_{x'=0}$$

(b) Evaluate the integral to obtain

$$y(x, t) = f(t - x/c)H(t - x/c)$$

4.7.3 *The Doppler Effect.* Beginning at $t = 0$ a source at a single frequency ω starts from the origin and moves in the positive direction with a constant velocity $V(<c)$ along an infinite string. The wave equation for the string is thus

$$\frac{\partial^2 y}{\partial x^2} - \frac{1}{c^2} \frac{\partial^2 y}{\partial t^2} = -\frac{F_0}{T} \delta(x - Vt)e^{-i\omega t}$$

where F_0 is the strength of the source.

One expects that the wave ahead of the source propagates in the positive direction with a frequency ω_+ that is greater than ω while the wave behind the source propagates in the negative direction with a frequency ω_- that is less than ω (cf. Fig. 4.4). We now obtain this result analytically.

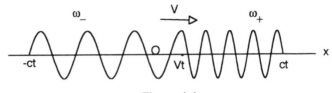

Figure 4.4.

(a) Show that the displacement of the string is given by

$$y(x, t) = \frac{1}{T} \int_0^{t^+} dt' \int_{-\infty}^{\infty} dx' \, \mathcal{F}(x', t') g(x, t | x', t')$$

where $\mathcal{F}(x, t) = F_0 \delta(x - Vt) e^{-i\omega t}$ and g is the one-dimensional free-space Green's function given in (4.7.25).

(b) Show that

$$\frac{2T}{F_0 c} y(x, t) = \int_0^{x/V} dt' \, e^{-i\omega t'} H\left(\frac{t - x/c}{1 - V/c} - t'\right)$$

$$+ \int_{x/V}^{t^+} dt' \, e^{-i\omega t'} H\left(\frac{t + x/c}{1 + V/c} - t'\right)$$

$$= I_< + I_>$$

Consider only the case $x > 0$ and note that $H(a - bt) = H(a/b - t)$.

(c) Evaluate the integrals in the three regions shown in Figure 4.5 by considering the interplay between the time intervals determined by the step functions and by the limits of integration.

(d) In region 1, show that $I_< = I_> = 0$.

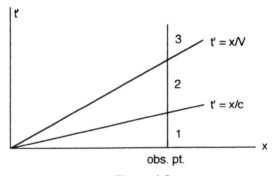

Figure 4.5.

(e) In region 2, show that $(t - x/c)/(1 - V/c) < x/V$, so

$$I_< = \int_0^{(t-x/c)/(1-V/c)} dt'\, e^{-i\omega t'}, \qquad I_> = 0$$

(f) In region 3, show that $(t + x/c)/(1 + V/c) < t$, so

$$I_< + I_> = \int_0^{(t+x/c)/(1+V/c)} dt'\, e^{-i\omega t'}$$

(g) Evaluate the integrals, take the real part, and obtain

$$y(x, t) = \frac{F_0 c}{2\omega T} \sin \omega_+ \left(t - \frac{x}{c} \right) \qquad \text{(region 2)}$$

$$y(x, t) = \frac{F_0 c}{2\omega T} \sin \omega_- \left(t + \frac{x}{c} \right) \qquad \text{(region 3)}$$

where $\omega_+ = \omega/(1 - V/c)$ and $\omega_- = \omega/(1 + V/c)$.

4.7.4 Waves on a string that is embedded in a material providing an elastic restoring force density $K(x)$ are governed by Eq. (1.8.9). The Green's function for this equation (assuming a constant tension T and no resistance) satisfies

$$g_{xx} - c^{-2}g_{tt} - \mu^2 g = -\delta(x - x')\delta(t - t'), \qquad \mu^2 = K(x)/T$$

(a) Obtain the Laplace transform of this equation and show that when $K(x)$ has the constant value K_0, the resulting ordinary differential equation has solutions

$$u_1(x) = e^{-st' + x\sqrt{(s/c)^2 + \mu_0^2}}, \qquad u_2(x) = e^{-st' - x\sqrt{(s/c)^2 + \mu_0^2}}$$

where $\mu_0^2 = K_0/T$.

(b) Use the general relation for constructing a Green's function given in (4.2.10) to show that

$$G(x, s | x', t') = \frac{e^{-st' - |x - x'|\sqrt{(s/c)^2 + \mu_0^2}}}{2\sqrt{(s/c)^2 + \mu_0^2}}$$

(c) Use a transform pair given in Table F.2 to obtain

$$g(x, t | x', t')$$

$$= \begin{cases} (c/2)\, J_0(\mu_0 c\sqrt{(t - t')^2 - |x - x'|^2/c^2}), & t - t' > |x - x'|/c \\ 0, & t - t' < |x - x'|/c \end{cases}$$

4.8   TWO AND THREE SPACE DIMENSIONS—
GREEN'S THEOREM

The techniques already developed in this chapter for problems having one space dimension are readily extended to accommodate problems in more space dimensions. The exact differential that was constructed in the one-dimensional problem will now be obtained by an application of Green's theorem. The multiply-and-subtract procedure used previously can also be viewed as a one-dimensional version of Green's theorem. We first summarize the notions that lead to Green's theorem.

In vector analysis one encounters the divergence theorem

$$\int_v \nabla \cdot \mathbf{V} \, dv = \int_S \mathbf{V} \cdot \mathbf{n} \, dS \qquad (4.8.1)$$

where S is a *closed* surface that bounds the volume v. The unit vector $\mathbf{n}$ points *outward* from the enclosed volume. We shall only consider vectors $\mathbf{V}$ that are smoothly varying and thus possess the derivatives required for the calculation of $\nabla \cdot \mathbf{V}$. We now consider the application of the divergence theorem to the two different vectors $\mathbf{V}_1 = p \, \nabla G$ and $\mathbf{V}_2 = G \, \nabla p$ where at present p and G are arbitrary differentiable functions. Then, since $\nabla \cdot \mathbf{V}_1 = \nabla \cdot (p \, \nabla G) = \nabla p \cdot \nabla G + p \, \nabla^2 G$ and similarly for $\mathbf{V}_2 = G \, \nabla p$, the divergence theorem (4.8.1), when applied to $\mathbf{V}_1$ and $\mathbf{V}_2$, yields the two integral relations

$$\int_v (\nabla p \cdot \nabla G + p \, \nabla^2 G) \, dv = \int_S p \, \nabla G \cdot \mathbf{n} \, dS \qquad (4.8.2)$$

and

$$\int_v (\nabla G \cdot \nabla p + G \, \nabla^2 p) \, dv = \int_S G \, \nabla p \cdot \mathbf{n} \, dS \qquad (4.8.3)$$

On subtracting one result from the other, we obtain

$$\int_v (p \, \nabla^2 G - G \, \nabla^2 p) \, dv = \int_S (p \, \nabla G - G \, \nabla p) \cdot \mathbf{n} \, dS \qquad (4.8.4)$$

a result known as Green's theorem. It relates the volume integral of a certain quantity containing Laplacians of p and G to the value of another function of p and G that is evaluated on the bounding surface. The Laplacian enters in many partial differential equations and Green's theorem is found to provide a way of expressing the solution of these partial differential equations in terms of values of the solution on the bounding surface. It should be expected that some such relation would exist since one frequently encounters situations in which a field quantity is determined *throughout* a region by information specified only on the *surface* of the region (such as the sound field produced by a loudspeaker located on the wall of a room).

In these applications, one of the functions, say p, is the quantity of physical interest while the other function, G, is the Green's function. It is chosen to satisfy boundary conditions in a way that simplifies the solution of a given problem as much as possible. The technique is, for the most part, an extension to higher dimensions of the methods used already in one dimension. As an example, consider the three-dimensional wave equation.

$$\nabla^2 p - \frac{1}{c^2}\frac{\partial^2 p}{\partial t^2} = -Q(\mathbf{r}, t) \tag{4.8.5}$$

If $Q(\mathbf{r}, t) = F(\mathbf{r})e^{-i\omega t}$, that is, a source at a single frequency ω, then we can expect, at least after an initial transient period, that the entire wave motion is also at this one frequency and write $p(\mathbf{r}, t) = p(\mathbf{r})e^{-i\omega t}$. We then obtain

$$\nabla^2 p + k^2 p = -F(\mathbf{r}), \quad k = \omega/c \tag{4.8.6}$$

This equation is to be satisfied within some volume v on which p satisfies some specified boundary conditions. Note that in the limit $\omega \to 0$ so that $k \to 0$ and with $F(\mathbf{r})$ interpreted as a heat source $-s(\mathbf{r})/K$, (4.8.6) would become Poisson's equation (3.1.2). The function p would then represent the steady state temperature distribution in v due to both the heat source $s(\mathbf{r})$ and the specified temperature or heat flow at the boundary. In general, p is interpreted as the function of physical interest while for G we use a solution of

$$\nabla^2 G + k^2 G = -\delta^3(\mathbf{r} - \mathbf{r}') \tag{4.8.7}$$

In the notation developed in Section 4.3, we thus set $G = G_k(\mathbf{r}|\mathbf{r}')$. The boundary conditions satisfied by G_k are determined by each particular problem. Then, multiplying (4.8.6) by G_k and (4.8.7) by p, subtracting, and integrating over the volume v, we obtain

$$\int_v (G_k \nabla^2 p - p\nabla^2 G_k)\, dv = -\int_v GF\, dv + \int_v dv\, \delta^3(\mathbf{r} - \mathbf{r}')p(\mathbf{r}') \tag{4.8.8}$$

We can now rewrite the left-hand side of this result as a surface integral by using Green's theorem. Using the reciprocity relation $G_k(\mathbf{r}|\mathbf{r}') = G_k(\mathbf{r}'|\mathbf{r})$ obtained below in (4.8.14), we can interchange primed and unprimed coordinates and obtain

$$\int_S (G_k \nabla' p - p \nabla' G_k) \cdot \mathbf{n}\, dS' + \int_v F(\mathbf{r}')G(\mathbf{r}|\mathbf{r}')\, dv'$$

$$= \begin{cases} p(r), & r \text{ inside } v \\ 0 & r \text{ outside } v \end{cases} \tag{4.8.9}$$

in direct analogy with the one-dimensional case (4.1.22). This result has a very natural physical interpretation. It specifies the field $p(\mathbf{r})$ in terms of the presumably known source $F(\mathbf{r})$ and the value of p and/or ∇p on the boundary S. If, for example, p were known on the boundary, then the unknown ∇p could be eliminated from (4.8.9) by constructing $G_k(\mathbf{r}|\mathbf{r}')$ [i.e., imposing boundary conditions on $G_k(\mathbf{r}|\mathbf{r}')$], so that it vanishes on S. The function $p(r)$ would then be expressed solely in terms of known quantities. Similarly, if ∇p were known on the boundary, a function $G_k(\mathbf{r}|\mathbf{r}')$ (with certain modifications for the case $k = 0$ that are discussed in Section 4.13) satisfying the homogeneous boundary condition $\nabla' G_k(\mathbf{r}|\mathbf{r}') = 0$ for $\mathbf{r}'$ on the boundary would again express the solution $p(r)$ in terms of known quantities. If p were known on part of the boundary and ∇p on another part, however, the problem, known as a mixed boundary value problem, would be more complicated. Such a problem would most likely entail an integral of an unknown function on part of the surface. We would then obtain an integral equation for p. An example of this situation will be considered in Section 9.3.

A source in free space, that is in the absence of all boundaries, may be considered to be enclosed by a surface at infinity. In Section 4.9, the entire contribution from the surface integral will be shown to vanish in this case and the solution to $\nabla^2 p + k^2 p = -F$ given in (4.8.9) then reduces to just the volume integral. We then have

$$p = \int_v F(\mathbf{r}')G_k(\mathbf{r}|\mathbf{r}')\, dv' \qquad (4.8.10)$$

where v is the volume of the source $F(\mathbf{r})$.

The reciprocity relation for $G_k(\mathbf{r}|\mathbf{r}')$ is readily developed as an extension of the one-dimensional case considered in Section 4.1. We consider the two equations

$$\nabla^2 G_k(\mathbf{r}|\mathbf{r}_1) + k^2 G_k(\mathbf{r}|\mathbf{r}_1) = -\delta^3(\mathbf{r} - \mathbf{r}_1)$$

$$\nabla^2 G_k(\mathbf{r}|\mathbf{r}_2) + k^2 G_k(\mathbf{r}|\mathbf{r}_2) = -\delta^3(\mathbf{r} - \mathbf{r}_2) \qquad (4.8.11)$$

By carrying out the multiply-and-subtract procedure and using Green's theorem, we obtain

$$\int_S [G_k(\mathbf{r}_s|\mathbf{r}_2)\nabla G_k(\mathbf{r}_s|\mathbf{r}_1) - G_k(\mathbf{r}_s|\mathbf{r}_1)\nabla G_k(\mathbf{r}_s|\mathbf{r}_2)] \cdot \mathbf{n}\, dS$$

$$= G_k(\mathbf{r}_2|r_1) - G_k(\mathbf{r}_1|r_2) \qquad (4.8.12)$$

Here $\mathbf{r}_s$ refers to the evaluation of $\mathbf{r}$ on the boundary S.

As long as $G_k(\mathbf{r}|\mathbf{r}')$ satisfies a homogeneous boundary condition of the general form

$$f(\mathbf{r}_s)G_k(\mathbf{r}_s|\mathbf{r}) + g(\mathbf{r}_s)\nabla G_k(\mathbf{r}_s|\mathbf{r}) = 0 \qquad (4.8.13)$$

the integrand of the surface integral will vanish identically and we obtain the reciprocity relation

$$G_k(\mathbf{r}_1|\mathbf{r}_2) = G_k(\mathbf{r}_2|\mathbf{r}_1) \tag{4.8.14}$$

4.9 GREEN'S FUNCTION IN FREE SPACE

We now obtain a particular solution of (4.8.7), the equation satisfied by $G_k(\mathbf{r}|\mathbf{r}')$. For $\mathbf{r} \neq \mathbf{r}'$, the equation is homogeneous. Since the solution has the significance of being the radiation away from a point source located at $\mathbf{r}'$, we expect the solution to be spherically symmetric about $\mathbf{r}'$. Thus, on setting $R = |\mathbf{r} - \mathbf{r}'|$, we expect $G_k(\mathbf{r}|\mathbf{r}')$ to be of the form $G_k(|\mathbf{r} - \mathbf{r}'|) = G_k(R)$. The proper form of the three-dimensional Laplacian, when expressed in terms of R, thus contains only a radial part, and from (3.5.16) we have

$$\frac{1}{R^2}\frac{d}{dR}\left(R^2\frac{dG_k}{dR}\right) + k^2 G_k = 0, \qquad R \neq 0 \tag{4.9.1}$$

or equivalently, by using Eq. (3.5.17).

$$\frac{d^2}{dR^2}(RG_k) + k^2(RG_k) = 0, \qquad R \neq 0 \tag{4.9.2}$$

In this latter form, the solution is recognized to be $RG_k = Ae^{ikR} + Be^{-ikR}$. To have only outgoing waves (i.e., a function of $R - ct$ in conjunction with $e^{-i\omega t}$ time dependence), we must set $B = 0$ and obtain

$$G_k(R) = A\frac{e^{ikR}}{R} \tag{4.9.3}$$

where $R^2 = (x - x')^2 + (y - y')^2 + (z - z')^2$. The constant A is determined by the strength of the source in (4.8.7), the equation satisfied by G_k. If we integrate that equation over a small sphere of radius ϵ about $R = 0$, we have

$$\int_v \nabla \cdot \nabla G_k \, dv + k^2 \int_v G_k \, dv = -\int_v \delta^3(\mathbf{R}) \, dv = -1 \tag{4.9.4}$$

In a spherical coordinate system centered at $R = 0$ we take $dv = R^2 \, dR \, d\Omega$ where $d\Omega = \sin\theta \, d\theta \, d\varphi$. Note that the second integral in (4.9.4) vanishes as $\epsilon \to 0$, i.e.,

$$k^2 \int_v G_k \, dv = k^2 \int_{R \le \epsilon} A\frac{e^{ikR}}{R} R^2 \, dR \, d\Omega \cong 4\pi Ak^2 \int_0^\epsilon R \, dR \to 0 \tag{4.9.5}$$

where we have replaced e^{ikR} by 1 since $k\epsilon \ll 1$. For the first integral in (4.9.4) we use the divergence theorem, which reduces (4.9.4) to the form

$$\int_{R=\epsilon} \nabla G_k \, \epsilon^2 \, d\Omega = -1 \qquad (4.9.6)$$

Since $\nabla G = -Ae^{ikR}(1 + ikR)/R^2 \approx -A/\epsilon^2$ for $k\epsilon \ll 1$, we obtain

$$\frac{A}{\epsilon^2} \int_{R=\epsilon} \epsilon^2 \, d\Omega = -4\pi A = -1 \qquad (4.9.7)$$

so that $A = \frac{1}{4}\pi$ and we have the solution

$$G_k(\mathbf{r}|\mathbf{r}') = \frac{e^{ik|\mathbf{r}-\mathbf{r}'|}}{4\pi|\mathbf{r} - \mathbf{r}'|} \qquad (4.9.8)$$

Note that the reciprocity relation $G_k(\mathbf{r}|\mathbf{r}') = G_k(\mathbf{r}'|\mathbf{r})$ obtained in (4.8.14) is displayed by this result.

The equation satisfied by the free-space Green's function for Poisson's equation is immediately obtained by setting $k = 0$ in (4.8.7). We then have

$$\nabla^2 G_0(\mathbf{r}|\mathbf{r}') = -\delta^3(\mathbf{r} - \mathbf{r}') \qquad (4.9.9)$$

and from the corresponding limit in the solution (4.9.8), we obtain

$$G_0(\mathbf{r}|\mathbf{r}') = \frac{1}{4\pi|\mathbf{r} - \mathbf{r}'|} = \frac{1}{4\pi\sqrt{(x - x')^2 + (y - y')^2 + (z - z')^2}} \qquad (4.9.10)$$

As will be seen in Section 4.10, this passage to the zero-frequency limit is not as immediate for two-dimensional problems or, as shown in Sections 4.13 and 9.5, when Neumann conditions are imposed on the boundary.

4.9.1 Boundary Condition at Infinity

To apply Green's theorem to a region of infinite extent, such as the half space above the plane $z = 0$, we first consider a hemispherical region of finite radius and then examine the value of the surface integral on the hemispherical surface in the limit that the radius of this surface increases indefinitely. As an example, consider the application of Green's theorem to the half space $0 \le \rho = \sqrt{x^2 + y^2} < \infty$, $z \ge 0$. We first apply Green's theorem to the hemispherical region of radius R shown in Figure 4.6. Using (4.8.9) with $F = 0$ we have

$$u = -\int_{\text{plane}} \left(G \frac{\partial u}{\partial z'} - r \frac{\partial G}{\partial z'} \right) dS' + \int_{S_R} \left(G \frac{\partial u}{\partial R'} - u \frac{\partial G}{\partial R'} \right) dS' \qquad (4.9.11)$$

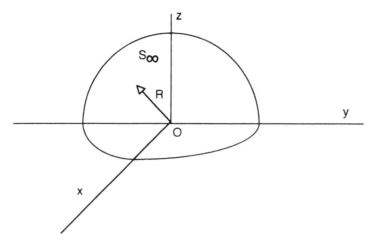

Figure 4.6. Two octants of hemisphere for applying Green's theorem.

The minus sign before the first integral arises because on the plane $z = 0$ the outward normal from the upper hemisphere is in the $-\mathbf{k}$ direction.

Since there is no source at infinity, we expect the contribution from the surface integral over S_R to vanish. We now formulate conditions required of the solution that are sufficient to have this take place. The condition is somewhat different for static problems (the Laplace or Poisson equations) than for wave problems (the Helmholtz equation). We begin with the static problem.

We are interested in the value of $p = \int (G \nabla p - p \nabla G) \cdot \mathbf{n} \, dS$ on the surface S_R shown in Figure 4.6. For the static problem we use $G = 1/4\pi R$ and thus have

$$4\pi p = \int_S \left[\frac{1}{R} \frac{\partial p}{\partial R} - p \frac{\partial}{\partial R} \left(\frac{1}{R} \right) \right] R^2 \, d\Omega \qquad (4.9.12)$$

in which the surface element dS has been written as $R^2 d\Omega$. This expression is equivalent to

$$4\pi p = \int_S \left(R \frac{\partial p}{\partial R} + p \right) d\Omega \qquad (4.9.13)$$

Note that the integrand will vanish if we require

$$\lim_{R \to \infty} \left(R \frac{\partial p}{\partial R} + p \right) = 0 \qquad (4.9.14)$$

This condition is certainly satisfied if p decreases at least as rapidly as $1/R$ for large R. As will be seen later in Section 5.2, this is the case whenever p is evaluated at distances from a source that are large compared with its size.

For wave problems there is a disturbance that propagates to infinity and the requirement is somewhat different. Now we use $G = e^{ikR}/4\pi R$ and a similar calculation yields

$$4\pi p = \int_S \left[\frac{e^{ikR}}{R} \frac{\partial p}{\partial R} - p \frac{\partial}{\partial R} \left(\frac{e^{ikR}}{R} \right) \right] R^2 \, d\Omega$$

$$= \int_S e^{ikR} R \left(\frac{\partial p}{\partial R} - ikp \right) d\Omega + \int_S e^{ikR} p \, d\Omega \qquad (4.9.15)$$

Since $|p|$ goes to zero as R^{-1} as R goes to infinity, the last integral vanishes in this limit. From the first integral we obtain the requirement

$$\lim_{R \to \infty} \left[R \left(\frac{\partial p}{R} - ikp \right) \right] = 0 \qquad (4.9.16)$$

if the surface integral at infinity is to vanish. Again, it will be found that this condition, frequently referred to as the Sommerfeld radiation condition, is satisfied for radiation from localized sources. For a problem in which there is no surface at all, that is, a localized source in free space, the surface S in (4.9.13) or (4.9.15) may be taken to be a complete sphere enclosing the source. Note that the radiation condition is satisfied if p is the point source expression given in (4.9.8).

As an example of the Green's function formulation of a problem, consider the temperature distribution in the half space above the plane $z = 0$ on which the heat flow, that is, the normal derivative u_z into the region $z \geq 0$, is specified. To express the temperature in the half space in terms of an integral over a Green's function, we use the multiply-and-subtract procedure and combine the steady state heat equation $\nabla^2 u = 0$ with a Green's function satisfying $\nabla^2 G = -\delta^3(\mathbf{r} - \mathbf{r}')$. This procedure yields

$$u(\mathbf{r}) = \int_{S_{XY} + S_\infty} dS' \, \mathbf{n}' \cdot (G \nabla' u - u \nabla' G) \qquad (4.9.17)$$

where S_{xy} and S_∞ together comprise a surface that encloses the half space $z > 0$ (cf. Fig. 4.6). In the present problem, the only heat source is on the surface S_{xy}. There is no volume source. Thus (4.9.17) for u could be obtained from (4.8.9) by setting $F = 0$ and using some solution of (4.9.9) for G.

For the Green's function, we introduce one that satisfies boundary conditions that are appropriate for this problem. The temperature in the half space is determined by the heat flux at $z = 0$. The contribution from S_∞ can be expected to vanish. It will be shown below that $u(\mathbf{r})$ decreases as R^{-1} away from a localized heat source at the boundary and thus condition (4.9.13) for the vanishing of the integral over S_∞ will be satisfied. Since the heat flux $J_z = -K$

$\nabla_z u$ is specified, the quantity $\nabla' u$ under the integral sign in (4.9.17) is known while u is unknown. We thus choose G to satisfy the boundary condition $\nabla' G = 0$ on $z' = 0$. If this can be accomplished, the temperature in the upper half space will be given by

$$u(\mathbf{r}) = \int_{S_{XY}} dS \; \mathbf{n}' \cdot G \nabla' u \qquad (4.9.18)$$

that is, in terms of an integral in which the entire integrand is known.

The Green's function with vanishing z' derivative is readily constructed by the method of images. Setting $\mathbf{r} = \mathbf{i}x + \mathbf{j}y + \mathbf{k}z$, $\mathbf{r}' = \mathbf{i}x' + \mathbf{j}y' + \mathbf{k}z'$, and $\mathbf{r}_i' = \mathbf{i}x' + \mathbf{j}y' - \mathbf{k}z'$, we have

$$G(\mathbf{r}|\mathbf{r}') = G_0(\mathbf{r}|\mathbf{r}') + G_0(\mathbf{r}|\mathbf{r}_i')$$

$$= \frac{1}{4\pi} \left[\frac{1}{\sqrt{(x-x')^2 + (y-y')^2 + (z-z')^2}} \right.$$

$$\left. + \frac{1}{\sqrt{(x-x')^2 + (y-y')^2 + (z+z')^2}} \right] \qquad (4.9.19)$$

A simple calculation of the z' derivative yields $\partial G/\partial z'|_{z'=0} = 0$ for this form of G. Note that the image term has a source at $x = 0$, $y = 0$, $z = -z'$, which is *outside* the upper half space where we are determining the temperature field. Thus the image term satisfies a *homogeneous* equation in the *upper* half space. The image term may be viewed as the appropriate solution of the homogeneous equation $\nabla^2 G = 0$ that must be added to $G_0(\mathbf{r}|\mathbf{r}')$ to satisfy the boundary conditions of this problem.

In the present problem it will be expedient to introduce cylindrical coordinates. We thus set $\mathbf{r} = \boldsymbol{\rho} + \mathbf{k}z$, $\mathbf{r}' = \boldsymbol{\rho}' + \mathbf{k}z'$, and $\mathbf{r}_i = \boldsymbol{\rho}' - \mathbf{k}z'$. Since $\mathbf{n}' = -\mathbf{k}'$, we have $\mathbf{n}' \cdot \nabla' u = -\partial u/\partial z'$ and $\mathbf{n}' \cdot \nabla' G = -\partial G/\partial z'$. The latter term, as noted above, vanishes on the surface $z' = 0$ and (4.9.18) thus reduces to

$$u(\boldsymbol{\rho}, z) = \int_{\substack{z'=0 \\ \text{plane}}} G(\boldsymbol{\rho}, z | \boldsymbol{\rho}', 0) \frac{\partial u}{\partial z'}\bigg|_{z'=0} dS' \qquad (4.9.20)$$

With $u_z(\boldsymbol{\rho}', 0) = -J(\boldsymbol{\rho}')/K$, and noting that on the surface $z' = 0$ the Green's function (4.9.19) reduces to

$$G(\boldsymbol{\rho}, z | \boldsymbol{\rho}, 0) = \frac{1}{2\pi} \frac{1}{\sqrt{|\boldsymbol{\rho} - \boldsymbol{\rho}'|^2 + z^2}} \qquad (4.9.21)$$

we obtain the temperature $u(\mathbf{\rho}, z)$ at any point in the upper half space in the form

$$u(\mathbf{\rho}, z) = \frac{1}{2\pi K} \int_{\substack{z'=0 \\ \text{plane}}} \frac{dS' J(\mathbf{\rho}')}{\sqrt{|\mathbf{\rho} - \mathbf{\rho}'|^2 + z^2}} \qquad (4.9.22)$$

In order to understand this result better, we consider a simple example in which the source has no angular dependence, i.e.,

$$J(\rho) = \begin{cases} F_0, & \rho < a \\ 0, & \rho > a \end{cases} \qquad (4.9.23)$$

Then, writing $\mathbf{\rho} = \rho(\mathbf{i}\cos\varphi + \mathbf{j}\sin\varphi)$ and $\mathbf{\rho}' = \rho'(\mathbf{i}\cos\varphi' + \mathbf{j}\sin\varphi')$, we have $|\mathbf{\rho} - \mathbf{\rho}'|^2 = \rho^2 + \rho'^2 - 2\rho\rho'\cos(\varphi - \varphi')$. Since the source is independent of φ, we expect this same symmetry in the field. This may be shown in various ways. We may integrate on φ' from $\varphi' = \varphi$ to $\varphi' = \varphi + 2\pi$. When the new integration variable $\alpha = \varphi' - \varphi$ is used in the angular integration in (4.9.22), and dS' is written as $\rho' \, d\rho' \, d\alpha$, we obtain

$$u(\rho, z) = \frac{F_0}{2\pi K} \int_0^a \rho' \, d\rho' \int_0^{2\pi} \frac{d\alpha}{\sqrt{\rho^2 + \rho'^2 - 2\rho\rho'\cos\alpha + z^2}} \qquad (4.9.24)$$

which is independent of φ. Another approach is to merely evaluate the field at $\varphi = 0$, in which case φ does not appear in the integrand at all, and then to recognize that due to symmetry, the result so obtained will hold for all φ.

We now consider the evaluation of the integral in (4.9.22) for various special cases. At fixed points far from the heated region, that is, for $R = \sqrt{\rho^2 + z^2} \gg a$, we may write

$$\frac{1}{\sqrt{R^2 + \rho'^2 - 2\rho\rho'\cos\alpha}} = \frac{1}{R}\left[1 + \left(\frac{\rho'}{R}\right)^2 - \frac{2\rho\rho'}{R^2}\cos\alpha\right]^{-1/2}$$

$$\cong \frac{1}{R}, \qquad R \gg \rho' \qquad (4.9.25)$$

since the largest value that ρ' can attain is $\rho' = a$. We then obtain

$$u(\rho, z) = \frac{F_0}{2\pi K} \int_0^a \rho' \, d\rho' \int_0^{2\pi} \frac{d\alpha}{R} = \frac{F_0 a^2}{2K} \frac{1}{R} \qquad (4.9.26)$$

It should be noted that this dependence upon R is in accordance with that required to be able to neglect the surface integral at infinity. From the temperature dependence obtained here in (4.9.26) we may determine the total amount of heat that flows out through a large hemispherical surface of radius

R. We have

$$H = \int \mathbf{J} \cdot \mathbf{n}\, dS = \int - K \frac{\partial u}{\partial R} R^2\, d\Omega$$

$$= -K \int_0^{2\pi} d\varphi \int_0^{\pi} R^2 \sin\theta\, d\theta \left(-\frac{F_0}{K} \frac{a^2}{2R^2} \right)$$

$$= \pi a^2 F_0 \tag{4.9.27}$$

As expected, this is the same amount of heat as that flowing in through the circular region of radius a on the plane $z = 0$.

It is also quite easy to determine the field on axis above the center of the circular source region. Here we set $\rho = 0$ in (4.9.24) and obtain

$$u(0, z) = \frac{F_0}{2\pi K} \int_0^a \rho'\, d\rho' \int_0^{2\pi} \frac{d\alpha}{\sqrt{\rho'^2 + z^2}}$$

$$= \frac{F_0}{K} (\sqrt{a^2 + z^2} - z) \tag{4.9.28}$$

At the surface we obtain $u(0, 0) = F_0 a/K$. For $z \gg a$ the radical may be expanded to yield $u(0, z) \approx F_0 a^2/2Kz$, which agrees with (4.9.26) when $\rho = 0$.

Problems

4.9.1 The plane $z = 0$ is insulated except for the circular region $r < a$ where there is a constant heat flow $J_0 = -Ku_z(\boldsymbol{\rho}, 0)$ into the upper half space $z > 0$.

   (a) Use Green's function methods to obtain an expression for the steady state temperature in the half space $z > 0$.

   (b) Show that on a hemispherical surface of radius $R \gg a$ the total heat flow outward is $J_0 \pi a^2$.

   (c) Show that $u(0, z) = J_0 \rho_0^2/2Kz$, $z \gg \rho_0$.

4.9.2 The steady heat flow into the region $z > 0$ is given by $J_z(\rho) = -Ku_z(\rho, 0) = J_0 \exp[-(\rho/\rho_0)^2]$.

   (a) Write the steady state temperature in the region $z > 0$ in terms of an integral over an appropriate Green's function and show that

$$u(\boldsymbol{\rho}, z) = \frac{J_0}{2\pi K} \int dS' \frac{e^{-(\rho'/\rho_0)^2}}{\sqrt{|\boldsymbol{\rho} - \boldsymbol{\rho}'|^2 + z^2}}$$

   (b) Simplify the integral for the case of the temperature along the z axis and show that for $z \gg \rho_0$ the result is

$$u(0, z) \approx \frac{J_0 \rho_0^2}{2Kz}$$

4.9.3 The temperature in the plane $z = 0$ is $u(\rho, 0) = u_0 e^{-(\rho/\rho_0)^2}$ and governed by $\nabla^2 u = 0$ for $z > 0$.

   (a) Express the temperature $u(\rho, z)$ in the half space $z > 0$ in terms of an integral over an appropriate Green's function.

   (b) Simplify the integral for the case $\rho = 0$, that is, the temperature on the z axis.

   (c) Evaluate this integral for z large (i.e., $z \gg \rho_0$) and obtain $u(0, z) \approx u_0 (\rho_0/z)^2$.

4.9.4 **(a)** A heated ring of radius a is embedded in a half space at a distance H above an insulated plane. Express the temperature field above the plane in terms of a Green's function.

   (b) Determine $u(0, 0)$, the temperature at the point on the plane that is directly below the center of the ring. The source term may be expressed as

$$s(\rho, z) = \frac{Q_0}{2\pi\rho} \delta(\rho - a)\delta(z - H)$$

4.9.5 The temperature in an upper half space $z \geq 0$, $-\infty < x, y < \infty$, is determined by a heat influx Q calories per square centimeter per second in the form of a heated ring located at the base of the half space. We thus set

$$J_z = -K \left.\frac{\partial u}{\partial z}\right|_{z=0} = \frac{Q}{2\pi a} \delta(\rho - a)$$

   (a) Show that the total heat input to the region is $\int_{\text{plane} z=0} J_z \, dS = Q$.

   (b) Show that the temperature in the upper half space is given by

$$u(\rho, z) = \int dS' \, g(\rho, z \mid \rho, 0) \left.\frac{\partial u}{\partial z'}\right|_{z'=0}$$

   where

$$g(\rho, z \mid \rho', z') = \frac{1}{4\pi} \left(\frac{1}{|\mathbf{r} - \mathbf{r}'|} + \frac{1}{|\mathbf{r} - \mathbf{r}'_i|} \right)$$

   and $\mathbf{r} = \boldsymbol{\rho} + \mathbf{k}z$, $\mathbf{r}' = \boldsymbol{\rho}' + \mathbf{k}z'$.

   (c) Show that

$$u(\rho, z) = \frac{Q}{4\pi^2 K} \int_0^{2\pi} \frac{d\varphi'}{\sqrt{r^2 + a^2 - 2\rho a \cos\varphi'}}$$

   where $r^2 = \rho^2 + z^2$.

(d) For $r \gg a$, show that

$$u(\rho, z) = \frac{Q}{2\pi K r}$$

(e) At large distances from the ring ($r \gg a$), the heat flow is radially outward from the ring and given by $J_r = -K \, \partial u / \partial r$. Show that

$$\int_{\substack{\text{upper} \\ \text{hemisphere}}} r^2 \, d\Omega J_r = Q$$

that is, all heat entering through the ring, as calculated in (a), leaves through a spherical surface at $r = \infty$.

4.9.6 A ring source of strength Q_0 and radius a heats an infinite medium. If the source is located in an XY plane, show that the cylindrically symmetric temperature distribution is given by

$$u(\rho, z) = \frac{Q_0}{8\pi^2 K} \int_0^{2\pi} \frac{d\varphi'}{\sqrt{\rho^2 + a^2 - 2a\rho \cos\varphi' + z^2}}$$

and that on axis the temperature is

$$u(0, z) = \frac{Q_0}{4\pi K} \frac{1}{\sqrt{a^2 + z^2}}$$

4.10 TWO-DIMENSIONAL PROBLEMS

Three-dimensional problems that have no spatial variation along one direction (say along a z coordinate) are readily envisioned. If all sources in the problem are line sources of infinite extent parallel to the z axis and all boundaries are surfaces of infinite extent that are also parallel to the z axis, then there will be no z dependence in the problem at all and we are dealing with a problem that is essentially two dimensional. The appropriate elementary source that provides the free space Green's function is the field about a *line* source parallel to the z axis. This solution will have neither z nor z' dependence. Green's theorem as developed previously for three-dimensional problems may be taken over directly to the two-dimensional situation by applying the divergence theorem to a cylindrical region parallel to the z axis. When this cylinder is closed at top and bottom, there is no contribution from surface integrals associated with these two surfaces since they contain terms involving $\partial\varphi / \partial z$ and $\partial G / \partial z$ that are zero due to the lack of z dependence in either φ or G.

For steady wave problems, that is, solutions of the scalar Helmholtz equation in two dimensions, the Green's function represents a circular (i.e., cylindrical) wave emanating from a source at some point $\boldsymbol{\rho}' = \mathbf{i}x' + \mathbf{j}y'$. Such a disturbance is expressed in terms of a Hankel function $H_0^{(1)}(kR)$ where $R = |\boldsymbol{\rho} - \boldsymbol{\rho}'|$ (cf. Appendix D). The appropriate amplitude for a source of unit strength is the solution of

$$\nabla^2 G_k + k^2 G_k = -\delta^2(\boldsymbol{\rho} - \boldsymbol{\rho}') \tag{4.10.1}$$

that represents outgoing waves. The amplitude may be determined by following a procedure analogous to that used in Section 4.8 for the three-dimensional case. Integration of (4.10.1) over a cylindrical region of radius ϵ about $\boldsymbol{\rho}'$ yields

$$\int_{R=\epsilon} d^2 S(\nabla^2 G_k + k^2 G_k) = -1 \tag{4.10.2}$$

Just as for the three-dimensional case considered in Section 4.8, the integral of G over the region of radius ϵ will vanish as ϵ goes to zero. Applying the divergence theorem to the first integral in (4.10.2), we obtain

$$\int \nabla G \bigg|_{R=\epsilon} \cdot \mathbf{a}_R \epsilon \, d\varphi = -1 \tag{4.10.3}$$

where φ is an angular coordinate on the cylindrical surface of radius ϵ. For small kR we have

$$H_0^{(1)}(kR) = J_0(kR) + i Y_0(kR)$$

$$\approx 1 + i\frac{2}{\pi} \ln kR \tag{4.10.4}$$

where we have used the approximate forms given in Eq. (D.26). Then, with $G = AH_0^{(1)}(kR)$, (4.10.4) yields the small kR approximation $\nabla G \cdot \mathbf{a}_R = \partial G/\partial R = -2i/\pi R$ and the integral in (4.10.3) reduces to

$$A \int_0^{2\pi} \frac{2i}{\pi R} R \, d\varphi = -1 \tag{4.10.5}$$

so that $A = i/4$. Thus, in two dimensions we have

$$G_k(\boldsymbol{\rho}|\boldsymbol{\rho}') = \frac{i}{4} H_0^{(1)}(k|\boldsymbol{\rho} - \boldsymbol{\rho}'|) \tag{4.10.6}$$

This result may also be obtained by using the idea introduced above that in two dimensions the Green's function represents the field due to a line source.

We should thus be able to calculate the Green's function in two dimensions by solving the three-dimensional problem of the field due to a uniform line source of unit strength. When located along the z axis, such a source may be written as $F(\boldsymbol{\rho}, z) = \delta^2(\boldsymbol{\rho})$. Since there are no surfaces in the example being considered, (4.8.9) without the surface integral is appropriate and the field due to the line source is

$$u(\rho, z) = \int_{-\infty}^{\infty} dz'\, G(\rho, z \mid \rho', z)$$

$$= \frac{1}{4\pi} \int_{-\infty}^{\infty} dz'\, \frac{e^{ik\sqrt{|\rho - \rho'|^2 + (z - z')^2}}}{\sqrt{|\rho - \rho'|^2 + (z - z')^2}} \qquad (4.10.7)$$

Introducing the new integration variable $\zeta = z - z'$ and setting $R = |\rho - \rho'|$, we obtain

$$u(R) = G_k(\rho \mid \rho') = \frac{1}{4\pi} \int_{-\infty}^{\infty} d\zeta\, \frac{e^{ik\sqrt{R^2 + z^2}}}{\sqrt{R^2 + z^2}} \qquad (4.10.8)$$

This result should be equivalent to $G(\boldsymbol{\rho} \mid \boldsymbol{\rho}')$. To show that the Hankel function may be expressed as the integral obtained here would require techniques developed in the theory of complex variables and will not be considered. However, the identification of this result with the Hankel function can at least be made plausible by evaluating this integral approximately for the case $R \gg \zeta$. Writing $\sqrt{R^2 + \zeta^2} \approx R[1 + (\frac{1}{2})(\zeta/R)^2] = R + \zeta^2/2R$, we obtain the approximate expression

$$G(R) \approx \frac{e^{ikR}}{4\pi R} \int_{-\infty}^{\infty} d\zeta\, e^{ik\zeta^2/2R} \qquad (4.10.9)$$

On setting $t^2 = k\zeta^2/2R$ we obtain[1]

$$G(R) \approx \frac{e^{ikR}}{\pi\sqrt{2kR}} \int_{-\infty}^{\infty} dt\, e^{it^2}$$

$$= \frac{i}{4} \sqrt{\frac{2}{\pi k R}}\, e^{i(kR - \pi/4)} \qquad (4.10.10)$$

which is $i/4$ times the large-argument form of the Hankel function given in Eq. (D.19). We can thus infer the plausibility of the result $G(R) = (i/4)H_0^{(1)}(kR)$, in agreement with (4.10.6).

The two-dimensional version of the Sommerfeld radiation condition (4.9.16) is obtained by examining the surface integral

$$\int_{S_\infty} R\, d\theta \left(G\, \frac{\partial p}{\partial R} - p\, \frac{\partial G}{\partial R} \right) \qquad (4.10.11)$$

---

[1]The relations $\int_0^\infty \cos(x^2)\, dx = \int_0^\infty \sin(x^2)\, dx = \frac{1}{2}\sqrt{\pi/2}$ are employed.

which corresponds to the surface integral in (4.9.11) for the three-dimensional case. Using the large-argument form of $G(R)$ given in (4.10.10) and proceeding as in the derivation of the three-dimensional case in (4.9.15), we obtain

$$\lim_{R \to \infty} \left[\sqrt{R} \left(\frac{\partial r}{\partial R} - ikp \right) \right] = 0 \qquad (4.10.12)$$

as the Sommerfeld radiation condition in two dimensions. When satisfied, it guarantees that outgoing waves will provide no contribution to the surface integral an infinity.

As in the three-dimensional case, we can again expect to obtain the Green's function for the two-dimensional *Laplace* equation by taking the limit of (4.10.6) as $kR \to 0$. Using the small-argument form for $J_0(kR)$ and $Y_0(kR)$ as in (4.10.4), we obtain

$$G_k(R) \approx \frac{i}{4} \left(1 + i \frac{2}{\pi} \ln kR \right) = -\frac{1}{2\pi} \ln R + \text{const} \qquad (4.10.13)$$

The constant in the result is customarily ignored and the final result written

$$G_0(|\boldsymbol{\rho} - \boldsymbol{\rho}'|) = -\frac{1}{2\pi} \ln |\boldsymbol{\rho} - \boldsymbol{\rho}'| \qquad (4.10.14)$$

The dimensional argument in the logarithm does not cause any difficulty when this Green's function is used to solve problems.

In order to understand this limiting process better, we now consider the three-dimensional problem of determining the steady state temperature in the vicinity of a line source of finite length. The strength of the line source is q_0 calories per centimeter per second and of length $2h$ located on a z axis within $-h \le z \le h$. Confinement of the source to the z axis may be expressed in terms of the two-dimensional delta function $q_0 \delta^2(\boldsymbol{\rho})$. According to the general solution given in (4.8.10), with $F = q_0 \delta^2(\boldsymbol{\rho})/K$, the temperature is

$$u(\boldsymbol{\rho}, z) = \frac{q_0}{4\pi K} \int_{-h}^{h} dz' \int d^2\boldsymbol{\rho}' \frac{\delta^2(\boldsymbol{\rho}')}{[|\boldsymbol{\rho} - \boldsymbol{\rho}'|^2 + (z - z')^2]^{1/2}}$$

$$= \frac{q_0}{4\pi K} \int_{-h}^{h} dz' \frac{1}{[\rho^2 + (z - z')^2]^{1/2}} \qquad (4.10.15)$$

The integration is relatively elementary and leads to

$$u(\rho, z) = -\frac{q_0}{4\pi K} \ln \left[\sqrt{\rho^2 + (z - z')^2} + (z - z') \right] \Big|_{-h}^{h} \qquad (4.10.16)$$

For our purposes it is sufficient to examine this result on the midplane $z = 0$ where we obtain

$$u(\rho, 0) = \frac{q_0}{4\pi K} \ln \frac{\sqrt{\rho^2 + h^2} + h}{\sqrt{\rho^2 + h^2} - h} \qquad (4.10.17)$$

We now investigate this result in the two limiting cases $\rho/h \gg 1$, far from the source where the problem is three dimensional, and $\rho/h \ll 1$, near the source where the temperature should in some sense be equivalent to the two-dimensional result associated with the field about a line source of infinite extent.

Far from the source ($\rho/h \gg 1$) the temperature on the midplane may be written

$$u(\rho, 0) \cong \frac{q_0}{4\pi K} \ln \frac{1 + \delta}{1 - \delta}$$

$$\cong \frac{q_0}{4\pi K} 2 \left(\delta + \frac{\delta^3}{3} + \cdots \right)$$

$$\cong \frac{(2h)q_0}{4\pi K} \frac{1}{\rho} \left[1 - \frac{1}{6} \left(\frac{h}{\rho} \right)^2 + \cdots \right], \qquad \delta = \frac{h/\rho}{\sqrt{1 + (h/\rho)^2}}$$

$$(4.10.18)$$

Since $R = \rho$ in the midplane $z = 0$, this decrease in u as ρ^{-1} is the expected result at large distances from a localized source. Note also that the strength of the localized source is the linear source density q_0 times $2h$ the length of the source.

For $\rho/h \ll 1$, that is, near the midpoint of the source, the field is essentially two dimensional. In this case we have

$$u(\rho, 0) = \frac{q_0}{4\pi K} \ln \frac{2 + \delta}{\delta}$$

$$\cong \frac{q_0}{4\pi K} \left[\ln \frac{2}{\delta} + \ln \left(1 + \frac{\delta}{2} \right) \right]$$

$$\cong \frac{q_0}{4\pi K} \left[-2 \ln \frac{\rho}{2h} + \frac{1}{4} \left(\frac{\rho}{h} \right)^2 \right], \qquad \delta = \sqrt{1 + (\rho/h)^2} - 1$$

$$(4.10.19)$$

When only the logarithmic term is retained, we obtain

$$u(\rho, 0) = -\frac{q_0}{2\pi K} \ln \frac{\rho}{2h} \qquad (4.10.20)$$

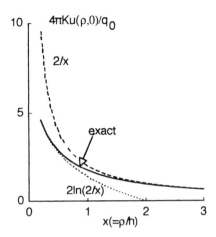

Figure 4.7. Exact and approximate temperature variations due to finite line source.

It is seen that the temperature depends upon the length h in an additive way since $\ln(\rho/2h) = \ln \rho - \ln 2h$. The heat flow, however, being proportional to $\partial u/\partial \rho$, is independent of the constant h. The relation of these two approximate solutions to the exact solution is shown in Figure 4.7.

Problems

4.10.1 Solve Laplace's equation in the quarter plane $x \geq 0$, $y \geq 0$ having boundary conditions $u(x, 0) = \varphi(x)$ and $u_x(0, y) = \psi(y)$ imposed at the rim.

(a) Using **n** to represent an outward normal, show that

$$u(x, y) = \int dS' \left(g \frac{\partial u}{\partial n'} - u \frac{\partial g}{\partial n'} \right)$$

$$= - \int_0^\infty dy' \left(g\psi - u \frac{\partial g}{\partial x'} \right) \bigg|_{x'=0}$$

$$- \int_0^\infty dx' \left(g \frac{\partial u}{\partial y'} - \varphi \frac{\partial g}{\partial y'} \right) \bigg|_{y'=0}$$

(b) We may eliminate the unknown functions $\partial u/\partial y'|_{y'=0}$ and $u(0, y')$ by constructing a Green's function that vanishes on $y' = 0$ and satisfies $\partial g/\partial x'|_{x'=0} = 0$. Use the method of images to show that

$$g(\mathbf{r}|\mathbf{r}') = -\frac{1}{2\pi} \ln \frac{|\mathbf{r} - \mathbf{r}'||\mathbf{r} - \mathbf{r}_1'|}{|\mathbf{r} - \mathbf{r}_2'||\mathbf{r} - \mathbf{r}_3'|}$$

where $\mathbf{r}' = \mathbf{i}x' + \mathbf{j}y'$, $\mathbf{r}_1' = -\mathbf{i}x' + \mathbf{j}y'$, $\mathbf{r}_2' = \mathbf{i}x' - \mathbf{j}y'$, and $\mathbf{r}_3' = -\mathbf{i}x' - \mathbf{j}y'$.

(c) Show that if $\varphi(x) = 0$, the solution is

$$u(x, y) = \frac{1}{2\pi} \int_0^\infty dy' \, \psi(y') \ln \frac{x^2 + (y - y')^2}{x^2 + (y + y')^2}$$

4.11 INVERSION AND THE METHOD OF IMAGES FOR A CIRCLE

For problems involving boundaries, as most problems do, success with the Green's function method depends upon our ability not only to obtain a particular solution to the differential equation for a unit point source, but also to obtain a solution to the corresponding homogeneous equation that will enable us to deal with whatever conditions are prescribed at the boundaries. If the boundary is a plane, then as shown previously in Section 4.9, the appropriate homogeneous solution can frequently be determined by using the method of images. As noted there, the image source is *outside* the region of interest for the problem being solved. The image source thus provides a solution to a *homogeneous* equation *inside* the region of interest. For the Laplace equation (but not for the Helmholtz equation) there is a corresponding image technique available for circular and spherical boundaries. Here we shall develop the circular case. For spherical geometry, see Section 5.3.

The method follows from the observation that for a radial coordinate $\rho = a^2/\underline{\rho}$, where for now a is an arbitrary constant, we have the relation

$$\rho \frac{\partial f}{\partial \rho} = \frac{a^2}{\rho} \frac{\partial f}{\partial \underline{\rho}} \frac{d\underline{\rho}}{d\rho} = -\underline{\rho} \frac{\partial f}{\partial \underline{\rho}}, \qquad \rho \neq 0 \tag{4.11.1}$$

This is precisely the type of differential expression that occurs twice in Laplace's equation in polar coordinates, that is,

$$\nabla^2 \psi(\rho, \theta) = \frac{1}{\rho^2} \left[\rho \frac{\partial}{\partial \rho} \left(\rho \frac{\partial \psi}{\partial \rho} \right) + \frac{\partial^2 \psi}{\partial \theta^2} \right] \tag{4.11.2}$$

If $\psi(\rho, \theta)$ is assumed to be a solution of $\nabla^2 \psi = 0$, then $\varphi(\rho, \theta) = \psi(a^2/\rho, \theta)$ will also satisfy Laplace's equation in the ρ, θ coordinate system. This result follows from the identity

$$\rho \frac{\partial}{\partial \rho} \left[\rho \frac{\partial \psi(a^2/\rho, \theta)}{\partial \rho} \right] + \frac{\partial^2 \psi(a^2/\rho, \theta)}{\partial \theta^2}$$

$$= \underline{\rho} \frac{\partial}{\partial \underline{\rho}} \left[\underline{\rho} \frac{\partial \psi(\underline{\rho}, \theta)}{\partial \underline{\rho}} \right] + \frac{\partial^2 \psi(\underline{\rho}, \theta)}{\partial \theta^2} \tag{4.11.3}$$

and this latter differential expression, being proportional to the Laplacian of $\psi(\rho, \theta)$ in "its own" coordinates, has been assumed to vanish.

The points ρ and $\bar{\rho}$ are referred to as the images of each other in a circle of radius a. Note that $\bar{a}$ is the geometric mean of ρ and $\bar{\rho}$, since $a = \sqrt{\rho\bar{\rho}}$.

We have already seen in (4.10.14) that the Green's function for Laplace's equation in two dimensions is $G(\mathbf{\rho}|\mathbf{\rho}') = -(\frac{1}{2}\pi)\ln|\mathbf{\rho} - \mathbf{\rho}'|$. For $\mathbf{\rho} \neq \mathbf{\rho}'$, this is a solution of Laplace's equation that we will indicate by $\psi(\rho, \theta)$, and we write

$$\psi(\rho, \theta) = -\frac{1}{2\pi} \ln \sqrt{\rho^2 + \rho'^2 - 2\rho\rho' \cos(\theta - \theta')}, \qquad \rho \neq \rho' \quad (4.11.4)$$

We may now construct another solution $\varphi(\rho, \theta)$ by using the prescription just introduced, namely,

$$\varphi(\rho, \theta) = \psi\left(\frac{a^2}{\rho}, \theta\right) = -\frac{1}{2\pi} \ln \sqrt{\frac{a^4}{\rho^2} + \rho'^2 - 2\frac{a^2\rho'}{\rho} \cos(\theta - \theta')}$$

$$(4.11.5)$$

Note that at $\rho = a$ the two solutions φ and ψ reduce to the same expression. Thus the difference between φ and ψ will vanish on a circular boundary of radius a. We can thus think of $\varphi(\rho, \theta)$ as a function to be subtracted from the Green's function $\psi(\rho, \theta)$ in order to satisfy Dirichlet boundary conditions at $\rho = a$.

However, $\varphi(\rho, \theta)$ is singular at the origin $\rho = 0$. It is thus not the solution to a homogeneous equation within the circle unless we can subtract out this divergent term. As $\rho \to 0$ we find $\varphi(\rho, a) \to (\frac{1}{4}\pi)\ln(\rho/a)$. The Green's function that vanishes on the boundary of a circle of radius a may thus be written

$$G(\mathbf{\rho}|\mathbf{\rho}') = \varphi(\rho, \theta) - \left[\varphi\left(\frac{a^2}{\rho}, \theta\right) - \frac{1}{4\pi} \ln \frac{\rho}{a}\right]$$

$$= -\frac{1}{4\pi} \ln \frac{\rho^2 + \rho'^2 - 2\rho\rho' \cos(\theta - \theta')}{a^2 + (\rho\rho'/a)^2 - 2\rho\rho' \cos(\theta - \theta')} \quad (4.11.6)$$

Note that the term $(\frac{1}{4}\pi)\ln(\rho/a)$ that was subtracted to eliminate the source at the origin also vanishes at $\rho = a$ so that $G(\mathbf{\rho}|\mathbf{\rho}')$ still vanishes there.

Since the argument of the logarithm reduces to unity for either ρ or $\rho' = a$, $G(\mathbf{\rho}|\mathbf{\rho}')$ vanishes on the boundary in both primed and unprimed coordinates. It is also immediately evident that the reciprocity principle is satisfied.

The radial derivative of G is

$$\left.\frac{\partial G}{\partial \rho'}\right|_{\rho'=a} = -\frac{1}{2\pi} \frac{a^2 - \rho^2}{a^2 + \rho^2 - 2\rho a \cos(\theta - \theta')} \quad (4.11.7)$$

Consequently, if the temperature is specified on the rim of a circular region, then the temperature inside the region is

$$u(\rho, \theta) = \int_0^{2\pi} a \, d\theta' u(a, \theta) \left.\frac{\partial G}{\partial \rho'}\right|_{\rho' = a}$$

$$= \frac{a^2 - \rho^2}{2\pi} \int_0^{2\pi} \frac{u(a, \theta') \, d\theta'}{a^2 + \rho^2 - 2a\rho \cos(\theta - \theta')} \qquad (4.11.8)$$

a result known as Poisson's integral. Determination of the solution within a circular region is thus reduced to the evaluation of an integral. A number of properties of this integral as well as examples of its evaluation for specific functions $u(a, \theta)$ may be found elsewhere (Edwards 1930, pp. 278, 305).

For the region outside the circle the outward normal is in the negative ρ' direction and the temperature is given by the negative of the result given in (4.11.8). The singular integral that occurs when the field is evaluated on the boundary may be treated by using the delta function (cf. Problem 4.11.5).

The Green's function for a circle that satisfies homogeneous Neumann boundary conditions does not follow as a minor variation of the calculation presented in this section. The Neumann Green's function for the sphere is considered in Section 9.5.

Problems

4.11.1 A circular disk of radius a and conductivity K is kept at zero temperature at the rim and is heated by a point source of strength q_0 located on the disk at a point $\rho = \rho_0$.

(a) Show that the temperature on the disk is given by

$$u(\rho) = \frac{q_0}{K} g(\rho | \rho_0)$$

where $g(\rho | \rho_0)$ is given by (4.11.6).

(b) Use the result $\int_0^{2\pi} (a - b \cos \theta)^{-1/2} \, d\theta = 2\pi (a^2 - b^2)^{-1/2}$, $a > b > 0$, to show that heat flow is conserved, that is,

$$-K \int_0^{2\pi} a \, d\theta \left.\frac{\partial u}{\partial \rho}\right|_{\rho = a} = q_0$$

4.11.2 The temperature on the rim of a circular disk of radius a is given by

$$u(a, \theta) = \begin{cases} -u_0, & -\pi < \theta < 0 \\ u_0, & 0 < \theta < \pi \end{cases}$$

Use the Poisson integral (4.11.8) to show that

$$\frac{u(\rho, \theta)}{u_0} = \frac{\alpha}{\pi} \int_0^\pi d\theta' \, \frac{\cos(\theta - \theta')}{1 + \alpha^2 \sin^2(\theta - \theta)}$$

$$= \frac{2}{\pi} \tan^{-1}(\alpha \sin \theta), \qquad \alpha = \frac{2a\rho}{a^2 - \rho^2}$$

4.11.3 The temperature on the rim of a circular disk of radius a is given by

$$u(a, \theta) = u_0 \sin \frac{\theta}{2}, \qquad 0 < \theta < 2\pi$$

Use the Poisson integral to show that the temperature within the disk is given by

$$\frac{2\pi u(\rho, \theta)}{u_0(a^2 - \rho^2)} = \frac{\cos(\theta/2)}{(a + \rho)^2} \int_{-\theta}^{2\pi - \theta} \frac{d\varphi \, \sin(\varphi/2)}{1 - \beta^2 \cos^2(\varphi/2)}$$

$$+ \frac{\sin(\theta/2)}{(a - \rho)^2} \int_{-\theta}^{2\pi - \theta} \frac{d\varphi \, \sin(\varphi/2)}{1 + \alpha^2 \sin^2(\varphi/2)}$$

where $\alpha = 2\sqrt{(a\rho)}/(a - \rho)$ and $\beta = 2\sqrt{(a\rho)}/(a + \rho)$. Evaluate the integrals and obtain

$$\frac{u(\rho, \theta)}{u_0} = \frac{2}{\pi} \left[\alpha^{-1} \cos \frac{\theta}{2} \tanh^{-1} \left(\beta \cos \frac{\theta}{2} \right) \right.$$

$$\left. + \beta^{-1} \sin \frac{\theta}{2} \tan^{-1} \left(\alpha \sin \frac{\theta}{2} \right) \right]$$

Show that this result reduces to that given in Problem 3.7.5.

4.11.4 The temperature on the rim of a circular disk of radius a is given by

$$u(a, \theta) = u_0 \sin \frac{\theta}{2}, \qquad -\pi < \theta < \pi$$

Note that with this specification of the range for θ there is a discontinuity in the rim temperature across $\theta = \pm \pi$.

Show that the temperature is expressed in terms of the same integrands as those of Problem 4.11.3 but the limits of integration are now $-\pi - \theta < \varphi < \pi - \theta$. Evaluate the integrals and obtain

$$\frac{u(\rho, \theta)}{u_0} = \frac{2}{\pi}\left[-\alpha^{-1}\cos\frac{\theta}{2}\tanh^{-1}\left(\beta\sin\frac{\theta}{2}\right)\right.$$

$$\left. + \beta^{-1}\sin\frac{\theta}{2}\tan^{-1}\left(\alpha\cos\frac{\theta}{2}\right)\right]$$

Show that near the point at which the rim temperature has the discontinuity, that is, for $\rho = a(1 - \epsilon)$ and $\theta = \pi - \delta$, where δ may be positive or negative, one has $\alpha \approx 2/\epsilon$ and $\beta \approx 1$. Simplify the general result for the temperature given above and obtain

$$\frac{u(\rho, \theta)}{u_0} \approx \frac{2}{\pi}\tan^{-1}\frac{\delta}{\epsilon}$$

Examine this limit as $\delta \to 0^{\pm}$ for fixed ϵ and as $\epsilon \to 0$ for fixed δ.

4.11.5 Show that the integrand in the Poisson integral, (4.11.8), has the properties of a delta function as ρ approaches the rim. Set $\theta - \theta' = \varphi$ and define

$$D(\rho, a, \varphi) = \frac{a^2 - \rho^2}{2\pi(a^2 + \rho^2 - 2a\rho\cos\varphi)}$$

Show that

$$\lim_{\rho \to a} D(\rho, a, \varphi) = 0, \qquad \varphi \neq 0$$

$$\lim_{\rho \to a} D(\rho, a, \varphi) = \infty, \qquad \varphi = 0$$

$$\int_0^{2\pi} D(\rho, a, \varphi)\, d\varphi = 1$$

In evaluating the integral, the relation

$$\int_0^{\pi/2} \frac{d\theta}{A + B\sin^2\theta} = \frac{\pi}{2\sqrt{A(A + B)}}$$

is useful.

4.12 EIGENFUNCTION EXPANSION METHODS

Use of the method of images to satisfy boundary conditions is only available in a few simple geometries. In other cases, the Green's function can frequently be expressed as an expansion in terms of the eigenfunctions appropriate to the geometry being considered. There is more than one way to set up an expansion, and the ideas involved in making a proper choice are perhaps best illustrated by an example. It was found in Problem 3.2.1 that when all edges of a rectangular region $0 \leq x \leq L$, $0 \leq y \leq H$ are kept at zero temperature except along the edge $y = 0$, where it is required that $u(x, 0) = u_0 \sin \pi x/L$, the temperature at any interior point is given by $u(x, y) = u_0 \sin(\pi x/L) \sinh[\pi(H - y)/L] \operatorname{csch} \pi H/L$. This result may be recovered by Green's function methods by considering the temperature to be expressed in the form

$$u(x, y) = \int \left(G \frac{\partial u}{\partial n'} - u \frac{\partial G}{\partial n'} \right) ds' \tag{4.12.1}$$

Since u is known on the boundary, we require that G vanish on the boundary. For the particular problem considered here, the temperature within the rectangle will then be expressed as

$$u(x, y) = \int_0^L dx' \, u(x', 0) \left. \frac{\partial G(x, y | x', y')}{\partial y'} \right|_{y'=0} \tag{4.12.2}$$

Note that since n' is an outward normal, it is pointed in the negative y' direction. Because of the rectangular geometry, we use an expansion of G in terms of some type of Fourier series. Three possibilities suggest themselves:

$$G_1(\boldsymbol{\rho} | \boldsymbol{\rho}') = \sum_{n=1}^{\infty} g_{1n}(y | \boldsymbol{\rho}') \sin \frac{n\pi x}{L}$$

$$G_2(\boldsymbol{\rho} | \boldsymbol{\rho}') = \sum_{n=1}^{\infty} g_{2n}(x | \boldsymbol{\rho}') \sin \frac{n\pi y}{H} \tag{4.12.3}$$

$$G_3(\boldsymbol{\rho} | \boldsymbol{\rho}') = \sum_{n=1}^{\infty} \sum_{m=1}^{\infty} g_{3nm}(\boldsymbol{\rho}') \sin \frac{n\pi x}{L} \sin \frac{m\pi y}{H}$$

If either of the functions g_{1n} or g_{2n} that appear in G_1 and G_2, respectively, were to be expanded in a Fourier series, we would obtain the double sum given as G_3. The coefficients in each sum are determined by substitution into the equation for the Green's function, namely,

$$\frac{\partial^2 G}{\partial x^2} + \frac{\partial^2 G}{\partial y^2} = -\delta(x - x')\delta(y - y') \tag{4.12.4}$$

For $G = G_1$ we obtain

$$\sum_{n=1}^{\infty} \left[-\left(\frac{n\pi}{L}\right)^2 g_{1n} + \frac{d^2 g_{1n}}{dy^2} \right] \sin \frac{n\pi}{L} = -\delta(x - x')\delta(y - y') \quad (4.12.5)$$

A particular term in this series is isolated by using orthogonality. We obtain

$$\frac{d^2 g_{1n}}{dy^2} - \left(\frac{n\pi}{L}\right)^2 g_{1n} = -J\delta(y - y') \quad (4.12.6)$$

where $J = (2/L)\sin n\pi x'/L$. The appropriate solution of this equation can be obtained immediately by using the technique developed in Section 4.2. Since the Green's function is to vanish on the boundaries at $y = 0, H$, we use the two elementary solutions $u_1 = \sinh n\pi y/L$ and $u_2 = \sinh n\pi (H - y)/L$. The Wronskian for these solutions is $W = u_1 u_2' - u_2 u_1' = -(n\pi/L)\sinh n\pi H/L$. Since $p(x) = 1$ in the present example, (4.2.10) yields

$$g_{1n}(y|\boldsymbol{\rho}') = \frac{2}{L} \sin \frac{n\pi x'}{L} \frac{\sinh(n\pi y_</L)\sinh[n\pi(H - y_>)/L]}{(n\pi/L)\sinh n\pi H/L} \quad (4.12.7)$$

and the Fourier series for $G_1(\boldsymbol{\rho}|\boldsymbol{\rho}')$ is obtained from the first of (4.12.3). If a similar calculation is carried out for $G_2(\boldsymbol{\rho}|\boldsymbol{\rho}')$ we find

$$G_2(\boldsymbol{\rho}|\boldsymbol{\rho}') = \frac{2}{\pi} \sum_{n=1}^{\infty} \frac{\sin(n\pi y/H)\sin(n\pi y'/H)\sinh(n\pi x_</H)\sinh n\pi(L - x_>)/H}{n \sinh n\pi L/H}$$

$$(4.12.8)$$

When the third form, $G_3(\boldsymbol{\rho}|\boldsymbol{\rho}')$, is substituted into the defining equation for G, we obtain

$$\sum_{n,m=1}^{\infty} g_{3nm} \left[\left(\frac{n\pi}{L}\right)^2 + \left(\frac{m\pi}{H}\right)^2 \right] \sin \frac{n\pi x}{L} \sin \frac{m\pi y}{H} = -\delta(x - x')\delta(y - y')$$

$$(4.12.9)$$

Multiplying by $\sin(p\pi x/L)\sin(q\pi y/H)$ and integrating over the rectangle, we find

$$g_{3pq} = \frac{4}{LH} \frac{\sin p\pi x'/L \sin q\pi y'/H}{(p\pi/L)^2 + (q\pi/H)^2} \quad (4.12.10)$$

and the coefficients in the double Fourier series for G_3 have been determined.

We now consider the use of these three forms of $G(\boldsymbol{\rho}|\boldsymbol{\rho}')$ for solving the problem mentioned at the beginning of this section. When $G_1(\boldsymbol{\rho}|\boldsymbol{\rho}')$ is substituted into (4.12.2) and the orthogonality of the function $\sin n\pi x/L$ is used, we immediately recover the solution $u(x, y)$ in the form quoted at the beginning of this section. When $G_2(\boldsymbol{\rho}\,|\,\boldsymbol{\rho}')$ is used, however, the integration over x is somewhat cumbersome since we must use one form for $g_{2n}(x\,|\,\boldsymbol{\rho}')$ when $x' < x$ and a different form for $x' > x$. Clearly, this expansion is less convenient for the present example. It is the form to use when the nonvanishing temperature is imposed along a y boundary. Finally, substitution of the third form for G yields

$$u(x, y) = u_0 \left[\frac{2}{\pi} \sum_{m=1}^{\infty} a_m \frac{\sin m\pi y/H}{m^2 + (H/L)^2} \right] \sin \frac{\pi x}{L} \qquad (4.12.11)$$

The series in brackets is just the Fourier series for the expression $\sin \pi(H - y)/\sinh \pi H/L$ that occurred when G_1 was used. The double series is equally useful for integration along either x' or y'. In the present problem it has the disadvantage of displaying a rather simple solution as an infinite series.

As an example of a series representation for $G(\boldsymbol{\rho}\,|\,\boldsymbol{\rho}')$ in another coordinate system, consider the two-dimensional free-space Green's function for Laplace's equation in polar coordinates. In (4.10.14) it was given in closed form as

$$G(\boldsymbol{\rho}\,|\,\boldsymbol{\rho}') = -\frac{1}{2\pi} \ln|\boldsymbol{\rho} - \boldsymbol{\rho}'| \qquad (4.12.12)$$

We now recover this result by beginning with a Fourier series in θ. We introduce the expansion

$$G(\boldsymbol{\rho}\,|\,\boldsymbol{\rho}') = \sum_{n=0}^{\infty} g_n(\rho\,|\,\rho')\cos n(\theta - \theta') \qquad (4.12.13)$$

as an assumed form for obtaining the solution of

$$\frac{1}{\rho} \frac{\partial}{\partial \rho}\left(\rho \frac{\partial G}{\partial \rho}\right) + \frac{1}{\rho^2} \frac{\partial^2 G}{\partial \theta^2} = -\frac{\delta(\rho - \rho')\delta(\theta - \theta')}{\rho} \qquad (4.12.14)$$

Note that the solution will be symmetric in θ when it is measured with respect to the angle of the source term θ'. We have introduced this symmetry at the beginning of the calculation by writing the angular dependence of $G(\boldsymbol{\rho}\,|\,\boldsymbol{\rho}')$ in the form $\cos n(\theta - \theta')$. The terms $\sin n(\theta - \theta')$ are odd with respect to $\theta = \theta'$ and hence make no contribution to the solution to this problem. Substitution of the series into (4.12.14), the equation defining G, yields

$$\sum_{n=0}^{\infty} \left[\frac{1}{\rho} \frac{d}{d\rho}\left(\rho \frac{dg_n}{d\rho}\right) - \frac{n^2}{\rho^2} g_n \right] \cos n(\theta - \theta') = -\frac{\delta(\rho - \rho')\delta(\theta - \theta')}{\rho}$$

$$(4.12.15)$$

Multiplying this equation by $\cos p(\theta - \theta')$ and integrating from 0 to 2π, we obtain

$$\frac{d}{d\rho}\left(\rho\frac{dg_0}{d\rho}\right) = -\frac{1}{2\pi}\delta(\rho - \rho')$$

$$\frac{d}{d\rho}\left(\rho\frac{dg_p}{d\rho}\right) - \frac{p^2}{\rho}g_p = -\frac{1}{\pi}\delta(\rho - \rho'), \qquad p \neq 0$$

(4.12.16)

From the known form of the solution in (4.12.12), we know that $G(\mathbf{\rho}\,|\,\mathbf{\rho}') \to (1/2\pi)\ln\rho$ as $\rho \to \infty$ while $G(\mathbf{\rho}\,|\,\mathbf{\rho}')$ is bounded as $\rho \to 0$. We now rewrite the appropriate solutions of Eq. (4.12.16) by using the general procedure developed in Section 4.2. Noting that for $\mathbf{\rho} \neq \mathbf{\rho}'$ the homogeneous equations in (4.12.16) are of Cauchy-Euler type for $p \neq 0$, we choose as elementary solutions

$$n = 0, \qquad u_1 = 1, \qquad u_2 = \ln\rho, \qquad W = -1/\rho$$

$$n > 0, \qquad u_1 = \rho^n, \qquad u_2 = \rho^{-n}, \qquad W = -2n/\rho$$

(4.12.17)

In the notation of Section 4.2, $p(\rho) = \rho$ and $c = -1$ for $n = 0$ while $c = 2n$ for $n > 0$. Also, $J = 1/2\pi$ for $n = 0$ and $1/\pi$ for $n > 0$. We thus obtain

$$g_0 = -\frac{1}{2\pi}\ln\rho_>$$

(4.12.18)

$$g_n = \frac{1}{2\pi n}(\rho_>/\rho_<)^n$$

The series expansion for G is now found to be

$$G(\rho, \theta\,|\,\rho, \theta) = -\frac{1}{2\pi}\ln\rho_> + \frac{1}{2\pi}\sum_{n=1}^{\infty}\frac{1}{n}\left(\frac{\rho_<}{\rho_>}\right)^n \cos n(\theta - \theta') \quad (4.12.19)$$

Note that the result is consistent with the Fourier series 27 and 28 of Table F.1, which may be written

$$\ln\sqrt{1 + a^2 - 2a\cos\theta} = \begin{cases} \ln a - \displaystyle\sum_{n=1}^{\infty}\frac{1}{n}a^{-n}\cos n\theta, & |a| > 1 \\[2mm] -\displaystyle\sum_{n=1}^{\infty}\frac{1}{n}a^n\cos n\,\theta, & |a| < 1 \end{cases}$$

(4.12.20)

For $a = 1$, both series reduce to

$$\ln\sqrt{2(1 - \cos\theta)} = \ln\left|2\sin\frac{\theta}{2}\right| = -\sum_{n=1}^{\infty}\frac{1}{n}\cos n\theta, \qquad \theta \neq 2n\pi \quad (4.12.21)$$

As another example of the eigenfunction method, consider the construction of the Green's function for Laplace's equation that vanishes on all surfaces of a circular cylinder of height H and radius a. As a simplification, let us confine attention to the case that will be useful for problems that have no angular variation on the walls of the cylinder. In such situations it would simplify matters if we could also ignore the angular dependence in the Green's function from the beginning. This may be accomplished by considering the Green's function to be the temperature distribution due to a *ring* source of unit strength rather than to a *point* source.

To obtain the source term for a ring, we need merely average the point source over the angle variable φ to obtain

$$\nabla^2 G = \frac{1}{2\pi} \int_0^{2\pi} d\varphi \left[-\frac{\delta(\rho - \rho')\delta(z - z')\delta(\varphi - \varphi')}{\rho} \right]$$

$$= -\frac{\delta(\rho - \rho')\delta(z - z')}{2\pi\rho'} \qquad (4.12.22)$$

Since there is now no angular dependence in G, we write

$$\frac{1}{\rho}\frac{\partial}{\partial\rho}\left(\rho\frac{\partial G}{\partial\rho}\right) + \frac{\partial^2 G}{\partial z^2} = -\frac{1}{2\pi\rho'}\delta(\rho - \rho')\delta(z - z') \qquad (4.12.23)$$

As in the case of rectangular coordinates considered earlier in (4.12.3), there are three choices for our series expansion. We choose the expansion

$$G(\rho, z | \rho', z') = \sum_{n=1}^{\infty} g_n(\rho, \rho', z') \sin\frac{n\pi z}{H} \qquad (4.12.24)$$

which will be useful for satisfying boundary conditions on the vertical wall of the cylinder since it contains a Fourier series in z. Substitution of the expansion for G into the governing equation (4.12.23) and use of the orthogonality relations for $\sin n\pi z/H$ yield

$$\frac{d}{d\rho}\left(\rho\frac{dg_n}{dz}\right) - k_n^2\rho g_n = -J_n\delta(\rho - \rho') \qquad (4.12.25)$$

where $k_n = n\pi/H$ and $J_n = (\pi H)^{-1}\sin n\pi z'/H$. As shown in Appendix D, the homogeneous equation has solutions $I_0(k_n\rho)$ and $K_0(k_n\rho)$. In using the method of Section 4.2 to construct a Green's function with appropriate elementary solutions, note that at $\rho = 0$ the solution must remain finite while at the outer edge $\rho = a$ it must vanish. Therefore we take

$$u_1(\rho) = I_0(k_n\rho)$$

$$u_2(\rho) = I_0(k_n\rho) - I_0(k_na)K_0(k_n\rho)/K_0(k_na) \qquad (4.12.26)$$

Defining $R_n = I_0(na)/K_0(na)$ and noting that the Wronskian of a solution with itself is zero, we obtain

$$W(u_1, u_2) = W(I_0, I_0 - R_n K_0)$$

$$= W(I_0, I_0) - R_n W(I_0, K_0)$$

$$= -k_n R_n (I_0 K_0' - I_0' K_0) \qquad (4.12.27)$$

where the primes refer to differentiation with respect to $k_n \rho$, the argument that has been suppressed. Using the large-argument approximation for I_0 and K_0 [cf. Eq. (D.32)] we find

$$W(u_1, u_2) = \frac{1}{\rho'} R_n = \frac{1}{\rho'} \frac{I_0(k_n a)}{K_0(k_n a)} \qquad (4.12.28)$$

Comparison of (4.12.25) with the standard Sturm-Liouville form (4.2.1) shows that the term $p(\rho')$ is equal to ρ'. From the development in Section 4.2, we then see that $c = R_n = I_0(k_n a)/K_0(k_n a)$ and finally obtain

$$g_n(\rho, \rho', z') = -\frac{J_n}{c} u_1(\rho_<) u_2(\rho_>)$$

$$= -\frac{\sin k_n z'}{\pi H} \frac{1}{R_n} [I_0(k_n \rho_>) - R_n K_0(k_n \rho_>)] \qquad (4.12.29)$$

The form of the Green's function in this case is therefore

$$G(\rho, z | \rho', z') = -\frac{1}{\pi H} \sum_{n=1}^{\infty} \frac{I_0(k_n \rho_<)}{I_0(k_n a)} [I_0(k_n \rho_>) K_0(k_n a)$$

$$- I_0(k_n a) K_0(k_n \rho_>)] \sin k_n z \sin k_n z' \qquad (4.12.30)$$

We now determine the ρ' derivative of this form for G. Since we are evaluating this derivative on the surface $\rho' = a$ for field points $\rho < a$, we choose $\rho_< = \rho$ and $\rho_> = \rho'$ in (4.12.30). The expression $-I_0'(k_n a) K_0(k_n a) + I_0(k_n a) K_0'(k_n a)$ is found to occur. This combination of terms is the Wronskian of these functions and, as noted above, has the value $(k_n a)^{-1}$. Thus

$$\frac{\partial G}{\partial \rho'}\bigg|_{\rho' = a} = -\frac{1}{\pi H a} \sum_{n=1}^{\infty} \frac{I_0(k_n \rho)}{I_0(k_n a)} \sin k_n z \sin k_n z' \qquad (4.12.31)$$

We now consider the use of these expressions for recovering a result that was obtained earlier without use of the Green's function method.

Example 4.7. Use the Green's function given in (4.12.30) to recover the result obtained in (3.7.4) where the temperature on the vertical wall of a cylinder was a specified function of z while the temperatures on the top and base were set equal to zero.

Using Green's theorem (Section 4.8), we obtain

$$
u = \int_{\substack{\text{top +} \\ \text{base}}} \left(G \frac{\partial u}{\partial n'} - u \frac{\partial G}{\partial n'} \right) dS' + \int_{\substack{\text{vert} \\ \text{wall}}} \left(G \frac{\partial u}{\partial n'} - u \frac{\partial G}{\partial n'} \right) dS' \quad (4.12.32)
$$

Since both G and u vanish on the top and base, the first integral vanishes. Since G has been constructed so as to vanish on the vertical wall as well, the second integral reduces to only an integral over the known temperature distribution on the wall multiplied by $\partial G/\partial n'$. Since $dS' = a\, d\varphi'\, dz'$ on the vertical wall, we obtain

$$
u(\rho, z) = -2\pi a \int_0^H dz'\, u(a, z') \frac{\partial}{\partial \rho'}\, G(\rho, z \mid \rho', z')\big|_{\rho' = a}
$$

$$
= \frac{2}{H} \int_0^H dz'\, u(a, z') \sum_{n=1}^{\infty} \frac{I_0(k_n \rho)}{I_0(k_n a)} \sin k_n z \sin k_n z' \quad (4.12.33)
$$

which agrees with the result obtained in (3.7.75).

The expansion of G that is appropriate when the temperature is specified on either the top or base of the cylinder is considered in Problem 4.12.3.

Problems

4.12.1 Determine the solution of

$$
\nabla^2 G(x, y \mid x'y') = -\delta(x - x')\delta(y - y')
$$

that vanishes on $x = 0, L, y = 0, H$ in the form

$$
G = \sum_{n=1}^{\infty} g_n(x, x', y') \sin \frac{n\pi y}{H}
$$

4.12.2 The boundaries of a cylinder of radius a and height H are maintained at zero temperature. A steady heat source Q_0 is located at the center of the cylinder so that the equation governing $u(r, z)$, the temperature distribution within the cylinder, is

$$
\nabla^2 u = -\frac{Q_0}{K} \frac{\delta(\rho)\delta(z - H/2)}{2\pi\rho}
$$

(a) Use the Green's function $G(\rho, z \mid \rho', z')$ given in (4.12.30) to show that the temperature at any point within the cylinder is

$$
u(\rho, z) = \frac{Q_0}{K} G(\rho, z \mid 0, H/2)
$$

(b) Show that the rate at which heat flows out through the vertical wall of the cylinder is given by

$$\int_{\substack{\text{vert.} \\ \text{wall}}} \mathbf{J} \cdot d\mathbf{S} = -K \int dS \left.\frac{\partial u}{\partial \rho}\right|_{\rho = a} = \frac{4Q_0}{\pi} \sum_{n=1}^{\infty} \frac{\sin(n\pi/2)}{nI_0(n\pi a/H)}$$

(c) For a long thin cylinder ($a/H \ll 1$), one expects that nearly all heat flows out through the vertical wall and that therefore the heat out is nearly equal to Q_0. We can see that the sum provides this approximate inequality by noting that in this limit we may set $I_0(n\pi a/H) \approx 1$ until n becomes very large. Although there is thus a nonuniform limiting process involved, if one sets $I(n\pi a/H) = 1$ for all n, the sum reduces to $\pi/4$ and one obtains a total heat flow out equal to Q_0, as expected.

4.12.3 For a cylinder $0 \le \rho \le a$, $0 \le z \le H$, determine a Green's function that vanishes at $z = 0, H$ in the form

$$G(\rho, z | \rho', z') = \sum_{n=1}^{\infty} g_n(z, \rho', z')J_0(\alpha_{0n}\rho/a), \qquad J_0(\alpha_{0n}) = 0$$

where

$$G(\rho, z | \rho', z') = \frac{1}{2\pi} \int_0^{2\pi} d\varphi\, g(\rho, \varphi, z | \rho', \varphi', z')$$

(a) Show that g_n satisfies

$$\frac{d^2 g_n}{dz^2} - k_n^2 g_n = -J_n \delta(z - z')$$

where $J_n = J_0(k_n \rho')/[\pi a^2 J_1^2(k_n a)]$.

(b) Use the general result (4.2.10) for constructing a Green's function to obtain

$$G = \frac{1}{\pi a^2} \sum_{n=1}^{\infty} \frac{\sinh k_n z_< \sinh k_n(H - z_>)}{k_n J_1^2(k_n a) \sinh k_n H} J_0(k_n \rho')J_0(k_n \rho)$$

4.12.4 A cylinder has temperature zero on all boundaries. A point heat source of strength Q_0 is located at the center.

(a) Show that the temperature at any internal point is given by

$$u(\mathbf{r}) = \frac{1}{K} \int d^3\mathbf{r}'\, s(\mathbf{r}')g(\mathbf{r}|\mathbf{r}')$$

where g vanishes on all boundaries of the cylinder.

(b) Since there is no angular dependence, we may average over the angle φ, as in Problem 4.12.3, and obtain

$$u(\rho, z) = \frac{1}{K} \int d^3\mathbf{r}' \, s(\mathbf{r}') G(\rho, z|\rho', z')$$

where G is the result obtained in Problem 4.12.3.

(c) Since the source term is the same as that used in Problem 4.12.2, obtain

$$u(\rho, z) = \frac{Q_0}{K} G(\rho, z|0, H/2)$$

(d) Show that

$$\left.\frac{\partial G}{\partial z}\right|_{z=H} = -\frac{1}{\pi a^2} \sum_{n=1}^{\infty} \frac{k_n \sinh k_n H/2}{J_1^2(k_n a)} \frac{J_0(k_n \rho)}{k_n \sinh k_n H}$$

and that the total heat flow out through the top of the cylinder is

$$\int_{\text{top}} \mathbf{J} \cdot d\mathbf{S} = -K \int dS \left.\frac{\partial u}{\partial z}\right|_{z=H}$$

$$= \frac{Q_0}{a} \sum_{n=1}^{\infty} \frac{1}{k_n \cosh k_n H/2 J_1(k_n a)}$$

(e) For $H/a \ll 1$, one expects nearly all heat to flow out equally through the top and base. Set $\cosh k_n H/2 = \cosh \alpha_{0n} H/2a = 1$ in this limit and obtain

$$\text{Heat flow at top} = Q_0 \sum_{n=1}^{\infty} \frac{1}{\alpha_{0n} J_1(\alpha_{0n})}$$

This sum has the value $\frac{1}{2}$, as found in Problem C.7. Hence one's intuition is confirmed. As in Problem 4.12.2, the sum is actually nonuniform since n is summed to infinity.

4.12.5 Obtain the eigenfunction expansion for the Green's function for the two-dimensional wave equation given in (4.9.6) by introducing an expansion of the form

$$g(\rho, \theta|\rho', \theta') = \sum_{n=0}^{\infty} g_n(\rho|\rho') \cos n(\theta - \theta')$$

Show, by substitution of this expansion into (4.9.1), that

$$g_0(\rho, \rho') = \frac{i}{4} J_0(k\rho_<) H_0^{(1)}(k\rho_>)$$

$$g_n(\rho, \rho') = \frac{i}{2} J_n(k\rho_<) H_n^{(1)}(k\rho_>)$$

Hence, obtain the result

$$g(\rho, \theta \,|\, \rho', \theta') = \frac{i}{4} \sum_{n=0}^{\infty} \epsilon_n J_n(k\rho_<) H_n^{(1)}(k\rho_>) \cos n(\theta - \theta')$$

where ϵ_n equals 1 for $n = 0$ and 2 for $n > 0$. Separate real and imaginary parts of this result to obtain

$$J_0(k|\boldsymbol{\rho} - \boldsymbol{\rho}'|) = \sum_{1}^{\infty} \epsilon_n J_n(k\rho) J_n(k\rho') \cos n(\theta - \theta')$$

$$Y_0(k|\boldsymbol{\rho} - \boldsymbol{\rho}'|) = \sum_{1}^{\infty} \epsilon_n J_n(k\rho_<) Y_n(k\rho_>) \cos n(\theta - \theta')$$

4.12.6 According to the two-dimensional Sommerfeld radiation condition, there is no contribution to surface integrals at infinity from fields that have the large argument radial dependence $\rho^{-1/2} e^{ik\rho}$. However, if the total wave field also contains an incident plane wave, as would be the case in a scattering problem in which an incident plane wave is scattered by a cylindrical obstacle, then S_∞ does provide a contribution, namely the incident plane wave. This result is the two-dimensional version of the one-dimensional result obtained in (4.3.16).

   Use the series obtained for the Green's function in Problem 4.12.5, the series for a plane wave $p_0 = e^{ik\rho\cos\theta}$ considered in Eq. (D.47) et seq., as well as appropriate Wronskian relations to show that

$$\int_{S_\infty} ds' \left(g \frac{\partial p_0}{\partial \rho'} - p_0 \frac{\partial g}{\partial \rho'} \right) = e^{ik\rho\cos\theta}$$

where $ds' = a\, d\theta'$ and a is a constant radius of the surface. Show, therefore, that the surface integral at infinity yields the incident plane wave.

4.12.7 A source of cylindrical waves of strength Q_0 is located along the axis of a circular cylinder of radius a. Assume that the boundary condition $p(a, \theta) = 0$ is satisfied.

(a) Combine

$$\nabla^2 p + k^2 p = -Q_0 \delta^2(\mathbf{p})$$

$$\nabla^2 g + k^2 g = -\delta^2(\mathbf{p} - \mathbf{p}')$$

by the multiply-and-subtract procedure to obtain an integral equation for waves within the cylinder in the form

$$p(\mathbf{p}) = Q_0 g(\mathbf{p}|0) + \int_{\text{cyl.}} dS' \, g(\mathbf{p}|\mathbf{p}'_s) \left.\frac{\partial p}{\partial \rho'}\right|_{\rho'=a}, \qquad \rho \leq a$$

where $g(\mathbf{p}|\mathbf{p}')$ is the free-space Green's function considered in Problem 4.12.4. As a result of symmetry, $\partial p/\partial \rho'|_{\rho'=a}$ is a constant and can be taken outside the integral to yield

$$p(\mathbf{p}) = Q_0 g(\mathbf{p}|0) + \left.\frac{\partial p}{\partial \rho'}\right|_{\rho'=a} \int_{\text{cyl.}} dS' \, g(\mathbf{p}|\mathbf{p}'_s)$$

(b) Use the series expansion for $g(\mathbf{p}|\mathbf{p}')$ derived in Problem 4.12.4 to show that

$$p(\rho) = Q_0 \frac{i}{4} H_0^{(1)}(k\rho) + \left.\frac{\partial p}{\partial \rho'}\right|_{\rho'=a} (2\pi a) \frac{i}{4} J_0(k\rho) H_0^{(1)}(ka)$$

and since $p(a) = 0$, obtain the result

$$\left.\frac{\partial p}{\partial \rho'}\right|_{\rho'=a} = -\frac{Q_0}{2\pi a J_0(ka)}$$

(c) Show that for *any* value of $\rho > a$ the integral equation for $p(\mathbf{p})$ obtained in (a) reduces to

$$p(\rho) = Q_0 \frac{i}{4} H_0^{(1)}(k\rho) + \int_0^{2\pi} a\,d\varphi' \, \frac{i}{4} H_0^{(1)}(k\rho) J_0(ka) \left.\frac{\partial p}{\partial \rho'}\right|_{\rho'=a}$$

$$= 0$$

The vanishing of $p(\rho)$ in this instance is an example of the general statement contained in (4.8.9) for the case in which the field point is located *outside* the region to which Green's theorem is applied.

4.12.8 Apply the method of the previous problem to the case in which $\partial p / \partial \rho |_{\rho = a} = 0$ on the boundary and show that

$$p(a) = -\frac{Q_0}{2\pi k a J_0'(ka)}$$

4.13 MODIFIED GREEN'S FUNCTIONS—ONE DIMENSION

The Green's function has been interpreted as the response of a system to a point source of unit strength. For problems involving time independent heat conduction, the response is the steady state temperature established by a heat source that is localized at one point and has been acting long enough so that all transient effects are negligible. If the system is of finite extent and the boundary condition is the homogeneous Neumann condition that the normal derivative vanish on all boundaries, then a contradiction presents itself. Vanishing normal derivative corresponds to an insulated boundary and therefore no heat can flow out of the system. If a source is located within such a region, the temperature will continue to increase and no steady state temperature will even occur. Consequently, a Green's function of the type considered previously cannot be associated with this situation. Since the formulation of problems in terms of Green's functions has proven to be useful, it is worth attempting to introduce a modification of the Green's function concept that will also enable us to treat insulated boundaries and Neumann boundary conditions in general.

In order to motivate a method for treating this situation, let us first consider the one-dimensional *Helmholtz* equation $u'' + k^2 u = 0$ since a Green's function for this equation does exist [and has already been obtained in (4.2.14)] for the boundary condition of vanishing normal derivative. We shall then examine the solution of this equation in the limit that k approaches zero. The limiting process will provide insight into how to treat the steady state problem. In (4.2.14) the solution of

$$G_{xx} + k^2 G = -\delta(x - x') \tag{4.13.1}$$

with boundary conditions $G_x(0|x') = G_x(L|x') = 0$ was given as

$$G(x|x') = -\frac{\cos k(L - x_>)\cos kx_<}{k \sin kL} \tag{4.13.2}$$

In the limit $k \rightarrow 0$ this solution reduces to $-1/k^2 L$, which diverges in this limit. The divergence is, of course, a manifestation of the difficulties referred to in the previous paragraph.

We now introduce the expression

$$G_m(x|x') = G(x|x') + 1/k^2 L \tag{4.13.3}$$

that is, an expression from which the divergent part of $G(x|x')$ has been subtracted. Substitution into (4.13.1) shows that G_m satisfies.

$$\frac{d^2 G_m}{dx^2} + k^2 \left(G_m - \frac{1}{kL^2} \right) = -\delta(x - x')$$

(4.13.4)

or

$$\frac{d^2 G_m}{dx^2} + k^2 G_m = -\delta(x - x') + \frac{1}{L}$$

(4.13.5)

Note that the strength of the *entire* source term in this latter equation is

$$\int_0^L dx' \left[-\delta(x - x') + \frac{1}{L} \right] = 0$$

(4.13.6)

The physical significance of G_m is now clear. It is the response of the system to both a *localized* source and a *uniformly distributed* sink (i.e., a source of negative strength) that is of sufficient magnitude to make the total source strength equal to zero. Such a source is, of course, compatible with the presence of insulated boundaries in a steady state heat problem.

To obtain the form of G_m in the limit $k \to 0$, we need merely introduce a series expansion for the trigonometric expressions occurring in (4.13.2) and retain only those terms that do not vanish in the limit $k \to 0$. We find

$$G_m(x|x') \cong \frac{[1 - \frac{1}{2}k^2(L - x_>)^2][1 - \frac{1}{2}(kx_<)^2]}{k^2 L[1 - \frac{1}{6}(kL)^2]} + \frac{1}{k^2 L}$$

(4.13.7)

$$\cong \frac{1}{2L}(x_<^2 + x_>^2 - 2Lx_>) + \frac{L}{3}$$

Since $k^2 G_m$ vanishes as $k \to 0$, the equation satisfied by G_m in this limit is seen from (4.13.5) to be

$$\frac{d^2 G_m}{dx^2} = -\delta(x - x') + \frac{1}{L}$$

(4.13.8)

This result may be used as a defining equation for G_m. Note that it has the expected physical significance of being the temperature distribution due to both a point source and a uniform background source of opposite sign. Since G_m occurs only under a derivative sign in both the defining equation (4.13.8) and in the Neumann boundary conditions, the value of the additive constant in

$G_m(x|x)$ is usually of no interest. However, as indicated below, it has the value in (4.13.7) such that the average value of $G_m(x|x')$ over the length of the system is zero, that is,

$$\int_0^L dx \, G_m(x|x') = \int_0^L dx \, G_m(x'|x) = 0 \qquad (4.13.9)$$

Note that the reciprocity principle is again satisfied in that $G_{m>}(x|x') = G_{m<}(x'|x)$.

The expression for $G_m(x|x')$ derived above may also be obtained by noting that the solution of (4.13.8), the differential equation defining G_m, will be composed of three parts:

1. A term $-\frac{1}{2}|x - x'|$ that provides the required unit discontinuity in slope at $x = x'$ [cf. (4.1.32)].
2. A particular solution $x^2/2L$ that supplies the inhomogeneous term $1/L$.
3. A homogeneous solution $ax + b$.

Thus

$$G_m(x|x') = -\frac{1}{2}|x - x'| + \frac{x^2}{2L} + ax + b \qquad (4.13.10)$$

The constants a and b may be functions of the parameter x'. To satisfy the boundary condition $dG_m/dx = 0$ at $x = 0, L$, we must require $a = -\frac{1}{2}$. The constant term b may be determined by requiring that the average value of G_m be zero. One finds $b = x'(x' - L)/2L + L/3$ and the final result is equivalent to that given above in (4.13.7).

When the modified Green's function is combined with the steady state heat conduction equation $u_{xx} = -s(x)/K$ by the usual multiply-and-subtract procedure, we obtain

$$G_m u_{xx} - u G_{mxx} = -\frac{1}{K} s(x) G_m - \delta(x - x')u(x) + \frac{1}{L} u(x) \qquad (4.13.11)$$

As noted above, in order for a steady state situation to be possible, we must impose the conservation condition (1.3.3) on the source term and require

$$-Ku_x(L) + Ku_x(0) = \int_0^L s(x) \, dx \qquad (4.13.12)$$

For homogeneous Neumann conditions, this condition reduces to $\int_0^L s(x) \, dx = 0$.

Interchanging the x and x' coordinates in (4.13.11) (recall that G_m is symmetric), integrating over x' from 0 to L, and noting that dG_m/dx vanishes at the end points, we obtain

$$u(x) - \langle u \rangle = G_{m<}(x|L)u_x(L) - G_{m>}(x|0)u_x(0)$$

$$+ \frac{1}{K} \int_0^L G_m(x|x')s(x')\, dx' \qquad (4.13.13)$$

where $\langle u \rangle = L^{-1} \int_0^L u(x)\, dx$, the average temperature on the beam. The temperature itself is never determined by the information provided in this problem. Note that this arbitrariness appears here in the form of an unspecified average temperature $\langle u \rangle$.

Example 4.8. The ends of a beam of length L are insulated. A uniform positive heat source density of strength s_0 calories per second per centimeter extends over the region $0 \le x \le L/2$ and a point heat sink (refrigerator) of strength q_0 calories per second is located at $x = 3L/4$. (a) Determine the relation between s_0 and q_0 for a steady state to exist. (b) Determine the temperature difference $u(x) - \langle u \rangle$ where $\langle u \rangle$ is the average temperature on the beam.

(a) For a steady state, the total source strength must vanish, that is,

$$\int_0^L dx\, s(x) = \int_0^L dx \left[s_0 H \left(\frac{L}{2} - x \right) + \delta \left(x - \frac{3L}{4} \right) \right] = 0 \qquad (4.13.14)$$

Evaluation of the integral yields $s_0 L/2 + q_0 = 0$.

(b) Combining the diffusion equation

$$\frac{d^2 u}{dx^2} = -\frac{1}{K} s(x), \qquad 0 \le x \le L \qquad (4.13.15)$$

with (4.13.8) for the modified Green's function, we obtain

$$u(x) - \langle u \rangle = \frac{1}{K} \int_0^L dx'\, G_m(x|x')s(x'), \qquad 0 \le x \le L \qquad (4.13.16)$$

Using the source term for the present problem and the appropriate form of G_m from (4.13.7) in each region, we obtain

$$u - \langle u \rangle = \frac{1}{K} \begin{cases} s_0 \left[\int_0^x dx'\, G_{m>} + \int_x^{L/2} dx'\, G_{m<} \right] + q_0 G_{m<} \left(x \Big| \frac{3L}{4} \right), & 0 \le x < \frac{L}{2} \\[2ex] s_0 \int_0^{L/2} dx'\, G_{m>} + q_0 G_{m<} \left(x \Big| \frac{3L}{4} \right), & \frac{L}{2} < x < \frac{3L}{4} \\[2ex] s_0 \int_0^{L/2} dx'\, G_{m>} + q_0 G_{m>} \left(x \Big| \frac{3L}{4} \right), & \frac{3L}{4} < x \le L \end{cases}$$

$$(4.13.17)$$

With G_m given by (4.13.7), the integrals are readily evaluated to yield

$$u(x) - \langle u \rangle = \frac{q_0 L}{K} \begin{cases} \left(\dfrac{x}{L}\right)^2 - \dfrac{25}{96}, & 0 \le x < \dfrac{L}{2} \\[2ex] \dfrac{x}{L} - \dfrac{49}{96}, & \dfrac{L}{2} < x \le \dfrac{3L}{4} \\[2ex] \dfrac{23}{96}, & \dfrac{3L}{4} < x \le L \end{cases} \qquad (4.13.18)$$

Application of the method of this section to the sphere is considered in Section 9.5.

Problems

4.13.1 A heat source and a heat sink of equal (and opposite) strengths are located a distance Δx apart at $x = a + \Delta x/2$ and $x = a - \Delta x/2$, respectively, on a beam of length L. Both ends of the beam are insulated. The equation governing the temperature distribution on the beam is

$$\frac{d^2 u}{dx^2} = -\frac{q_0}{K} \left[\delta\left(x - a - \frac{\Delta x}{2}\right) - \delta\left(x - a + \frac{\Delta x}{2}\right) \right]$$

In the limit $\Delta x \to 0$, $q_0 \to \infty$, such that $q_0 \Delta x = m_0$ remains fixed, we may write

$$\frac{d^2 u}{dx^2} = \frac{m_0}{K} \delta'(x - a)$$

where $\delta'(x - a)$ refers to the derivative of the delta function. Such a doublet source is referred to as a dipole with moment m_0.

Combine this latter equation for u with (4.13.5) for the modified Green's function and, after an integration by parts similar to that employed in obtaining Eq. (1.5.10), show that the temperature on the beam is

$$\frac{u(x) - \bar{u}}{m_0} = \begin{cases} a - L, & 0 < x < a \\ a, & a < y < L \end{cases}$$

Average the right-hand side of this expression over the length of the beam and show that the result is zero.

5 Spherical Geometry

The techniques for solving partial differential equations that have been developed thus far, namely separation of variables, Laplace transforms, and the use of Green's functions, can be applied to other geometries besides the polar and cylindrical configurations considered up until now. In Section 3.5 the temperature around a heated disk was obtained quite easily by using oblate spheroidal coordinates. A full development of the problems that can be solved with these other coordinate systems is beyond the scope of the present volume. One additional system that can be readily analyzed, however, is that of spherical coordinates. In the present chapter we take up this topic but restrict the development to problems with azimuthal symmetry, that is, problems that vary with latitude but not longitude.

5.1 SOLUTION OF LAPLACE'S EQUATION

As shown in Section 3.5, the Laplacian of $f(r, \theta, \varphi)$ in a spherical coordinate system has the form

$$\nabla^2 f = \frac{1}{r^2} \frac{\partial}{\partial r} \left(r^2 \frac{\partial f}{\partial r} \right) + \frac{1}{r^2 \sin \theta} \frac{\partial}{\partial \theta} \left(\sin \theta \frac{\partial f}{\partial \theta} \right) + \frac{1}{r^2 \sin^2 \theta} \frac{\partial^2 f}{\partial \varphi^2} \quad (5.1.1)$$

and as noted in that earlier section, the radial part of this expression may be rewritten by using the identity

$$\frac{1}{r^2} \frac{\partial}{\partial r} \left(r^2 \frac{\partial f}{\partial r} \right) = \frac{1}{r} \frac{\partial^2}{\partial r^2} (rf) \quad (5.1.2)$$

The coordinate system being used is shown in Figure 5.1. In Section 3.6 we have already considered problems possessing complete spherical symmetry, that is neither θ nor φ dependence.

Since the φ dependence is not being considered, Laplace's equation $\nabla^2 f = 0$ reduces to

$$\frac{1}{r} \frac{\partial^2}{\partial r^2} (rf) + \frac{1}{r^2 \sin \theta} \frac{\partial}{\partial \theta} \left(\sin \theta \frac{\partial f}{\partial \theta} \right) = 0 \quad (5.1.3)$$

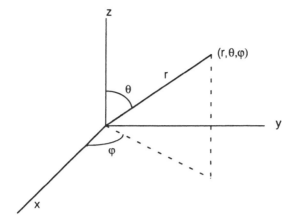

Figure 5.1. Spherical coordinate system.

and when the separated form $rf(r, \theta) = R(r)\Theta(\theta)$ is introduced, we obtain

$$\frac{r^2 R''}{R} = -\frac{(\sin\theta\,\Theta')'}{\Theta\sin\theta} = k^2 \qquad (5.1.4)$$

where k^2 is the separation constant. At this stage in the solution it is assumed that the reader is familiar with the contents of Appendix E, where the solution of the equation for Θ in terms of Legendre polynomials is developed. In particular, it is shown there that the solution cannot be finite at θ equal to both 0 and π unless k^2 is of the form $n(n + 1)$ where $n = 0, 1, 2, \ldots$. We thus impose this condition on k^2 and write the two separated equations in the form

$$r^2 R'' - n(n + 1)R = 0$$
$$(\Theta'\sin\theta)' + n(n + 1)\sin\theta\,\Theta = 0 \qquad (5.1.5)$$

The equation for R is of the Cauchy-Euler form and has solutions $R_n(r)$ that are linear combinations of r^{n+1} and r^{-n} while, as shown in Appendix E, the solution for Θ is expressed in terms of the Legendre functions $P_n(\cos\theta)$ and $Q_n(\cos\theta)$. The solutions $P_n(\cos\theta)$ are polynomials in $\cos\theta$, the first seven of which are listed in Eq. (E.14). Employing a superposition of these basic product solutions as in previous developments, we obtain the solution of Laplace's equation as a summation of the elementary solutions $f_n(r, \theta) = r^{-1}R_nP_n(\cos\theta)$ and $r^{-1}R_nQ_n(\cos\theta)$. The general solution thus takes the form

$$f(r) = \sum_{n=0}^{\infty} \{r^n[A_nP_n(\cos\theta) + B_nQ_n(\cos\theta)]$$
$$+ r^{-n-1}[C_nP_n(\cos\theta) + D_nQ_n(\cos\theta)]\} \qquad (5.1.6)$$

As was the case for polar and cylindrical coordinates, this solution must be specialized in various ways in order to avoid divergences. If the solutions must remain finite at the origin, the terms involving r^{-n-1} must be eliminated by choosing $C_n = D_n = 0$. Divergences in the limit $r \to \infty$ are avoided by choosing $A_n = B_n = 0$ so that the terms containing r^n do not appear. Finally, a knowledge of the solutions P_n and Q_n is required to determine which of these solutions can be included. As noted in Appendix E, the Q_n solutions have a logarithmic singularity at $\theta = 0, \pi$. If the solution is to remain finite at these angles, we must eliminate the Q_n solutions by setting $B_n = D_n = 0$.

For the common problem of determining a solution of Laplace's equation either inside or outside of a complete sphere we then have

$$f(r, \theta) = \sum_{n=0}^{\infty} A_n r^n P_n(\cos \theta), \qquad 0 < r < a$$

$$f(r, \theta) = \sum_{n=0}^{\infty} C_n r^{-n-1} P_n(\cos \theta), \qquad a < r < \infty \qquad (5.1.7)$$

In either case, the unknown coefficients A_n or C_n can be expressed in terms of $f(a, \theta)$, the value of $f(r, \theta)$ on the boundary, by using the orthogonality relations for the Legendre polynomials $P_n(\cos \theta)$, as shown in Appendix E. For $r \le a$ we multiply the first of (5.1.7) by $\sin \theta P_m(\cos \theta)$ and integrate over θ from 0 to π. The result yields the coefficient A_m in the form

$$A_m = \frac{2m+1}{2a^m} \int_0^{\pi} f(a, \theta) P_m(\cos \theta) \sin \theta \, d\theta \qquad (5.1.8)$$

For $r \ge a$, a similar procedure applied to the second of (5.1.7) yields

$$C_m = \frac{2m+1}{2a^{-m-1}} \int_0^{\pi} f(a, \theta) P_m(\cos \theta) \sin \theta \, d\theta \qquad (5.1.9)$$

With the coefficients A_m or C_m thus determined, the expressions (5.1.7) provide the solution $f(r, \theta)$ for all values of r and θ either inside or outside the sphere $r = a$.

Although the representation of the surface temperature as a sum of Legendre polynomials usually requires an infinite series of such terms, simple problems involving a series with only a finite number of terms may be constructed. The following example is of this type.

Example 5.1. The temperature on the surface of a sphere of radius a is given by $u(a, \theta) = u_0 \cos^2 \theta$. Determine the temperature within the sphere.

Since $P_2(\cos \theta) = \frac{1}{2}(3 \cos^2 \theta - 1)$ and $P_0(\cos \theta) = 1$, the surface temperature may be expressed as a finite sum of Legendre polynomials by writing $\cos^2 \theta = \frac{2}{3} P_2(\cos \theta) + \frac{1}{3} P_0(\cos \theta)$. The temperature within the sphere is given by the first of (5.1.7) with the coefficients A_n obtained from (5.1.8). We find

$$A_0 = \frac{1}{2} \int_0^\pi u_0 \left[\frac{2}{3} P_2(\cos\theta) + \frac{1}{3} P_0(\cos\theta) \right] P_0(\cos\theta) \sin\theta \, d\theta$$

$$= \frac{1}{2} \frac{1}{3} 2u_0 = \frac{1}{3} u_0$$

$$A_2 = \frac{5}{2a^2} \int_0^\pi u_0 \left[\frac{2}{3} P_2(\cos\theta) + \frac{1}{3} P_0(\cos\theta) \right] P_2(\cos\theta) \sin\theta \, d\theta$$

$$= \frac{5}{2a^2} u_0 \frac{2}{3} \frac{2}{5} = \frac{2}{3} \frac{u_0}{a^2} \tag{5.1.10}$$

while all other coefficients are zero by virtue of the orthogonality of the Legendre polynomials. The temperature within the sphere is thus

$$u(r, \theta) = u_0 \left[\frac{1}{3} + \frac{2}{3} \left(\frac{r}{a} \right)^2 P_2(\cos\theta) \right] \tag{5.1.11}$$

Note that since heat flows from hotter to cooler regions, if we assume $u_0 > 0$, the specified surface temperature implies an influx of heat near the north and south poles of the sphere and an efflux of heat near the equator. To determine the values of θ that divide these various regions, we need merely set $u_r(a, \theta) = 0$ and solve for θ. From the solution (5.1.11) we readily obtain $P_2(\cos\theta) = 0$, which is equivalent to $3\cos^2\theta - 1 = 0$. Solving for θ we find $\theta = \pm 1/\sqrt{3}$ or θ is $54.7°$ and $125.3°$.

As a more extensive usage of the method, consider the surface of a sphere of radius a that is kept at temperature $u(a, \theta) = 0$ except for a cap $0 \leq \theta \leq \alpha$ on which $u(a, \theta) = u_0$. We again determine the temperature within the sphere and obtain the temperature at the center of the sphere as a function of the size of the cap as measured by the angle α as well as the temperature along the axis $\theta = 0, \pi$ when $\alpha = \pi/2$.

Using the solution for $r < a$ given in (5.1.7) with the coefficients A_n given by (5.1.8) with

$$u(a, \theta) = \begin{cases} u_0, & 0 \leq \theta < \alpha \\ 0, & \alpha < \theta \leq \pi \end{cases} \tag{5.1.12}$$

we have

$$A_n = \frac{2n+1}{2a^n} u_0 \int_0^\alpha P_n(\cos\theta) \sin\theta \, d\theta \tag{5.1.13}$$

The integral is evaluated by using the result obtained in Problem E.5. We have either of the two equivalent forms

$$A_n = \frac{u_0}{2a^n} \frac{2n+1}{n+1} [P_{n-1}(\cos \alpha) - \cos \alpha P_n(\cos \alpha)]$$

$$A_n = \frac{u_0}{2a^n} [P_{n-1}(\cos \alpha) - P_{n+1}(\cos \alpha)] \qquad (5.1.14)$$

The temperature may thus be written

$$u_\alpha(r, \theta) = \frac{u_0}{2} \sum_{n=0}^{\infty} \left(\frac{r}{a}\right)^n [P_{n-1}(\cos \alpha) - P_{n+1}(\cos \alpha)] P_n(\cos \theta) \qquad (5.1.15)$$

Note that since the equation defining Legendre polynomials contains the term $n(n + 1)$, which is unchanged by replacing n by $-n - 1$, we have $P_{-n-1}(\cos \theta) = P_n(\cos \theta)$ and, in particular, $P_{-1}(\cos \theta) = P_0(\cos \theta) = 1$.

At the center of the sphere ($r = 0$), where only the $n = 0$ term in (5.1.15) is nonvanishing, the temperature is

$$u_\alpha(0) = \frac{u_0}{2} [1 - P_1(\cos \alpha)] = \frac{u_0}{2} (1 - \cos \alpha) = u_0 \sin^2 \frac{\alpha}{2} \qquad (5.1.16)$$

For $\alpha = \pi/2$, the first of the two expressions for A_n given in (5.1.14) is more convenient. It reduces to $A_n = (u_0/2a^n)[(2n + 1)/(n + 1)]P_{n-1}(0)$. Along the axis in the upper hemisphere ($\theta = 0$) and the lower hemisphere ($\theta = \pi$), the temperature is

$$u_{\pi/2}(r, 0) = \frac{u_0}{2} \left[1 + \sum_{n=1}^{\infty} \frac{2n+1}{n+1} \left(\frac{r}{a}\right)^n P_{n-1}(0)\right]$$

$$u_{\pi/2}(r, \pi) = \frac{u_0}{2} \left[1 - \sum_{n=1}^{\infty} (-1)^n \frac{2n+1}{n+1} \left(\frac{r}{a}\right)^n P_{n-1}(0)\right] \qquad (5.1.17)$$

where we have again used $P_{-1}(\cos \theta) = 1$ as well as the relations $P_n(1) = 1$ and $P_n(-1) = (-1)^n$. As noted in (E.19), $P_{n-1}(0)$ is only nonzero for $n - 1$ even, that is, n odd, all factors $(-1)^n$ thus reduce to -1, and both summations are of the same form. Writing

$$S = \sum_{n=1}^{\infty} \frac{2n+1}{n+1} \left(\frac{r}{a}\right)^n P_{n-1}(0) \qquad (5.1.18)$$

we have

$$u_{\pi/2}(r, 0) = \frac{u_0}{2} (1 + S)$$

$$u_{\pi/2}(r, \pi) = \frac{u_0}{2} (1 - S) \qquad (5.1.19)$$

The sums may be evaluated by first writing $(2n + 1)/(n + 1) = 2 - 1/(n + 1)$ and then setting $n = 2k + 1$. This procedure yields

$$S = 2\frac{r}{a} \sum_{k=0}^{\infty} \left(\frac{r}{a}\right)^{2k} P_{2k}(0) - \frac{a}{r} \sum_{k=0}^{\infty} \left(\frac{r}{a}\right)^{2k+2} \frac{P_{2k}(0)}{2k + 2} \qquad (5.1.20)$$

These sums may be expressed in closed form by using (E.18) and (E.22). The result is

$$S = \frac{2r/a}{\sqrt{1 + (r/a)^2}} - \frac{a}{r}[\sqrt{1 + (r/a)^2} - 1] \qquad (5.1.21)$$

The temperature on axis is now given by using this result for S in (5.1.19).

For the azimuthally symmetric problems being considered in this chapter, there is a simple relation between the solution of Laplace's equation along the polar (z) axis and the solution throughout the entire spherical region. The relation is obtained by noting that on the positive z axis, where $\theta = 0$, and according to the definition of $P_n(\cos \theta)$ introduced in Appendix E, where $P_n(1) = 1$, the solutions reduce to

$$f(r, 0) = \sum_{n=0}^{\infty} A_n z^n, \qquad 0 \le z \le a$$

$$f(r, 0) = \sum_{n=0}^{\infty} C_n z^{-n-1}, \qquad a \le z < \infty \qquad (5.1.22)$$

Hence, for an azimuthally symmetric problem, if the solution is known as a series in powers of z on the positive z axis, then the solution may be obtained for all θ by merely replacing z by r and multiplying the nth term in the series by $P_n(\cos \theta)$. It should be emphasized that this is only true for problems that are independent of φ.

As a simple example of this result, note that when specialized to the z axis, the solution in Example 5.1 takes the form

$$u(r, 0) = f(z) = \frac{u_0}{3}\left[1 + 2\left(\frac{z}{a}\right)^2\right], \qquad z > 0 \qquad (5.1.23)$$

If this information on the temperature were available initially and we also knew that the temperature were symmetric about the z axis, we could use the principle referred to above to immediately infer (5.1.11) throughout the entire sphere.

An example of this principle that will be of use in other problems is provided by the field about a unit point source in an infinite region. The expression is actually the free-space Green's function for Laplace's equation, namely,

$$g(\mathbf{r}|\mathbf{r}') = \frac{1}{4\pi|\mathbf{r} - \mathbf{r}'|} = \frac{1}{4\pi\sqrt{r^2 + r'^2 - 2rr'\cos\gamma}} \qquad (5.1.24)$$

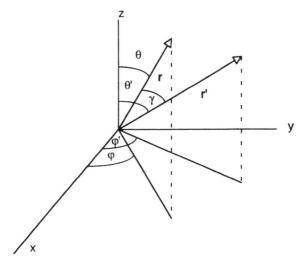

Figure 5.2. Angles determining observation point **r** and field point **r′**.

where γ is the angle between the position vectors **r** and **r′**, as shown in Figure 5.2. In terms of the unit vectors **i**, **j**, and **k** of rectangular coordinates we have

$$\mathbf{r} = r(\mathbf{i}\cos\varphi\sin\theta + \mathbf{j}\sin\varphi\sin\theta + \mathbf{k}\cos\theta)$$

$$\mathbf{r'} = r'(\mathbf{i}\cos\varphi'\sin\theta' + \mathbf{j}\sin\varphi'\sin\theta' + \mathbf{k}\cos\theta') \qquad (5.1.25)$$

Thus, we obtain

$$\cos\gamma = \frac{\mathbf{r}\cdot\mathbf{r'}}{rr'} = \cos\theta\cos\theta' + \sin\theta\sin\theta'\cos(\varphi - \varphi') \qquad (5.1.26)$$

The application of the principle that we now consider is to the integral of $g(\mathbf{r}|\mathbf{r'})$ over all φ, that is, $\int_0^{2\pi} g(\mathbf{r}|\mathbf{r'})\,d\varphi$. It should be immediately evident (by integrating from φ' to $\varphi' + 2\pi$ and changing the integration variable to $\varphi - \varphi'$) that the result will be independent of φ'. Since $\int_0^{2\pi} g(\mathbf{r}|\mathbf{r'})\,d\varphi$ also satisfies Laplace's equation for $\mathbf{r} \neq \mathbf{r'}$ and is independent of φ, we expect it to have an expansion of the form

$$g(r,\theta|r',\theta') \equiv \int_0^{2\pi} d\varphi\, g(\mathbf{r}|\mathbf{r'}) = \sum_{n=0}^{\infty} f(r,r',\theta')P_n(\cos\theta) \qquad (5.1.27)$$

To determine the coefficients $f_n(r,r',\theta')$, we note that on the positive z axis, where $\theta = 0$, we have, from (5.1.24),

$$g(r,0|r',\theta') = \frac{1}{4\pi}\int_0^{2\pi} \frac{d\varphi}{\sqrt{r^2 + r'^2 - 2rr'\cos\gamma}}\bigg|_{\theta=0} \qquad (5.1.28)$$

Since $\cos \gamma = \cos \theta'$ for $\theta = 0$, we may introduce the customary expansion for $1/\sqrt{r^2 + r'^2 - 2rr' \cos \theta'}$ as given in (E.25) and immediately integrate over φ to obtain

$$g(r, 0|r', \theta') = \frac{1}{2r_>} \sum_{n=0}^{\infty} \left(\frac{r_<}{r_>}\right)^n P_n(\cos \theta') \qquad (5.1.29)$$

where $r_>$ and $r_<$ refer, respectively, to the larger and smaller of r and r'. Thus we have

$$\frac{1}{4\pi} \int_0^{2\pi} \frac{d\varphi}{|\mathbf{r} - \mathbf{r}'|}\bigg|_{\theta=0} = \frac{1}{2r_>} \sum_{n=0}^{\infty} \left(\frac{r_<}{r_>}\right)^n P_n(\cos \theta') \qquad (5.1.30)$$

By employing the symmetry principle we can now write the result off the z axis (i.e., for $\theta \neq 0$) in the form

$$g(r, \theta|r', \theta') = \frac{1}{4\pi} \int_0^{2\pi} \frac{d\varphi}{|\mathbf{r} - \mathbf{r}'|} = \frac{1}{2r_>} \sum_{n=0}^{\infty} \left(\frac{r_<}{r_>}\right)^n P_n(\cos \theta') P_n(\cos \theta)$$

$$(5.1.31)$$

This result is quite useful for evaluating surface integrals that possess azimuthal symmetry.

Finally, note that if the z axis is aligned so as to pass through the origin and the point $\mathbf{r}'$, then $\theta' = 0$. In this case there is no φ dependence in the unaveraged Green's function (5.1.24) and it becomes

$$g(\mathbf{r}|\mathbf{r}') = \frac{1}{4\pi|\mathbf{r} - \mathbf{r}'|} = \frac{1}{4\pi \sqrt{r^2 + r'^2 - 2rr' \cos \theta}} \qquad (5.1.32)$$

According to the expansion given in (E.25) we may write

$$g(\mathbf{r}|\mathbf{r}') = \frac{1}{4\pi r_>} \sum_{n=0}^{\infty} \left(\frac{r_<}{r_>}\right)^n P_n(\cos \theta), \qquad \theta' = 0 \qquad (5.1.33)$$

When the point $\mathbf{r}'$ is not on the z axis, the angle θ in (5.1.33) is replaced by the angle γ defined in (5.1.26). The expression of $P_n(\cos \gamma)$ in terms of θ, φ, θ', and φ', a relation known as the addition theorem for spherical harmonics, requires the use of the associated Legendre polynomials and will not be considered here.

Problems

5.1.1 A spherical shell has an inner radius a that is insulated. At the outer radius b the temperature is $u(b, \theta) = u_0 \sin^2\theta$. Show that the heat flow out at the north pole is

$$J = -Ku_r(b, 0) = \frac{4Ku_0}{b} \frac{a^5 + b^5}{2a^5 + 3b^5}$$

5.1.2   The surface of radius a is maintained at a temperature u_0 over the upper hemisphere and a temperature $-u_0$ over the lower hemisphere.

(a) Use the relation obtained in Problem E.5 to show that

$$u(r, \theta) = \sum_{n=0}^{\infty} A_{2k+1} r^{2k+1} P_{2k+1}(\cos \theta)$$

where

$$A_{2k+1} = \frac{u_0}{2a^{2k+1}} \frac{4k+3}{k+1} P_{2k}(0)$$

(b) Use relations (E.20) and (E.22) to show that

$$u(r, 0) = \frac{u_0 a}{r} \left(1 - \frac{a^2 - r^2}{a\sqrt{a^2 + r^2}} \right)$$

5.1.3   A sphere of radius a and conductivity K is irradiated over its northern hemisphere so that the heat input is proportional to latitude, that is,

$$-Ku_r(a, \theta) = \begin{cases} -J_0\cos\theta, & 0 \le 0 < \pi/2 \\ 0, & \pi/2 < \theta \le \pi \end{cases}$$

where J_0 is a constant.

(a) Show that the temperature within the sphere (determined only to within a constant) is given by

$$u(r, \theta) - A_0 = \sum_{1}^{\infty} A_n r^n P_n(\cos \theta)$$

where

$$A_n = \frac{J_0}{K} \frac{2n+1}{2na^{n-1}} I_n, \qquad n = 1, 2, 3, \ldots$$

in which (using the first of E.16 and Problem E.5)

$$I_n = \int_0^1 d\mu \, \mu P_n(\mu) = \frac{1}{2n+1} \left[\frac{n+1}{n+2} P_n(0) + P_{n+2}(0) \right]$$

(b) Show that

$$u(r, 0) - A_0 = \frac{J_0 a}{2K}(1 + S)$$

$$u(a, \pi) - A_0 = \frac{J_0 a}{2K}(-1 + S)$$

where $S = \sum_1^\infty (4k + 1)I_{2k}/k$. Thus show that the difference in temperature between the north and south poles is $J_0 a/K$.

5.2 SOURCE TERMS AND THE MULTIPOLE EXPANSION

As shown in Section 4.8, the solution of

$$\nabla^2 u = -s(\mathbf{r}) \tag{5.2.1}$$

when no surfaces are present may be written

$$u(\mathbf{r}) = \int d^3\mathbf{r}'\, G(\mathbf{r}|\mathbf{r}')s(\mathbf{r}')$$

$$= \frac{1}{4\pi}\int d^3\mathbf{r}'\, \frac{s(\mathbf{r}')}{|\mathbf{r} - \mathbf{r}'|} \tag{5.2.2}$$

where the region of integration extends throughout the source, that is, the region for which $s(\mathbf{r}') \neq 0$. The expansion of $|\mathbf{r} - \mathbf{r}'|^{-1}$ in terms of Legendre polynomials provides a convenient approach to the evaluation of this integral. We shall only consider the case in which $s(\mathbf{r}')$ symmetric about the polar axis so that $s(\mathbf{r}') = s(r', \theta')$. Then the field $u(\mathbf{r})$ also depends only upon r and θ and we have $u(\mathbf{r}) = u(r, \theta)$. We can thus use the result noted in the previous section that a solution on the z axis of the form $A_n z^n$ or $C_n z^{-n-1}$ may be extended to all angles by simply replacing z^n by $r^n P_n(\cos \theta)$ or z^{-n-1} by $r^{-n-1}P_n(\cos \theta)$.

The distance between the field point $\mathbf{r}$ and a point $\mathbf{r}'$ within the source region may be written

$$|\mathbf{r} - \mathbf{r}| = \sqrt{r^2 + r'^2 - 2rr'\cos \gamma} \tag{5.2.3}$$

with

$$\cos \gamma = \frac{\mathbf{r} \cdot \mathbf{r}'}{rr'} = \cos \theta \cos \theta' + \sin \theta \sin \theta'\cos(\varphi - \varphi') \tag{5.2.4}$$

as shown in Section 5.1. As noted there, we need only evaluate the integral in (5.2.2) at $\theta = 0$. In this case, $\cos \gamma = \cos \theta'$ and $r = z$. Then $|\mathbf{r} - \mathbf{r}'| =$

$\sqrt{z^2 + r'^2 - 2zr'\cos\theta'}$ and the temperature $u(r, 0)$ takes the form

$$u(r, 0) = u(z, 0)$$

$$= \frac{1}{4\pi} \int_0^{2\pi} d\varphi' \int_0^{\pi} \sin\theta' \, d\theta' \int_0^{r_{max}} \frac{r'^2 \, dr' \, s(r', \theta')}{\sqrt{z^2 + r'^2 - 2zr'\cos\theta'}}$$

$$(5.2.5)$$

where $r_{max}(\theta)$ is the largest value taken on by r' when one integrates over the source region. For field points $r = z > r_{max}(\theta')$ one has

$$u(z, 0) = \frac{1}{4\pi} \int_0^{2\pi} d\varphi' \int_0^{\pi} \sin\theta' \, d\theta' \int_0^{r_{max}} r'^2 \, dr'^2 \, s(r', \theta')$$

$$\cdot \frac{1}{z} \sum_{n=1}^{\infty} \left(\frac{r'}{z}\right)^n P_n(\cos\theta')$$

$$= \frac{1}{4\pi} \sum_{n=0}^{\infty} \frac{m_n}{z^{n+1}} \qquad (5.2.6)$$

in which

$$m_n = 2\pi \int_0^{\pi} \sin\theta' \, d\theta' \int_0^{r_{max}} dr' \, s(r', \theta') r'^{n+2} P_n(\cos\theta') \quad (5.2.7)$$

The field at an arbitrary angle is then

$$u(r, \theta) = \frac{1}{4\pi} \sum_{n=0}^{\infty} \frac{m_n}{r^{n+1}} P_n(\cos\theta) \qquad (5.2.8)$$

5.2.1 Axial Multipoles

A simple physical interpretation may be given to the sequence of constants m_n that appears in the expansion (5.2.8). To this end, we first consider the temperature in an infinite solid that results from a localized heat source that we model by a point source. In spherical coordinates a point source of strength q_0 and located at $\mathbf{r} = \mathbf{r}_0$ having spherical coordinates $r = a$, $\theta = \theta_0$, and $\varphi = \varphi_0$ is written

$$s_{point} = q_0 \delta^3(\mathbf{r} - \mathbf{r}_0)$$

$$= q_0 \frac{\delta(r - a)\delta(\theta - \theta_0)\delta(\varphi - \varphi_0)}{r^2 \sin\theta} \qquad (5.2.9)$$

This expression has the required form for a delta function in spherical coordinates since

$$\int s_{point}(r)\, d^3r = \int_0^{2\pi} d\varphi \int_0^{\pi} \sin\theta\, d\theta \int_{-0}^{\infty} r^2\, dr\, s_{point}(r, \theta, \varphi)$$

$$= q_0 \int_0^{2\pi} d\varphi\, \delta(\varphi - \varphi_0) \int_0^{\pi} d\theta\, \delta(\theta - \theta_0) \int_0^{\infty} dr\, \delta(r - a)$$

$$= q_0 \qquad\qquad (5.2.10)$$

Special cases of this expression are readily constructed. It is only necessary that the integral over all space be unity. As an example, consider the azimuthally symmetric case of a point source on the z axis at a distance a above the origin. The source term is then

$$s(r, \theta) = q_0 \frac{\delta(r - a)\delta(\theta)}{2\pi r^2 \sin\theta} \qquad\qquad (5.2.11)$$

When the source at the origin as given in (5.2.9) is introduced into the solution (5.2.2), we obtain

$$u(r) = \frac{1}{4\pi} \int \frac{d^3r'\, q_0 \delta^3(\mathbf{r}')}{|\mathbf{r} - \mathbf{r}'|} = \frac{q_0}{4\pi r} \qquad\qquad (5.2.12)$$

This integral may be evaluated without any recourse to angular decompositions by using the general result $\int d^3r'\, f(\mathbf{r}, \mathbf{r}')\delta^3(\mathbf{r}' - \mathbf{a}) = f(\mathbf{r}, \mathbf{a})$ and with $\mathbf{a} = (0, 0, 0)$ in the present instance. On setting $q_0 = m_0$ we have the first term in the general solution (5.2.8). The first term in this expansion is therefore equivalent to a point source of strength m_0 located at the origin. We now introduce considerations that will lead to an interpretation of the constant m_1.

The temperature field obtained in (5.2.12) displays the general space dependence associated with a point source in an unbounded medium. If we place a source on the z axis at a distance a above the origin and measure distance from that point by R, then (cf. Fig. 5.3)

$$u(r, \theta) = \frac{q_0}{4\pi R} = \frac{q_0}{4\pi} \frac{1}{\sqrt{a^2 + r^2 - 2ar\cos\theta}}$$

$$= \frac{q_0}{4\pi r} \frac{1}{\sqrt{1 + (a/r)^2 - 2(a/r)\cos\theta}} \qquad\qquad (5.2.13)$$

The radical is seen to be the generating function for Legendre polynomials given in (E.17). If we make the assumption $r \gg a$ and retain only the two

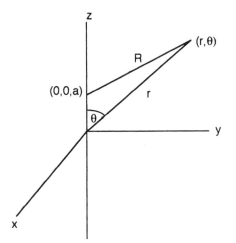

Figure 5.3. Location of field point with respect to a point source on the z axis.

lowest order terms in the expansion, we have

$$u(r, \theta) = \frac{q_0}{4\pi r} \left(1 + \frac{a}{r} \cos \theta \right) \qquad (5.2.14)$$

We now construct a source composed of two point sources. The first source is the one at $z = a$ that was just considered while the second is a negative source (i.e., a refrigeration unit) at the origin. Then, combining the results just obtained in (5.2.12) and (5.2.14), we obtain

$$u(r, \theta) = \frac{q_0}{4\pi r} \left(1 + \frac{a}{r} \cos \theta \right) - \frac{q_0}{4\pi r}$$

$$= \frac{q_0 a}{4\pi} \frac{1}{r^2} \cos \theta \qquad (5.2.15)$$

If we now set $q_0 a = m_1$ and note that $\cos \theta = P_1(\cos \theta)$, we see that the result corresponds to the second term in the expansion (5.2.8). This doublet source is referred to as a dipole and the strength of the source m_1 is referred to as the dipole moment of the source. It might be assumed that this result can only be approximate since we have neglected higher order terms in writing (5.2.14). However, in a mathematical sense, the result may be made exact for all values of r by considering the dipole moment m_1 to result from application of the limiting process $a \rightarrow 0$ and $q_0 \rightarrow \infty$ in such a way that m_1 remains constant. In general, by considering more elaborate groupings of positive and negative point sources (quadrupole, octopole, etc.), it is possible to represent any more complicated source by an equivalent multipole expansion located at the origin. The quadrupole is considered in Problem 5.2.2. It should be noted that we are only considering those source distributions and multipoles that are symmetric about the z axis, usually referred to as longitudinal multipoles.

As noted in Appendix E, we may also write

$$u(\mathbf{r}) = \frac{1}{4\pi} \sum_{n=0}^{\infty} m_n \frac{(-1)^n}{n!} \frac{\partial^n}{\partial r^n} \frac{1}{r} \tag{5.2.16}$$

so that the moments m_n have the interpretation of being the coefficients in an expansion of R^{-1} about the origin.

Example 5.2. Determine the temperature about a ring source described by $s(r', \theta')$ $= (Q_0/K)\delta(r' - a)\delta(\theta' - \alpha)/(2\pi a^2 \sin \alpha)$, that is, a circular ring of radius $a\sin \alpha$ at a distance $a\cos \alpha$ above the xy plane.

From (5.2.7), the various multipole moments are

$$m_n = (Q_0/K)a^n(\cos \alpha) \tag{5.2.17}$$

and the temperature at a field point $r > a$ is

$$u_\alpha(r, \theta) = \frac{Q_0}{4\pi Kr} \sum_{n=0}^{\infty} \left(\frac{a}{r}\right)^n P_n(\cos \alpha)P_n(\cos \theta) \tag{5.2.18}$$

If we ask for the field on axis when the source is in the xy plane, that is, set $\theta = 0$ and $\alpha = \pi/2$, we obtain

$$u_{\pi/2}(r, 0) = \frac{Q_0}{4\pi Kr} \sum_{n=0}^{\infty} \left(\frac{a}{r}\right)^n P_n(0) \tag{5.2.19}$$

As noted in (E.19), $P_n(0) = 0$ for $n = 1, 3, 5, \ldots$. Hence the sum is of the form obtained in (E.20) and we have

$$u_{\pi/2}(r, 0) = \frac{Q_0}{4\pi K} \frac{1}{\sqrt{a^2 + r^2}} \tag{5.2.20}$$

This result agrees with that obtained in Problem 4.9.6 where the field due to such a source in the XY plane was solved by using cylindrical coordinates.

Problems

5.2.1 (a) A bar of length $2a$ centered at the origin and lying along a z axis has a linear density λ so that the mass of the bar is $2\lambda a$. Obtain this result by integrating $M = \int \rho(\mathbf{r}) \, dv$ where $\rho(\mathbf{r}) = \lambda H(a - r)[\delta(\theta) - \delta(\theta - \pi)]/2\pi r^2 \sin \theta$ in which H is the Heaviside step function.

(b) A disk of radius a centered at the origin and lying in the xy plane has a surface density σ so that the mass of the disk is $\pi \sigma a^2$. Obtain this result by integrating $M = \int \rho(\mathbf{r}) \, dv$ where $\rho(\mathbf{r}) = \sigma H(a - r)\delta(\theta - \pi/2)/2\pi r^2 \sin \theta$.

5.2.2   A medium of infinite extent has heat sources of strength q_0 embedded at $z = \pm a$ and a negative source $- 2q_0$ located at the origin. Show that at large distances from the origin ($r \gg a$), the temperature in the medium is given approximately by

$$u = \frac{q_0 a^2}{2\pi r^3} P_2(\cos \theta)$$

The source is referred to as an axial quadrupole.

5.3   INVERSION—GREEN'S FUNCTION FOR A SPHERE

Consider one of the elementary solutions of the Laplace equation (5.1.3) that remains finite at the origin, that is,

$$f(r, \theta) = A_n r^n P_n(\cos \theta) \tag{5.3.1}$$

Note that the expression

$$\frac{1}{r} f\left(\frac{a^2}{r}, \theta\right) = C_n r^{-n-1} P_n(\cos \theta) \tag{5.3.2}$$

where $C_n = A_n a^{2n}$ is in the form of the associated *outer* solution, that is, it remains finite, in fact goes to zero, as r approaches infinity. Note also that if $r < a$, then $\underline{r} \equiv a^2/r$ is greater than a. This relation between inner and outer solutions is analogous to the one considered for the circle in Section 4.10 and is used here to introduce the notion of inversion for a sphere. The constant a will play the role of the radius of the sphere while r and $\underline{r}$ refer to field points inside and outside the sphere, respectively.

   We begin by noting that derivatives with respect to r are related to those with respect to the coordinate $\underline{r}$ according to $r f_r = -\underline{r} f_{\underline{r}}$ or, equivalently,

$$\frac{\partial f}{\partial r} = -\frac{r^2}{a^2} \frac{\partial f}{\partial \underline{r}} \tag{5.3.3}$$

Let us assume that a function $\psi(r, \theta)$ satisfies Laplace's equation in the r, θ coordinate system [or, by a mere change of notation, that $\psi(\underline{r}, \theta)$ satisfies Laplace's equation in an $\underline{r}, \theta$ system] and evaluate the Laplacian of the function $\varphi(r, \theta) = (a/r)\psi(a^2/r, \theta) = a^{-1}\underline{r}\psi(\underline{r}, \theta)$. We will find that $\nabla^2 \varphi(r, \theta) = 0$. Note that the function f introduced in (5.3.2) is an example of such a function.

   It will prove convenient to exploit the two different forms for the radial part of the Laplacian that were noted in (5.1.2). Then, abbreviating the angular derivatives by

$$L = \frac{1}{\sin \theta} \frac{\partial}{\partial \theta} \left(\sin \theta \frac{\partial}{\partial \theta} \right) \tag{5.3.4}$$

we begin with

$$\nabla^2 \varphi(r, \theta) = \frac{1}{r} \frac{\partial^2 [r\varphi(r, \theta)]}{\partial r^2} + \frac{1}{r^2} L\varphi(r, \theta) \tag{5.3.5}$$

Using (5.3.3) to transform to $\underline{r}$ derivatives, as well as $r\varphi = a\psi$ and $r = a^2/\underline{r}$, we obtain

$$\nabla^2 \varphi(r, \theta) = \frac{\underline{r}}{a^2} \left\{ \frac{\underline{r}^2}{a^2} \frac{\partial}{\partial \underline{r}} \left[\frac{\underline{r}^2}{a^2} \frac{\partial}{\partial \underline{r}} (a\psi) \right] \right\} + \frac{\underline{r}^3}{a^5} L\psi(\underline{r}, \theta) \tag{5.3.6}$$

On extracting the term $\underline{r}^5/a^5$, the Laplacian of φ becomes

$$\nabla^2 \varphi = \frac{\underline{r}^5}{a^5} \left[\frac{1}{\underline{r}^2} \frac{\partial}{\partial \underline{r}} \left(\underline{r}^2 \frac{\partial \psi}{\partial \underline{r}} \right) + \frac{1}{\underline{r}^2} L\psi \right] = \frac{\underline{r}^5}{a^5} \nabla^2 \psi = 0 \tag{5.3.7}$$

The final expression is zero since we have assumed that the Laplacian of $\psi(\underline{r}, \theta)$ vanishes in the $\underline{r}, \theta$ coordinate system. We therefore have the result that the Laplacian of $\varphi(r, \theta) = (a/r)\psi(a^2/r, \theta)$ also vanishes in the original r, θ coordinate system if $\nabla^2 \psi(\underline{r}, \theta)$ vanishes in the $\underline{r}, \theta$ coordinate system.

We have previously noted that, for $\mathbf{r} \neq \mathbf{r}_0$, a function $\psi(r, \theta) = \psi(\mathbf{r}|\mathbf{r}_0) = 1/(4\pi|\mathbf{r} - \mathbf{r}_0|)$ is a solution of Laplace's equation and is in fact the free-space Green's function for that equation. The corresponding second solution $(a/r)\psi(a^2/r, \theta)$ just introduced may be combined with the first solution to construct a Green's function that satisfies boundary conditions in spherical coordinates. In particular, note that on the surface $r = a$ we have $\psi(a, \theta) = \varphi(a, \theta) = 1/(4\pi|\mathbf{a} - \mathbf{r}_0|)$ where $\mathbf{a}$ is a vector from the origin to a point on the surface of the sphere of radius a. Thus, the function

$$g(\mathbf{r}|\mathbf{r}_0) = \psi(r, \theta) - \varphi(r, \theta)$$

$$= \psi(r, \theta) - \frac{a}{r} \psi \left(\frac{a^2}{r}, \theta \right) \tag{5.3.8}$$

is a Green's function for Laplace's equation that vanishes on the spherical surface $r = a$. This result is conveniently rearranged by noting that

$$4\pi \frac{a}{r} \psi \left(\frac{a^2}{r}, \theta \right) = \frac{a}{r} \frac{1}{|\mathbf{u}(a^2/r) - \mathbf{u}_0 r_0|} = \frac{a}{r_0 |\mathbf{u}(a^2/r_0) - \mathbf{u}_0 r|} \tag{5.3.9}$$

Since $|\mathbf{u}a^2/r_0 - \mathbf{u}_0 r| = |\mathbf{u}r - \mathbf{u}_0 a^2/r_0|$, as is evident from the geometry of this problem, we have

$$\frac{a}{r}\,\psi\left(\frac{a^2}{r},\,\theta\right) = \frac{1}{4\pi}\frac{a/r_0}{|\mathbf{r} - \mathbf{u}_0 a^2/r_0|} \qquad (5.3.10)$$

which has the interpretation of an image source of strength a/r_0 located along the same radius vector from the origin as the original source but outside the sphere at the point $r_i = a^2/r_0$. Introducing this result into (5.3.8) we have

$$g(\mathbf{r}|\mathbf{r}_0) = \frac{1}{4\pi|\mathbf{r} - \mathbf{r}_0|} - \frac{a/r_0}{4\pi|\mathbf{r} - \mathbf{u}_0 a^2/r_0|} \qquad (5.3.11)$$

As an application of this Green's function, let us reexamine the problem considered in Section 5.1 in which the temperature field was determined within a sphere on which a cap of polar angle α was maintained at a temperature u_0. Combining Laplace's equation $\nabla^2 u(\mathbf{r}) = 0$ with $\nabla^2 g(\mathbf{r}|\mathbf{r}_0) = -\delta^3(\mathbf{r} - \mathbf{r}_0)$ and using Green's theorem in the usual way, we obtain

$$u(\mathbf{r}) = \int dS'\left(g\frac{\partial u}{\partial n'} - u\frac{\partial g}{\partial n'}\right)\Bigg|_{r'=a} \qquad (5.3.12)$$

where n', the outward normal used in the divergence theorem, is in the outward radial direction. Since the temperature $u(\mathbf{r})$ is specified on the surface, it is appropriate to use the Green's function introduced above since it vanishes on the surface of the sphere. We then have

$$u(r, \theta) = -a^2\int_0^{2\pi} d\varphi'\int_0^\alpha \sin\theta'\, d\theta'\,\frac{\partial g}{\partial r'}\Bigg|_{r'=a} \qquad (5.3.13)$$

A simple way of evaluating the radial derivative in this result is to express the Green's function (5.3.11) in terms of Legendre polynomials. We then have

$$g(r, \theta|r'\theta') = \frac{1}{4\pi}\frac{1}{r'}\sum_{n=0}^\infty\left(\frac{r}{r'}\right)^n P_n(\cos\gamma) - \frac{1}{4\pi}\frac{a}{r'}\sum_{n=0}^\infty\left(\frac{rr'}{a^2}\right)^2 P_n(\cos\gamma)$$

$$(5.3.14)$$

where γ is the angle between $\mathbf{r}$ and $\mathbf{r}'$ as given in (5.2.4). Since $\mathbf{r}'$ is to be placed on the surface $r' = a$, we have used the choice for $r_>$ and $r_<$ that conforms to r being less than either r' or a^2/r'. As before, we first evaluate the field on the z axis and then extend the field off axis by using the symmetry principle introduced in Section 5.1. On the z axis, where $r = z$, we have $\theta' = 0$ and thus $\cos\gamma = \cos\theta$. We then find

$$\frac{\partial g}{\partial r'}\Bigg|_{r'=a} = -\frac{1}{4\pi}\sum_{n=0}^\infty\frac{2n+1}{a^{n+2}}z^n P_n(\cos\theta') \qquad (5.3.15)$$

and (5.3.13) becomes

$$u(z, 0) = u_0 2\pi a^2 \int_0^\alpha \sin \theta' \, d\theta' \, \frac{1}{4\pi} \sum_{n=1}^\infty \frac{2n+1}{a^{n+2}} z^n P_n(\cos \theta')$$

$$= \sum_{n=0}^\infty A_n z^n \qquad\qquad (5.3.16)$$

where

$$A_n = u_0 \frac{2n+1}{2a^n} \int_0^\alpha P_n(\cos \theta') \sin \theta' \, d\theta' \qquad\qquad (5.3.17)$$

which agrees with (5.1.13). The field at any point within the sphere may again be obtained by extending the on-axis result in (5.3.16) to yield

$$u(r, \theta) = \sum_{n=0}^\infty A_n r^n P_n(\cos \theta) \qquad\qquad (5.3.18)$$

It should be noted that this result was obtained much more directly in Section 5.1. As noted previously, the Green's function method is relatively cumbersome for simple problems.

Problems

5.3.1 (a) Show that the radial derivative of the image Green's function (5.3.11) is

$$\left.\frac{\partial g}{\partial r_0}\right|_{r_0 = a} = \frac{r^2 - a^2}{4\pi a (r^2 + a^2 - 2ar \cos \gamma)^{3/2}}$$

where $\cos \gamma = \cos \theta \cos \theta' + \sin \theta \sin \theta' \cos(\varphi - \varphi')$.

(b) Thus obtain Poisson's integral formula for a sphere

$$u(\mathbf{r}) = \frac{a^2 - r^2}{4\pi a} \int \frac{dS' u(\mathbf{r}'_s)}{(r^2 + a^2 - 2ar \cos \gamma)^{3/2}}$$

(c) Evaluate the integral for $u_1(r'_s) = u_0$ and show that $u(\mathbf{r}) = u_0$ throughout the sphere.

5.3.2 Use the generating function for the Legendre polynomials to show that

$$\frac{a^2 - r^2}{(r^2 + a^2 - 2ar \cos \gamma)^{3/2}} = \frac{1}{a} \sum_{n=0}^\infty (2n+1) \left(\frac{r}{a}\right)^n P_n(\cos \gamma)$$

and thus recover (5.3.16) from the Poisson integral formula obtained in Problem 5.3.1.

5.3.3 Obtain the result given in Problem 5.1.2b by using the Poisson integral formula.

5.3.4 The surface of a uniform sphere of radius a and conductivity K is kept at temperature $u(a, \theta) = 0$. A point source of heat of strength q_0 is located within the sphere at $r = r_0$. Use Green's theorem to show that

$$u(\mathbf{r}) = \frac{q_0}{K} g(\mathbf{r}|\mathbf{r}_0) + \int_S ds' \, g(\mathbf{r}|\mathbf{r}'_s) \frac{\partial u}{\partial r'_s}$$

and conclude that if the image Green's function (5.3.11) is used, then $u(\mathbf{r}) = (q_0/K) g(\mathbf{r}|\mathbf{r}_0)$, where $g(\mathbf{r}|\mathbf{r}_0)$ is this image Green's function with source term at the location of the source in the physical problem.

5.3.5 If we were to use merely the free-space Green's function $g(\mathbf{r}|\mathbf{r}') = \frac{1}{4}\pi|\mathbf{r} - \mathbf{r}'|$ in Problem 5.3.4, the surface integral over the unknown $\partial u/\partial r'_s$ would still remain and determination of $u(\mathbf{r})$ would require solution of an integral equation. Solve this integral equation as follows:

(a) Set the field point $\mathbf{r}$ on the boundary where $u(\mathbf{r}) = 0$ to obtain

$$0 = \frac{q_0}{4\pi} g(\mathbf{r}_s|\mathbf{r}_0) + \int_S ds' \, g(\mathbf{r}_s|\mathbf{r}'_s) \frac{\partial u}{\partial r'_s}$$

The field within the sphere will be symmetric about an axis passing through both the origin and the source point $\mathbf{r}_0$. Use this line as a z axis and expand $\partial u/\partial r'_s$ in the form

$$\frac{\partial u}{\partial r'_s} = \sum_{n=0}^{\infty} f_n P_n(\cos \theta')$$

where the coefficients f_n are to be determined. Integrate over φ' by using (5.1.30) and expand the Green's function in Legendre polynomials to show that

$$f_n = -(2n + 1) \frac{q_0}{Ka} \left(\frac{r_0}{a}\right)^n$$

(b) With the f_n now known, perform the surface integration over θ' to show that

$$u(\mathbf{r}) = \frac{q_0}{K} g(\mathbf{r}|\mathbf{r}_0) - \frac{q_0/r_0}{K} \frac{1}{r} \sum_{n=0}^{\infty} \left(\frac{r}{r}\right)^n P_n(\cos \theta)$$

where $r = a^2/r_0$. The series has the interpretation of being the expansion for the appropriate image source for this problem. The result thus agrees with that obtained in Problem 5.3.4.

5.4 SPHERICAL WAVES

Waves having complete spherical symmetry have been considered in Section 3.6. We now introduce a generalization of those results. When θ dependence is included, the wave equation $\nabla^2 u - c^{-2} u_{tt} = 0$ becomes

$$\frac{1}{r} \frac{\partial^2 (ru)}{\partial r^2} + \frac{1}{r^2 \sin \theta} \frac{\partial}{\partial \theta} \left(\sin \theta \frac{\partial u}{\partial \theta} \right) - \frac{1}{c^2} \frac{\partial^2 u}{\partial t^2} = 0 \qquad (5.4.1)$$

For waves at a single frequency ω we introduce the factored form $u(r, \theta, t) = R(r) \Theta(\theta) e^{-i\omega t}$ and obtain the separated equations

$$\frac{r}{R} (rR)'' + r^2 \left(\frac{\omega}{c} \right)^2 = -\frac{(\Theta' \sin \theta)'}{\Theta \sin \theta} = \lambda \qquad (5.4.2)$$

where λ is the separation constant. The equation for Θ is the same as the one that arose in solving Laplace's equation in Section 5.1. To obtain nonsingular solutions we again choose $\lambda = n(n + 1)$ with $n = 0, 1, 2, \ldots$. The nonsingular solutions are again the Legendre polynomial $P_n(\cos \theta)$. The second solution $Q_n(\cos \theta)$, which is singular at $\theta = 0, \pi$, is not of interest unless the points $\theta = 0, \pi$ are excluded. On noting that $n(n + 1) = (n + \frac{1}{2})^2 - \frac{1}{4}$, the radial equation can be written as

$$(rR)'' + \left\{ \left(\frac{\omega}{c} \right)^c + \frac{1}{r^2} \left[\frac{1}{4} - \left(n + \frac{1}{2} \right)^2 \right] \right\} rR = 0 \qquad (5.4.3)$$

which is of the same form as (D.24), an equation that was shown to have a solution equal to $r^{1/2}$ times a linear combination of Bessel functions. On comparison with (D.24) it is seen that in the present instance the constant $n + \frac{1}{2}$ plays the role of the order of the Bessel function. Denoting any Bessel function of order $n + \frac{1}{2}$ by $Z_{n+1/2}$, a solution of (5.4.3) for rR may be written as $rR = \sqrt{r} Z_{n+1/2}(kr)$. Since n is an integer, $n + \frac{1}{2}$ is nonintegral and thus two linearly independent solutions are provided by $Z_{n+1/2}(kr)$ and $Z_{-n-1/2}(kr)$. Hence the radial solutions for $R(r)$ are of the form

$$R(r) = \frac{1}{\sqrt{r}} Z_{\pm(n+1/2)}(kr) \qquad (5.4.4)$$

The additional factor of $r^{-1/2}$ is to be expected since Bessel functions only decrease as $r^{-1/2}$ for large kr whereas, as noted in Section 3.6, we expect a spherical disturbance to decrease as r^{-1}.

These radial solutions are actually much simpler than one might expect. From (5.4.3), we see that for $n = 0$ the equation reduces to

$$(rR)'' + \left(\frac{\omega}{c}\right)^2 rR = 0 \tag{5.4.5}$$

which is the equation associated with waves in one dimension and has solutions rR expressible as a combination of $\cos kr$, $\sin kr$, or $e^{\pm ikr}$. Thus, the Bessel functions $Z_{n+1/2}$ obtained here must be proportional to trigonometric functions for *all* values of kr and not merely for $kr \gg 1$. To associate these trigonometric solutions of the present problem with the appropriate Bessel functions, we need only examine the approximate expression for $J_{n+1/2}(kr)$ at large kr. From (D.25) we have

$$J_{n+1/2}(kr) \cong \sqrt{\frac{2}{\pi kr}} \cos \left\{ kr - \left[2\left(n + \frac{1}{2}\right) + 1\right]\frac{\pi}{4} \right\}$$

$$\cong \sqrt{\frac{2}{\pi kr}} \cos \left[kr - (n + 1)\frac{\pi}{2} \right] \tag{5.4.6}$$

A similar result obtains for $J_{-n-1/2}(kr)$. Then, for $n = 0$ we find $J_{1/2}(kr) = (2/\pi kr)^{1/2} \cos(kr - \pi/2) = (2/\pi kr)^{1/2} \sin kr$ and $J_{-1/2}(kr) = (2/\pi kr)^{1/2} \cos kr$. These expressions are *exact* for all kr. Using these results we find that a linear combination of the two solutions given in (5.4.4) reduces to

$$R(r) = r^{-1/2}[AJ_{1/2}(kr) + BJ_{-1/2}(kr)]$$

$$= a\frac{\sin kr}{r} + b\frac{\cos kr}{r} \tag{5.4.7}$$

where $a = A\sqrt{2/\pi}$ and $b = B\sqrt{2/\pi}$.

We may use the recurrence relations for Bessel functions (D.41) to obtain the trigonometric expressions for higher values of n. For instance, setting $\nu = \frac{1}{2}$ in the relation $J_{\nu+1}(z) = (2\nu/z)J_\nu(z) - J_{\nu-1}(z)$, we obtain

$$J_{3/2}(z) = \frac{1}{z}J_{1/2}(z) - J_{-1/2}(z)$$

$$= \sqrt{\frac{2}{\pi}}\left(\frac{\sin z}{z^{3/2}} - \frac{\cos z}{z^{1/2}}\right) \tag{5.4.8}$$

For extensive calculations with these functions it is convenient to introduce spherical Bessel functions defined as

$$j_n(z) = \sqrt{\frac{\pi}{2z}} J_{n+1/2}(z) \tag{5.4.9}$$

with similar definitions applying for the other Bessel functions. In the following sections we consider these functions further as well as some examples of their use.

5.4.1 Spherical Bessel Functions

As just noted, extensive calculations with Bessel functions of half-odd-integer order are conveniently carried out in terms of the spherical Bessel functions. The two linearly independent solutions of (5.4.3) are usually defined by

$$j_n(z) = \sqrt{\frac{\pi}{2z}} J_{n+1/2}(z)$$

$$n_n(z) = (-1)^n \sqrt{\frac{\pi}{2z}} J_{-n-1/2}(z) \tag{5.4.10}$$

Since the order of the Bessel function is noninteger, the Y-type solution introduced for integer order Bessel functions in Appendix D is not needed. The extra factor $(-1)^n$ in the definition of $n_n(z)$ is convenient when dealing with the large-argument form. From (5.4.6) we find that

$$j_n(z) \xrightarrow[z \to \infty]{} \frac{1}{z} \cos\left[z - \left(n + \frac{1}{2}\right)\frac{\pi}{2}\right]$$

$$n_n(z) \xrightarrow[z \to \infty]{} \frac{1}{z} \sin\left[z - \left(n + \frac{1}{2}\right)\frac{\pi}{2}\right] \tag{5.4.11}$$

Small-argument forms may be shown to be (Morse and Feshbach, 1953, p. 1573)

$$j_n(z) \to \frac{z^n}{1 \cdot 3 \cdot 5 \cdots (2n + 1)}$$

$$n_n(z) \to \frac{1 \cdot 1 \cdot 3 \cdot 5 \cdots (2n - 1)}{z^{n+1}} \tag{5.4.12}$$

Also, as is to be expected from our previous development of Bessel functions, the spherical Hankel functions are defined as

$$h_n^{(1)}(z) = j_n(z) + i n_n(z)$$

$$h_n^{(2)}(z) = j_n(z) - i n_n(z) \tag{5.4.13}$$

From the definition of $J_{\pm 1/2}(z)$ given above, one finds $j_0(z) = (\sin z)/z$, $n_0(z) = (\cos z)/z$ and $h_0^{(1)} = e^{iz}/z$, $h_0^{(2)} = e^{-iz}/z$. Thus, in conjunction with $e^{-i\omega t}$ time dependence, $h_0^{(1)}(z)$ represents an outgoing spherical wave while $h_0^{(2)}(z)$ represents an incoming spherical wave. It should be noted that $j_0(z)$ is the only solution that remains finite at the origin.

The properties of spherical Bessel functions parallel those of the regular Bessel functions summarized in Appendix D. In particular, the Wronskians $W(h_n, j_n) = h_n(x)j_n'(x) - j_n(x)h_n'(x)$ can be evaluated by using the large-argument forms as was done in Problem C.3, and one finds

$$W[h_n, j_n] = i/z^2 \tag{5.4.14}$$

5.4.2 Radiation from a Point Source

Some of the details associated with wave propagation in spherical coordinates can be displayed by considering radiation from a point source located a distance r_0 from the origin, as shown in Figure 5.4. In order to maintain azimuthal symmetry in our development, the source is located on the negative z axis at $r = r_0$, $\vartheta = \pi$. The wave equation to be solved is

$$\nabla^2 p + k^2 p = -q_0 \delta^3(\mathbf{r} - \mathbf{r}_0) = -q_0 \frac{\delta(r - r_0)\delta(\theta - \pi)}{2\pi r^2 \sin \theta} \tag{5.4.15}$$

where

$$\nabla^2 p = \frac{1}{r^2}\frac{\partial}{\partial r}\left(r^2 \frac{\partial p}{\partial r}\right) + \frac{1}{r^2} Lp \tag{5.4.16}$$

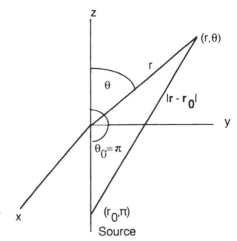

Figure 5.4. Coordinate system for source on the negative z axis.

with L given by (5.3.4). Due to the absence of azimuthal (φ) dependence in the problem, the solution may be expanded in the form

$$p(r, \theta) = \sum_{n=0}^{\infty} p_n(r, r_0) P_n(\cos \theta) \qquad (5.4.17)$$

Since we are considering a point source, the coefficients $p_n(r, r_0)$ can be determined by using the same procedure as that used previously to determine a Green's function. The solution may in fact be written as $p(r, \theta) = q_0 G(r, \theta | r_0, \pi)$. Substituting the form of the solution (5.4.17) into the wave equation (5.4.15) and noting that $LP_n(\cos \theta) = n(n + 1)P_n(\cos \theta)$, we have

$$Lp = - \sum_{n=0}^{\infty} p_n(r, r_0) n(n + 1) P_n(\cos \theta) \qquad (5.4.18)$$

Hence we obtain

$$\sum_{n=0}^{\infty} \left\{ \frac{1}{r^2} \frac{d}{dr} \left(r^2 \frac{dp_n}{dr} \right) + \left[k^2 - \frac{n(n + 1)}{r^2} \right] p_n \right\} P_n(\cos \theta)$$

$$= -q_0 \frac{\delta(r - h)\delta(\theta - \pi)}{2\pi r^2 \sin \theta} \qquad (5.4.19)$$

By using the orthogonality properties of the Legendre polynomials in the usual way, one finds that the individual coefficients $p_n(r, r_0)$ satisfy

$$\frac{d}{dr} \left(r^2 \frac{dp_n}{dr} \right) + [(kr)^2 - n(n + 1)] p_n = -J\delta(r - r_0) \qquad (5.4.20)$$

where the amplitude of the delta function is

$$J = \frac{q_0}{4\pi} (2n + 1) P_n(\cos \pi) \qquad (5.4.21)$$

In the notation used for constructing Green's functions that was introduced in Section 4.2, the inner and outer solutions $p_{n<}$ and $p_{n>}$ are now chosen on the basis of physical considerations. Since the region $r < r_0$ contains the origin, and the field is finite there, we must choose $p_{n<}$ proportional to $j_n(kr)$. For $r > r_0$ we expect radiation to be outgoing from the source region and hence choose $p_{n>}$ proportional to $h_n^{(1)}(kr)$. The Wronskian for this pair of solutions is given in (5.4.14). We can now write the solution for $p_n(r, r_0)$ by employing the Green's function expression (4.2.11) with $p(r) = r^2$, and J as given in

(5.4.21) with $P_n(\cos \pi)$ set equal to $(-1)^n$. We obtain

$$
\begin{aligned}
p_n(r, r_0) &= \frac{1}{p(r)\,W}\,p_{n<}p_{n>} \\
&= \frac{ikq_0}{4\pi}(2n + 1)(-1)^n j_n(kr_<)h_n^{(1)}(kr_>)
\end{aligned}
\tag{5.4.22}
$$

The final result is now seen from (5.4.17) to be

$$
p(r, \theta \,|\, r_0, \pi) = \frac{ikq_0}{4\pi}\sum_{n=0}^{\infty}(2n + 1)(-1)^n j_n(kr_<)h_n^{(1)}(kr_>)P_n(\cos \theta)
\tag{5.4.23}
$$

Since we also know that the solution is merely a spherical wave emanating from the point $r_0 = (r_0, \theta_0)$, the final result provides us with an expansion of a spherical wave in the form

$$
\begin{aligned}
p(r, \theta \,|\, r_0, \pi) &= q_0\frac{e^{ik|\mathbf{r} - \mathbf{r}_0|}}{4\pi|\mathbf{r} - \mathbf{r}_0|} \\
&= \frac{q_0}{4\pi}ik\sum_{n=0}^{\infty}(2n + 1)(-1)^n j_n(kr_<)h_n^{(1)}(kr_>)P_n(\cos \theta)
\end{aligned}
\tag{5.4.24}
$$

It should be emphasized that this result applies only when the source is located on the negative z axis and there is no azimuthal (φ) dependence in the expression for the field. When the source is located on the positive z axis, the factor $(-1)^n$ is absent.

5.4.3   Reduction to a Plane Wave

In the limit that $r_0 \to \infty$, the source location recedes to $z = -\infty$ and the wave motion described by $p(r, \theta \,|\, r_0, \pi)$ at locations such that $r \ll r_0$, for example, near the origin, should appear as a spherical wave with a large radius of curvature, that is, as nearly a plane wave. It should then be possible to use the above result (5.4.24) to obtain an expression for a plane wave in spherical coordinates by merely examining the limit $r_0 \to \infty$. For $r \ll r_0$ we have (cf. Fig. 5.4)

$$
\begin{aligned}
|\mathbf{r} - \mathbf{r}_0| &= \sqrt{r^2 + r_0^2 - 2rr_0\cos(\pi - \theta)} \\
&= r_0\left[1 + \left(\frac{r}{r_0}\right)^2 + 2\frac{r}{r_0}\cos \theta\right]^{1/2} \\
&\approx r_0\left(1 + \frac{r}{r_0}\cos \theta\right) = r_0 + r\cos \theta
\end{aligned}
\tag{5.4.25}
$$

where we have only kept the term in the expansion of the radical that is of first order in r/r_0. The dominant contribution to the first form for the spherical wave given in (5.4.24) is now obtained by using this approximation for $|r - r_0|$ in the phase and r_0 in the amplitude. We then obtain

$$\frac{q_0}{4\pi} \frac{e^{ik|r - r_0|}}{|r - r_0|} \approx \frac{q_0}{4\pi r_0} e^{ikr + ikr\cos\theta} \tag{5.4.26}$$

To simplify the second form in (5.4.24), we use the large-argument form for the spherical Hankel function obtainable from (5.4.11) and (5.4.13) to obtain

$$h_n^{(1)}(kr_0) = \frac{1}{r_0} e^{i[kr_0 - (n + 1)\pi/2]} \tag{5.4.27}$$

When this result is used in (5.4.24) and the common factor $(q_0/4\pi r_0) e^{ikr_0}$ is suppressed, we obtain

$$e^{ikr\cos\theta} = \sum_{n=0}^{\infty} i^n (2n + 1) j_n(kr) P_n(\cos\theta) \tag{5.4.28}$$

as the expansion of a plane wave in spherical coordinates. This result is analogous to the corresponding expression for polar coordinates given in (D.47).

The plane-wave expansion obtained here provides a simple introduction to the interrelations between spherical Bessel functions and Legendre polynomials. Some of them are outlined in Problems 5.4.1–5.4.4.

5.4.4 Scattering of a Plane Wave by a Sphere

The expression for a plane wave given in (5.4.28) may be used to satisfy boundary conditions on a spherical surface. As a simple example, consider the scattering of a plane wave that is incident upon a sphere. The total wave field may be thought of as being composed of that due to the incident wave plus that due to the scattered field. Since the wave equation is linear, these two components of the field may be added together to form the total field. The scattered field, since it emanates from the scatter, must be in the form of a wave propagating radially outward from the scatter and having some angular dependence that is to be determined. Thus, the total field may be written as $p(r, \theta) = p_{\text{inc}} + p_{\text{sc}}$ where

$$p_{\text{inc}} = p_0 e^{ikz} = p_0 \sum_{n=0}^{\infty} i^n (2n + 1) j_n(kr) P_n(\cos\theta)$$

$$p_{\text{sc}} = \sum_{n=0}^{\infty} A_n h_n^{(1)}(kr) P_n(\cos\theta) \tag{5.4.29}$$

in which p_0 represents the amplitude of the incident wave and the coefficients A_n in the second expansion are to be determined.

If the surface of the scatterer is such that the total field must vanish on the surface of a sphere of radius a, then we require $p(a, \theta) = 0$. Due to the orthogonality of the Legendre polynomials, the total coefficient of each $P_n(\cos \theta)$ must vanish separately and we have, from (5.4.29),

$$p_0 i^n (2n + 1) j_n(ka) + A_n h_n^{(1)}(ka) = 0 \tag{5.4.30}$$

Solving for the A_n we have the scattered wave in the form

$$p_{sc}(r, \theta) = p_0 \sum_0^\infty i^n (2n + 1) \frac{j_n(ka)}{h_n^{(1)}(ka)} h_n^{(1)}(kr) P_n(\cos \theta) \tag{5.4.31}$$

Since scattered waves are customarily observed at distances that are many wavelengths from the scatterer, we may approximate the scattered wave by using the large-argument form of the spherical Hankel function. From (5.4.11) and (5.4.13) we have

$$h_n^{(1)}(kr) \cong \frac{1}{kr} e^{i\{kr - [(n + 1)/2]\pi\}} \tag{5.4.32}$$

and the first terms in the series (5.4.31) may be written

$$p_{sc} \cong -p_0 \left[\frac{j_0}{h_0^{(1)}} + 3 \frac{j_1}{h_1^{(1)}} P_1(\cos \theta) + \cdots \right] \frac{e^{ikr}}{ikr} \tag{5.4.33}$$

where the argument of each spherical Bessel function is ka.

An additional simplification may be introduced if we now confine attention to wavelengths that are large compared with the radius a. Then $\lambda \gg a$, and since $k = 2\pi/\lambda$, we have $ka \ll 1$. In this limit the small-argument forms for the various Bessel function of argument ka are appropriate. From (5.4.12) we find that, to lowest order in ka, $j_0(ka)/h_0^{(1)}(ka) \approx ika$ while $j_1(ka)/h_1^{(1)}(ka)$ is proportional to $(ka)^3$ and thus provides a much smaller contribution to the scattered field than that of the first term. Higher order terms are even smaller and thus at low frequencies only the first term in the series (5.4.33) need be retained. We finally obtain

$$p_{sc} = -p_0 \left(\frac{a}{r} \right) e^{ikr} \tag{5.4.34}$$

The scattering in this limit is thus spherically symmetric and does not go to zero as ka becomes arbitrarily small. This result is perhaps counter to one's intuitive ideas concerning wave scattering processes since one expects that a wave will not "see" a scatterer that is much smaller than the wavelength. One might say that the Dirichlet boundary condition $p(a, \theta) = 0$ is so extreme that it offsets this expected result. Scattering by Neumann boundary conditions is considered in Problem 5.4.5.

Problems

5.4.1 Replace n by $n + \frac{1}{2}$, and similarly for m in Problem D.4, and show that the spherical Bessel function satisfy the orthogonality relation

$$\int_0^\infty j_n(kr)j_m(kr) \, dr = \begin{cases} \dfrac{\pi}{2k}\dfrac{1}{2n+1}, & n = m \\ 0, & n \neq m \end{cases}$$

5.4.2 Use the orthogonality of the Legendre polynomials to show that

$$\int_0^\pi \sin \theta d\theta e^{ikr\cos\theta} P_n(\cos \theta) = 2i^n j_n(kr)$$

5.4.3 Set $\theta = \pi/2$ in the plane wave expansion (5.4.28) and separate real and imaginary terms to obtain

$$1 = \sum_{n=0}^\infty (-1)^n (4n + 1) j_{2n}(kr) P_{2n}(0)$$

5.4.4 Use the result of Problem 5.4.3 and the orthogonality of spherical Bessel functions obtained in Problem 5.4.1 to show that

$$\int_0^\infty j_{2n}(kr) \, dr = \frac{\pi}{2k}(-1)^n P_{2n}(0)$$

5.4.5 Consider the scattering of a plane wave of amplitude p_0 by a sphere of radius a on which the total field satisfies the homogeneous Neumann boundary condition $p_r(a, \theta) = 0$.

(a) Show that the scattered field is given by

$$p_{sc}(r, \theta) = -p_0 \sum_{n=0}^\infty (2n + 1) \frac{j_n'(ka)}{h_n^{(1)}(ka)} P_n(\cos \theta) \frac{e^{ikr}}{ikr}$$

(b) Use the small-argument forms given in (5.4.12) to show that for $ka \ll 1$ the first two terms are of the same order in ka and hence must both be retained. Show that the scattered field takes the form

$$p_{sc}(r, \theta) = -\frac{p_0}{3} (ka)^2 \left(1 - \frac{3}{2} \cos \theta \right) \left(\frac{a}{r} \right) e^{ikr}$$

This result should be compared with that given in (5.4.34).

5.4.6 Show that in the limit $kr \ll 1$ the spherical wave expansion (5.4.24) reduces to the version of (E.25) that is appropriate for a static point source at $r = (r_0, \pi)$.

6 Fourier Transform Methods

In this chapter we introduce the Fourier sine and cosine transforms and show that they can provide a convenient approach to solving partial differential equations in a semi-infinite region. Guided by the success of the Laplace transform technique in solving problems on a semi-infinite *temporal* interval, it might be expected that problems of a semi-infinite *spatial* extent could equally well be solved by this method. However, the convenience of the Laplace transform method in solving initial-value problems stems from the circumstance that both initial values $f(x, 0)$ and $f_t(x, 0)$ are customarily provided in the statement of such problems. Thus, all the initial information on a function and its time derivative that are required for the Laplace transform of the second order differential equation are known. In a boundary value problem, however, the spatial Laplace transform of the second derivative, $f_{xx}(x, t)$, would involve the two boundary conditions $f(0, t)$ and $f_x(0, t)$ whereas only one of these quantities, or a single relation between them, is usually specified by the boundary condition at $x = 0$. As indicated in Problem 2.2.8, where the Laplace transform of the spatial variable was used in a simple example of this type, the unknown boundary condition would have to be carried through the entire calculation in a somewhat cumbersome fashion and determined only after a solution of the problem has been obtained. We shall find that when second spatial derivatives are subjected to Fourier sine and cosine transforms, however, this difficulty is obviated.

Just as the Fourier sine and cosine series could be extended to treat periodic functions that were neither even nor odd, so also the Fourier sine and cosine transforms may be used on the infinite interval as well. Transforms in more than one spatial dimension will also be developed. It is found that the case of circular symmetry leads to a transform that can be expressed in terms of Bessel functions. These topics are also developed in this chapter.

6.1 FOURIER SINE AND COSINE TRANSFORMS

As shown in Appendix A, the Fourier sine and cosine transform pairs are obtained as limiting cases of the corresponding Fourier sine and cosine series representations of periodic functions. In the present chapter it is assumed that the reader is familiar with Fourier series and, in particular, with the limiting procedure used in Section A.5 to obtain the Fourier transform relations. For the Fourier sine transform the relations are

$$\mathcal{F}_s\{f(x)\} = \int_0^\infty f(x)\sin\ kx\ dx = F_s(k)$$

$$\mathcal{F}_s^{-1}\{F_s(k)\} = \frac{2}{\pi} \int_0^\infty F_s(k)\sin\ kx\ dk = f(x) \tag{6.1.1}$$

while for the cosine transform we have

$$\mathcal{F}_c\{f(x)\} = \int_0^\infty f(x)\cos\ kx\ dx = F_c(k)$$

$$\mathcal{F}_c^{-1}\{F_c(k)\} = \frac{2}{\pi} \int_0^\infty f(x)\cos\ kx = f(x) \tag{6.1.2}$$

Only positive values of x and k occur in these definitions. Since $\sin\ kx$ is an odd function of either x or k, the sine transform may be extended to negative values of these variables as an odd function, that is, $f(-x) = -f(x)$ and $F_s(-k) = -F_s(k)$. Similarly, when a function is expressed as a cosine transform, it is to be interpreted as an even function so that $f(-x) = f(x)$ and $F_c(-k) = F_c(k)$.

As an example, the function $f(x) = e^{-ax}$, $x > 0$, has the Fourier cosine transform[1]

$$\mathcal{F}_c\{e^{-ax}\} = \int_0^\infty e^{-ax}\cos\ kx\ dx = \frac{a}{k^2 + a^2} \tag{6.1.3}$$

which is obviously an even function of k. The inverse transform is given by

$$\mathcal{F}_c^{-1}\left\{\frac{a}{k^2 + a^2}\right\} = \frac{2}{\pi} \int_0^\infty \frac{a}{k^2 + a^2} \cos\ kx\ dk = \begin{cases} e^{-ax}, & x > 0 \\ e^{ax}, & x < 0 \end{cases} \tag{6.1.4}$$

where the integration is customarily performed by using complex variable techniques. Since this background is not assumed of the reader, results for inverse transforms will usually be obtained from tables of sine and cosine transforms in a manner analogous to that used for the Laplace transform.

The corresponding sine transform of $f(x) = e^{-ax}$, $x > 0$, is

$$\mathcal{F}_s\{e^{-ax}\} = \frac{k}{k^2 + a^2} \tag{6.1.5}$$

[1]Note that this integral has been encountered previously as the Laplace transform of the cosine.

which, as expected, is an odd function of k. The inverse is

$$\mathcal{F}_s^{-1}\left\{\frac{k}{k_2 + a^2}\right\} = \frac{2}{\pi}\int_0^\infty \frac{k}{k^2 + a^2}\sin kx\,dk = \begin{cases} e^{-ax}, & x > 0 \\ -e^{ax}, & x < 0 \end{cases}$$

$$= \mathrm{sgn}(x)e^{-a|x|} \tag{6.16}$$

where $\mathrm{sgn}(x) = x/|x|$, that is, $\mathrm{sgn}(x)$ is $+1$ for $x > 0$ and -1 for $x < 0$.

The class of functions for which the Fourier transform integrals converge is more restricted than that for the Laplace transform since the integrand for the Fourier transform does not contain an exponentially decreasing factor. As a sufficient condition for the existence of the Fourier transform, it is customary to require the existence of the integral $\int_0^\infty |f(x)|\,dx$. However, it should be noted that this requirement is unnecessarily restrictive since, for example, both $x^{-1/2}$ and $\tan^{-1}x$ possess Fourier transforms while the integral $\int_0^\infty |f(x)|\,dx$ diverges in each of these instances.

The usefulness of these transforms of a function $f(x)$ becomes evident when we consider the associated transforms for a second derivative. After performing two partial integrations and assuming that $f(x)$ and its derivatives decay sufficiently rapidly so that there is no contribution from the upper limit, we obtain

$$\mathcal{F}_s\left\{\frac{d^2f}{dx^2}\right\} = -k^2 F_s(k) + kf(0)$$

$$\mathcal{F}_c\left\{\frac{d^2f}{dx^2}\right\} = -k^2 F_c(k) - f'(0) \tag{6.1.7}$$

Whether a sine or cosine transform should be used in a given problem is determined by the boundary condition that is specified at $x = 0$. In contrast to the Laplace transform, where the transform of a second derivative yields an expression involving the value of both the function and its derivative on the boundary, one of which will be unknown in a boundary value problem, the Fourier sine or cosine transform of the second derivative is related to the transform of the function plus a *single* known term. We thus avoid the difficulty that would occur if a Laplace transform were used. The more general boundary condition in which $f'(0) - af(0)$ is specified is treated in Section 9.9.

Sine and cosine transforms are given in Tables F.3 and F.4. For more extensive tables, Erdelyi et al. (1954) may be consulted.

Problems

6.1.1 Solve $f'' - \gamma^2 f = -e^{-\alpha x}$, $f(0) = 1$, $f(\infty) = 0$, $0 \le x < \infty$, by using the Fourier sine transform.

6.1.2 Solve $f'' - \gamma^2 f = -e^{-\alpha x}$, $f'(0) = 1$, $f(\infty) = 0$, $0 \le x < \infty$, by using the Fourier cosine transform.

6.1.3 Obtain the results in Problems 6.1.1 and 6.1.2 by using the cosine transform in (6.1.1) and the sine transform in (6.1.2).

6.1.4 **(a)** Set $\alpha = 0$ in the differential equations in Problems 6.1.1 and 6.1.2 and solve by methods familiar from some previous course in differential equations.

(b) Set $\alpha = 0$ in the *solutions* you obtained in Problems 6.1.1 and 6.1.2 and show that the results obtained in Problem 6.1.4a are recovered.

(c) What happens if you set $\alpha = 0$ in the *differential equations* in Problems 6.1.1 and 6.1.2 and then attempt to solve by using Fourier transforms?

6.2 EXAMPLES

In this section we consider the use of sine and cosine transforms in solving both ordinary and partial differential equations. The problems considered here could also be solved by using image Green's function techniques for semi-infinite regions, as was developed in Section 4.4. It is of interest to note how the present approach incorporates the method of images.

6.2.1 Green's Function for Steady Waves on a Semi-infinite String

The problem considered in this first example has already been analyzed in Section 4.4 by using the method of images. It is considered here to indicate the somewhat more delicate nature of the convergence of the integrals that occur in Fourier transform theory.

We take the equation for the string to be

$$\rho \frac{\partial^2 w}{\partial t^2} + R \frac{\partial w}{\partial t} - T \frac{\partial^2 w}{\partial x^2} = f(x, t) = \mathcal{F}(x)e^{-i\omega t}, \qquad x \geq 0 \quad (6.2.1)$$

with the boundary condition $w(0, t) = 0$. The damping term $R\, \partial w/\partial t$ will be found to play a crucial role in the construction of the solution. Since the motion of the string $w(x, t)$ is at the single frequency ω, we write $w(x, t) = y(x)e^{-i\omega t}$ and obtain the ordinary differential equation

$$\frac{d^2 y}{dx^2} + \left[\left(\frac{\omega}{c}\right)^2 + i\,\frac{\omega R}{T} \right] y = -\frac{1}{T}\mathcal{F}(x), \qquad y(0) = 0 \quad (6.2.2)$$

The Green's function $g(x|x')$ associated with this equation satisfies

$$\frac{d^2 g}{dx^2} + \left[\left(\frac{\omega}{c}\right)^2 + i\,\frac{\omega R}{T} \right] g = -\delta(x - x') \quad (6.2.3)$$

with the boundary condition $g(0|x') = g(x|0) = 0$. According to the transforms for second derivatives given in (6.1.7), it is the Fourier sine transform that is expressed in terms of the function evaluated at $x = 0$. Since this is the information that is provided by the boundary condition in the present problem, we use the Fourier sine transform in obtaining the solution. We thus introduce the Fourier sine transform pair

$$G_s(k|x') = \int_0^\infty g(x|x')\sin kx \, dx$$

$$g(x|x') = \frac{2}{\pi} \int_0^\infty G_x(k|x')\sin kx \, dk$$

(6.2.4)

When we take the sine transform of the equation for $g(x|x')$, we find

$$-k^2 G_s + \left[\left(\frac{\omega}{c}\right)^2 + i\frac{\omega R}{T} \right] G_s = -\sin kx'$$

(6.2.5)

Solving this algebraic equation for G_s and using the expression (6.2.4) for the inverse transform, we obtain

$$g(x|x') = \frac{2}{\pi} \int_0^\infty \frac{\sin kx \sin kx' \, dk}{k^2 - [(\omega/c)^2 + i\omega R/T]}$$

(6.2.6)

The crucial role played by the damping term is now evident since the integrand would be singular at $k = \omega/c$ if the damping term were not present. This integral is readily related to an entry in Table F.4 when we write $\sin kx \sin kx' = \frac{1}{2}[\cos k(x - x') - \cos k(x + x')]$. It is with this identity that we express the solution in such a way that a contribution from a source and its image begins to become apparent. On writing $\epsilon = \omega R/T$ and introducing the presumably negligible term $(\epsilon c/2\omega)^2$ into the denominator of (6.2.6) so as to complete the square, we find that the denominator may be written in the form

$$k^2 - \left(\frac{\omega}{c}\right)^2 - i\epsilon \approx k^2 - \left(\frac{\omega}{c}\right)^2 - i\epsilon + \left(\frac{\epsilon c}{2\omega}\right)^2$$

$$= k^2 + \left(\frac{\epsilon c}{2\omega} - i\frac{\omega}{c}\right)^2$$

(6.2.7)

Now using entry 3 in Table F.4 with $a = \epsilon c/2\omega - i\omega/c$, we obtain

$$g(x|x') = \frac{e^{-\epsilon x}}{2(\epsilon - i\omega/c)} [e^{i(\omega/c)|x - x'|} - e^{i(\omega/c)(x + x')}]$$

(6.2.8)

At this stage we need no longer retain the term ϵ that describes the damping. Letting ϵ vanish we obtain the result for undamped steady waves given in (4.4.3).

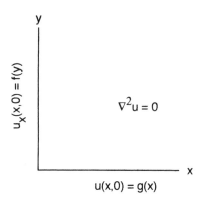

Figure 6.1. Boundary conditions determining temperature distribution in quarter plane.

6.2.2 Temperature Distribution in a Quarter Plane

To give a more extensive indication of the choices available in using the Fourier transform method, we consider the solution of Laplace's equation in a quarter plane with boundary conditions as shown in Figure 6.1. Using the notation for transforms introduced in Section 6.1, we find that the Fourier sine transform on y of Laplace's equation is

$$\int_0^\infty dy \, \sin ky \left(\frac{\partial^2 u}{\partial x^2} + \frac{\partial^2 u}{\partial y^2} \right) = 0$$

$$\frac{d^2 U_s(x, k)}{dx^2} - k^2 U_s(x, k) + ku(x, 0) = 0$$

(6.2.9)

while the cosine transform on x would yield

$$\frac{d^2 U_c(k, y)}{dy^2} - k^2 U_c(k, y) - u_x(0, y) = 0$$

(6.2.10)

Since both $u(x, 0) = f(x)$ and $u_x(0, y) = g(y)$ are assumed known in this problem, either of these two transforms would be appropriate. The choice to be made is based on the relative simplicity of the known functions $f(y)$ and $g(x)$. Certainly, if either boundary has a homogeneous boundary condition, for example, $g(x) = 0$, then the transform that introduces this function would be the appropriate one to employ since it would merely entail the solution of a *homogeneous* ordinary differential equation. Note that either of the other two possibilities, a cosine transform on y or a sine transform on x, would contain either $u_y(x, 0)$ or $u(x, 0)$ which would remain unknown until the problem has been solved.

We now pursue this matter further by considering the case of homogeneous Dirichlet conditions on the boundary $y = 0$, and we set $u(x, 0) = 0$ in (6.2.9). We thus use a Fourier sine integral on y and obtain from (6.2.9) the homogeneous equation

$$\frac{d^2 U_s(x, k)}{dx^2} - k^2 U_s(x, k) = 0 \tag{6.2.11}$$

The solution that remains bounded as x goes to $+\infty$ for positive k is

$$U_s(x, k) = A(k)e^{-kx} \tag{6.2.12}$$

so that

$$u(x, y) = \frac{2}{\pi} \int_0^\infty U_s(x, k)\sin ky \, dk$$

$$= \frac{2}{\pi} \int_0^\infty A(k)e^{-kx}\sin ky \, dk \tag{6.2.13}$$

To impose the boundary conditions along the edge at $x = 0$, and thus determine $A(k)$, we must first calculate $u_x(x, y)$ from this result. At $x = 0$ we obtain

$$u_x(0, y) = \frac{2}{\pi} \int_0^\infty A(k)(-k)\sin ky \, dk \tag{6.2.14}$$

The inverse transform gives

$$-kA(k) = \int_0^\infty u_x(0, y')\sin ky' \, dy' \tag{6.2.15}$$

For a specific function $u_x(0, y)$ we could evaluate this integral and obtain $A(k)$ for substitution into the second of (6.2.13). The solution would then be expressed as an integral over k. We may also proceed in a more general way by substituting the general expression for $A(k)$ given in (6.2.15) into (6.2.13). It will be seen that the result may then be interpreted in terms of the Green's function for this problem. With this latter substitution for $A(k)$ we have

$$u(x, y) = \frac{2}{\pi} \int_0^\infty dk \, e^{-kx}\sin ky \left(-\frac{1}{k}\right) \int_0^\infty d\xi \, u_x(0, \xi)\sin k\xi$$

$$= -\frac{2}{\pi} \int_0^\infty d\xi \, u_x(0, \xi) \int_0^\infty dk \, e^{-kx} \frac{\sin ky \sin k\xi}{k} \tag{6.2.16}$$

The integral over k may be obtained from a table of Laplace transforms (Table F.2, entry 13). The final form of the solution is thus

$$u(x, y) = \int_0^\infty dy' \, u_x(0, y') \frac{1}{2\pi} \ln \frac{(y - y')^2 + x^2}{(y + y')^2 + x^2} \tag{6.2.17}$$

The expression multiplying $u_x(0, y)$ in the integrand is just the negative of the Green's function for this problem (cf. Problem 4.10.1).

6.2.3   d'Alembert Solution for a Semi-infinite String

A string having an equilibrium position along the positive x axis has a fixed end point at $x = 0$. The string displacement $y(x, t)$ is assumed to satisfy the wave equation

$$y_{xx} - c^{-2}y_{tt} = 0 \qquad (6.2.18)$$

with initial conditions $y(x, 0) = \varphi(x)$ and $y_t(x, 0) = \psi(x)$. The fixed end at $x = 0$ provides the boundary condition $y(0, t) = 0$ while the slope $y_x(0, t)$ remains unknown until the problem has been solved. Since the known quantity $y(0, t)$ appears in the Fourier sine transform for $y_{xx}(x, t)$, we use this transform in solving this problem and introduce a transform $Y_s(k, t)$ defined as

$$Y_s(k, t) = \int_0^\infty y(x, t)\sin kx \, dx \qquad (6.2.19)$$

Unlike the steady state string problem considered in Section 6.2.1, no damping term is required here to provide convergence for the transform integral. In a steady state problem the wave motion is assumed to have been taking place for an infinitely long time. Thus an undamped wave will have propagated infinitely far along the x axis and lead to convergence difficulties for the transform integral. In the present problem, however, an initially localized wave profile propagates along the string. One can always consider regions of the string that are remote enough so that no disturbance has yet reached that region. The displacement $y(x, t)$ is zero in that section of the string and the integral in (6.2.19) may be assumed to converge.

The Fourier sine transform of the wave equation (6.2.18) is thus seen to be appropriate and we have

$$-k^2 Y_s(k, t) - c^{-2}\frac{d^2 Y_s}{dt^2} = 0 \qquad (6.2.20)$$

The solution of this transformed equation, and the first time derivative of that solution, $Y_{st}(k, t)$, are

$$Y_s(k, t) = A(k)\cos kct + B(k)\sin kct$$
$$Y_{st}(x, t) = kc[-A(k)\sin kct + B(k)\cos kct] \qquad (6.2.21)$$

At $t = 0$ the solution $Y_s(k, 0)$ as well as $Y_{st}(k, 0)$ must reduce, respectively, to the Fourier sine transform of the initial displacement and velocity, which we define as

$$\Phi_s(k) = \int_0^\infty \varphi(x)\sin kx \, dx$$

$$\Psi_s(k) = \int_0^\infty \Psi(x)\sin kx \, dx \tag{6.2.22}$$

Evaluation of the above expressions for $Y_s(k, t)$ and $Y_{st}(k, t)$ at $t = 0$ yields

$$Y_s(k, 0) = \Phi_s(k) = A(k)$$

$$Y_{st}(k, 0) = \Psi_s(k) = kcB(k) \tag{6.2.23}$$

With the transform of the displacement now expressed in terms of the transform of the initial conditions, we can use the inverse sine transform to write

$$y(x, t) = \frac{2}{\pi} \int_0^\infty \left[\Phi_s(k)\cos kct + \Psi_s(k) \frac{\sin kct}{kc} \right] \sin kx \, dk \tag{6.2.24}$$

This result may be rewritten in the form of a d'Alembert solution similar to that introduced in Section 1.9 for an infinite string. This transformation will be carried out by considering separately the two cases $y(x, 0) = \varphi(x)$, $y_t(x, 0) = 0$ and $y(x, 0) = 0$, $y_t(x, 0) = \Psi(x)$. Due to linearity of the wave equation, the two results may be added to give the general solution.

Case 1: $y(x, 0) = \varphi(x)$, $y_t(x, 0) = 0$
Since $\Psi_s(k) = 0$, the string displacement is now given by

$$y(x, t) = \frac{2}{\pi} \int_0^\infty \Phi_s(k)\cos kct \sin kx \, dk$$

$$= \frac{2}{\pi} \int_0^\infty \Phi_s(k) \frac{1}{2} [\sin k(x - ct) + \sin k(x + ct)] \, dk$$

$$= \frac{1}{2} [\varphi(x - ct) + \varphi(x + ct)] \tag{6.2.25}$$

Note that at $x = 0$ this solution reduces to $y(0, t) = \frac{1}{2}[\varphi(-ct) + \varphi(ct)]$, which is zero, since $\varphi(x)$, being a Fourier sine transform, is an odd function of its argument. Another interpretation of this result is that it is the same solution as that obtained for an infinite string as given in (1.9.3) if the initial displacement of the semi-infinite string is assumed to be extended into the region $x < 0$ as an odd function of x.

Case 2: $y(x, 0) = 0$, $y_t(x, 0) = \psi(x)$
From (6.2.24), the string displacement is

$$y(x, t) = \frac{2}{\pi} \int_0^\infty \Psi_s(k) \frac{\sin kct}{kc} \sin kx \, dk \tag{6.2.26}$$

Using the relation

$$\frac{\sin kct}{kc} = \frac{1}{c} \int_0^{ct} \cos k\xi \, d\xi \qquad (6.2.27)$$

we have

$$y(x, t) = \frac{2}{\pi c} \int_0^{ct} d\xi \int_0^{\infty} \Psi_s(k) \cos k\xi \, \sin kx \, dk \qquad (6.2.28)$$

The integral over k is now of the same form as that considered in Case 1 and we have

$$y(x, t) = \frac{1}{c} \int_0^{ct} d\xi \, \frac{1}{2} [\psi(x - \xi) + \psi(x + \xi)] \qquad (6.2.29)$$

This result may also be rewritten in the form obtained for the infinite string by noting that, since we are using a sine transform for ψ, it must be an odd function of its argument. We consider separately the cases $x > ct$ and $x < ct$.

For $x > ct$ the value of x is beyond the range of integration for ξ in (6.2.29) and so the argument $x - \xi$ is always positive. Setting $x - \xi = \eta$ in the first integrand and $x + \xi = \eta$ in the second, we obtain

$$y(x, t) = \frac{1}{2c} \left[-\int_x^{x-ct} \psi(\eta) \, d\eta + \int_x^{x+ct} \psi(\eta) \, d\eta \right]$$

$$= \frac{1}{2c} \int_{x-ct}^{x+ct} \psi(\eta) \, d\eta \qquad (6.2.30)$$

For $x < ct$ the first integral in (6.2.31) may be decomposed into integrals over the subregions $0 \leq \xi < x$ and $x < \xi \leq ct$. Since ψ is an odd function, so that $\psi(x - \xi) = -\psi(\xi - x)$ for $x < \xi$, we have

$$y(x, t) = \frac{1}{2c} \left[\int_0^x \psi(x - \xi) \, d\xi - \int_x^{ct} \psi(\xi - x) \, d\xi + \int_0^{ct} \psi(x + \xi) \, d\xi \right]$$

$$= \frac{1}{2c} \left[\int_0^x \psi(\eta) \, d\eta + \int_{ct-x}^0 \psi(\eta) \, d\eta + \int_x^{x+ct} \psi(\eta) \, d\eta \right]$$

$$= \frac{1}{2c} \int_{ct-x}^{ct+x} \Psi(\eta) \, d\eta \qquad (6.2.31)$$

which differs from (6.2.30) only in the form of the lower limit. Both results may be combined into the following solution for Case 2:

$$y(x, t) = \frac{1}{2c} \int_{|x-ct|}^{x+ct} \psi(\eta) \, d\eta \qquad (6.2.32)$$

It should also be observed that since $\psi(\eta)$ is odd,

$$\int_{x-ct}^{ct-x} \psi(\eta)\, d\eta = 0 \tag{6.2.33}$$

and therefore the lower limit in (6.2.32) may be replaced by $x - ct$. When both φ and ψ are nonzero, the results of both cases are added to give the d'Alembert solution (1.9.16).

Problems

6.2.1 Use the Fourier sine transform on x to solve

$$\nabla^2 w + k^2 w = 0, \qquad w(0, y) = w_0 \sin \frac{n\pi y}{H}$$

$$0 \le x < \infty, \qquad 0 \le y < H$$

where

$$\nabla^2 w = \frac{\partial^2 w}{\partial x^2} + \frac{\partial^2 w}{\partial y^2}$$

6.2.2 A semi-infinite string $(0 \le x \le \infty)$ is free to move vertically at $x = 0$ and is driven by a point source of frequency ω_0 and strength F_0 at $x = x_0$. The equation governing the string is thus

$$\frac{\partial^2 y}{\partial x^2} - \frac{1}{c^2}\frac{\partial^2 y}{\partial t^2} = -F_0 \delta(x - x_0)e^{-i\omega t}, \qquad \left.\frac{\partial y}{\partial x}\right|_{x=0} = 0$$

Set $y(x, t) = f(x)e^{-i\omega t}$ and solve the resulting equation for $f(x)$ by using an appropriate transform.

6.2.3 A semi-infinite beam is initially at zero temperature. For $t > 0$, $u(0, t) = u_0 f(t)$.
(a) Show that $U(k, t)$, the Fourier sine transform of $u(x, t)$, satisfies

$$\frac{dU}{dt} + (\gamma k)^2 U = k\gamma^2 u_0 f(t)$$

(b) Solve the differential equation for $U(k, t)$ and obtain

$$u(x, t) = \frac{x}{2\sqrt{\pi}\gamma} \int_0^t dt' \frac{f(t')}{(t - t')^{3/2}} e^{-x^2/4\gamma^2(t - t')}$$

(c) For $f(t) = H(t)$, the Heaviside step function, obtain

$$u(x, t) = u_0 \operatorname{erfc} \frac{x}{2\gamma \sqrt{t}}$$

6.3 CONVOLUTION THEOREMS

When a Fourier sine or cosine transform can be interpreted as the product of two simpler functions that are themselves transforms of known functions of x, then, as is the case for the Laplace transform (cf. Section B.2), it is possible to express the inversion in terms of a convolution. As has been seen in the problems that were solved in the previous section, the sine and cosine transforms already incorporate information on boundary conditions that has been associated with the method of images. This information is also displayed by the convolution theorems that we are about to consider. Hence the various forms of these convolutions are somewhat more cumbersome than the one associated with the Laplace transform.

Since the sine transform is associated with odd functions of k and the cosine transform with even functions of k, there will be more than one convolution theorem to be developed. In particular, we shall obtain a convolution theorem for the sine transform of an odd function of k that is the product of an odd and even function of k. The even function of k is itself a cosine transform. We thus consider the integral $(2/\pi)\int_0^\infty F_s(k)G_c(k)\sin kx \, dk$.

There are two possibilities for the cosine transform. Since a cosine transform is associated with an even function of k, it can be composed of the product of either two odd or two even functions of k. Hence for the cosine transform we must consider the two forms $(2/\pi)\int_0^\infty F_c(k)G_c(k)\cos kx \, dk$ and $(2/\pi)\int_0^\infty F_s(k)G_s(k)\cos kx \, dk$. Since the development of the various convolution theorems are all quite similar, we shall only consider the sine transform in detail and then list the two results for the cosine transform.

Returning to a consideration of the sine transform, recall that for the Laplace transform, there were two forms in which the convolution theorem could be expressed depending upon which term is to have the argument that contains the difference of two variables. To obtain one of the two possible forms in the present instance, we write $F_s(k)$ as the transform of a function $f(u)$ and obtain

$$\frac{2}{\pi}\int_0^\infty F_s(k)G_c(k)\sin kx \, dk$$

$$= \frac{2}{\pi}\int_0^\infty G_c(k)\left[\int_0^\infty f(u)\sin ku \, du\right]\sin kx \, dk$$

$$= \frac{2}{\pi}\int_0^\infty du \, f(u)\int_0^\infty dk \, G_c(k)\sin ku \, \sin kx$$

$$= \frac{2}{\pi} \int_0^\infty du\, f(u) \int_0^\infty dk\, G_c(k) \frac{1}{2} [\cos k(x - u) + \cos k(x + u)]$$

$$= \frac{1}{2} \int_0^\infty du\, f(u) [\, g(|u - x|) - g(u + x)] \tag{6.3.1}$$

The result is one form for the convolution theorem for the sine transform. The magnitude sign is used in writing $g(|x - u|)$ since g, being associated with a cosine transform, must be treated as an even function of its argument. While the form of this convolution expression may appear somewhat cumbersome, it is seen to contain information expected on the basis of the method of images.

If the substitution in (6.3.1) had been made for the function $G_c(k)$ rather than $F_s(k)$, we would have obtained the result

$$\frac{2}{\pi} \int_0^\infty F_s(k)G_c(k) \sin kx\, dx = \frac{1}{2} \int_0^\infty du\, g(u)[f(u + x) - f(u - x)] \tag{6.3.2}$$

In a similar way we may derive the following convolution theorems for the cosine transform:

$$\frac{2}{\pi} \int_0^\infty F_c(k)G_c(k)\cos kx\, dx = \frac{1}{2} \int_0^\infty g(u)[f(u + x) + f(|u - x|)]\, du$$

$$= \frac{1}{2} \int_0^\infty f(u)[g(u + x) + g(|u - x|)]\, du \tag{6.3.3}$$

as well as

$$\frac{2}{\pi} \int_0^\infty F_s(k)G_s(k)\cos kx\, dx = \frac{1}{2} \int_0^\infty g(u)[f(u - x) + f(u + x)]\, du$$

$$= \frac{1}{2} \int_0^\infty f(u)[g(u - x) + g(u + x)]\, du \tag{6.3.4}$$

Both pairs of relations are seen to be symmetric in the interchange of f and g. Before using some of these expressions to recover the solutions obtained in the previous section, we note that the convolution theorems for the cosine transform provide useful results known as Parseval relations when the special case $x = 0$ is considered. We obtain

$$\frac{2}{\pi} \int_0^\infty F_c(k)G_c(k)\, dk = \int_0^\infty g(u)f(u)\, du$$

$$\frac{2}{\pi} \int_0^\infty F_s(k)G_s(k)\, dk = \int_0^\infty g(u)f(u)\, du \tag{6.3.5}$$

Finally, if $g(u)$ is the complex conjugate of $f(u)$, we obtain

$$\frac{2}{\pi} \int_0^\infty |F_c(k)|^2 \, dk = \int_0^\infty |f(u)|^2 \, du$$

$$\frac{2}{\pi} \int_0^\infty |F_s(k)|^2 \, dk = \int_0^\infty |f(u)|^2 \, du$$

(6.3.6)

Since the expression for a transform of a function may have a much simpler analytical form than the function itself, the Parseval relations can provide a useful way of evaluating certain integrals. As an example, note that the cosine transform of the Bessel function $K_0(ax)$ is $(\pi/2)(k^2 + a^2)^{-1/2}$. Thus (6.3.6) yields

$$\int_0^\infty [K_0(ax)]^2 \, dx = \frac{\pi}{2} \int_0^\infty \frac{dk}{k^2 + a^2} = \frac{\pi^2}{4}$$

(6.3.7)

As examples of the use of the convolution theorems we now reconsider the three problems that were solved in the previous section.

1. *Green's Function for a Semi-infinite String.* For this example we obtained the Fourier sine transform

$$G_s(k) = \frac{\sin kx'}{k^2 - [(\omega/c)^2 + i\epsilon]}$$

(6.3.8)

For application of the convolution theorem, we factor the sine transform G_s into the product form $F_s(k)G_c(k)$. Recalling that $F_s(k)$ must be an odd function of k and $G_c(k)$ must be an even function of k, we write

$$F_s(k) = \sin kx', \qquad G_c(k) = \frac{1}{k^2 - [(\omega/c)^2 + i\epsilon]}$$

(6.3.9)

Then, on modifying the denominator of G_c as was done in (6.2.7) and using entries 8 in Table F.3 and 3 in Table F.4, we have

$$f(x) = \delta(x - x'), \qquad g(x) = \frac{ic}{\omega} e^{i\omega x/c}$$

(6.3.10)

where we have set $\epsilon = 0$ in the expression for $g(x)$. Using the convolution theorem given in (6.3.1) we obtain

$$g(x|x') = \frac{1}{2} \int_0^\infty du \, f(u)[g(|u - x|) - g(u + x)]$$

$$= \frac{1}{2} \int_0^\infty du \, \delta(u - x') \frac{ic}{\omega} [e^{i\omega|u - x|/c} - e^{i\omega(x + u)/c}]$$

$$= \frac{ic}{2\omega} [e^{i\omega|x - x'|/c} - e^{i\omega(x + x')/c}] \tag{6.3.11}$$

which agrees with the form given in (6.2.8) when we also let ϵ vanish in that result.

2. *Temperature Distribution in a Quarter Plane.* In Section 6.2.2 the temperature distribution $u(x, y)$ in the quarter plane with boundary conditions $u(x, 0) = 0$ for $x > 0$ and $u_x(0, y) = \varphi(y)$ for $y > 0$ was determined. The solution was given in terms of the Fourier sine transform

$$u(x, y) = \frac{2}{\pi} \int_0^\infty A(k)e^{-kx}\sin ky \, dk \tag{6.3.12}$$

where

$$A(k) = -\frac{1}{k} \int_0^\infty u_x(0, \eta)\sin k\eta \, d\eta \tag{6.3.13}$$

The result obtained previously in (6.2.17) is recovered here by using the convolution theorem for the sine transform given in (6.3.1). Since $A(k)$ is seen from (6.3.13) to be an even function of k, we associate it with $G_c(k)$ in the convolution theorem (6.3.1). The remaining term e^{-kx} is then identified with $F_s(k)$. Since the transform is with respect to the variable y, we obtain

$$f(y) = \frac{2}{\pi} \int_0^\infty F_s(k)\sin ky \, dk$$

$$= \frac{2}{\pi} \int_0^\infty e^{-kx}\sin ky \, dk = \frac{2}{\pi} \frac{y}{x^2 + y^2} \tag{6.3.14}$$

and

$$g(y) = \frac{2}{\pi} \int_0^\infty G_c(k)\cos ky \, dk$$

$$= \frac{2}{\pi} \int_0^\infty dk \cos ky \left(-\frac{1}{k}\right) \int_0^\infty d\eta \, u_x(0, \eta)\sin k\eta$$

$$= -\frac{2}{\pi} \int_0^\infty d\eta \, u_x(0, \eta) \int_0^\infty dk \frac{\cos ky \sin k\eta}{k} \tag{6.3.15}$$

According to Table F.3, 10, the integral over k is equal to $\pi/2$ for $|y| < \eta$ and vanishes for $|y| > \eta$. Hence we obtain

$$g(y) = -\int_y^\infty u_x(0, \eta)\, d\eta \qquad (6.3.16)$$

The form of the convolution theorem given in (6.3.2) now yields

$$u(x, y) = \frac{1}{2} \int_0^\infty d\xi\,(-) \int_\xi^\infty d\eta\, u_x(0, \eta)$$

$$\frac{2}{\pi}\left[\frac{(\xi + y)}{(\xi + y)^2 + x^2} - \frac{(\xi - y)}{(\xi - y)^2 + x^2} \right] \qquad (6.3.17)$$

To recover the result given previously in (6.2.17) we perform an integration by parts on the variable ξ. Noting that

$$\frac{dg(y)}{dy} = u_x(0, y) \qquad (6.3.18)$$

according to (6.3.16), and from (6.3.14) that

$$\int f(\xi)\, d\xi = \frac{1}{\pi} \ln(x^2 + \xi^2) \qquad (6.3.19)$$

we obtain, since the integrated term vanishes at the limits 0 and ∞,

$$u(x, y) = \frac{1}{2\pi} \int_0^\infty d\xi\, u_x(0, \xi) \ln \frac{(y - \xi)^2 + x^2}{(y + \xi)^2 + x^2} \qquad (6.3.20)$$

in agreement with (6.2.17).

3. *D'Alembert Solution for a Semi-infinite String.* In Section 6.2.3 we obtained the sine transform of the string displacement in the form

$$Y_s(k, t) = \Phi_s(k)\cos kct + \Psi_s(k)\frac{\sin kct}{kc} \qquad (6.3.21)$$

To obtain the string displacement $y(x, t)$, we now apply the convolution theorem for the sine transform to each of these two terms separately.

For the first term we choose $F_s(k) = \Phi_s(k)$ and $G_c(k) = \cos kct$ so that $f(x) = \varphi(x)$, the initial displacement. Also, according to entry 6 in Table F.4, we have $g(x) = \delta(x - ct)$. Application of the convolution theorem (6.3.1) now yields

$$\frac{2}{\pi} \int_0^\infty \Phi_s(k)\cos kct \sin kx \, dk$$

$$= \frac{1}{2} \int_0^\infty d\xi \, \varphi(\xi) [\delta(|x - \xi| - ct) - \delta(x + \xi - ct)]$$

$$= \frac{1}{2} \left\{ \int_0^x d\xi \, \varphi(\xi)\delta(x - \xi - ct) + \int_x^\infty d\xi \, \varphi(\xi)\delta(\xi - x - ct) \right.$$

$$\left. + \int_0^\infty d\xi \, \varphi(\xi)\delta(x + \xi - ct) \right. \tag{6.3.22}$$

On examining each integral to determine whether or not the contribution from the delta function is in the range of integration, we find that for $x > ct$ only the first and second integrals contribute while for $x < ct$ only the second and third integrals contribute. The string displacement is found to be

$$\frac{2}{\pi} \int_0^\infty \phi_s(k)\cos kct \sin kx \, dk$$

$$= \frac{1}{2} [\varphi(x + ct) + \varphi(x - ct)], \qquad x > ct$$

$$\frac{1}{2} [\varphi(x + ct) + \varphi(ct - x)], \qquad x < ct$$

$$\frac{1}{2} [\varphi(x + ct) - \varphi(x - ct)] \tag{6.3.23}$$

For the second term in (6.3.21) we choose $F_s = \Phi_s(k)$ and $G_c(k) = (\sin kct)/kc$. We then have $f(x) = \psi(x)$, the initial string velocity. Also, from entry 7 of Table F.4 we have $g(x) = H(ct - x)$. The convolution theorem (6.3.1) now yields

$$\frac{2}{\pi} \int_0^\infty \Psi_s(k) \frac{\sin kct}{kc} \sin kx \, dk$$

$$= \frac{1}{2} \int_0^\infty d\xi \, \psi(\xi) \{H(ct - |x - \xi|) - H[ct - (x + \xi)]\} \tag{6.3.24}$$

Again separating the integrand for the first two terms into subregions for $0 < \xi < x$ and $x < \xi < \infty$ and examining the relation of the step functions to the range of integration, we find.

$$\frac{2}{\pi} \int_0^\infty \Psi_s(k) \frac{\sin kct}{kc} \sin kx = \int_{|x - ct|}^{x + ct} \psi(\xi) \, d\xi \tag{6.3.25}$$

On combining (6.3.23) and (6.3.25) we obtain the d'Alembert solution for the semi-infinite string. As noted previously in the discussion of (6.2.32), an odd extension of $\psi(x)$ to negative values of x enables us to replace the lower limit in (6.3.25) by $x - ct$. Clearly, the convolution theorems associated with the sine and cosine transforms give rise to somewhat tedious calculations.

Problems

Results obtained in the problems of Section 6.1 and 6.2 may be recovered by using appropriate convolution theorems.

6.4 COMPLEX FOURIER TRANSFORMS

The Fourier transform method can also be applied to functions that are specified over the entire x axis ($-\infty < x < \infty$) as being neither even nor odd. As in the case of Fourier series, the appropriate relations may be obtained by combining those for the sine and cosine transforms. The results are most conveniently expressed in a complex exponential notation.

In order to use the results already developed for sine and cosine transforms, we decompose some given function, say $f(x)$, into even and odd parts and write

$$f(x) = f_e(x) + f_o(x) \tag{6.4.1}$$

Assuming that $f_e(x)$ and $f_o(x)$ have cosine and sine transforms, respectively, we write

$$f_e(x) = \frac{2}{\pi} \int_0^\infty F_c(k)\cos kx \, dk$$
$$\tag{6.4.2}$$
$$f_o(x) = \frac{2}{\pi} \int_0^\infty F_s(k)\sin kx \, dk$$

as well as the accompanying relations for $F_c(k)$ and $F_s(k)$. For $F_c(k)$ we have

$$F_c(k) = \int_0^\infty f_e(x)\cos kx \, dk = \frac{1}{2} \int_{-\infty}^\infty f_e(x)\cos kx \, dx$$

$$= \frac{1}{2} \int_{-\infty}^\infty f(x)\cos kx \, dx \tag{6.4.3}$$

The final form has been obtained by adding the odd function $f_o(x)\cos kx$ to the integrand and then replacing $f_e + f_o$ by f. Since the integrand is between symmetric limits, the added term, which is an odd function of x, contributes nothing to the integral. Proceeding in a similar way for $F_s(k)$ we have

$$F_s(k) = \int_0^\infty f_o(x)\sin kx\, dx = \frac{1}{2}\int_{-\infty}^\infty f_o(x)\sin kx\, dx \qquad (6.4.4)$$

Multiplying (6.4.4) by $-i$ and adding to (6.4.3), we obtain

$$2F(k) = \int_{-\infty}^\infty f(x)e^{-ikx}\, dx \qquad (6.4.5)$$

where $F(k) = F_c(k) - iF_s(k)$. In a similar way we can write

$$f_e(x) = \frac{2}{\pi}\int_0^\infty F_c(k)\cos kx\, dk$$

$$= \frac{1}{\pi}\int_{-\infty}^\infty F_c(k)(\cos kx + i\sin kx)\, dk$$

$$= \frac{1}{\pi}\int_{-\infty}^\infty F_c(k)e^{ikx}\, dx \qquad (6.4.6)$$

and

$$f_o(x) = \frac{1}{\pi}\int_{-\infty}^\infty -iF_s(k)(\cos kx + i\sin kx)\, dk$$

$$= \frac{-i}{\pi}\int_{-\infty}^\infty F_s(k)e^{ikx}\, dx \qquad (6.4.7)$$

Combining the results for f_e and f_o, we have

$$f_e(x) + f_o(x) = \frac{1}{\pi}\int_{-\infty}^\infty e^{ikx}(F_c - iF_s)\, dk \qquad (6.4.8)$$

We thus have

$$f(x) = \frac{1}{2\pi}\int_{-\infty}^\infty e^{ikx}F(k)\, dk \qquad (6.4.9)$$

As an example, consider the function

$$f(x) = e^{-\gamma|x-a|} \qquad (6.4.10)$$

A simple integration of (6.4.5) for this function yields

$$F(k) = \frac{2\gamma}{k^2 + \gamma^2}e^{-ika} \qquad (6.4.11)$$

To obtain the convolution theorem for the complex Fourier transform, we consider

$$\int_{-\infty}^{\infty} \frac{dk}{2\pi} F(k)G(k)e^{ikx} = \int_{-\infty}^{\infty} \frac{dk}{2\pi} e^{ikx}F(k) \int_{-\infty}^{\infty} d\xi \; e^{-ik\xi}g(\xi)$$

$$= \int_{-\infty}^{\infty} d\xi \; g(\xi) \int_{-\infty}^{\infty} \frac{dk}{\pi} F(k)e^{ik(x-\xi)}$$

$$= \int_{-\infty}^{\infty} d\xi \; g(\xi)f(x-\xi) \tag{6.4.12}$$

The result is, of course, symmetric in the functions $f(x)$ and $g(x)$ and, by substituting the transform for $F(k)$ instead of that for $G(k)$, could be written with the integrand in the form $f(\xi)g(x-\xi)$. On setting $x = 0$ in (6.4.12) we obtain the Parseval relation

$$\int_{-\infty}^{\infty} F(k)G(k) \; dk = \int_{-\infty}^{\infty} f(x)g(x) \; dx \tag{6.4.13}$$

A shift theorem for the complex Fourier transform can also be developed. Provided the integral of $e^{-ax}f(x)$ over all x converges, we can write

$$\int_{-\infty}^{\infty} e^{-ikx}e^{-ax}f(k) = \int_{-\infty}^{\infty} e^{-i(k-ia)}f(x) \; dx$$

$$= F(k-ia) \tag{6.4.14}$$

where $F(k)$ is the transform of $f(x)$. This result proves useful in evaluating a number of transforms and inverse transforms.

Assuming that $f(x)$ and its derivative vanish as $|x| \to \infty$, we may use partial integration to obtain the following results for the transforms of derivatives:

$$\int_{-\infty}^{\infty} e^{-ikx} \frac{df}{dx} \; dx = ikF(k)$$

$$\int_{-\infty}^{\infty} e^{-ikx} \frac{d^2f}{dx^2} \; dx = -k^2F(k) \tag{6.4.15}$$

As an example of the method, we determine the solution of Laplace's equation in the half plane $-\infty < x < \infty$, $y \geq 0$, when the solution is specified on the boundary $y = 0$.

We write Laplace's equation in the form

$$\frac{\partial^2 u}{\partial x^2} + \frac{\partial^2 u}{\partial y^2} = 0 \tag{6.4.16}$$

and introduce the complex Fourier transform

$$U(k, y) = \int_{-\infty}^{\infty} e^{-ikx} u(x, y) \, dx \qquad (6.4.17)$$

Using (6.1.5) for the transform of the second derivative, we obtain

$$\frac{d^2 U}{dy^2} - k^2 U = 0 \qquad (6.4.18)$$

which has the general solution

$$U(k, y) = A(k)e^{-ky} + B(k)e^{ky} \qquad (6.4.19)$$

This solution must, of course, remain bounded as $y \to +\infty$ for both positive and negative values of k. We thus write the solution in the form

$$U(k, y) = \begin{cases} A(k)e^{-ky}, & k > 0 \\ B(k)e^{ky}, & k < 0 \end{cases} \qquad (6.4.20)$$

a result more conveniently expressed as

$$U(k, y) = F(k)e^{-|k|y} \qquad (6.4.21)$$

where $F(k) = A(k)$ for $k > 0$ and $F(k) = B(k)$ for $k < 0$. Using the inverse transform, we express the solution in the form

$$u(x, y) = \int_{-\infty}^{\infty} \frac{dk}{2\pi} e^{ikx} U(k, y)$$

$$= \int_{-\infty}^{\infty} \frac{dk}{2\pi} F(k)e^{-|k|y + ikx} \qquad (6.4.22)$$

To determine the function $F(k)$, we evaluate this expression for $y = 0$ and obtain

$$u(x, 0) = \int_{-\infty}^{\infty} \frac{dk}{2\pi} F(k)e^{ikx} \qquad (6.4.23)$$

and then use the inverse expression

$$F(k) = \int_{-\infty}^{\infty} d\xi \, e^{-ik\xi} u(\xi, 0) \qquad (6.4.24)$$

For a specific boundary condition $u(\xi, 0)$ this integral may be evaluated and the resulting expression for $F(k)$ introduced into (6.4.23). This procedure may also be carried out in general. Replacing $F(k)$ in (6.4.23) by the integral in (6.4.24), we obtain

$$u(x, y) = \int_{-\infty}^{\infty} \frac{dk}{2\pi} e^{ikx - |k|y} \int_{-\infty}^{\infty} d\xi \, u(\xi, 0)e^{-ik\xi} \tag{6.4.25}$$

On interchanging orders of integration, we have

$$u(x, y) = \int_{-\infty}^{\infty} d\xi \, u(\xi, 0) \int_{-\infty}^{\infty} \frac{dk}{2\pi} e^{ik(x - \xi) - |k|y} \tag{6.4.26}$$

The integration over k is of the type that was employed in obtaining (6.4.11) with k and x interchanged. For the present choice of variables the result is

$$u(x, y) = \frac{y}{\pi} \int_{-\infty}^{\infty} \frac{u(\xi, 0) \, d\xi}{(x - \xi)^2 + y^2} \tag{6.4.27}$$

6.4.1 Approach to the Boundary

It should be noted that as y approaches the boundary, the integrand in this result becomes singular. However, the singularity arises in a form that is a standard representation for a delta function [cf. (1.5.1)]. Thus, as the boundary is approached, the expression should be interpreted as yielding an identity, that is,

$$\lim_{y \to 0^+} u(x, y) = u(x, 0) = \int_{-\infty}^{\infty} d\xi \, u(\xi, 0) \lim_{y \to 0^+} \left| \frac{1}{\pi} \frac{y}{(x - \xi)^2 + y^2} \right|$$

$$= \int_{-\infty}^{\infty} u(\xi, 0)\delta(x - \xi) = u(x, 0) \tag{6.4.28}$$

If the temperature on the boundary, $u(\xi, 0)$, is discontinuous at a point $\xi = x_0$ where

$$\lim_{\xi \to x_0^+} u(\xi, 0) = u_+, \qquad \lim_{\xi \to x_0^-} u(\xi, 0) = u_- \tag{6.4.29}$$

the temperature at field points $u(x_0, y)$ just about the point of discontinuity will approach the average of these two values as the boundary is approached, that is,

$$\lim_{y \to 0^+} (x_0, y) = \tfrac{1}{2}(u_+ + u_-) \tag{6.4.30}$$

This result may be readily obtained by first converting the solution (6.4.27) to an integral over the angle ϑ shown in Figure 6.2.

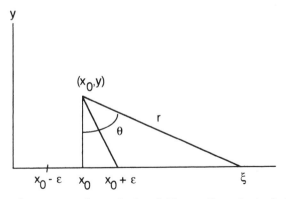

Figure 6.2. Coordinate system for analyzing field near discontinuity in boundary conditions.

Since $\xi - x_0 = y \tan \vartheta$, we have $d\xi = y \sec^2 \vartheta \, d\vartheta$ and the solution (6.4.27) becomes

$$u(x_0, y) = \frac{1}{\pi} \int_{-\pi/2}^{\pi/2} u(x_0 + y \tan \vartheta, 0) \, d\vartheta \qquad (6.4.31)$$

Setting $\vartheta = -\vartheta$ in the integration from $-\pi/2$ to 0, we have

$$u(x_0, y) = \frac{1}{\pi} \int_0^{\pi/2} [u(x_0 + y \tan \vartheta, 0) - u(x_0 - y \tan \vartheta, 0)] \, d\vartheta \qquad (6.4.32)$$

Now consider an arbitrarily small region of the boundary about the point of discontinuity, that is, $x_0 - \delta < \xi < x_0 + \delta$. For values of ξ in this region, the integration variable ϑ satisfies $0 < y \tan \vartheta < \delta$ and we may write

$$u(x_0 + y \tan \vartheta, 0) = u_+ + u_\xi(x_0^+, 0)y \tan \vartheta + \cdots$$

$$u(x_0 - y \tan \vartheta, 0) = u_- - u_\xi(x_0^-, 0)y \tan \vartheta + \cdots \qquad (6.4.33)$$

When the integration along the surface $0 < \xi < \infty$ is decomposed into regions $x_0 \leq \xi < x_0 + \delta$ and $x_0 + \delta < \xi < \infty$, the integral over the angle ϑ in (6.4.32) becomes

$$u(x_0, y) = \frac{1}{\pi} \int_0^{\tan^{-1}(\delta/y)} [u_+ + u_- + (u_{\xi+} - u_{\xi-})y \tan \vartheta + \cdots] \, d\vartheta$$

$$+ \frac{1}{\pi} \int_{\tan^{-1}(\delta/y)}^{\pi/2} [u(x_0 + y \tan \vartheta, 0) + u(x_0 - y \tan \vartheta, 0) + \cdots] \, d\vartheta$$

$$(6.4.34)$$

where $u_{\xi\pm} = u_{\xi}(x_0^{\pm})$. As the field point coordinate y approaches 0, $\tan^{-1}(\delta/y)$ approaches $\pi/2$. Also, since $\tan^{-1}\delta/y = \cos^{-1}y/r$, we have

$$\int_0^{\tan^{-1}(\delta/y)} \tan \vartheta \, d\vartheta = y \ln (y/r) \qquad (6.4.35)$$

which vanishes as y approaches zero. The second integral in (6.4.34) also vanishes as long as $u(\xi, 0)$ remains finite since the lower limit of integration approaches the upper limit, $\pi/2$. The only remaining term in (6.4.34) is the integral over $u_+ + u_-$. As y approaches 0, the upper limit approaches $\pi/2$ and we obtain the result quoted in (6.4.30). When the temperature on the boundary is continuous, so that $u^+ = u^-$, we, of course, recover the result given in (6.4.28). Examples of this result occur in Problems 4.11.2 and 4.11.4.

Returning to consideration of application of the method, we determine the temperature distribution on an infinite strip $-\infty < x < \infty, 0 \le y \le h$. We choose the situation for which the heat flow is specified along the lower edge [i.e., $u_y(x, 0)$ is known] while the temperature along the upper edge $y = h$ is maintained at zero temperature, $u(x, h) = 0$. We ask for the spatial variation of the heat flux at the upper boundary and the total amount of heat flowing out through the boundary at $y = h$.

Writing the Fourier transform of the temperature distribution $u(x, y)$ in the form

$$U(k, y) = \int_{-\infty}^{\infty} dx \, e^{-ikx}u(x, y) \qquad (6.4.36)$$

we obtain the transform of Laplace's equation in the form

$$\frac{d^2U}{dy^2} - k^2U = 0 \qquad (6.4.37)$$

The solution of this equation that vanishes at the edge $y = h$ is

$$U(k, y) = A(k)\sin k(h - y) \qquad (6.4.38)$$

Using the inverse of the definition of $U(k, y)$ that was given above in (6.4.36), we have

$$U(k, y) = \int_{-\infty}^{\infty} \frac{dk}{2\pi} e^{ikx}A(k)\sin k(h - y) \qquad (6.4.39)$$

To relate $A(k)$ to the known quantity $u_y(x, 0)$, we differentiate this result for u to obtain

$$U_y(k, y) = \int_{-\infty}^{\infty} \frac{dk}{2\pi} e^{ikx}(-k)A(k)\cosh k(h - y) \qquad (6.4.40)$$

On applying the inverse transform, we have

$$-kA(k)\cosh kh = \int_{-\infty}^{\infty} dx\, e^{-ikx} u_y(x, 0) \tag{6.4.41}$$

Proceeding as in the previous example, we use this expression for $A(k)$ in $u(x, y)$ and interchange the orders of integration. The result is

$$u(x, y) = -\int_{-\infty}^{\infty} d\xi\, u_y(\xi, 0) \int_{-\infty}^{\infty} \frac{dk}{2\pi} e^{ik(x - \xi)} \frac{\sinh k(h - y)}{k\cosh kh} \tag{6.4.42}$$

Although the integral over k can be performed, we shall here consider only the evaluation of the simpler expression for $u_y(x, h)$. It is proportional to the heat flow at the upper edge. We find

$$u_y(x, h) = \int_{-\infty}^{\infty} d\xi\, u_y(\xi, 0) \int_{-\infty}^{\infty} \frac{dk}{2\pi} e^{ik(x - \xi)} \operatorname{sech} kh \tag{6.4.43}$$

The complex Fourier transform of sech kh is given in Table F.5. Using the result given there, we have

$$u_y(x, h) = \frac{1}{2h} \int_{-\infty}^{\infty} d\xi\, u_y(\xi, 0)\operatorname{sech}\left(\frac{\pi}{2} \frac{\xi - x}{h}\right) \tag{6.4.44}$$

The form of this expression shows how the heat influx at any point $x = \xi$ along the base is "smeared out" as it diffuses across the strip. Note also that as $h \to 0$, there is no opportunity for a smearing out to take place and the term multiplying $u_y(\xi, 0)$ becomes a representation for a delta function. In this case we find

$$\delta(\xi - x) = \lim_{h \to 0} \frac{1}{2h} \operatorname{sech}\left(\frac{\pi}{2} \frac{\xi - x}{h}\right) \tag{6.4.45}$$

Problems

6.4.1 The base of the half space $z > 0$ has two narrow strips that are maintained at a constant temperature and separated by a distance $2a$. This source is modeled by the boundary condition

$$u(x, 0) = A[\delta(x - a) + \delta(x + a)]$$

Significance of the constant A will be considered later.

(a) Use the complex Fourier transform in x to show that the temperature $u(x, z)$ in the half space is given by

$$u(x, z) = \frac{Az}{2\pi} \left[\frac{1}{z^2 + (x - a)^2} + \frac{1}{z^2 + (x + a)^2} \right]$$

(b) Given a qualitative sketch of $u(x, z)$ as a function of x for fixed z both for $z \ll a$ and $z \gg a$.

(c) The two qualitatively different curves in (b) are distinguished by having $u_{xx}(0, z) = 0$ for $z = a\sqrt{3}$. What do these two regions means physically?

(d) If $u(0, a)$, that is, the temperature on axis at a distance a above the XY plane, is set equal to a reference temperature u, show that the source amplitude A is given as $A = 2\pi a u_0$.

(e) Another common way of relating a source amplitude to physically significant quantities is to relate it to the total amount of heat in the region of the problem. If one attempts to determine A in terms of H_0, the total heat in the half space $z > 0$ by using the definition

$$H_0 = \rho c \int_{-\infty}^{\infty} dx \int_0^{\infty} dz\, u(x, z)$$

what happens?

6.4.2 The base of the half space $z \geq 0$ is heated by a filament at $x = -a$ and also cooled by a parallel heat sink at $x = a$. The situation may be represented by the boundary condition

$$J_z(x, 0) = -K \left. \frac{\partial u}{\partial z} \right|_{z=0} = c[\delta(x + a) - \delta(x - a)]$$

The significance of the constant c will be considered later.

(a) Use the complex Fourier transform in x to show that

$$u(x, z) = \frac{2ic}{K} \int_{-\infty}^{\infty} \frac{dk}{2\pi} \frac{\sin ka}{|k|} e^{ikx - |k|z}$$

$$= \frac{c}{\pi} \int_0^{\infty} dk\, k^{-1}(-2 \sin ka \sin kx)e^{-kz}$$

(b) Evaluate by using the table of Laplace transforms to obtain

$$u(x, z) = \frac{c}{2\pi K} \ln \frac{z^2 + (x - a)^2}{z^2 + (x + a)^2}$$

(c) If the total rate of heat flow across the plane $x = 0$ is known to be H_0, that is,

$$\int_0^\infty dz \left(-k \frac{\partial u}{\partial x} \right)\Bigg|_{x=0} = H_0$$

show that the constant c has the value H_0.

6.5 FOURIER TRANSFORMS IN TWO AND THREE DIMENSIONS

The transform techniques previously developed for one-space dimension are readily extended to two and three dimensions. As a two-dimensional example, consider the function $f(x, y)$ and introduce the transform $F(k_x, k_y)$ according to the definition

$$F(k_x, k_y) = \int_{-\infty}^{\infty} dx \int_{-\infty}^{\infty} dy \, e^{-i(k_x x + k_y y)} f(x, y) \tag{6.5.1}$$

It is convenient to refer to the two transform variables k_x and k_y as the components of a two-dimensional vector $\mathbf{k} = \mathbf{i}k_x + \mathbf{j}k_y$. Also introducing the vector $\boldsymbol{\rho} = \mathbf{i}x + \mathbf{j}y$, we can write the above definition in the form

$$F(\mathbf{k}) = \int d^2\boldsymbol{\rho} \, e^{-i\mathbf{k}\cdot\boldsymbol{\rho}} f(\boldsymbol{\rho}) \tag{6.5.2}$$

The associated two-dimensional inverse transform is also a natural extension of the one-dimensional case. We have

$$f(x, y) = \int_{-\infty}^{\infty} \frac{dk_x}{2\pi} \int_{-\infty}^{\infty} \frac{dk_y}{2\pi} \, e^{i(k_x x + k_y y)} F(k_x, k_y) \tag{6.5.3}$$

or equivalently

$$f(\boldsymbol{\rho}) = \int \frac{d^2\mathbf{k}}{(2\pi)^2} \, e^{i\mathbf{k}\cdot\boldsymbol{\rho}} F(\mathbf{k}) \tag{6.5.4}$$

As a simple but useful example of this extension of notation, the two-dimensional delta function $\delta^2(\boldsymbol{\rho} - \boldsymbol{\rho}') = \delta(x - x')\delta(y - y')$ may be written

$$\delta^2(\boldsymbol{\rho} - \boldsymbol{\rho}') = \int_{-\infty}^{\infty} \frac{dk_x}{2\pi} \, e^{ik_x(x - x')} \int_{-\infty}^{\infty} \frac{dk_y}{2\pi} \, e^{ik_y(y - y')}$$

$$= \int \frac{d^2\mathbf{k}}{(2\pi)^2} \, e^{i\mathbf{k}\cdot(\boldsymbol{\rho} - \boldsymbol{\rho}')} \tag{6.5.5}$$

Other techniques used for the one-dimensional transform are readily extended to the multidimensional case. If the function $f(\mathbf{\rho})$ as well as its derivatives $f_x(\mathbf{\rho})$ and $f_y(\mathbf{\rho})$ vanish as $|x|, |y| \to \infty$, then partial integration yields

$$\int d^2\rho\, e^{-i\mathbf{k}\cdot\mathbf{\rho}} \frac{\partial f}{\partial x} = \int_{-\infty}^{\infty} dy\, e^{-ik_y y} \int_{-\infty}^{\infty} dx\, e^{-ik_x x} ik_x f(x, y)$$

$$= ik_x F(\mathbf{k}) \tag{6.5.6}$$

$$\int d^2\rho\, e^{-i\mathbf{k}\cdot\mathbf{\rho}} \frac{\partial^2 f}{\partial x^2} = -k_x^2 F(\mathbf{k})$$

and similarly for the y derivatives. Hence, for the transform of the two-dimensional Laplacian we obtain

$$\int d^2\rho\, e^{-i\mathbf{k}\cdot\mathbf{\rho}} \nabla^2 f = -k^2 F(\mathbf{k}) \tag{6.5.7}$$

where $k^2 = |\mathbf{k}|^2 = k_x^2 + k_y^2$.

As an example of the use of a multidimensional transform, we consider the determination of the solution of Laplace's equation in the half space $z > 0$, $-\infty < x, y < \infty$ when the normal derivative $u_z(x, y, 0)$ is specified over the plane $z = 0$. The problem is thus equivalent to determining the temperature above a plane on which the heat flux has been specified.

Since the x and y values extend to $\pm\infty$, they are conveniently treated with the two-dimensional transform developed here. Writing this transform as

$$U(k_x, k_y, z) = \int_{-\infty}^{\infty} dx \int_{-\infty}^{\infty} dy\, e^{-i(k_x x + k_y y)} u(x, y, z) \tag{6.5.8}$$

we obtain the transform of the three-dimensional Laplace equation in the form

$$-(k_x^2 + k_y^2)U + \frac{d^2 U}{dz^2} = 0 \tag{6.5.9}$$

where $U = U(k_x, k_y, z)$ as defined above. As in the case of the single-variable transform obtained in (6.2.12), we use the solution of this ordinary differential equation that decreases with increasing z and write

$$U(k_x, k_y, z) = A(k_x, k_y)e^{-z\sqrt{k_x^2 + k_y^2}} \tag{6.5.10}$$

The solution $u(x, y, z)$ can now be written

$$u(x, y, z) = \int \frac{dk_x\, dk_y}{(2\pi)^2} e^{i(k_x x + k_y y)} U(k_x, k_y, z)$$

$$= \int \frac{dk_x\, dk_y}{(2\pi)^2}\, e^{i(k_x x + k_y y)} A(k_x,\, k_y)\, e^{-z\sqrt{k_x^2 + k_y^2}} \qquad (6.5.11)$$

Calculation of $u_z(x,\, y,\, z)$ will enable us to express $A(k_x,\, k_y)$ in terms of the known quantity $u_z(x,\, y,\, 0)$. We obtain

$$u_z(x,\, y,\, 0) = -\int \frac{dk_x\, dk_y}{(2\pi)^2}\, e^{i(k_x x + k_y y)}\, \sqrt{k_x^2 + k_y^2}\, A(k_x,\, k_y) \qquad (6.5.12)$$

The inverse of this expression is

$$-\sqrt{k_x^2 + k_y^2}\, A(k_x,\, k_y) = \int_{-\infty}^{\infty} dx'\, dy'\, e^{i(k_x x' + k_y y')} u_z(x',\, y',\, 0) \qquad (6.5.13)$$

With this result for $A(k_x,\, k_y)$, the value of $u(x,\, y,\, z)$ is given by (6.5.11) in the form

$$u(x,\, y,\, z) = -\int_{-\infty}^{\infty} dx'\, dy'\, u_z(x',\, y',\, 0)$$

$$\cdot \int_{-\infty}^{\infty} \frac{dk_x\, k_y}{(2\pi)^2}\, \frac{e^{ik_x(x - x') + ik_y(y - y') - z\sqrt{k_x^2 + k_y^2}}}{\sqrt{k_x^2 + k_y^2}} \qquad (6.5.14)$$

In terms of the vector notation introduction previously, this result is

$$u(\boldsymbol{\rho},\, z) = -\int d^2\rho'\, u(\boldsymbol{\rho}',\, 0) \int \frac{d^2k}{(2\pi)^2}\, \frac{e^{i\mathbf{k}\cdot(\boldsymbol{\rho} - \boldsymbol{\rho}') - kz}}{k} \qquad (6.5.15)$$

where $k = (k_x^2 + k_y^2)^{1/2}$.

The two-dimensional integral over k is readily performed. If we set $\mathbf{R} = \boldsymbol{\rho} - \boldsymbol{\rho}'$ and choose the k_x axis along the vector $\mathbf{R}$, so that $\mathbf{k} \cdot \mathbf{R} = kR \cos \vartheta$, where ϑ is the angle between $\mathbf{k}$ and $\mathbf{R}$, the integration proceeds as follows:

$$\int \frac{d^2k}{(2\pi)^2}\, \frac{e^{i\mathbf{k}\cdot\mathbf{R} - kt}}{k} = \int_0^{\infty} \frac{k\, dk}{(2\pi)^2} \int_0^{2\pi} d\vartheta\, \frac{e^{ikR\cos\vartheta - kz}}{k}$$

$$= \frac{1}{2\pi} \int_0^{\infty} dk\, e^{-kz} J_0(kR)$$

$$= \frac{1}{2\pi\sqrt{z^2 + R^2}} \qquad (6.5.16)$$

We have used the result (D.49) that the integral over ϑ is a representation of the Bessel function $J_0(kR)$ and that the resulting integral over k is in the form of a Laplace transform of a Bessel function (Table F.2, 20).

The final expression for $u(\mathbf{\rho}, z)$ is now

$$u(\rho_s, z) = -\int d^2\mathbf{\rho}' \, u_z(\mathbf{\rho}', 0) \left[\frac{1}{2\pi} \frac{1}{\sqrt{(x - x')^2 + (y - y')^2 + z^2}} \right]$$

$$(6.5.17)$$

The term in brackets will be recognized as $G(x, y, z|x', y', 0)$, the Green's function for Laplace's equation that has a vanishing normal derivative at $z = 0$. The result obtained here is, of course, equivalent to that obtainable from Eq. (4.9.20).

Problems

6.5.1 The half space $z > 0$ is heated by the temperature $u(x, 0) = C\delta(x)$ where C is a constant.

   (a) Use a two-dimensional Fourier transform to obtain the temperature for $z > 0$.

   (b) Show that the result reduces to a standard representation of $\delta(x)$ as $y \to 0$.

6.5.2 The temperature $u(x, y, z)$ in the half space $z > 0$ is determined by the boundary condition

$$u(x, y, 0) = \begin{cases} u_0, & |x| < a \\ 0, & |x| > a \end{cases}$$

and the requirement that it vanish as $z \to \infty$.

   (a) Use the two-dimensional Fourier transform to obtain

$$u(x, y, z) = \int \frac{dk_x \, dk_y}{(2\pi)^2} e^{ik_x x + k_y y} A(k_x, k_y) e^{-\sqrt{k_x^2 + k_y^2}\, z}$$

   where

$$A(k_x, k_y) = 4\pi u_0 \delta(k_y) \frac{\sin k_x a}{k_x}$$

   (b) The integral over k_x may be evaluated as a Laplace transform (Table F.2) to yield

$$u(x, z) = \frac{u_0}{\pi} \tan^{-1} \frac{2az}{x^2 + z^2 - a^2}$$

   There is, of course, no y dependence in the problem.

6.5.3 In the previous problem we considered the two-dimensional problem of determining the steady state temperature distribution in the half space $z > 0$ when $u(x, y, 0) = u_0 H(a^2 - x^2)$ in which H is the unit step function. There is thus a strip of width $2a$ that is maintained at temperature u_0 while the remainder of the plane $z = 0$ is kept at temperature zero.

(a) Consider the corresponding problem in which the strip of width $2a$ has a constant heat flux $J(x, y) = -Ku_z(x, y, 0) = J_0 H(a^2 - x^2)$ when J_0 is a constant. Use the complex Fourier transform to obtain

$$u(x, y, z) = \frac{2J_0}{K} \int_{-\infty}^{\infty} \frac{dk}{2\pi} e^{ikx - |k|z} \frac{\sin ka}{k^2}$$

$$= \frac{J_0}{K} \frac{2}{\pi} \int_0^{\infty} dk\, e^{-kz} \frac{\cos kx \sin ka}{k^2}$$

(b) Comment on this result.

6.5.4 The half space $z \geq 0$ is heated by a square region such that

$$J(x, y, 0) = -K \left.\frac{\partial u}{\partial z}\right|_{z=0} = \begin{cases} J_0, & |x|, |y| < a \\ 0, & \text{outside the square} \end{cases}$$

(a) Use the complex Fourier transform to show that the solution of Laplace's equation, which provides the steady state temperature distribution in the half space, may be written as

$$u(x, y, z) = \int_{-\infty}^{\infty} \frac{dk_x}{2\pi} \int_{-\infty}^{\infty} \frac{dk_y}{2\pi} U(k_x, k_y, z) e^{i(k_x x + k_y y)}$$

where

$$U(k_x, k_y, z) = \frac{4J_0}{K} \frac{1}{\sqrt{k_x^2 + k_y^2}} \frac{\sin k_x a}{k_x} \frac{\sin k_y a}{k_y} e^{-z\sqrt{k_x^2 + k_y^2}}$$

(b) Decouple the k_x and k_y dependence in the integrand by using the integral representation obtainable from Eq. (B.60), namely,

$$\frac{e^{-2\sqrt{ab}}}{\sqrt{a}} = \frac{2}{\sqrt{\pi}} \int_0^{\infty} d\theta\, e^{-(a\theta^2 + b/\theta^2)}$$

to obtain the temperature an axis (i.e., $x = y = 0$) in the form

$$u(0, 0, z) = \frac{8J_0}{\sqrt{\pi} K} \int_0^{\infty} d\theta\, e^{-z^2/4\theta^2} \left(\int_{-\infty}^{\infty} \frac{dk}{2\pi} \frac{\sin ka}{k} e^{-k^2\theta^2} \right)^2$$

where the integration variables k_x and k_y have each been written as k.

(c) Use the relation Table F.3, 7,

$$\int_0^\infty dx\, e^{-\beta x^2}\frac{\sin \alpha x}{x} = \frac{\pi}{2}\operatorname{erf}\frac{\alpha}{2\sqrt{\beta}}$$

to obtain

$$u(0, 0, z) = \frac{2}{\sqrt{\pi}}\frac{J_0}{K}\int_0^\infty d\theta\, e^{-z^2/4\theta^2}\operatorname{erf}\left(\frac{a}{2\theta}\right)^2$$

(d) Although this integral may be done exactly, merely set $z/2\theta = \zeta$ and expand the error function as

$$\operatorname{erf}(x) \cong \frac{2x}{\sqrt{\pi}}\left(1 - \frac{x^2}{3}\right)$$

to obtain

$$u(0, 0, z) \cong \frac{4J_0 a^2}{\pi^{3/2} Kz}\int_0^\infty d\zeta\, e^{-\zeta^2}\left(1 - \frac{2}{3}\frac{a^2}{z^2}\zeta^2\right)$$

which is a valid approximation for $a/z \ll 1$.

(e) Use standard results for the integrals over ζ to finally obtain

$$U(0, 0, z) \cong \frac{J_0 A}{2\pi Kz}\left(1 - \frac{A}{12z^2}\right)$$

in which $A = 4a^2$, the area of the source region.

6.6 CIRCULAR SYMMETRY, FOURIER-BESSEL TRANSFORM

Of special interest are two-dimensional problems with no angular dependence and three-dimensional problems that are symmetric about some axis. In the two-dimensional case $f(\boldsymbol{\rho})$ then depends only on $\rho = |\boldsymbol{\rho}|$ while in three dimensions $f(\boldsymbol{\rho}, z)$, where z is the axis of symmetry, is of the form $f(\rho, z)$. In these situations the angular integration in the transform relations can be performed once and for all. The transform is then found to depend only on $k = |\mathbf{k}|$. In two dimensions (6.5.2) reduces to

$$F(k) = \int_0^\infty \rho\, d\rho\, f(\rho)\int_0^{2\pi} d\vartheta\, e^{-ik\rho\cos\vartheta} \tag{6.6.1}$$

Following the method used in obtaining (6.5.16), the x axis has been chosen along $\mathbf{k}$ so that $\mathbf{k} \cdot \mathbf{R} = kR \cos \vartheta$. For the inverse relation we proceed in a similar way to obtain

$$
f(\rho) = \int_0^\infty \frac{k \, dk}{(2\pi)^2} F(k) \int_0^{2\pi} d\vartheta \, e^{ik\rho\cos\vartheta}
$$

$$
= \frac{1}{2\pi} \int_0^\infty k \, dk \, J_0(k\rho) F(k) \tag{6.6.2}
$$

It is customary to absorb the factor of 2π into $F(k)$ and define a Fourier-Bessel transform pair by the relations

$$
F_B(k) = \int_0^\infty \rho \, d\rho \, J_0(k\rho) f(\rho)
$$

$$
f(\rho) = \int_0^\infty k \, dk \, J_0(k\rho) F_B(k) \tag{6.6.3}
$$

where $F_B(k) = F(k)/2\pi$.

As is to be expected, this transform leads to a simple result when it is applied to the radial part of the Laplace operator in polar or cylindrical coordinates. Performing two partial integrations and using relations (D.34) for differentiation of the Bessel functions, we obtain

$$
\int_0^\infty \rho \, d\rho \, J_0(k\rho) \frac{1}{\rho} \frac{d}{d\rho} \left(\rho \frac{df}{d\rho} \right) = -k^2 F_B(k) + J_0(k\rho) \rho f_\rho(k\rho) |_0^\infty
$$

$$
+ k\rho J_1(k\rho) f(\rho) |_0^\infty \tag{6.6.4}
$$

where $f_\rho(k\rho) = \partial f/\partial\rho$. Unless $f(\rho)$ is singular at the origin, which could be the case if there were a source located there, the integrated terms vanish and only the first term, $-k^2 F_B(k)$, remains.

A polar coordinate representation for the two-dimensional delta function may be obtained by carrying out the angular integration in the definition (6.5.5). Using as angular integration variable the angle between $\mathbf{k}$ and $\boldsymbol{\rho} - \boldsymbol{\rho}'$, we have

$$
\delta^2(\boldsymbol{\rho} - \boldsymbol{\rho}') = \int \frac{d^2k}{(2\pi)^2} e^{i\mathbf{k}\cdot(\boldsymbol{\rho}-\boldsymbol{\rho}')} = \frac{1}{(2\pi)^2} \int_0^\infty k \, dk \int_0^{2\pi} d\vartheta \, e^{ik|\boldsymbol{\rho}-\boldsymbol{\rho}'|\cos\vartheta}
$$

$$
= \frac{1}{\pi} \int_0^\infty k \, dk \, J_0(k|\boldsymbol{\rho} - \boldsymbol{\rho}'|) \tag{6.6.5}
$$

On using the series expansion for $J_0(k|\boldsymbol{\rho} - \boldsymbol{\rho}'|)$ given in Problem 4.12.5 we have

$$\delta^2(\boldsymbol{\rho} - \boldsymbol{\rho}') = \frac{\delta(\rho - \rho')\delta(\varphi - \varphi')}{\rho}$$

$$= \frac{1}{2\pi} \int_0^\infty k\, dk \sum_{n=0}^\infty \epsilon_n J_n(k\rho) J_n(k\rho') \cos n(\varphi - \varphi') \quad (6.6.6)$$

Upon integrating this result over φ from 0 to 2π, we find that only the $n = 0$ term in the series remains and thus

$$\frac{\delta(\rho - \rho')}{\rho} = \int_0^\infty k\, dk\, J_0(k\rho) J_0(k\rho') \quad (6.6.7)$$

This result is also implied by the Fourier-Bessel transform pair introduced above. Substituting the integral for $F_B(k)$ into the second of (6.6.3) and interchanging the order of integration, we obtain

$$f(\rho) = \int_0^\infty \rho'\, d\rho'\, f(\rho') \int_0^\infty k\, dk\, J_0(k\rho) J_0(k\rho') \quad (6.6.8)$$

Replacing this integral over k by the delta function according to (6.6.7) and then integrating over ρ', we obtain the expected identity for $f(\rho)$.

As an example of the method, recall that in Section 4.9 the temperature distribution in a half space was determined for the case in which the base of the region was heated uniformly by a circular region of radius a. We now use Fourier-Bessel transform techniques to recover the result obtained there.

We require a solution of Laplace's equation in the half space $z > 0$ that reduces to the boundary value

$$-Ku_z(\rho, 0) = \begin{cases} F_0, & 0 \le \rho < a \\ 0, & \rho > a \end{cases} \quad (6.6.9)$$

Since the problem has no angular dependence about the z axis, we write the Laplacian in cylindrical coordinates and obtain

$$\frac{1}{\rho} \frac{\partial}{\partial \rho} \left(\rho \frac{\partial u}{\partial \rho} \right) + \frac{\partial^2 u}{\partial z^2} = 0 \quad (6.6.10)$$

Introducing the Fourier-Bessel transform of $u(\rho, z)$ as

$$U_B(k, z) = \int_0^\infty \rho\, d\rho\, J_0(k\rho) u(\rho, z) \quad (6.6.11)$$

and transforming Laplace's equation with the aid of (6.6.4), we have

$$\frac{d^2 U_B}{dz^2} - k^2 U_B = 0 \quad (6.6.12)$$

The solution that remains bounded as $z \to +\infty$ is $A(k)e^{-kz}$. Hence the solution for $u(\rho, z)$ is of the form

$$u(\rho, z) = \int_0^\infty k\, dk\, J_0(k\rho) A(k) e^{-kz} \qquad (6.6.13)$$

To relate $A(k)$ to the boundary condition, we now calculate $u_z(\rho, 0)$ and find

$$u_z(\rho, 0) = -\int_0^\infty k\, dk\, J_0(k\rho) kA(k) \qquad (6.6.14)$$

The inverse of this relation yields

$$
\begin{aligned}
-kA(k) &= \int_0^\infty \rho\, d\rho\, J_0(k\rho) u_z(\rho, 0) \\
&= \frac{F_0}{K} \int_0^a \rho\, d\rho\, J_0(k\rho) \\
&= -\frac{F_0}{K} \frac{aJ_1(ka)}{k} \qquad (6.6.15)
\end{aligned}
$$

where we have used (D.34) in performing the integration over $J_0(k\rho)$. Thus $A(k) = (F_0/k) aJ_1(ka)/k^2$ and the temperature distribution is

$$u(\rho, z) = \frac{F_0 a}{K} \int_0^\infty \frac{dk}{k} J_0(k\rho) J_1(ka) e^{-kz} \qquad (6.6.16)$$

The temperature on the z axis is

$$u(0, z) = \frac{F_0 a}{K} \int_0^\infty \frac{dk}{k} J_1(ka) e^{-kz} = \frac{F_0}{K} (\sqrt{z^2 + a^2} - z) \quad (6.6.17)$$

which agrees with the result obtained in (4.9.28). The integral over k may be viewed as a Laplace transform and its value obtained from entry 21 in Table F.2.

Problems

6.6.1 (a) Determine the steady state temperature in the half space $z > 0$ when the plane $z = 0$ has temperature

$$
u(\rho, 0) = \begin{cases} u_0, & 0 \le \rho < a \\ 0, & a < \rho < \infty \end{cases}
$$

(b) Use your answer to write the integral for $u(\rho, 0)$ and infer from it the value of a certain discontinuous integral.

6.6.2 The half space $z > 0$ having conductivity K is heated in the plane $z = 0$ by the ring source

$$J_z(r, 0) = \frac{Q_0}{2\pi a} \delta(r - a)$$

(a) Show that the temperature may be written

$$u(r, z) = \frac{Q_0}{2\pi K} \int_0^\infty dk \, e^{-kz} J_0(kr) J_0(ka)$$

(b) For the temperature on the axis of symmetry, obtain

$$u(0, z) = \frac{Q_0}{2\pi K} \frac{1}{\sqrt{z^2 + a^2}}$$

$$= \frac{Q_0}{2\pi Kz} \left[1 - \frac{1}{2} \left(\frac{a}{z} \right)^2 + \cdots \right], \quad \frac{a}{z} \ll 1$$

6.6.3 Let $u(\rho, z)$ denote steady state temperature in a *slab* $\rho > 0, 0 \leq \varphi < 2\pi, 0 \leq z \leq L$, where ρ, φ, z are cylindrical coordinates. The face $z = L$ is kept at temperature $u = 0$ and the face $z = 0$ is insulated except that heat is supplied through a circular region such that

$$J_z = \begin{cases} F_0, & 0 \leq \rho < a \\ 0, & a < \rho < \infty \end{cases}$$

Determine $u(\rho, z)$ within the slab.

6.6.4 Determine the steady temperature in the half space $z > 0, -\infty < x, y < \infty$, when the plane $z = 0$ has the heat input

$$J_z(r) = \begin{cases} Q_0/\pi a^2, & r < a \\ 0, & r > a \end{cases}$$

(a) Show that

$$u(r, z) = \frac{Q_0}{\pi a K} \int_0^\infty dk \, e^{-kz} \frac{J_0(kr) J_1(ka)}{k}$$

(b) Since this result is in the form of a Laplace transform, use Table F.2 to obtain

$$u(0, z) = \frac{Q_0}{\pi K} \frac{1}{z + \sqrt{z^2 + a^2}} = \frac{Q_0}{\pi a^2 K} (\sqrt{z^2 + a^2} - z)$$

$$= \frac{Q_0}{2\pi Kz} \left[1 - \frac{1}{4} \left(\frac{a}{z} \right)^2 + \cdots \right], \qquad \frac{a}{z} \ll 1$$

Note that for a heated *ring* (Problem 6.6.2), we obtained

$$u_{\text{ring}}(r, z) = \frac{Q_0}{2\pi K} \frac{1}{\sqrt{z^2 + a^2}}$$

$$\cong \frac{Q_0}{2\pi Kz} \left[1 - \frac{1}{2} \left(\frac{a}{z} \right)^2 + \cdots \right], \qquad \frac{a}{z} \ll 1$$

6.6.5 A half space $z > 0$ is heated by a ring source that satisfies the boundary condition

$$u(\rho, 0) - \gamma u_z(\rho, 0) = C\delta(\rho - a)$$

Obtain an integral representation for $u(\rho, z)$.

6.7 GREEN'S FUNCTIONS FOR TIME DEPENDENT WAVE EQUATION IN ONE, TWO, AND THREE DIMENSIONS

By way of recapitulating, we now use various integral transforms to determine the Green's functions for the time dependent wave equation in one, two, and three dimensions. The result for one dimension has already been given in (4.7.25) where it was obtained by using a Laplace transform on time. Here we shall use appropriate spatial transforms. In each case the resulting differential equation for the time dependence will be found to be that encountered in determining the Green's function for the undamped oscillator considered in Section 4.6.

6.7.1 One Dimension

Here we solve

$$\frac{\partial^2 g_1}{\partial x^2} - \frac{1}{c^2} \frac{\partial^2 g_1}{\partial t^2} = -\delta(x - x')\delta(t - t') \qquad (6.7.1)$$

Since the coefficients in the wave equation are all constants, the system is spatially homogeneous and time independent. We therefore expect a solution that depends only upon the differences $\xi = x - x'$ and $\tau = t - t'$, and therefore set $g(x, t | x', t') = g(\xi, \tau)$. Furthermore, we expect the solution to be symmetric about the source point $\xi = 0$. We may thus assume that $g(\xi, \tau)$ is an even function of ξ and also that $g_\xi(0, \tau) = 0$ (at least for $\tau > 0$). We thus need only determine $g(\xi, \tau)$ in the region $\xi \geq 0$. In the ξ, τ coordinates we ask for the solution of

$$\frac{\partial^2 g_1}{\partial \xi^2} - \frac{1}{c^2} \frac{\partial^2 g_1}{\partial \tau^2} = -\delta(\xi)\delta(\tau), \qquad g_{1\xi}(0, \tau) = 0, \qquad \xi, \tau \geq 0 \quad (6.7.2)$$

With the indicated boundary condition at $\xi = 0$, a Fourier cosine transform is appropriate and we introduce the transform

$$G_1(k, \tau) = \int_0^\infty d\xi \, g_1(\xi, \tau) \cos k\xi \qquad (6.7.3)$$

The transformed wave equation takes the form

$$\frac{d^2 G_1}{d\tau^2} + (kc)^2 G_1 = \frac{1}{2} c^2 \delta(\tau) \qquad (6.7.4)$$

In obtaining this result, we have used the relation

$$\int_0^\infty d\xi \, \delta(\xi) \cos k\xi = \tfrac{1}{2} \qquad (6.7.5)$$

since, on the basis of symmetry, only half of the source is being considered.

Equation (6.7.4) is just the equation for the Green's function for the oscillator that was solved in Section 4.6 except that the strength of the source is now $c^2/2$ and the frequency is expressed as kc. We also have no damping. Using the solution given in (4.6.5) with these modifications we obtain

$$G_1(k, \tau) = \left(\frac{1}{2} c^2\right) \frac{1}{kc} \sin kc\tau H(\tau) \qquad (6.7.6)$$

The Green's function itself is obtained by using the inverse transform. We have

$$g_1(\xi, \tau) = \frac{c}{\pi} \int_0^\infty dk \cos k\xi \, \frac{\sin kc\tau}{k} \qquad (6.7.7)$$

Using entry 7 in Table F.4, we obtain

$$g_1(\xi, \tau) = \frac{c}{2} H(c\tau - \xi) \qquad (6.7.8)$$

Due to the symmetry inherent in this problem, the result may be used for negative ξ as well by merely replacing ξ by $|\xi|$. The result then agrees with (4.7.25). The step function $H(\tau)$ could also be included, but it is less restrictive than the one arising in the evaluation of the transform and has therefore been ignored.

6.7.2 Two Dimensions

Using the two-dimensional Laplacian (3.7.1), we now solve

$$\nabla^2 g_2 - \frac{1}{c^2} \frac{\partial^2 g_2}{\partial t^2} = -\delta^2(\mathbf{\rho} - \mathbf{\rho}')\delta(t - t') \qquad (6.7.9)$$

We again introduce the time difference $\tau = t - t'$, and since the solution may be expected to be symmetric about the source point at $\mathbf{\rho} = \mathbf{\rho}'$, we introduce the radial coordinate $R = |\mathbf{\rho} - \mathbf{\rho}'|$ to measure distance from the source. Ignoring the angular derivative in the Laplacian, we have

$$\frac{1}{R} \frac{\partial}{\partial R} \left(R \frac{\partial g_2}{\partial R} \right) - \frac{1}{c^2} \frac{\partial^2 g_2}{\partial \tau^2} = -\frac{1}{2\pi R} \delta(R)\delta(\tau) \qquad (6.7.10)$$

As a result of the circular symmetry we introduce a Fourier-Bessel transform

$$G_{2B}(k, \tau) = \int_0^\infty R \, dR \, J_0(kR) g_2(R, \tau) \qquad (6.7.11)$$

The transform of (6.7.10), in which (6.6.4) has been used for the second derivative, is

$$\frac{d^2 G_{2B}}{d\tau^2} + (kc)^2 G_{2B} = \frac{c^2}{2\pi} \delta(\tau) \int_0^\infty dR \, \delta(R) J_0(kR)$$

$$= \frac{c^2}{2\pi} \delta(\tau) \qquad (6.7.12)$$

In evaluating the integral over the delta function we have used the limiting value $J_0(0) = 1$. Also, since R is never negative, we use $\int_0^\infty dR \, \delta(R) = 1$ rather than $\frac{1}{2}$ as in the one-dimensional case. One could also obtain this result by introducing a model for the two-dimensional delta function such as

$$\delta^2(\mathbf{\rho}) = \lim_{a \to 0} \frac{H(a - r)}{\pi a^2} \qquad (6.7.13)$$

The source term in (6.7.12) would then be

$$\frac{c^2 \delta(\tau)}{\pi a^2} \int_0^a rd \, rJ_0(kr) = \frac{c^2 \delta(\tau)}{\pi a^2} \frac{a}{k} J_1(ka) \qquad (6.7.14)$$

where the integral has been evaluated by using (D.34). When the limit $ka \to 0$ is taken by using (D.26), one recovers the result $(c^2/2\pi)\delta(\tau)$ given in (6.7.12).

The equation for $G_{2B}(k, \tau)$ in (6.7.12) is again of the form encountered for the oscillator. The solution is

$$G_{2B}(k, \tau) = \frac{c}{2\pi k} \sin kc\tau H(\tau) \tag{6.7.15}$$

and the inverse Fourier-Bessel transform yields

$$
\begin{aligned}
g_2(R, \tau) &= \int_0^\infty k \, dk \, J_0(kR) \left[\frac{cH(\tau)}{2\pi k} \sin kc\tau \right] \\
&= \frac{c}{2\pi} H(\tau) \int_0^\infty dk \, J_0(kR) \sin kc\tau
\end{aligned}
\tag{6.7.16}
$$

From entry 4 in Table F.6 we have

$$
\begin{aligned}
g_2(R, \tau) &= \frac{c}{2\pi} \frac{H(c\tau - R)}{\sqrt{(c\tau)^2 - R^2}} \\
&= \frac{1}{2\pi} \frac{H(\tau - R/c)}{\sqrt{\tau^2 - R^2/c^2}}
\end{aligned}
\tag{6.7.17}
$$

as the Green's function for waves in two dimensions.

6.7.3   Three Dimensions

We now solve

$$\nabla^2 g_3 - \frac{1}{c^2} \frac{\partial^2 g_3}{\partial t^2} = -\delta^3(\mathbf{r} - \mathbf{r}')\delta(t - t') \tag{6.7.18}$$

where the Laplacian is that given in (3.5.16). Only the radial part is relevant due to spherical symmetry about the source point at $\mathbf{r}'$. Introducing $R = |\mathbf{r} - \mathbf{r}'|$ and writing the radial Laplacian in accordance with (3.5.17), we have

$$\frac{1}{R} \frac{\partial^2}{\partial R^2} (Rg_3) - \frac{1}{c^2} \frac{\partial^2 Rg_3}{\partial \tau^2} = -\frac{\delta(R)}{4\pi R^2} \delta(\tau) \tag{6.7.19}$$

On setting $\psi(R, \tau) = Rg_3(R, \tau)$, we obtain

$$\frac{\partial^2 \psi}{\partial R^2} - \frac{1}{c^2} \frac{\partial^2 \psi}{\partial \tau^2} = -\frac{1}{4\pi R} \delta(R)\delta(\tau) \tag{6.7.20}$$

The radial derivative is appropriately treated by using a Fourier transform. Since the cosine transform of the right-hand side of (6.7.19) would be divergent, we use a sine transform and set

$$\Psi_s(k, \tau) = \int_0^\infty dR \, \psi(R, \tau) \sin kR \tag{6.7.21}$$

Using (6.1.7) for the transform of the second derivative, we have

$$-k^2 \Psi_s(k, \tau) + k\psi(0, \tau) - \frac{1}{c^2} \frac{d^2 \Psi_s}{d\tau^2} = -\frac{\delta(\tau)}{4\pi} \int_0^\infty dR \, \frac{\delta(R) \sin kR}{R}$$

$$= -\frac{k}{4\pi} \delta(\tau) \tag{6.7.22}$$

where we have used the limiting expression $(\sin kR)/kR \rightarrow 1$ and evaluated the delta function according to $\int_0^\infty dR \, \delta(R) = 1$ as in the two-dimensional case.

Since $g_3(0, \tau)$ must remain finite (at least for $\tau > 0$), we must impose the boundary condition $\psi(0, \tau) = 0$ and (6.7.22) is again of the form of the equation for the oscillator. The solution this time is

$$\Psi_s(k, \tau) = \frac{c}{4\pi} H(\tau) \sin kc\tau \tag{6.7.23}$$

and the inverse transform, provided by entry 8 in Table F.3, yields

$$g_3(R, \tau) = \frac{1}{R} \psi(R, \tau) = \frac{1}{4\pi R} \delta(\tau - R/c) \tag{6.7.24}$$

Problems

6.7.1 Evaluate the integral over $\delta(R)$ in (6.7.22) by representing the three-dimensional delta function as

$$\delta^3(\mathbf{R}) = \lim_{a \to 0} \frac{3H(a - R)}{4\pi a^3}$$

6.7.2 Use the methods developed in this section to determine the Green's function for the diffusion equation in one, two, and three dimensions. Show that the solution of

$$\nabla_N^2 g_N - \frac{1}{\gamma^2} \frac{\partial g_N}{\partial t} = -\delta^N(\mathbf{r} - \mathbf{r}')\delta(\tau), \qquad N = 1, 2, 3$$

is

$$g_N(R, \tau) = \frac{\gamma^2 H(\tau)}{(2\gamma \sqrt{\pi\tau})^N} e^{-R^2/4\gamma^2\tau}$$

where R is the distance between $\mathbf{r}$ and $\mathbf{r}'$ in N dimensions.

6.7.3 Determine the Green's function for the equation governing the elastically braced string and its generalizations to two and three dimensions, that is,

$$\nabla_N^2 g_N - \frac{1}{c^2}\frac{\partial^2 g_N}{\partial \tau^2} - \mu^2 g_N = -\delta^N(R)\delta(\tau), \qquad N = 1, 2, 3$$

All required transforms are available in Tables F.3–F.5. Show that

$$g_1(\xi, \tau) = \frac{c}{2} J_0(\mu\sqrt{(c\tau)^2 - \xi^2})H(c\tau - \xi)$$

$$g_2(R, \tau) = \frac{c}{2\pi}\frac{\cos(\mu\sqrt{(c\tau)^2 - R^2})}{\sqrt{(c\tau)^2 - R^2}}H(c\tau - R)$$

$$g_3(R, \tau) = \frac{c}{4\pi}\frac{H(\tau)}{R}\frac{2}{\pi}\int_0^\infty dk \frac{k\sin(c\tau\sqrt{k^2 + \mu^2})}{\sqrt{k^2 + \mu^2}}\sin kR$$

$$= -\frac{c}{4\pi}\frac{H(\tau)}{R}\frac{2}{\pi}\frac{\partial}{\partial R}\int_0^\infty dk \frac{\sin(c\tau\sqrt{k^2 + \mu^2})}{\sqrt{k^2 + \mu^2}}\cos kR$$

$$= \frac{H(\tau)}{4\pi R}\left\{\delta(\tau - R/c) - \frac{\mu c J_1(\mu\sqrt{(c\tau)^2 - R^2})}{\sqrt{(c\tau)^2 - R^2}}H(c\tau - R)\right\}$$

7 Perturbation Methods

It should be clear that the techniques developed in previous chapters are quite limited. Restrictions on their use are imposed either by the geometric configurations for which separability may be employed or by the inhomogeneities for which the separated equations may be solved. In some instances it is the boundary conditions that prohibit the construction of solutions by the methods developed previously. For situations that differ only slightly from some exactly soluble problem, it is frequently possible to express the solution as a slight modification of the soluble case by calculating a correction term. The term that constitutes the difference between the problem of interest and the exactly soluble problem is usually referred to as a *perturbation*. For situations that differ greatly from any soluble case, it is sometimes possible to construct expressions for quantities of interest (resonant frequencies, reflection coefficients, etc.) that are relatively insensitive to our lack of detailed understanding of the exact solution to the problem. Such expressions, known as variational principles, are said to be stationary with respect to variations in the solution about its correct value. They can frequently be used to obtain surprisingly accurate information about the system with only a qualitative understanding of the solution to the problem.

The subject of perturbation theory is a large one and contains a number of subtle issues. In the present chapter we will only touch upon some of the simpler aspects of this topic.

7.1 FIRST ORDER CORRECTIONS

When the governing equations for a problem cannot be solved exactly because of the presence of a term that, in some sense, is small, it is frequently possible to obtain an approximate solution in terms of a solution to the problem in which this small term has been neglected plus a correction term that is determined by the small perturbation. The general ideas are readily conveyed by an example. We consider the changes in the resonant frequencies and mode shapes of a membrane that are due to the change in the density of the membrane from a constant value σ_0, for which the problem is readily solved (cf. Section 3.4), to $\sigma(\mathbf{r}) = \sigma_0 + \epsilon \sigma_1(\mathbf{r})$. The parameter ϵ is a constant that characterizes the smallness of the density variation $\sigma_1(\mathbf{r})$. One may think of the perturbation as being turned on by increasing the value of ϵ from 0 to some final value that then provides the perturbation amplitude of interest. From the derivation in Section

3.4, the equation governing wave motion on the membrane will now be of the form

$$[\sigma_0 + \epsilon\sigma_1(\mathbf{r})] \frac{\partial^2 w}{\partial t^2} - T \nabla^2 w = 0 \tag{7.1.1}$$

It will be assumed that the boundary conditions remain unchanged as the perturbation is turned on. When we factor out the dominant term σ_0 and write $\sigma_0 + \epsilon\sigma_1(\mathbf{r}) = \sigma_0[1 + \epsilon f(\mathbf{r})]$, the wave equation becomes

$$\nabla^2 w - \frac{1}{c_0^2}[1 + \epsilon f(\mathbf{r})] \frac{\partial^2 w}{\partial t^2} = 0, \qquad f(\mathbf{r}) = \frac{\sigma(\mathbf{r})}{\sigma_0} \tag{7.1.2}$$

where $c_0^2 = T/\sigma_0$. The natural frequencies and associated orthogonal vibrational modes for the *unperturbed* membrane will be written $\omega_{mn}^{(0)}$ and $\varphi_{mn}(\mathbf{r})$, respectively, that is, we write $w_{mn}(\mathbf{r}, t) = \varphi_{mn}(\mathbf{r})\exp(-i\omega_{mn}^{(0)}t)$ when no perturbation is present. At times it is convenient to choose the amplitude of the spatial modes φ_{mn} so that $\int dS\, \varphi_{mn}^2 = 1$. For a rectangular membrane with fixed edges and dimensions $L \times H$, one would then use $\varphi_{mn} = 2/\sqrt{(LH)}$ $\sin m\pi x/L \sin n\pi y/H$. Introducing the unperturbed wave numbers $k_{mn}^{(0)} = \omega_{mn}^{(0)}/c_0$, we can write the equation satisfied by a mode φ_{mn} in the form

$$\nabla^2 \varphi_{mn} + k_{mn}^{(0)2} \varphi_{mn} = 0 \tag{7.1.3}$$

In the presence of the perturbation we expect changes to new frequencies and mode shapes that may be expressed in the form

$$\begin{aligned} \omega_{mn} &= \omega_{mn}^{(0)} + \epsilon\omega_1 \\ u_{mn} &= \varphi_{mn} + \epsilon u_1 \end{aligned} \tag{7.1.4}$$

It will be seen that higher order corrections proportional to higher powers of ϵ can enter the calculation in a natural way. However, the present development is only an attempt to convey the simplest ideas of the perturbation method and therefore the expansion will not be carried to higher order in ϵ. All terms proportional to ϵ^2, ϵ^3, ... will therefore be discarded.

Let us now concentrate on some one of the modes of the perturbed membrane and write $w(\mathbf{r}, t) = u_{mn}(\mathbf{r})e^{-i\omega_{mn}t}$. The equation governing the mode u_{mn} is seen from the wave equation (7.12) to be

$$\nabla^2 u_{mn} + k_{mn}^2 u_{mn} = -\epsilon k_{mn}^2 f(\mathbf{r})u_{mn} \tag{7.1.5}$$

where $k_{mn} = \omega_{mn}/c_0$. When the representations of u_{mn} and k_{mn}, available from (7.1.4), are introduced here, we have

$$\nabla^2(\varphi_{mn} + \epsilon u_1) + (k_{mn}^{(0)} + \epsilon k_1)^2(\varphi_{mn} + \epsilon u_1) = -\epsilon k_{mn}^{(0)2} f(\mathbf{r})\varphi_{mn} \tag{7.1.6}$$

in which the terms ϵk_1 and ϵu_1 have been neglected on the right-hand side since their contributions are proportional to ϵ^2 and, as mentioned above, such terms are being neglected. When the terms on the left-hand side are multiplied out and only terms of first order in ϵ are retained, we have

$$\nabla^2(\varphi_{mn} + \epsilon u_1) + k_{mn}^{(0)2}(\varphi_{mn} + \epsilon u_1) + 2\epsilon k_{mn}^{(0)} k_1 \varphi_{mn}$$

$$= -\epsilon k_{mn}^{(0)2} f(\mathbf{r}) \varphi_{mn} \tag{7.1.7}$$

Since φ_{mn} is one of the modes of the unperturbed system, it satisfies the Helmholtz equation $\nabla^2 \varphi_{mn} + k_{mn}^{(0)2} \varphi_{mn} = 0$, and thus the two terms in (7.1.6) that are independent of ϵ are seen to cancel out. All remaining terms contain a common factor of ϵ that may be discarded, and (7.1.7) reduces to

$$\nabla^2 u_1 + k_{mn}^{(0)2} u_1 + 2k_{mn}^{(0)} k_1 \varphi_{mn} = -k_{mn}^{(0)2} f(\mathbf{r}) \varphi_{mn} \tag{7.1.8}$$

To solve this equation for the correction terms $u_1(\mathbf{r})$ and k_1, we expand u_1 in terms of the presumably known modes of the unperturbed system; we therefore set

$$u_{mn} = \varphi_{mn} + \epsilon \sum a_{\alpha\beta} \varphi_{\alpha\beta}$$

$$= (1 + \epsilon a_{mn}) \varphi_{mn} + \epsilon \sum' a_{\alpha\beta} \varphi_{\alpha\beta} \tag{7.1.9}$$

The term in the perturbation that corresponds to the unperturbed mode being considered, namely, $\epsilon a_{mn} \varphi_{mn}$, has been extracted from the sum and combined with the unperturbed mode. Whether or not this correction term is displayed explicitly is of no concern since it merely changes the amplitude of φ_{mn}, which can be eventually set by the normalization of u_{mn}. Absence of the term φ_{mn} in the summation in (7.1.9) is indicated by a prime on the summation sign.

The mode perturbation term u_1 now takes the form

$$u_1 = \sum' a_{\alpha\beta} \varphi_{\alpha\beta} \tag{7.1.10}$$

Since, as noted above, the unperturbed modes satisfy a Helmholtz equation, we have

$$\nabla^2 u_1 = \nabla^2 \sum' a_{\alpha\beta} \varphi_{\alpha\beta} = \sum' a_{\alpha\beta} k_{\alpha\beta}^{(0)2} \varphi_{\alpha\beta} \tag{7.1.11}$$

and (7.18) becomes

$$-\sum' a_{\alpha\beta} k_{\alpha\beta}^{(0)2} \varphi_{\alpha\beta} + k_{mn}^{(0)2} \sum' a_{\alpha\beta} \varphi_{\alpha\beta} + 2k_{mn}^{(0)} k_1 \varphi_{mn}$$

$$= -k_{mn}^{(0)2} f(\mathbf{r}) \varphi_{mn} \tag{7.1.12}$$

To obtain the correction term k_1 and thus the frequency correction ω_1 as well as the coefficients $a_{\alpha\beta}$ in the expansion of u_1, we first multiply (7.1.12) by φ_{mn}

and then integrate over the membrane surface. Since the $\varphi_{\alpha\beta}$ are orthogonal, and the term φ_{mn} is absent from the sum over $(\alpha\beta)$, all terms in the sum vanish and we obtain

$$2k_{mn}^{(0)} k_1 \int \varphi_{mn}^2 \, dS = -k_{mn}^{(0)2} \int \varphi_{mn}^2 f(\mathbf{r}) \, dS \qquad (7.1.13)$$

The frequency correction can now be written as

$$k_1 = -\frac{k_{mn}^{(0)}}{2} \frac{\displaystyle\int \varphi_{mn}^2 f(r) \, dS}{\displaystyle\int \varphi_{mn}^2 \, dS} \qquad (7.1.14)$$

Note that if $f(\mathbf{r})$ is positive over the entire membrane surface, both integrals in (7.1.14) are positive and k_1 is therefore negative, indicating a decrease in the frequency of the membrane. This result conforms to our intuition that an increase in the mass of a vibrating system can be expected to lower its natural frequencies.

To determine the correction to the mode shape given by u_1, we must obtain the coefficients $a_{\alpha\beta}$ in (7.1.10). They are readily determined by multiplying (7.1.12) by one of the $\varphi_{\alpha\beta}$, say φ_{pq}, and integrating over the membrane surface. Since the mode φ_{mn} is not included among the $\varphi_{\alpha\beta}$, the integral $\int dS \, \varphi_{pq}\varphi_{mn}$ vanishes and we find

$$a_{pq}(-k_{pq}^{(0)2} + k_{mn}^{(0)2}) = -k_{mn}^{(0)2} \int \varphi_{pq}^2 f(\mathbf{r}) \varphi_{mn} \, dS \qquad (7.1.15)$$

The amplitude is therefore

$$a_{pq} = \frac{k_{mn}^{(0)2}}{k_{pq}^{(0)2} - k_{mn}^{(0)2}} \int \varphi_{pq} f(\mathbf{r}) \varphi_{mn} \, dS \qquad (7.1.16)$$

Note that if k_{pq} is close to k_{mn}, the denominator in this last result becomes small and the contribution of this term to the solution may become large. The simple first order perturbation expansion used here is then no longer adequate and higher order terms in the expansion or a modification of the procedure may be required. The special case when two frequencies are equal is considered in the next section.

The correction term for k_{mn} given in (7.1.14) may also be obtained by using a method that displays features similar to those to be used for obtaining a variational result in Section 7.3. As preparation for these later developments we now consider this alternate derivation of (7.1.14).

Returning to (7.1.5), the equation governing the exact expression for u_{mn}, we multiply that equation by φ_{mn} and integrate over the surface of the membrane to obtain

$$\int \varphi_{mn} \nabla^2 u_{mn} \, dS + k_{mn}^2 \int \varphi_{mn} u_{mn} \, dS = -\epsilon k_{mn}^2 \int \varphi_{mn} f(\mathbf{r}) u_{mn} \, dS \quad (7.1.17)$$

Applying Green's theorem to u_{mn} and φ_{mn}, we have

$$\int (\varphi_{mn} \nabla^2 u_{mn} - u_{mn} \nabla^2 \varphi_{mn}) \, dS = \oint (\varphi_{mn} \nabla u_{mn} - u_{mn} \nabla \varphi_{mn}) \cdot \mathbf{n} \, dl$$

$$(7.1.18)$$

where $\mathbf{n}$ is an outward unit vector in the plane of the membrane and normal to its rim. As in our consideration of Sturm-Liouville theory in Appendix C where we assume that the eigenfunctions satisfy boundary conditions such that integrated terms vanish, so now we assume that both φ_{mn} and u_{mn} satisfy boundary conditions such that the line integral around the rim will vanish. We then have $\int dS \, \varphi_{mn} \nabla^2 u_{mn} = \int dS \, u_{mn} \nabla^2 \varphi_{mn}$. This latter surface integral may be further simplified by noting that $\nabla^2 \varphi_{mn} = k_{mn}^{(0)2} \varphi_{mn}$. We may write (7.1.17) in the form

$$(-k_{mn}^{(0)2} + k_{mn}^2) \int \varphi_{mn} u_{mn} \, dS = -\epsilon k_{mn}^2 \int \varphi_{mn} f(\mathbf{r}) u_{mn} \, dS \quad (7.1.19)$$

Introducing the decomposition $u_{mn} = \varphi_{mn} + \epsilon u_1$ from (7.1.4) and recalling from (7.1.9) that u_1 is orthogonal to φ_{mn}, we have $\int dS \, \varphi_{mn} u_{mn} = \int \varphi_{mn}^2 \, dS$. Using the approximation $k_{mn}^2 - k_{mn}^{(0)2} \approx 2\epsilon k_1 k_{mn}^{(0)}$, which is valid to first order in ϵ, we find that (7.1.19) reduces to

$$k_1 = -\frac{k_{mn}^{(0)}}{2} \frac{\int \varphi_{mn} f(r) u_{mn} \, dS}{\int \varphi_{mn}^2 \, dS} \quad (7.1.20)$$

Finally, when u_{mn} is replaced by the lowest approximation φ_{mn}, we recover the first order result for k_1 given in (7.1.14).

Example 7.1. Determine the shift in the diagonal wave numbers k_{nn} of a square membrane of length L on a side when a mass M is placed at the center.

The perturbation may be written $\sigma_1(\mathbf{r}) = M\delta(x - L/2)\delta(y - L/2)$ and the unnormalized unperturbed modes as $\varphi_{nn}(\mathbf{r}) = \sin n\pi x/L \sin n\pi y/L$. Substituting these values into (7.1.14) and performing the integration, we immediately obtain

$$k_1 = -\frac{k_{nn}^{(0)}}{2}\frac{M}{\sigma_0}\frac{\sin^4(n\pi/2)}{(L/2)^2} \tag{7.1.21}$$

When this result is added to $k_{mn}^{(0)}$ to form the shifted value of the wave number, we obtain

$$\frac{k_{nn}}{k_{nn}^{(0)}} = 1 - \frac{2M}{\sigma_0 L^2}\sin^4\frac{n\pi}{2} \tag{7.1.22}$$

For even values of n, for which the displacement has a nodal line through the location of the mass, there is no shift in wave number.

The result obtained in the previous example can be expected to be valid as long as the correction term in (7.1.22) is much less than unity, that is, as long as M is much less than the mass of the membrane $\sigma_0 L^2$. Note that although the perturbation itself is singular, the "smallness" of the perturbation refers to the magnitude of the integral in the numerator of (7.1.20).

Problems

7.1.1 A uniform string of length L and total mass m has a mass M located at $x = x_0$. Use first order perturbation theory (7.1.14) to determine the perturbed values of $k_n L$. Show that the perturbed values of k_n are given by

$$\frac{k_n}{k_n^{(0)}} = 1 - \frac{M}{m}\sin^2\frac{n\pi x_0}{L}$$

where $k_n^{(0)} = n\pi/L$.

7.1.2 A square membrane with uniform density σ_0 and fixed edges has a tension T_1 in the x direction and a tension $T_2 = T_1 + \delta T$ in the y direction. The modes of the membrane are thus governed by

$$\sigma_0\frac{\partial^2 w}{\partial t^2} - T_1\frac{\partial^2 w}{\partial x^2} - T_2\frac{\partial^2 w}{\partial y^2} = 0$$

Setting $w(x, y, t) = f(x, y)e^{-i\omega t}$ for vibration at a single frequency ω, one then has

$$\nabla^2 f + \left(\frac{\omega}{c_0}\right)^2 f = -\frac{\delta T}{T_1}\frac{\partial^2 w}{\partial y^2}$$

Show that k_1, the perturbation in the diagonal wave number k_{nn}, is given by

$$k_1 = -\frac{1}{2k_{nn}^{(0)}} \frac{\delta T}{T_1} \frac{\int dS\, \varphi_{nn}\, (\partial^2 \varphi_{nn}/\partial y^2)}{\int dS\, \varphi_{nn}^2}$$

and obtain the result

$$k_1 = \frac{1}{2^{3/2}} \frac{\delta T}{T_1} \frac{n\pi}{L}$$

which is an increase (decrease) for δT positive (negative).

7.1.3 A mass M is placed at the center of a uniform circular membrane with fixed rim and total mass $m = \sigma_0 \pi a^2$. Show that the wave numbers for the radially symmetric modes are given by

$$\frac{k_{0n}}{k_{0n}^{(0)}} = 1 - \frac{1}{2J_1^2(\alpha_{0n})} \frac{M}{m}$$

The ratio of perturbed to unperturbed wave numbers equals $1 - 1.85M/m$ for $n = 1$.

7.2 EQUAL FREQUENCIES (DEGENERACY)

When the lengths L and H are incommensurate, no two modes will have the same frequency. When the lengths are commensurate, however, two distinct modal patterns can have the same frequency. Two such modes are then said to be degenerate. We shall consider only the case of a square membrane with fixed rim. The diagonal case, which is nondegenerate, has been considered above in Example 7.1. For $m = 1$ and $n = 2$ or $m = 2$ and $n = 1$ we have the two modal patterns $\varphi_{12}(x, y) = A_{12} \sin \pi x/L \sin 2\pi y/L$ and $\varphi_{21} = A_{21} \sin 2\pi x/L \sin \pi y/L$, respectively, with the same frequency $\omega_{12} = \omega_{21} = (\pi c/L) \sqrt{1^2 + 2^2} = (\pi c/L) \sqrt{5} = 1.58\omega_{11}$, where ω_{11}, the fundamental frequency, equals $\pi c \sqrt{2}/L$. In general, the displacement of the membrane at the frequency ω_{12} will be a linear combination of φ_{12} and φ_{21}. The amplitude of each mode is determined by the manner in which the membrane is being excited.

One means of separating the contributions of the degenerate modes is to introduce a perturbation in the properties of the membrane. In such cases, one assumes that the unperturbed mode is a linear combination of the two modal patterns that have this same frequency. As will be shown, the appropriate amount of each mode is determined by the perturbation.

For the two modes considered above, the perturbation u_1 is written

$$u_1 = a_{12}\varphi_{12} + a_{21}\varphi_{21} + \sum{}'' a_{\alpha\beta}\varphi_{\alpha\beta} \tag{7.2.1}$$

where the double prime indicates the extraction of two modes from the sum. Setting $a_{12} = a$, $a_{21} = b$, $\varphi_{12} = \varphi_a$, and $\varphi_{21} = \varphi_b$ for brevity and using (7.2.1) in (7.1.12), with the unperturbed mode φ_{mn} replaced by $a\varphi_a + b\varphi_b$, we have

$$-\sum{}'' a_{\alpha\beta}\varphi_{\alpha\beta}k_{\alpha\beta}^{(0)2} + k_{mn}^{(0)2} \sum{}'' a_{\alpha\beta}\varphi_{\alpha\beta} + 2k_1 k_{mn}^{(0)}(a\varphi_a + b\varphi_b)$$
$$= -k_{mn}^{(0)2} f(r)(a\varphi_a + b\varphi_b) \tag{7.2.2}$$

Since neither φ_a nor φ_b is present in the summation, these sums may again be eliminated by successively multiplying through by φ_a and φ_b, and then integrating over the membrane surface. We obtain the pair of equations

$$2k_{mn}^{(0)}k_1 a \int \varphi_a^2 \, dS = -k_{mn}^{(0)2} \left[a \int \varphi_a f \varphi_a \, dS + b \int \varphi_a f \varphi_b \, dS \right] \tag{7.2.3}$$
$$2k_{mn}^{(0)}k_1 b \int \varphi_b^2 \, dS = -k_{mn}^{(0)2} \left[a \int \varphi_b f \varphi_a \, dS + b \int \varphi_b f \varphi_b \, dS \right]$$

It is convenient to incorporate normalization constants into φ_a and φ_b so that $\int \varphi_a^2 \, dS = \int \varphi_b^2 \, dS = 1$. Setting $M_{ij} = \int dS\, \varphi_i f(r) \varphi_j$ and rearranging terms, one then finds

$$(M_{aa} - \kappa)a + M_{ab}b = 0 \tag{7.2.4}$$
$$M_{ba}a + (M_{bb} - \kappa)b = 0$$

where $\kappa = -2k_1/k_{mn}^{(0)}$. This homogeneous system of equations for a and b has nonzero solutions when the determinant of the system vanishes. We thus set

$$\begin{vmatrix} M_{aa} - \kappa & M_{ab} \\ M_{ba} & M_{bb} - \kappa \end{vmatrix} = 0 \tag{7.2.5}$$

and obtain a quadratic equation for κ. The solutions are

$$\kappa = -\frac{2k_1}{k_{mn}^{(0)}} = \frac{1}{2}[M_{aa} + M_{bb} \pm \sqrt{(M_{aa} + M_{bb})^2 - 4D}] \tag{7.2.6}$$

where $D = M_{aa}M_{bb} - M_{ab}M_{ba}$. As long as the term under the radical does not vanish, there are two distinct values for κ and hence of the wave number perturbation k_1. The perturbation is thus seen to have split the degeneracy. The ratio b/a is now obtainable from either relation in (7.2.4) and the amplitude of the remaining constant (b or a) is set by normalization of $u(\mathbf{r})$ as in the nondegenerate case.

Other less obvious instances of degeneracy may also be encountered. As an example, for a square membrane, the two modal patterns with indices (m, n) equal to $(1, 8)$ and $(4, 7)$ have frequencies proportional to $\sqrt{1^2 + 8^2}$ and $\sqrt{4^2 + 7^2}$, which both equal $\sqrt{65}$. There are now four degenerate modes, φ_{18}, φ_{81}, φ_{47}, and φ_{74}.

Example 7.2. For a square membrane, the normalized modes $\varphi_{12} = (2/L)\sin \pi x/L \sin 2\pi y/L$ and $\varphi_{21} = (2/L)\sin 2\pi x/L \sin \pi y/L$ are degenerate. Determine the shift in the wave numbers k_{12} and k_{21} when a thin strip of length L and linear density λ is placed on the membrane at the line $x = L/4$.

The density may be written $\sigma(\mathbf{r}) = \sigma_0 + \lambda\delta(x - L/4)$. Setting $\varphi_a = \varphi_{12}$ and $\varphi_b = \varphi_{21}$, the matrix elements M_{ab} are

$$M_{ab} = \frac{\lambda}{\sigma_0} \int_0^L dx \int_0^L dy \, \varphi_a \delta\left(x - \frac{L}{4}\right) \varphi_b \qquad (7.2.7)$$

and are readily evaluated to obtain

$$M_{12,12} = \frac{\lambda}{\sigma_0 L}, \qquad M_{21,21} = \frac{2\lambda}{\sigma_0 L}, \qquad M_{12,21} = M_{21,12} = 0 \qquad (7.2.8)$$

The values of κ given by (7.2.6) are $M_{12,12}$ and $M_{21,21}$. For the first of these values, (7.2.4) shows that $b = 0$ while a, the amplitude of φ_{12}, is undetermined. For the second choice of κ, one finds that a is zero and b is undetermined. The corresponding wave numbers are thus

$$k_{12} = k_{12}^{(0)} - \lambda/2\sigma_0 L \quad \text{and} \quad k_{21} = k_{21}^{(0)} - \lambda/\sigma_0 L \qquad (7.2.9)$$

Note that the perturbation is located along the line of maximum displacement of the mode φ_{21} and hence has a larger effect upon this mode.

Problems

7.2.1 Determine the shifts in wave number for the square membrane in Example 7.2 when the strip of linear density λ is replaced by one that extends along the line $x = L/4$ between $y = L/2$ to $y = 0$, that is, $\sigma(\mathbf{r}) = \sigma_0 + \lambda\delta(x - L/4)H(L/2 - y)$. Show that

$$M_{12,12} = \delta/2, \qquad M_{21,21} = \delta, \qquad M_{12,21} = M_{21,12} = \frac{4\sqrt{2}}{3\pi}\delta$$

where $\delta = \lambda/\sigma_0 L$ and that

$$\frac{k_1}{k_{12}^{(0)}} = -(0.375 \pm 0.329)\lambda/\sigma_0 L$$

7.2.2 A square membrane of uniform density σ_0 has tension T_1 in the x direction and $T_2 = T_1 + \delta T$ in the y direction. The wave equation is thus

$$\nabla^2 w - \frac{1}{c^2}\frac{\partial^2 w}{\partial t^2} = \frac{\delta T}{T_1} = -\frac{\delta T}{T_1}\frac{\partial^2 w}{\partial y^2}$$

where $c^2 = T/\sigma_0$.

Assume $\delta T/T_1 \ll 1$ and use first order perturbation theory to determine the shift in the off-diagonal modes $\varphi_{mn} = (2/L)\sin m\pi x/L \sin n\pi y/L$ and $\varphi_{nm} = (2/L)\sin n\pi x/L \sin m\pi y/L$.

(a) Show that the right-hand side of (7.2.2) is replaced by

$$-\frac{\delta T}{T_1}\, a\, \frac{\partial^2 \varphi_a}{\partial y^2} + b\, \frac{\partial^2 \varphi_b}{\partial y^2}$$

where $\varphi_a = \varphi_{mn}$ and $\varphi_b = \varphi_{mn}$. Equation (7.2.4) is retained with κ defined as previously with $M_{ij} = (\delta T/T_1)k_{mn}^{(0)\,-2}\int dS\, \varphi_i\, \partial^2 \varphi_j/\partial y^2$.

(b) Obtain the wave number shifts

$$\frac{k_{mn}}{k_{mn}^{(0)}} = 1 + \frac{1}{2}\frac{\delta T}{T_1}\frac{1}{k_{mn}^{(0)2}}\left(\frac{n\pi}{L}\right)^2, \qquad \frac{k_{nm}}{k_{nm}^{(0)}} = 1 + \frac{1}{2}\frac{\delta T}{T_1}\frac{1}{k_{nm}^{(0)2}}\left(\frac{m\pi}{L}\right)^2$$

where $k_{mn}^{(0)2} = k_{nm}^{(0)2} = (m^2 + n^2)\pi^2/L^2$. Note that an increase in tension corresponds to an increase in wave number and hence in frequency.

7.2.3 The modes of a circular membrane that do not possess circular symmetry are degenerate since the two modes $J_n(k_{nm}r)\cos m\vartheta$ and $J_n(k_{nm}r)\sin m\vartheta$ have the same frequency $\omega_{nm} = ck_{nm} = c\alpha_{mn}/a$ where $J_n(\alpha_{mn}) = 0$.

Assume a strip of total mass M is placed along the line $\vartheta = \vartheta_0$ so that the total density of the membrane is $\sigma(\mathbf{r}) = \sigma_0 + (2M/a^2)\delta(\vartheta - \vartheta_0)$.

(a) Show that the matrix elements for normalized modes $\varphi_c = NJ_n(k_{nm}r)\cos m\vartheta$ and $\varphi_s = NJ_n(k_{nm}r)\sin m\vartheta$ are

$$M_{cc} = \delta \cos^2 \vartheta_0, \qquad M_{ss} = \delta \sin^2 \vartheta_0,$$

$$M_{cs} = M_{sc} = \delta \sin \vartheta_0 \cos \vartheta_0$$

where $\delta = 2M/\sigma_0 a^2$.

(b) Show that κ has the two values 0 and δ so that the mode shifts are $k_1 a = 0$, $-M\alpha_{mn}/\sigma_0 a^2$.

(c) Show that the mode corresponding to $\kappa = 0$, the unshifted mode, is $J_n(k_{nm}r)\sin m(\vartheta - \vartheta_0)$, which has a node along the perturbation, while the mode corresponding to $\kappa = \delta$ is $J_n(k_{nm}r)\cos m(\vartheta - \vartheta_0)$, which has a maximum along the perturbation.

7.3 VARIATIONAL METHODS

When the perturbation is not small, it may still be possible to obtain analytical results for various quantities of experimental interest. One method for accomplishing this is to construct what is known as a variational principle for that quantity. As a preparation for introducing this concept, we first note that usually it is not the solution of the wave equation itself that is of concern, but rather certain experimentally interesting properties of the system such as resonant frequencies, reflection or transmission coefficients, angular radiation patterns, etc. In many instances the solution to the wave equation is merely an intermediary that enables us to calculate these experimentally interesting quantities. It would be extremely useful, therefore, if we could manipulate an expression for the quantity of interest, expressed in terms of the solution to the wave equation, into a form such that the value of the physically interesting quantity is nearly the same no matter what reasonable expression for the wave function is substituted into the formula. We could then not even bother to solve the wave equation but merely substitute into this formula an educated guess for the solution. If our formula is truly insensitive to the trial solution that we use and our guess is truly educated, we could hope that the resulting value of the experimentally interesting quantity would be fairly close to the value that would be obtained if we had substituted the correct wave function into the formula. Variational principles provide a quantitative framework for carrying out this program.

Two different approaches to this idea will be developed. One is based upon the wave equation in differential form, while the other approach is based upon the integral equation formulation of the wave equation that is obtained by introducing a Green's function.

7.3.1 Differential Equation Approach (Rayleigh's Method)

To have a specific physical model in mind, we continue with our consideration of the modes of an inhomogeneous membrane. We begin with the general equation for the membrane vibrating at one of its natural frequencies by setting $w(\mathbf{r}, t) = u(\mathbf{r})e^{-i\omega t}$ in the membrane equation (7.1.2). Since the membrane modes u_{mn} are now no longer close to any known unperturbed modes φ_{mn}, we abandon expansion in terms of them. Equation (7.1.8) becomes

$$\nabla^2 u + k^2 F(\mathbf{r})u = 0 \tag{7.3.1}$$

where $k = \omega/c_0$ and $F(\mathbf{r}) = 1 + f(\mathbf{r})$. Since no expansion in terms of ϵ is contemplated, we set the parameter ϵ equal to unity and express the inhomogeneity in terms of $F(\mathbf{r})$. Multiplying the wave equation (7.3.1) by u and integrating over the membrane, we have

$$\int u \, \nabla^2 u \, dS = -k^2 \int uFu \, dS \qquad (7.3.2)$$

This relation may now be solved for k^2. To appreciate the significance of the resulting expression for k^2, let us examine the change in k^2 that takes place when we make a slight change in the expression for the mode function u. When u is replaced by $u + \delta u$, k^2 can be expected to change to $k^2 + \delta k^2$. Change in the mode function u may be thought of as resulting from a change in some parameter contained in u. As an example, a change in the function $u = \sin ax$ can be expressed as $\sin(a + \delta a)x$. When the smallness of δa is incorporated into the expansion of the sine function, we obtain $\sin(z + \delta a)x = \sin ax + \delta a \, x \cos ax = u + \delta u$. It will be noted that δu may be obtained by merely performing a standard differential calculus operation on the parameter a contained in u. We now employ this procedure in (7.3.2). We assume that the δ operation may be interchanged with ∇^2 and thus write $\delta \, \nabla^2 u = \nabla^2 \, \delta u$. Application of the δ operation to (7.3.2) yields

$$\int \delta u \, \nabla^2 u \, dS + \int u \, \nabla^2 \, \delta u \, dS = -\delta k^2 \int uFu \, dS - 2k^2 \int u \, \delta u \, F \, dS$$

$$(7.3.3)$$

Using Green's theorem we may rewrite $\int u \, \nabla^2 \, \delta u \, dS$ as

$$\int u \, \nabla^2 \, \delta u \, dS = \int \delta u \, \nabla^2 u \, dS + \oint (u \, \nabla \, \delta u - \delta u \, \nabla u) \cdot \mathbf{n} \, dl \quad (7.3.4)$$

Assuming that both u and δu or that ∇u and $\delta(\nabla u) = \nabla \, \delta u$ vanish on the boundary we have $\int u \, \nabla^2 \, \delta u \, dS = \int \delta u \, \nabla^2 u \, dS$. The first two terms on the left of (7.3.3) are thus equal. Solving this equation of δk^2 we obtain

$$\delta k^2 \int u \, Fu \, dS = 2 \int \delta u (\nabla^2 u + k^2 Fu) \, dS \qquad (7.3.5)$$

If u satisfies the wave equation (7.3.1), the integral on the right in (7.3.5) will vanish and δk^2 will be zero. We conclude, then, that if changes in u are introduced in first order, changes in k^2 can only occur in second order and thus that k^2 is stationary with respect to first order changes in u about its correct value.

In the present example it is possible to show that the stationarity is in the form of a minimum, that is, substitution of a perturbed form of u into (7.3.2) yields a value of k^2 that is larger than the correct value. To see this, we make temporary use of the complete set of *exact* modes for the inhomogeneous membrane, although no attempt will be made to determine them. The orthogonality relation satisfied by these modes is obtained by using the multiply-and-subtract procedure used in the development of Sturm-Liouville theory. We select two modes $u_{\alpha\beta}$ and $u_{\gamma\delta}$ (ignoring the possibility of degeneracy) and consider the equations satisfied by these modes. For the sake of brevity of notation we label these two modes as u_a and u_b, respectively, and write

$$\nabla^2 u_a + k_a^2 F u_a = 0$$

$$\nabla^2 u_b + k_b^2 F u_b = 0$$

(7.3.6)

Multiplying the first by u_b and the second by u_a, subtracting, and integrating over the membrane, we obtain

$$\int (u_b \nabla^2 u_a - u_a \nabla^2 u_b)\, dS + (k_a^2 - k_b^2) \int u_a F u_b\, dS = 0 \quad (7.3.7)$$

Using Green's theorem as in the derivation of (7.1.19) and assuming boundary conditions on u_a and u_b such that the integral around the rim vanishes, we have the orthogonality relation

$$\int u_a F u_b\, dS = 0, \qquad a \neq b \qquad (7.3.8)$$

For $a = b$, the integral will be written as I_a. With this result concerning the orthogonality of the membrane modes, we return to (7.3.2) and rewrite it in the form

$$k^2 = \frac{\int u \nabla^2 u\, dS}{\int u F u\, dS} = -\frac{\int dS \sum c_a u_a \nabla^2 \sum c_b u_b}{\int dS\, F \sum c_a u_a \sum c_b u_b} \qquad (7.3.9)$$

where the expansion $u = \sum c_a u_a$ has been introduced. Writing $\nabla^2 u_b = -k_b^2 F u_b$, and using the orthogonality relation (7.3.8), we obtain

$$\frac{\int u \nabla^2 u\, dS}{\int u F u\, dS} = -\frac{\sum c_a^2 k_a^2 I_a}{\sum c_a^2 I_a} \qquad (7.3.10)$$

Applying the divergence theorem in the form $\int u\, \nabla u \cdot \mathbf{n}\, dl = \int \nabla \cdot (u\, \nabla u)\, dS$ $= \int |\nabla u|^2\, dS + \int u\, \nabla^2 u\, dS$, we have $\int u\, \nabla^2 u\, dS = -\int |\nabla u|^2\, dS$ as a result of assuming homogeneous boundary conditions at the rim. Equation (7.3.10) then becomes

$$\frac{\displaystyle\int |\nabla u|^2\, dS}{\displaystyle\int uFu\, dS} = -\frac{\sum c_a^2 k_a^2 I_a}{\sum c_a^2 I_a} \tag{7.3.11}$$

Although the various terms in the summation are unknown, an extremely useful inequality may be obtained from this result. Note that if we replace all k_a by the lowest value k_0, thereby *decreasing* the value of the numerator, the remaining series in both numerator and denominator become equal and cancel out. Equation (7.3.11) therefore provides the inequality

$$\frac{\displaystyle\int |\nabla u|^2\, dS}{\displaystyle\int uFu\, dS} \geq k_0^2 \tag{7.3.12}$$

The result in (7.3.12) provides a variational principle for the lowest resonance k_0. By evaluating the integrals in (7.3.2) for an assumed form for u, we obtain a value of k_0^2 that is above the correct value, and according to (7.3.5), the variation in the value is stationary about its correct value. Two one-dimensional examples are now considered. In both instances the exact solution is available so the accuracy of the approximate result may be ascertained.

Example 7.3. Use the trial function $u = x(L - x)$ to estimate the lowest frequency of a uniform string of length L with fixed ends.
   Here $F(x) = 1$ and $u_x = L - 2x$. The inequality for k_0^2 in (7.3.12) takes the form

$$k_0^2 < \frac{\displaystyle\int_0^2 (L - 2x)^2\, dx}{\displaystyle\int_0^2 x^2(L - x)^2\, dx} \tag{7.3.13}$$

The integrals in numerator and denominator are found to equal $L^3/3$ and $L^5/30$, respectively, so that $(k_0 L)^2 = 10$. The approximate value thus differs from the correct value of π^2 by about 1.3%.

Example 7.4. Use the trial function employed in the previous example to estimate the lowest value of $k_0 L$ for an inhomogeneous string of length L having density $\rho(x) =$

$\rho_0(1 + x/L)^{-2}$. In this case $F(x) = (1 + x/L)^2$ and (7.3.12) reduces to

$$k_0^2 < \frac{\displaystyle\int_0^L (L - 2x)^2 \, dx}{\displaystyle\int_0^L x^2 (L - x)^2 (1 + x/L)^{-2} \, dx} \tag{7.3.14}$$

The numerator has the same value as in the previous example and the denominator is found to equal $L^5(25/3 - 12 \ln 2)$. Hence,

$$(k_0 L)^2 < \frac{1}{25 - 36 \ln 2} \approx 21.413 \tag{7.3.15}$$

The exact result for this example was obtained in Problem 1.8.5 where it was shown that $(k_0 L)^2 = \frac{1}{4} + (\pi/\ln 2)^2 = 20.792$. The estimate obtained here is in error by 3%.

Since the correct value of k_0^2 is always lower than the value obtained from the variational expression (7.3.12) when an approximate form of u is employed, it is desirable to have a trial function containing a free parameter. The parameter may then be varied until the lowest value of k_0^2 is obtained. The trial function must satisfy the boundary conditions of the problem so some artistry may be required in constructing an appropriate expression.

An improvement in the trial function for Example 7.3 is provided by the choice $y = x(L - x) + \alpha x^2 (L - x)^2$. With this trial function the variational expression reduces to[1]

$$(k_0 L)^2 \le \frac{10 + 4\mu + 4\mu^2/7}{1 + 3\mu/7 + \mu^2/21} \tag{7.3.16}$$

where $\mu = \alpha L^2$. To obtain the values of μ that minimize this expression, we set $d(k_0 L)^2/d\mu = 0$ and solve the resulting quadratic equation for μ to obtain $\mu = 1.33140649, -4.633140649$. Use of the first of these values in (7.3.16) yields the extremely accurate value $(k_0 L)^2 \approx 9.86974962 = (1.0000147)\pi^2$. The second root yields $(k_0 L)^2 \cong 102.13$, which is a poor approximation to $(3\pi)^2 \approx 88.83$. Since the trial function is symmetric about the midpoint of the string and has two nodes for this latter value of μ, it is to be expected that the approximation will be to the third eigenvalue.

Clearly a trial function with additional parameters could be expected to add further improvement and a trial function with an infinite number of parameters, such as a Fourier series, could provide the exact result. Such extensions will not be considered here.

---

[1]The integrals $\int_0^L dx \, x^n (L - x)^n (L - 2x)^2$ are reduced to $(L^{2n+3}/2^n) \int_0^{\pi/2} d\vartheta \, \sin^{2n+1} \vartheta \cos^2 \vartheta$ by setting $x = L \sin^2(\vartheta/2)$.

7.3.2 Integral Equation Approach

Again we consider the determination of k^2 for an inhomogeneous membrane. Our first task is to convert the differential equation (7.3.1) into an integral form. The Green's function for the differential equation (7.3.1) would be difficult to determine since it contains the inhomogeneity $F(\mathbf{r})$. Even the Green's function for $\nabla^2 u + k^2 u$ would be cumbersome since it contains k^2, the constant being determined, in a transcendental form. Hence we use the Green's function for the corresponding *static* problem, that is, we combine the wave equation in the form

$$\nabla^2 u = -k^2 F(\mathbf{r}) u \qquad (7.3.17)$$

with the Green's function

$$\nabla^2 g = -\delta^2 (\mathbf{r} - \mathbf{r}') \qquad (7.3.18)$$

by the standard multiply-and-subtract procedure. Using a Green's function that satisfies the same homogeneous boundary conditions as u, we obtain

$$u(\mathbf{r}) = k^2 \int dS' \ F(\mathbf{r}') u(\mathbf{r}') g(\mathbf{r}|\mathbf{r}') \qquad (7.3.19)$$

On multiplying this equation by $F(\mathbf{r}) u(\mathbf{r})$ and then integrating over the membrane, we obtain

$$k^2 = \frac{\displaystyle\int dS \ u^2(\mathbf{r}) F(\mathbf{r})}{\displaystyle\int dS \int dS' \ u(\mathbf{r}) F(\mathbf{r}) g(\mathbf{r}|\mathbf{r}') F(\mathbf{r}') u(\mathbf{r}')} \qquad (7.3.20)$$

This expression for k^2 is stationary with respect to first order variations in u about the solution of the integral equation (7.3.19). To show this, we perform a variation of u and k^2 analogous to that used in obtaining (7.3.3). Using a prime to indicate dependence upon the primed variable, that is, $u' = u(\mathbf{r}')$ and $F' = F(\mathbf{r}')$, we obtain

$$\delta k^2 \int dS \int dS' \ uFgF'u + k^2 \iint dS \ dS' FF' (\delta u \ gu' + ug \ \delta u')$$

$$= 2 \int dS \ Fu \ \delta u \qquad (7.3.21)$$

The two double integrals multiplying k^2 are equal since we can interchange integration variables $\mathbf{r}$ and $\mathbf{r}'$ in one of them and then use the reciprocity relation

$g(\mathbf{r}|\mathbf{r}') = g(\mathbf{r}'|\mathbf{r})$. As a result, (7.3.21) becomes

$$\delta k^2 \int dS \int dS'\, uFgF'u' = 2 \int dS\, F\, \delta u \left(u - k^2 \int dS'\, F'gu' \right) \quad (7.3.22)$$

If u satisfies the integral equation (7.3.19), the integrand on the right vanishes and we have $\delta k^2 = 0$. We thus find that (7.3.20) is an expression that is stationary with respect to first order changes in u about the correct value.

From a computational standpoint, the evaluation of the double integral in (7.3.20) presents a drawback to the use of this method. However, it is frequently possible to circumvent some of the computational details by employing the methods outlined in the following example.

Example 7.5. We reconsider the one-dimensional situation examined previously in Example 7.3 and thus set $F(\mathbf{r}) = 1$. Using the same trial function $u = x(L - x)$, we evaluate (7.3.20), that is

$$k^2 = \frac{N}{D} \quad (7.3.23)$$

where

$$N = \int_0^L dx\, u^2 = \int_0^L dx\, x^2 (L - x)^2 = L^5/30 \quad (7.3.24)$$

and

$$D = \int_0^L dx\, u(x) \int_0^L dx'\, g(x|x')u(x') \quad (7.3.25)$$

A direct evaluation of D would entail use of the Green's function $g(x|x') = x_<(L - x_>)/L$ obtained in (4.1.7). The calculation would thus be cumbersome since the two different regions of integration $0 \le x' < x$ and $x < x' \le L$ would have to be considered separately. A simpler procedure is to note that the integral over x' in the double integral, that is,

$$\varphi(x) \equiv \int_0^L dx'\, g(x|x')u(x') \quad (7.3.26)$$

may be thought of as being the solution of the boundary value problem

$$\frac{d^2\varphi}{dx^2} = -u(x), \qquad \varphi(0) = \varphi(L) = 0 \quad (7.3.27)$$

The solution of this problem is more readily obtained by direct integration. Setting $u(x)$ equal to the trial function being used, that is, $u(x) = x(L - x)$, and integrating (7.3.27), we immediately find

$$\varphi(x) = -\frac{L}{6}x^3 + \frac{1}{12}x^4 + Ax + B \qquad (7.3.28)$$

The integration constants are readily shown to have the values $A = L^3/12$ and $B = 0$. Substitution of the resulting form of $\varphi(x)$ into the integrand in (7.3.25) yields[2]

$$D = \int_0^L dx\, x(L - x)\left(-\frac{L}{6}x^3 + \frac{1}{12}x^4 + \frac{L^3}{12}x\right) \qquad (7.3.29)$$
$$= \frac{17L^7}{5040}$$

From (7.3.23), the value of k^2 is

$$(kL)^2 = \frac{168}{17} = 1.00129\pi^2 \qquad (7.3.30)$$

a result that is seen to differ from the correct solution by about one-tenth of 1%. This result should be contrasted with the larger 1.3% error obtained by using the same trial function in the differential approach given in Example 7.3. In general we may expect the integral equation method to be more accurate since the trial function appears under an integral sign that is a smoothing operation. It is perhaps of interest to note that although a Green's function was used to construct the stationary expression (7.3.23), the evaluation of D never required the actual determination of that Green's function.

Another approach to evaluation of the double integral in D is to introduce a Fourier series for $g(x|x')$. Using the series for the Green's function given in Problem 4.1.9, we find that the x and x' integrations are now decoupled and obtain

$$D = \frac{2L}{\pi^2}\sum_{n=1}^{\infty}\frac{1}{n^2}\int_0^L dx\, u(x)\sin\frac{n\pi x}{L}\int_0^L dx'\, u(x')\sin\frac{n\pi x'}{L} \qquad (7.3.31)$$

For the trial function $u = x(L - x)$ employed here, the integrals over x and x' are each equal to $4/(n\pi)^3$ for odd n and vanish for even n. Therefore, we have

$$D = \frac{32L^7}{\pi^8}\sum_{1,3,5,\ldots n}\frac{1}{n^8} \qquad (7.3.32)$$

The sum of the series is given in (Table F.1, series 1.13) as $17\pi^8/(4 \cdot 8!)$. Thus $D = 17L^7/5040$ as obtained above and the result given in (7.3.30) is recovered.

Problems

7.3.1   A mass m is attached at $x = x_0$ to a homogeneous string of length L and fixed ends. Use the trial function $y = \sin n\pi x/L$ in Rayleigh's

---

[2]Evaluation is facilitated by first setting $x = L\zeta$ and then using $\int_0^1 d\zeta\, \zeta^m(1 - \zeta)^n = n!m!/(n + m)!$

method (7.3.12) to obtain the perturbed wave number

$$(k_0L)^2 = \frac{(n\pi)^2}{1 + \left(\dfrac{2M}{\lambda L}\right) \sin^2\left(\dfrac{n\pi x_0}{L}\right)}$$

Note that the first term in the expansion of the denominator yields a result in agreement with Problem 7.1.1.

7.3.2 Choice of coordinate system may be conducive to the construction of trial functions. Consider a homogeneous string stretched between $\pm L/2$ and use the trial function $y = (L/2)^\nu - |x|^\nu$ in Rayleigh's method.

(a) Show that

$$(k_0L)^2 = 2\,\frac{(\nu + 1)(2\nu + 1)}{2\nu - 1}$$

(b) Show that the minimum value of k_0L is obtained for $\nu = (\sqrt{6} + 1)/2 = 1.7247$ for which $(k_0L)^2 = 5 + 2\sqrt{6} = 1.00248\pi^2$.

7.3.3 **(a)** Show that the stationarity of k^2 obtained in (7.3.5) is still obtained when u satisfies the more general boundary condition $u + \gamma\,\nabla u = 0$ where γ is a constant.

(b) In one dimension obtain the result

$$k^2 = \frac{-yy'\Big|_0^L + \displaystyle\int_0^L dx\,(y')^2}{\displaystyle\int_0^L dx\,y^2}$$

(c) For a homogeneous string satisfying $y(0) = 0$, $y(L) + \zeta L y'(L) = 0$, use the trial function $y = x(\mu L - x)$ with $\mu = (1 + 2\zeta)/(1 + \zeta)$ to satisfy the boundary condition at L.

(d) Show that

$$(kL^2) = 10\,\frac{3\mu - 2}{10\mu^2 - 15\mu + 6}$$

(e) The exact solution is obtained by solving the transcendental equation $\tan kL + \zeta kL = 0$. A comparison of the variation result for kL with the exact solution is shown below.

ζ	0.1	0.5	1.0	5.0	10.0
Variational	2.877	2.300	2.041	1.700	1.643
Exact	2.863	2.289	2.029	1.688	1.631

7.3.4 Use the Green's function method (7.3.20) to determine the values of $k_n L$ when a mass M is located at the center of a homogeneous string of length L.

(a) Show that

$$k^2 = \frac{N}{D}$$

with

$$N = \frac{L}{2}(1 + \delta), \qquad D = \frac{L^3}{2\pi^2}\left(1 + 2\delta + \frac{\pi^2 \delta^2}{8}\right)$$

where $\delta = 2M/\rho_0 L$.

(b) Show that for $\delta = 2$, so that the mass M has a mass equal to that of the string, one obtains $kL = \pi\sqrt{6/(10 + \pi^2)} = 1.7264$. The exact value is the lowest root of $(kL/2)\tan(kL/2) = 1$, which is 1.7207. Thus one finds $kL_{\text{var}} = 1.0033\,(kL)_{\text{exact}}$.

7.3.5 Use the trial function $y(r) = a^\nu - r^\nu$ in the variational expression (7.3.20) to estimate the lowest resonance of a homogeneous circular membrane that is fixed at the rim.

(a) Show that $(ka)^2 = 8(\nu + 2)^2(\nu + 4)/[(\nu + 1)(\nu^2 + 10\nu + 20)]$.

(b) Show that the stationary value of $(ka)^2$ is given by the positive root of $3\nu^3 + 14\nu^2 + 4\nu - 40 = 0$, which is found to be $= 1.3789$, approximately. The value of ka is 2.4055, which is 1.00028 times the first root of $J_0(ka)$. Suggestion: Evaluate the double integral in the demonimator by using the first method employed in Example 7.5.

7.3.6 Apply the integral equation approach to the inhomogeneous string considered in Example 7.4. Use the same trial function, $y = x(L - x)$, and show that (7.3.20) yields

$$k_0^2 \le \frac{N}{D}$$

in which

$$N = L^5(\tfrac{25}{3} - 12\ln 2) = 0.01557\,L^5$$

$$D = L^7\,[24\ln 2 - 40(\ln 2)^2 + \tfrac{31}{12}] = 0.0007452\,L^7$$

so that $(k_0 L)^2 = 20.894 = 1.0048 \times$ (correct value).

8 Generalizations and First Order Equations

In this chapter we consider some aspects of our subject that have not been addressed to any extent in the earlier development. It was mentioned briefly in Section 1.1 that the three equations to be considered, the diffusion equation, the wave equation, and Laplace's equation, are each an example of one of three distinct types of partial differential equations. We now consider this topic in more detail and develop a cataloging procedure for partial differential equations. Then, in Section 8.2 we consider two of the more qualitative issues concerning the subject, namely, the uniqueness of solutions and the extreme values that the solutions can take on (maximum principles). In the third section we introduce first order equations and examine them primarily from the standpoint of providing approximate solutions to second order equations. Finally, in Section 8.4 we consider a nonlinear partial differential equation known as Burgers' equation and show how it enables us to avoid a nonphysical multiple valuedness that can occur in the solution of certain first order equations. We also show how it is related to the linear diffusion equation.

8.1 CLASSIFICATION OF SECOND ORDER EQUATIONS

The two equations encountered in Chapter 1, that is, the diffusion equation $u_{xx} - \gamma^{-2}u_t = 0$ and the wave equation $u_{xx} - c^{-2}u_{tt} = 0$, as well as the Laplace equation $u_{xx} + u_{yy} = 0$ considered in Chapter 3 are actually examples of three fundamentally different types of partial differential equations. We now introduce a method for characterizing this distinction.

We consider the general second order equation

$$Az_{xx} + 2Bz_{xy} + Cz_{yy} + Dz_x + Ez_y + Fz = 0 \qquad (8.1.1)$$

where A, B, ..., F can be functions of x and y, and show that there are three categories into which such equations can fall. Each of the three equations referred to above is found to be in a different one of these categories. The possibility of x, y dependence in the coefficients A, B, ..., F means that, in general, the character of an equation can be different in various regions of the xy plane.

To motivate our approach, we first recall the situation for an *ordinary* differential equation such as $y'' + f(x)y' + g(x)y = 0$ with specified conditions $y(0) = a_0$, $y'(0) = a_1$. We inquire into the possibility of constructing a power series representation of the solution away from the point $x = 0$ by using these conditions at $x = 0$ as well as the given differential equation plus derivatives of the differential equation. We consider only the simplest situation in which f, g, and their derivatives remain finite at $x = 0$. The coefficients in an assumed expansion of the form

$$y(x) = y(0) + y'(0)x + \frac{1}{2!}y''(0)x^2 + \frac{1}{3!}y'''(0)x^3 + \cdots \quad (8.1.2)$$

are then readily obtained. From the given initial conditions one has $y(0) = a_0$, $y'(0) = a_1$, and from the differential equation, $y''(0) = -[f(0)a_1 + g(0)a_0]$. Successive differentiation of the differential equation then yields $y'''(0)$, etc.

We now consider the possibility of using a similar procedure to obtain the solution of a partial differential equation when information along a boundary is provided. As a simple example of some of the issues that arise, we first consider the two situations provided by the equations $z_{xx} + z_{yy} = 0$ and $z_{xy} = 0$ when $z(x, 0) = f(x)$ and $z_y(x, 0) = g(x)$ are given. We again consider the determination of the coefficients in a series expansion. To obtain the solution away from the boundary $y = 0$, we assume a series of the form

$$z(x, y) = z(x, 0) + z_y(x, 0)y + \frac{1}{2!}z_{yy}(x, 0)y^2 + \frac{1}{3!}z_{yyy}(x, 0)y^3 + \cdots$$
$$(8.1.3)$$

The first two coefficients are provided by the boundary conditions. For the first of the two equations referred to above, we can obtain z_{yy} from the given equation and write $z_{yy}(x, 0) = -z_{xx}(x, 0) = -[f(x)]_x$. Similarly, $z_{yyy}(x, 0) = -(z_{xx})_y = -(z_y)_{xx} = -(g(x))_{xx}$, etc. The series expansion may now be developed in complete analogy with the situation for the ordinary differential equation.

For the second example mentioned above, however, we find that the partial differential equation does not provide us with the information required to calculate the higher derivatives. We are thus unable to develop the series expansion. In order to understand the difference between these two equations, we return to the general equation (8.1.1) and consider the more general situation in which the boundary is specified by some curve Γ in the xy plane. We assume that z and its derivative normal to Γ are given. Specifically, if distance along Γ is measured by a parameter s, we consider $z(s) = F(s)$ and $\partial z(s)/\partial n = G(s)$ to be given. Note first that we can use this information to obtain $z_x(s)$ and $z_y(s)$ since $\partial z(s)/\partial s$ is obtainable from the given information and

$$\frac{\partial z}{\partial x} = \frac{\partial z}{\partial s}\frac{\partial s}{\partial x} + \frac{\partial z}{\partial n}\frac{\partial n}{\partial x}$$
$$\frac{\partial z}{\partial y} = \frac{\partial z}{\partial s}\frac{\partial s}{\partial y} + \frac{\partial z}{\partial n}\frac{\partial n}{\partial y}$$
$$(8.1.4)$$

with the equations defining Γ being used to calculate $\partial s/\partial x$, $\partial s/\partial y$ and $\partial n/\partial x$, $\partial n/\partial y$.

Example 8.1. On the circle of radius a assume $z(a,\ \theta) = z(s) = a\ \cos(s/a)$ and $\partial z/\partial r = \partial z/\partial n = \cos(s/a)$, where $s = \theta/a$. Use (8.1.4) to obtain $\partial z/\partial x$ and $\partial z/\partial y$ on $r = a$.

From the equations defining polar coordinates we have $s = a\theta = a\ \tan^{-1}y/x$, $n = r = (x^2 + y^2)^{1/2}$, and thus

$$\frac{\partial s}{\partial x} = -\sin\theta, \quad \frac{\partial s}{\partial y} = \cos\theta, \quad \frac{\partial r}{\partial x} = \cos\theta, \quad \frac{\partial r}{\partial y} = \sin\theta \qquad (8.1.5)$$

Also $\partial z/\partial s = -\sin s/a$. Substitution into (8.1.4) yields

$$\frac{\partial z}{\partial x} = 1, \quad \frac{\partial z}{\partial y} = 0 \qquad (8.1.6)$$

which is the expected result since the original problem follows from the requirement $z = x$ on $r = a$.

We may thus assume that the conditions given on Γ are equivalent to

$$z_x[x,\ y(x)] = f(x), \quad z_y[x,\ y(x)] = g(x) \qquad (8.1.7)$$

where $y = y(x)$ is the equation defining Γ and $f(x)$ and $g(x)$ are known.

In considering the conditions under which we can obtain the higher derivatives from the information provided on Γ plus the partial differential equation, it is convenient to introduce the notation $p = z_x$, $q = z_y$, $r = z_{xx}$, $s = z_{xy}$, and $t = z_{yy}$. We wish to determine the conditions under which we can determine the second derivatives r, s, and t.

The original equation (8.1.1) can be written

$$Ar + 2Bs + Ct = \Phi(p,\ q,\ x,\ y,\ z) \qquad (8.1.8)$$

For $z_x(x,\ y)$ we may write the differential

$$dz_x = \frac{\partial z_x}{\partial x}\ dx + \frac{\partial z_x}{\partial y}\ dy \qquad (8.1.9)$$

On Γ we have $y = y(x)$ and can write

$$\frac{dz_x}{dx} = f'(x) = \frac{\partial z_x}{\partial x} + \frac{\partial z_x}{\partial y}\frac{dy}{dx} = r + sy' \qquad (8.1.10)$$

Similarly, from the differential for z_y we obtain

$$s + ty' = g' \qquad (8.1.11)$$

Equations (8.1.8), (8.1.10), and (8.1.11) provide the system

$$r + y's = f'$$
$$s + y't = g' \tag{8.1.12}$$
$$Ar + 2Bs + Ct = \Phi$$

for determining the second derivatives r, s, and t. To proceed, we must require that the determinant of this system be nonzero, that is, we would be *unable* to obtain r, s, and t if

$$\begin{vmatrix} 1 & y' & 0 \\ 0 & 1 & y' \\ A & 2B & C \end{vmatrix} = 0 \tag{8.1.13}$$

Expansion of the determinant yields

$$A(y')^2 - 2By' + C = 0 \tag{8.1.14}$$

The change in sign with respect to that in (8.1.8) should be noted. Equation (8.1.14) is a nonlinear first order ordinary differential equation whose solution $y = y(x)$ is a curve in the xy plane. As is typical of nonlinear equations, it may have more than one solution even though it is only of first order. If such a curve or curves should be used as the boundary Γ on which conditions are given, it will *not* be possible to obtain r, s, and t away from the curve. We note first the consistency of this result with the second problem considered above. There we had the equation $z_{xy} = 0$ with z and z_y specified on the line $y = 0$. For that equation we have $A = C = 0$ and $B = \frac{1}{2}$ for which (9.1.14) yields $y' = 0$ or $y = $ const. We were thus attempting to solve this equation by providing boundary information on the line $y = $ const $= 0$, the line on which we are unable to obtain the second derivatives r, s, and t.

Returning to the general situation given by (8.1.14), we note that the differential equation for y is actually a quadratic equation in y'. Solving for y' we have

$$y' = \frac{B \pm \sqrt{B^2 - AC}}{A} \tag{8.1.15}$$

The paths in the xy plane corresponding to the solutions of these two equations are usually referred to as the characteristics of the equation. We now distinguish three situations, depending upon the value of the discriminant $\Delta \equiv B^2 - AC$. For $\Delta > 0$ there are two real solutions for y' and hence two distinct differential equations to solve for y. Since each solution introduces an integration constant, there is actually a double infinity of solutions. For $\Delta = 0$ there is only one differential equation and a single infinity of solutions. For $\Delta < 0$, both values of y' are complex and hence no real solution in the xy plane is provided. As indicated below, however, the complex solutions may be combined to yield real solutions to the original partial differential equation. In analogy with the terminology used for these three situations in analytic geometry, the three types of partial differential equations are referred to as being hyperbolic, parabolic, and elliptic, respectively.

The three equations emphasized in our development, the wave, diffusion, and Laplace equations, provide an example of each of these three categories. For the sake of consistency with the notation being employed here, we temporarily represent the time variable by y.

1. *Wave Equation* (*Hyperbolic*) $z_{xx} - c^{-2}z_{yy} = 0$, $A = 1$, $B = 0$, $C = -c^{-2}$. Thus $\Delta = c^{-2} > 0$ and the two differential equations for the characteristics are

$$\frac{dy}{dx} = \pm \frac{1}{c} \qquad (8.1.16)$$

The solutions are $x + cy = u$ and $x - cy = v$ where u and v are constants of integration. On replacing y by t we have the two straight lines that were introduced in Section 1.9. It was shown there that the general solution of the wave equation could be written in terms of these characteristic coordinates. In the present notation the result given in Eq. (1.9.10) takes the form $z = f(x + cy) + g(x - cy)$ where f and g are the arbitrary functions introduced there.

2. *Diffusion Equation* (*Parabolic*) $z_{xx} - \gamma^{-2}z_y = 0$, $A = 1$, $B = C = 0$, and so $\Delta = 0$. We thus obtain the single characteristic $y = $ const. The solution may now be expressed in terms of y and any other convenient variable (cf. Example 8.3 below).

3. *Laplace Equation* (*Elliptic*) $z_{xx} + z_{yy} = 0$, $A = C = 1$, $B = 0$, and $\Delta = -1$. In this case, $dy/dx = \pm i$ and the solutions are $x + iy = u$, $x - iy = v$. The corresponding solution takes the form $z = f(x + iy) + g(x - iy)$. If g is the function that is obtained from f by changing i to $-i$, then z will be real. In this way many solutions of Laplace's equation may be constructed. Some examples are considered in the problems at the end of this section.

Example 8.2. Determine the characteristics of the equation $z_{xx} - a^{-2}z_{yy} - z_y = 0$ and consider them in the limit $a \to \infty$.

In the notation of (8.1.8) we have $A = 1$, $B = 0$, $C = -a^{-2}$, and $\Phi = u_y$. The characteristics are the solutions of $y' = \pm 1/a$, which are $y = \pm x/a + $ const. As $a \to \infty$ the slope of each of these straight lines approaches zero and the double infinity of straight-line characteristics degenerates into the single infinity of lines $y = $ const that were obtained for the diffusion equation.

Example 8.3. For the equation $z_{xx} + 2z_{xy} + z_{yy} + z_y = 0$ we have $A = B = C = 1$ and $\Delta = 0$. The differential equation (8.1.16) now takes the form $y' = 1$. Thus $y - x = u$ and as a second coordinate we may choose $v = x$ to obtain

$$\frac{\partial}{\partial x} = \frac{\partial}{\partial v} - \frac{\partial}{\partial u}, \qquad \frac{\partial}{\partial y} = \frac{\partial}{\partial u} \tag{8.1.17}$$

The original equation now takes the standard form for the one-dimensional diffusion equation $z_{vv} + z_u = 0$.

When the partial differential equation has variable coefficients, the classification can still be carried out. Since the value of the discriminant now depends upon location in the xy plane, the equation may take on a different character in different regions of this plane. A simple example is the equation $z_{xx} + xz_{yy} = 0$, which is elliptic for $x > 0$ and hyperbolic for $x < 0$.

Problems

8.1.1 Determine the characteristic coordinates for the following equations and then write both the original equation and the solution in terms of these coordinates:

    (a) $z_{xx} + 4z_{xy} + 3z_{yy} = 0$
    (b) $z_{xx} + 4z_{xy} + 4z_{yy} = 0$
    (c) $z_{xx} + 4z_{xy} + 5z_{yy} = 0$

8.1.2 (a) Determine characteristic coordinates u, v for the equation $z_{xx} + 6z_{xy} + 13z_{yy} = 0$.

    (b) Rewrite the equation in terms of the variables λ, μ such that $u = \lambda + i\mu$, $v = \lambda - i\mu$ and obtain $z_{\lambda\lambda} + z_{\mu\mu} = 0$.

8.1.3 (a) Determine the characteristics of the equation

$$(1 - M^2)z_{xx} + \frac{2M}{c}z_{xt} - \frac{1}{c^2}z_{tt} = 0$$

    where $M = V/c = \text{const.}$

    (b) Show that the transformation of variables $x' = x - Vt$, $t' = t$ yields

$$z_{x'x'} - c^{-2}z_{t't'} = 0$$

    (c) Give a physical interpretation of the coordinate transformation.

8.1.4 The solution of $z_{xx} + z_{yy} = 0$ was shown to be $z = f(x + iy) + g(x - iy)$. Assume that $g(\zeta)$ is the function obtained from $f(\zeta)$ by changing i to $-i$. Then, if $f(x + iy) = \theta + i\chi$ where θ and χ are real, we have the real solution $z = 2\theta$. Obtain the solutions that correspond to $z(\zeta) = \zeta^3$, $e^{i\zeta}$, and $\ln \zeta$.

8.1.5 Show that under the transformation $u = u(x, y)$, $v = v(x, y)$, the general second order homogeneous linear equation

$$Lz + F(x, y)z = 0$$

where

$$Lz = Az_{xx} + 2Bz_{xy} + Cz_{yy} + Dz_x + Ez_y$$

and the coefficients A, B, ... may depend upon x and y, takes the form

$$A'z_{uu} + 2B'z_{uv} + C'z_{vv} + (Lu)z_u + (Lv)z_v + Fz = 0$$

where

$$A' = A(u_x)^2 + 2Bu_x u_y + C(u_y)^2$$
$$B' = Au_x v_x + B(u_x v_y + u_y v_x) + Cu_y v_y$$
$$C' = A(v_x)^2 + 2Bv_x v_y + C(v_y)^2$$

8.1.6 The results obtained in Problem 8.1.5 may be rewritten in matrix form as

$$A' = U^T M U, \qquad B' = U^T M V, \qquad C' = V^T M V$$

where

$$U = \begin{pmatrix} u_x \\ u_y \end{pmatrix}, \qquad V = \begin{pmatrix} v_x \\ v_y \end{pmatrix}, \qquad M = \begin{pmatrix} A & B \\ B & C \end{pmatrix}$$

and the superscript T refers to the operation of transposing rows and columns.

(a) Write

$$\Delta' = A'C' - B'^2 = U^T M (UV^T - VU^T) M V$$

and show that

$$UV^T - VU^T = J \begin{pmatrix} 0 & 1 \\ -1 & 0 \end{pmatrix}$$

where $J = u_x v_y - u_y v_x$ is the (assumed nonvanishing) Jacobian of the transformation between x, y and u, v.

(b) Show that

$$\Delta' = J^2\Delta$$

which shows that the sign of Δ' is the same as that of Δ and thus the classification of an equation is unchanged by coordinate transformations.

The following three problems arise from specialization of Laplace's equation in various coordinate systems.

8.1.7   Set $\rho = x$, $\varphi = y$, and $z = $ const in Laplace's equation in cylindrical coordinates (3.5.14) to obtain

$$x^2 f_{xx} + f_{yy} + x f_x = 0$$

(a) Obtain the characteristic coordinates $u = xe^{iy}$ and $v = xe^{-iy}$.
(b) Show that in these coordinates Laplace's equation becomes $4uv f_{uv} = 0$ and thus obtain the general solution

$$f = \theta(u) + \chi(v), \qquad x \neq 0$$

(c) Obtain real solutions for f by setting $\chi(\zeta) = \theta(\zeta)$ and considering $\theta(\zeta) = \zeta^n$, $e^{i\zeta}$, and $\ln \zeta$.

8.1.8   Set $r = x$, $\theta = y$, and $\varphi = $ const in Laplace's equation in spherical coordinates (3.5.16) and obtain

$$x^2 f_{xx} + f_{yy} + 2x f_x + (\cot y) f_y = 0$$

(a) Obtain characteristic coordinates $u = xe^{iy}$, $v = xe^{-iy}$.
(b) Show that Laplace's equation now becomes

$$2(u - v) f_{uv} - f_u + f_v = 0$$

(c) Show that the subsequent introduction of the complex conjugate variables $u = \lambda + i\mu$, $v = \lambda - i\mu$ is equivalent to the use of the cylindrical coordinates $\lambda = \rho$, $\mu = z$.

8.1.9   Set $r = $ const, $\theta = x$, and $\varphi = y$ in Laplace's equation in spherical coordinates (3.5.16) and obtain

$$f_{xx} + (\csc^2 x) f_{yy} + (\cot x) f_x = 0$$

(a) Obtain the characteristic coordinates $u = e^{iy}\tan(x/2)$, $v = e^{-iy}\tan(x/2)$.

(b) Show that in these coordinates Laplace's equation becomes

$$(uv + 1)^2 f_{uv} = 0$$

and hence obtain

$$f = \theta(u) + \chi(v)$$

(c) Obtain real solutions for f by setting $\chi(\zeta) = \theta(\zeta)$ or $i\theta(\zeta)$. The solutions corresponding to $\theta(\zeta) = \zeta^\mu$ are $\cos \mu y \, \tan^\mu(x/2)$ and $\sin \mu y \, \tan^\mu(x/2)$. In the original θ, φ coordinates the θ dependent factor, $\tan^\mu(\theta/2)$, is proportional to the associated Legendre function $P_0^\mu(\cos \theta)$. Since φ dependent problems in spherical coordinates have not been introduced in this text, the properties of these functions have not been developed.

8.1.10 Rewrite $x^2 z_{xx} + xz_x - z_{yy} = 0$ in characteristic coordinates and obtain the general solution.

8.1.11 (a) Rewrite $x^2 z_{xx} - z_{yy} = 0$ in characteristic coordinates and obtain

$$4uv z_{uv} - uz_u - vz_v = 0$$

where $u = xe^y$, $v = xe^{-y}$.

(b) Set $p = \ln u$ and $q = \ln v$ to obtain

$$4z_{pq} - z_p - z_q = 0$$

(c) Introduce the new dependent variable $w = ze^{\alpha p + \beta q}$ and determine α, β so that $w_{pq} = \gamma w$ where γ is a constant.

8.2 UNIQUENESS AND GENERAL PROPERTIES OF SOLUTIONS

For the most part, our previous considerations have stressed the solution of specific problems. In a number of examples, we have seen how solutions of the Laplace equation and the diffusion equation, for example, exhibit expected properties of steady and non–steady state temperature distributions, respectively. In this section we summarize certain general properties of partial differential equations and examine some of the ways in which these equations imply these properties in a more universal way rather than merely as displayed through the solution of specific problems. We also summarize a method for proving the uniqueness of the solutions to these equations as well as those for the wave equation.

8.2.1   Laplace's Equation

Solutions of Laplace's equation in a region can be viewed as representing an equilibrium temperature distribution in that region and hence there can be no maximum temperature within the region. If this were not the case, there would be continual heat flow away from the point of maximum temperature. The temperature would decrease in time and thus violate the steady state assumption. Maximum temperatures can therefore only occur at the boundary (recall the steady state heat flow problems of Section 1.3). To examine this idea in a more quantitative fashion, we integrate Laplace's equation throughout a region S and then use the divergence theorem. In two dimensions (a similar calculation may be performed in three dimensions) the vanishing of the Laplacian implies

$$\int_S \nabla^2 u \, dS = \int_S \nabla \cdot \nabla u \, dS = \int_C dl \, \mathbf{n} \cdot \nabla u = 0 \qquad (8.2.1)$$

where C is the closed curve bounding S and $\mathbf{n}$ is an outward normal for C. If we specialize C to be a circle of radius ρ' centered at some point $\mathbf{\rho_0}$ within S so that a field point $\mathbf{\rho}$ on the rim of this circle is given by $\mathbf{\rho} = \mathbf{\rho_0} + \mathbf{\rho}'$, then since ρ' is constant, the final expression in (8.2.1) reduces to

$$\rho' \int_0^{2\pi} d\theta' \, \frac{\partial u(\mathbf{\rho_0}, \rho', \theta')}{\partial \rho'} = 0 \qquad (8.2.2)$$

The integral here, divided by 2π, may be interpreted as the average value of $\partial u(\mathbf{\rho_0}, \rho', \theta')/\partial\rho'$ on the circumference of the circle. The average value of $u(\mathbf{\rho_0}, \rho', \theta')$ itself may be written

$$\langle u(\mathbf{\rho_0}, \rho') \rangle = \frac{1}{2\pi} \int_0^{2\pi} d\theta' u(\mathbf{\rho_0}, \rho', \theta') \qquad (8.2.3)$$

which might be expected to depend upon ρ'. This is not the case, however, since differentiation with respect to ρ' and insertion of appropriate factors of ρ' yield

$$\frac{\partial}{\partial\rho'} \langle u(\mathbf{\rho_0}, \rho') \rangle = \frac{1}{\rho'} \left[\frac{1}{2\pi} \rho' \int_0^{2\pi} d\theta' \, \frac{\partial u}{\partial\rho'} \right] \qquad (8.2.4)$$

This result is zero according to (8.2.2), and we conclude that $\langle u(\mathbf{\rho_0}, \rho') \rangle$ is actually independent of ρ'. On choosing ρ' to be arbitrarily small, we conclude that $\langle u(\mathbf{\rho_0}, \rho') \rangle = \langle u(\mathbf{\rho_0}, 0) \rangle = u(\mathbf{\rho_0})$, that is, the value of u at any point is equal to its average value on any circle centered at that point and surrounding the point (i.e., not intersecting the boundary). This result is known as the mean value theorem for Laplace's equation.

The mean value theorem may be used to show that a maximum value of a solution of Laplace's equation in the region S cannot be located within the region and must therefore be located on the boundary C. Consider the point $\boldsymbol{\rho}_M$ at which u takes on the supposed maximum value and a neighboring point $\boldsymbol{\rho}_1$ at which u has some consequently smaller value. If one considers $\boldsymbol{\rho}_1$ to be on a circle of radius $|\boldsymbol{\rho}_M - \boldsymbol{\rho}_1|$ centered at $\boldsymbol{\rho}_M$, then the value of u at $\boldsymbol{\rho}_M$, according to the mean value theorem, is the average of values on the circle. Since u on the circle is to have a value less than that at $\boldsymbol{\rho}_M$ and $u(\boldsymbol{\rho}_M)$ is the average of u on the rim, there must be some other place on the rim with a value larger than $u(\boldsymbol{\rho}_M)$ to compensate for this lower value at $\boldsymbol{\rho}_1$. Hence $u(\boldsymbol{\rho}_M)$ cannot be the largest value. This result is known as the maximum principle for Laplace's equation. A similar consideration also applies to minimum values.

8.2.2 Uniqueness

Laplace's equation provides a simple context for developing a method for proving the uniqueness of the solution to a boundary value problem. Consider this equation as being satisfied by some quantity u in a three-dimensional region V bounded by a closed surface S. Multiplying Laplace's equation by u, integrating throughout V, and applying Green's theorem yield

$$\int_V dv\, u\, \nabla^2 u = \int_S dS\, \mathbf{n} \cdot u\, \nabla u - \int_V dv\, |\nabla u|^2 = 0 \qquad (8.2.5)$$

If either u or $\mathbf{n} \cdot \nabla u = \partial u / \partial n$ should be zero on S, the surface integral of $u\mathbf{n} \cdot \nabla u$ will also vanish and (8.2.5) yields $\int_V dv\, |\nabla u|^2 = 0$. Since $|\nabla u|^2$ is inherently positive, this surface integral can only vanish if $\nabla u = 0$, which implies $u = \text{const}$. But if $u = 0$ on the boundary, the constant is zero and u is zero throughout V. If $\partial u / \partial n = 0$ on the boundary, then the constant is arbitrary.

We can now use these results to obtain a uniqueness theorem for solutions of Laplace's equation. For if we have two solutions u_1 and u_2 in V that satisfy the same boundary conditions on S, then the difference $u = u_1 - u_2$ is zero on S. It thus has the properties of the function u in the previous discussion and so $u = 0$ throughout V. Therefore $u_1 - u_2 = 0$ or $u_1 = u_2$ and the solution is unique. If, on the other hand, $\partial u_1 / \partial n = \partial u_2 / \partial n$ on S, then the solutions can at most differ by a constant. If the solutions are equal at any one point, this constant is zero and the two solutions are equal everywhere.

8.2.3 Diffusion Equation

The maximum temperature for a time dependent temperature distribution must also occur on the boundary. Here the term boundary refers to either the physical boundary or the temporal boundary $t = 0$. For a one-dimensional system $0 \leq x \leq L$ and $t \geq 0$, the boundary, which we label Γ, is thus a U-shaped region in the xt plane that is bounded by $0 \leq x \leq L$ as well as $x = 0$, $t \geq 0$, and

$x = L$, $t \geq 0$. We proceed by assuming that u takes on its maximum value u_0 at some space-time point x_0, t_0 not on the boundary. Let u_Γ denote values of u on Γ and $u_{\Gamma M}$ the largest value that u takes on Γ. Then, by hypothesis, $u_0 > u_{\Gamma M}$ or equivalently $u_0 = u_{\Gamma M} + \Delta$ where Δ is positive. We now introduce a function $v(x, t) = u(x, t) - \epsilon(t - t_0)$. At $t = t_0$ we have $v(x, t_0) = u(x, t_0)$ and in particular $v(x_0, t_0) = u(x_0, t_0) = u_0$. On Γ we write $v_\Gamma(x, t) = u_\Gamma(x, t) - \epsilon(t - t_0)$. Let $v_{\Gamma M}$ be the largest value that v_Γ takes on the boundary. Then $v_{\Gamma M} = u_\Gamma - \epsilon(t - t_0)$, and since $u_\Gamma \leq u_{\Gamma M}$, we have $u_\Gamma = u_{\Gamma M} - \delta$ with $\delta \geq 0$. Substituting into the expression for $v_{\Gamma M}$ we have $v_{\Gamma M} = u_{\Gamma M} - \delta - \epsilon(t - t_0) = u_0 - \Delta - \delta - \epsilon(t - t_0)$. The total increment to u_0 in this result may be positive or negative depending on $\epsilon(t - t_0)$. For ϵ small enough, however, the total increment is negative and we have $u_0 \geq v_{\Gamma M}$. Since $u_0 = v_0$, we also have $v_{\Gamma M} < v_0$, that is, the maximum value of v is somewhere off the boundary. Let is be at x_1, t_1. Then at that point we must have $v_{xx} \leq 0$, the customary requirement for a spatial maximum, as well as $v_t \geq 0$, for otherwise v would be decreasing in time from some previous larger value. From the definition of v we obtain

$$u_{xx} \leq 0, \qquad u_t - \epsilon > 0 \tag{8.2.6}$$

or $u_t > 0$ while $u_{xx} \leq 0$. Since u satisfies the heat equation $u_t - u_{xx} = 0$, the inequalities provide a contradiction, and we conclude that the maximum value of u cannot take place off the boundary.

8.2.4   Wave Equation

Here we confine attention to uniqueness and consider the case of a semi-infinite string. We follow the approach used for Laplace's equation and consider two possible solutions u_1 and u_2 with the same initial conditions $u_1(x, 0) = u_2(x, 0) = f(x)$, $u_{1t}(x, 0) = u_{2t}(x, 0) = g(x)$ as well as the same boundary condition $u_1(0, t) = u_2(0, t) = h(t)$. Note that although we may specify the initial conditions $f(x)$ and $g(x)$, it is only after we have solved the problem that we actually know the boundary value $h(t)$ since waves may propagate to the boundary and influence the value of the field variable there (unless we are imposing homogeneous boundary conditions). Hence, we merely require that $u_1(0, t) = u_2(0, t)$ and not that their value be known. As in our previous consideration of Laplace's equation, we consider the problem of the wave field $u(x, t) = u_1 - u_2$ and show that it must equal zero.

For the wave equation a crucial role is played by the finite propagation velocity of disturbances on the string. At a point ξ, τ the solution is determined by only that part of the boundary and initial conditions that can reach ξ in the time τ. There are thus two situations, as shown in Figure 8.1. In Figure 8.1a, for $\tau < \xi/c$, the signal from the boundary, due to either a source at the boundary or a reflected disturbance, has not had time to reach ξ, τ. The energy on the string at time t such that $0 \leq t \leq \tau$ that can reach ξ, τ is given by an

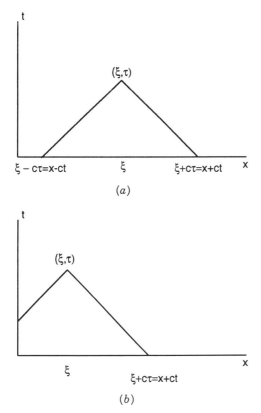

Figure 8.1. Figures for energy calculation on string: (a) $\tau < \xi/c$ and (b) $\tau > \xi/c$.

expression of the form

$$E(\xi, \tau) = \int_{\xi - c(\tau - t)}^{\xi + c(\tau - t)} dx \; [\tfrac{1}{2} \rho(u_t)^2 + \tfrac{1}{2} T(u_x)^2] \qquad (8.2.7)$$

To obtain an inequality from which to infer uniqueness, we multiply the wave equation by u_t and integrate as indicated to obtain

$$\int_{\xi - c(\tau - t)}^{\xi + c(\tau - t)} dx \; u_t(\rho u_{tt} - Tu_{xx}) = 0 \qquad (8.2.8)$$

Employing the relation $u_t u_{xx} = -\tfrac{1}{2}[(u_x)^2]_t + (u_t u_x)_x$, we may write

$$\int_{\xi - c(\tau - t)}^{\xi + c(\tau - t)} dx \; \frac{\partial}{\partial t} \left[\frac{\rho}{2}(u_t)^2 + \frac{T}{2}(u_x)^2 \right] - T(u_t u_x) \Big|_{\xi - c(\tau - t)}^{\xi + c(\tau - t)} = 0 \quad (8.2.9)$$

Since the limits depend upon time, we use the relation (Hildebrand, 1976, p. 365)

$$\frac{d}{dt} \int_{A(t)}^{B(t)} dx\, f(x,\, t) = \int_{A}^{B} dx\, \frac{\partial f}{\partial t} + f(B,\, t)\, \frac{dB}{dt} - f(A,\, t)\, \frac{dA}{dt} \qquad (8.2.10)$$

to extract the time integral from under the integral sign and obtain

$$\frac{d}{dt} \int_{\xi - c(\tau - t)}^{\xi + c(\tau - t)} dx \left[\frac{\rho}{2}\, (u_t)^2 + \frac{T}{2}\, (u_x)^2 \right] + c \left[\frac{\rho}{2}\, (u_t)^2 + \frac{T}{2}\, (u_x)^2 \right]\bigg|_{x = \xi + c(\tau - t)}$$

$$+ c \left[\frac{\rho}{2}\, (u_t)^2 + \frac{T}{2}\, (u_x)^2 \right]\bigg|_{x = \xi - c(\tau - t)} = T(u_t u_x)\bigg|_{\xi - c(\tau - t)}^{\xi + c(\tau - t)} \qquad (8.2.11)$$

On using the definition of $E(\xi,\, \tau)$ from (8.2.7) as well as the relation $T = \rho c^2$, we find that terms in (8.2.11) can be grouped into perfect squares to yield

$$\frac{d}{dt} E(t) = -\frac{\rho c}{2}\, [(u_t - cu_x)^2 + (u_t + cu_x)^2] \qquad (8.2.12)$$

The right-hand side is thus negative or zero and hence

$$\frac{dE}{dt} \le 0 \qquad (8.2.13)$$

Since E is either positive or zero, we see that if u is initially zero, it must remain so and hence $u_1 = u_2$.

For $\tau > \xi/c$ (cf. Fig. 8.1b) a signal originating at and/or reflected from the boundary can reach ξ, τ. In this case we examine the integral

$$E = \int_{0}^{\xi + c(\tau - t)} dx\, [\tfrac{1}{2}\, \rho\, (u_t)^2 + \tfrac{1}{2}\, T(u_x)^2] \qquad (8.2.14)$$

Proceeding as before, we have

$$\frac{dE}{dt} = -\frac{c}{2}\, [(u_t)^2 + c^2 (u_x)^2]\bigg|_{x = \xi + c(\tau - t)} + c^2 u_t u_x\bigg|_{x = \xi + c(\tau - t)} \qquad (8.2.15)$$

The terms evaluated at the lower limit of the integral vanish since the difference terms u and u_t are assumed to vanish at the boundary $x = 0$. We finally obtain

$$\frac{dE}{dt} = -\frac{c}{2}\, (u_t - cu_x)^2\bigg|_{\xi + c(\tau - t)}$$

$$\le 0 \qquad (8.2.16)$$

and the same conclusion may be drawn.

8.3 FIRST ORDER EQUATIONS

The shape of a wave frequently takes the form of a slowly varying amplitude superposed upon a much more rapidly varying "carrier" wave. An example is shown in Figure 8.2. When the wave represents a light pulse, the rapidly varying component determines the color of the light while the envelope determines the duration of the light pulse. If one is only interested in the shape of the pulse envelope, it is frequently possible to obtain an equation for the envelope alone and ignore the rapid oscillations associated with the underlying carrier wave. The approximate equation that describes the envelope is a first order partial differential equation. This description of the wave field is known as the slowly varying envelope approximation.

As an example, consider a one-dimensional wave satisfying

$$\frac{1}{c^2}\frac{\partial^2 w}{\partial t^2} + 2\gamma\frac{\partial w}{\partial t} - \frac{\partial^2 w}{\partial x^2} = p(x, t) \tag{8.3.1}$$

in which the source term has the form $p(x, t) = P(x, t)\cos(kx - \omega t)$ where $k = \omega/c$. It is convenient to consider this expression to be the real part of $P(x, t)e^{i\phi}$ where $\phi = kx - \omega t$. A slow variation of P in comparison with that of the oscillatory term $e^{i\phi}$ can be imposed by requiring $|\partial P/\partial x| \ll k|P|$ and $|\partial P/\partial t| \ll \omega|P|$. A corresponding complex variation in the wave amplitude $w(x, t)$ may be introduced by writing $w(x, t) = f(x, t)e^{i\phi}$. A slow variation in the phase term could also be included here, but such generality is unnecessary for our purposes. To obtain the equation for the slowly varying envelope $f(x, t)$ from (8.3.1), we first calculate the derivatives

$$w_{xx} = (-k^2 f + 2ikf_x + f_{xx})e^{i\phi}$$

$$w_t = (-i\omega f + f_t)e^{i\phi} \tag{8.3.2}$$

$$w_{tt} = (-\omega^2 f - 2i\omega f_t + f_{tt})e^{i\phi}$$

Substitution into (8.3.1) leads to a cancellation of the terms containing k^2 and ω^2. Assuming that spatial and temporal derivatives of f are related to f in the same way as those for P, that is, $|f_x| \ll k|f|$, $|f_t| \ll \omega|f|$, and similarly for higher derivatives, we may expect the terms f_{xx} and f_{tt} to be negligible in comparison with kf_x and ωf_t, respectively, and obtain from (8.3.1) the ap-

Figure 8.2. Slowly varying envelope of high-frequency wave.

proximate first order equation

$$\frac{\partial f}{\partial x} + \frac{1}{c}\frac{\partial f}{\partial t} + \gamma cf = \frac{i}{2k}P \tag{8.3.3}$$

The imaginary nature of the term on the right-hand side of this result merely indicates that f and P are out of phase by $90°$ (i.e., if P is the amplitude of $\cos\phi$, as indicated previously, then w is proportional to $\sin\phi$). Since we shall have no further concern for the underlying oscillation, this phase relation will be ignored. We shall, in fact, merely think of (8.3.3) as an example of a first order partial differential equation of the general form

$$\frac{\partial z}{\partial x} + \frac{1}{v}\frac{\partial z}{\partial t} + rz = q \tag{8.3.4}$$

We will henceforth consider the coefficients v, q, and r to be real but allow them to depend upon x, t, and z. As an aid in interpreting the solution, it will be found useful to retain the notion that v signifies a velocity at which some disturbance represented by z is propagating.

Equation (8.3.4) is a linear partial differential equation of the first order. When any of the coefficients depend upon z the equation is referred to as being quasi-linear. It will be found that the technique employed to solve this equation is still applicable when the coefficients include the dependent variable z.

The solution can be expected to be expressible in the form $z = z(x, t)$ or, in the more symmetric form, $u(x, t, z) = c$ where c is a constant. Either expression has the geometric significance of representing a surface in a three-dimensional x, t, z space. A typical condition that might be imposed upon the solution of the equation, such as the initial condition $z(x, 0) = f(x)$, where $f(x)$ is given, may be interpreted as a requirement that the curve $z = f(x)$, $t = 0$ be on the solution surface. In general, we seek the solution surface that contains a curve Γ on which $z(x, t)$ is specified at some initial time. It proves useful to consider the geometry of the surface and boundary curve Γ in some detail. We choose the representation $u(x, t, z) = c$ and recast the original equation (8.3.4) in terms of derivatives of u rather than of z. To accomplish this change, we first note that the variables x, t, and z must vary in such a way that u maintains the constant value c and therefore write

$$du = \frac{\partial u}{\partial x}dx + \frac{\partial u}{\partial t}dt + \frac{\partial u}{\partial z}dz = dc = 0 \tag{8.3.5}$$

The corresponding vector form is

$$\nabla u \cdot d\mathbf{r} = 0 \tag{8.3.6}$$

Here $\mathbf{r}$ is the position vector $\mathbf{r} = \mathbf{i}x + \mathbf{j}t + \mathbf{k}z$ from an origin to some point x, t, z on the surface $u(x, t, z) = c$. Since z itself is a function of x and t, we also have

$$dz = \frac{\partial z}{\partial x} dx + \frac{\partial z}{\partial t} dt \qquad (8.3.7)$$

and hence (8.3.5) may be written as

$$\left(\frac{\partial u}{\partial x} + \frac{\partial u}{\partial z} \frac{\partial z}{\partial x} \right) dx + \left(\frac{\partial u}{\partial t} + \frac{\partial u}{\partial z} \frac{\partial z}{\partial t} \right) dt = 0 \qquad (8.3.8)$$

Since x and t are independent variables, the coefficients of dx and dt in (8.3.8) must vanish separately. When the two resulting equations are solved individually so as to express $\partial z/\partial x$ and $\partial z/\partial t$ in terms of derivatives of u, we find that the original equation (8.3.4) takes the form

$$\frac{\partial u}{\partial x} + \frac{1}{v} \frac{\partial u}{\partial t} + (q - rz) \frac{\partial u}{\partial z} = 0 \qquad (8.3.9)$$

We now introduce a vector $\mathbf{T} = \mathbf{i} + v^{-1}\mathbf{j} + (q - rz)\mathbf{k}$ and write (8.3.9) as

$$\mathbf{T} \cdot \nabla u = 0 \qquad (8.3.10)$$

At any point on the surface $u(x, t, z) = c$ the gradient vector ∇u is normal to the surface, and according to (8.3.10), the vector $\mathbf{T}$ is therefore tangent to the surface at that point. It is thus tangent to a curve in that surface. Such a curve is referred to as a characteristic curve of the partial differential equation and the projection of the curve onto the xy plane is referred to as a characteristic trace. If position on the characteristic curve is parametrized by s, the distance along the characteristic away from the boundary curve Γ, then the position vector to a point on the characteristic is $\mathbf{r}(s) = \mathbf{i}x(s) + \mathbf{j}t(s) + \mathbf{k}z(s)$, and a unit tangent vector to the characteristic is $\mathbf{t} = d\mathbf{r}/ds$. The tangent vector $\mathbf{T}$ referred to above is parallel to $\mathbf{t}$, that is, $\mathbf{t} = \mu\mathbf{T}$, and on equating corresponding components of these two vectors, we have

$$1 = \frac{dx}{\mu \, ds}, \qquad v^{-1} = \frac{dt}{\mu \, ds}, \qquad q - rz = \frac{dz}{\mu \, ds} \qquad (8.3.11)$$

where μ may be a function of x, t, and z. Introducing a new parameter α by writing $d\alpha = \mu \, ds$, we obtain

$$\frac{dx}{d\alpha} = 1, \qquad \frac{dt}{d\alpha} = v^{-1}, \qquad \frac{dz}{d\alpha} = q - rz \qquad (8.3.12)$$

which is a system of three ordinary differential equations that may be solved to determine the characteristics and obtain z, x, and t in parametric form. The integration constants that arise in solving these equations are determined by requiring that for $\alpha = 0$ the quantities x, t, and z must be on the boundary curve Γ. If position on Γ is expressed by a parameter β, the information given on Γ may be expressed in the parametric form $x = x_0(\beta)$, $t = t_0(\beta)$, $z = z_0(\beta)$. We require that the solutions $x(\alpha, \beta)$, $t(\alpha, \beta)$, and $z(\alpha, \beta)$ on the characteristic reduce to $x(0, \beta) = x_0(\beta)$, $t(0, \beta) = t_0(\beta)$, and $z(0, \beta) = z_0(\beta)$ on Γ. Once the solutions $x(\alpha, \beta)$ and $t(\alpha, \beta)$ have been obtained, they are inverted to yield $\alpha(x, t)$ and $\beta(x, t)$. To carry out such an inversion, however, it is necessary that the Jacobian of the transformation be nonvanishing, that is

$$J\left(\frac{x, t}{\alpha, \beta}\right) = \frac{\partial x}{\partial \alpha}\frac{\partial t}{\partial \beta} - \frac{\partial t}{\partial \alpha}\frac{\partial x}{\partial \beta} \neq 0 \qquad (8.3.13)$$

Assuming the nonvanishing of the Jacobian, the two relations expressing α and β in terms of x and t are obtained and used in $z(\alpha, \beta)$ to yield the solution $z(x, t)$.

As an example (analyzed previously by the Laplace transform method in Problem 2.4.3), consider (8.3.4) for the case in which v and r are constants, q equals zero, and the solution satisfies the initial condition $z(x, 0) = \chi(x)$. Equations (8.3.12) yield

$$\frac{dx}{d\alpha} = 1, \qquad \frac{dt}{d\alpha} = v^{-1}, \qquad \frac{dz}{d\alpha} = -rz \qquad (8.3.14)$$

Since x, t, and z depend upon both α and β, the derivatives here are actually partial derivatives with respect to α and the solutions involve integration constants that are functions of β. We thus have

$$x = \alpha + \varphi_1(\beta), \qquad t = v^{-1}\alpha + \varphi_2(\beta), \qquad z = \varphi_3(\beta)e^{-\alpha r} \qquad (8.3.15)$$

At $\alpha = 0$, that is, on Γ, these solutions reduce to $x = x_0(\beta) = \varphi_1(\beta)$, $t = t_0(\beta) = \varphi_2(\beta) = 0$, and $z = z_0(\beta) = \varphi_3(\beta)$. The function $\varphi_1(\beta)$ allows for an arbitrary parametric representation of position on Γ. For simplicity, we choose $\varphi_1(\beta) = \beta$. Then, $\varphi_3(\beta) = \chi(\beta)$ and (8.3.15) yields $x(\alpha, \beta) = \alpha + \beta$, $t(\alpha, \beta) = v^{-1}\alpha$, and $z(\alpha, \beta) = \chi(\beta)e^{-\alpha r}$. Solving for α and β in terms of x and t, we obtain $\alpha = vt$ and $\beta = x - vt$. Substitution into $z(\alpha, \beta)$ gives the expected solution

$$z(x, t) = e^{-rvt}\chi(x - vt) \qquad (8.3.16)$$

It should be noted that in this example the Jacobian has a constant nonzero value of $-1/v$.

An alternate procedure that may be used to solve the general equation (8.3.4), or equivalently (8.3.9), is to eliminate μ from (8.3.11) by writing

$$\frac{dx}{1} = \frac{dt}{v^{-1}} = \frac{dz}{q - rz} = \mu \, ds \qquad (8.3.17)$$

to again obtain a system of ordinary differential equations. Solutions are now determined for two of the equations obtainable by equating various members of (8.3.17). The choice depends upon the x, t, and z dependence of the coefficients in any specific example and is usually dictated by convenience. Examples may be found in the problems at the end of this section. The two solutions will be of the general form

$$u_1(x, t, z) = c_1, \qquad u_2(x, t, z) = c_2 \qquad (8.3.18)$$

and represent two surfaces in x, t, z space. A solution that contains z is actually a particular solution of the partial differential equation (8.3.4). What we require for the general solution is the simultaneous satisfaction of both relations in (8.3.18), which will be the curve produced by the intersection of these two surfaces. When the constants c_1 and c_2 are varied, this intersection curve sweeps out a surface that is a possible solution surface $u(x, y, z) = c$ of the given partial differential equation. The solution, which is a relation between c_1 and c_2, may be expressed in either of the forms

$$F(c_1, c_2) = 0, \qquad c_2 = h(c_1) \qquad (8.3.19)$$

The x, t, z dependence of the solution is now exhibited by introducing $c_1 = u_1(x, t, z)$ and $c_2 = u_2(x, t, z)$ from (8.3.18). Our task now is to obtain the particular relation between c_1 and c_2 that provides the solution to a specific problem in which z has a prescribed dependence along some boundary curve Γ.

The boundary curve Γ may also be considered to be the intersection of two surfaces, say,

$$\psi_1(x, t, z) = 0, \qquad \psi_2(x, t, z) = 0 \qquad (8.3.20)$$

The general procedure employed is to eliminate x, t, and z between the four equations in (8.3.18) and (8.3.20) to obtain a relation between c_1 and c_2.

As an example, we reconsider the problem treated above and in Problem 2.4.3. Equating the first and second members of (8.3.17) as well as the second and third and then integrating, we obtain (8.3.18) in the form $u_1 = x - vt = c_1$ and $u_2 = ze^{rvt} = c_2$. The initial condition $z(x, 0) = \chi(x)$ is recast in terms of ψ_1 and ψ_2 by writing $\psi_1(x, t, z) = t = 0$ and $\psi_2(x, t, z) = x - \chi(x) = 0$. Imposing this initial condition on u_1 and u_2, we find $x = c_1$ and $\chi(x) = c_2$. The relation between c_1 and c_2 is therefore $c_2 = f(c_1)$. Since $c_1 = u_1 = x - vt$ and $c_2 = u_2 = ze^{rvt}$ at later times, we have the solution in the form $ze^{rvt} = \chi(x - vt)$, which is equivalent to the result obtained in (8.3.16).

We see elsewhere (Appendix B, Problem B.3) that the initial conditions imposed upon the solution of an ordinary differential equation may be incorporated as source terms. This equivalence may be introduced when solving partial differential equations as well. We may recover the results just obtained by considering

$$\frac{\partial z}{\partial x} + \frac{1}{v}\frac{\partial z}{\partial t} + rz = A(x)\delta(t - \epsilon) \tag{8.3.21}$$

subject to the *homogeneous* initial condition $z(x, 0) = 0$. The amplitude $A(x)$ is determined by integrating on t across the singularity at $t = \epsilon$. One obtains $v^{-1}[z(x, \epsilon^+) - z(x, \epsilon^-)] = A(x)$. Since $z(x, \epsilon^+) = \chi(x)$ and $z(x, \epsilon^-) = 0$, one finds $A(x) = v^{-1}\chi(x)$. With this expression for $A(x)$ and homogeneous boundary conditions, Eq. (8.3.21) represents the situation in the problem just considered. The two ordinary differential equations obtainable from (8.3.17), with q now set equal to $v^{-1}\chi(x)\delta(t - \epsilon)$, lead to the solution $x - vt = c_1$, as before, as well as the inhomogeneous equation

$$\frac{dz}{dt} + rvz = \chi(x)\delta(t - \epsilon) = \chi(c_1 + vt)\delta(t - \epsilon) \tag{8.3.22}$$

Integrating from 0 to t with $z(x, 0) = 0$, we obtain

$$z(x, t)e^{rvt} = \int_0^t dt' \, \chi(c_1 + vt')\delta(t' - \epsilon) = \chi(c_1) \tag{8.3.23}$$

in which the limit $\epsilon \to 0$ has been taken. Since $c_1 = x - vt$, this result may be rewritten in the form $z = e^{-rvt}\chi(x - vt)$ obtained above.

A situation deserving special attention is that in which the curve Γ on which data are specified can coincide with a characteristic curve. If the values of z that we intend to specify on Γ are written as $z_0(x, t)$, then in general $dz_0 = z_{0x}\, dx + z_{0t}\, dt$ and

$$\frac{dz_0}{dt} = \frac{\partial z_0}{\partial x}\frac{dx}{dt} + \frac{\partial z_0}{\partial t} \tag{8.3.24}$$

From the first equation of (8.3.12) we see that on the characteristic trace $dx/dt = v$ and (8.1.24) is then

$$\frac{dz_0}{dt} = v\left(\frac{\partial z_0}{\partial x} + v^{-1}\frac{\partial z_0}{\partial t}\right) \tag{8.3.25}$$

But since the solution of (8.3.4) is to reduce to z_0 on Γ, we may also write (8.3.25) as

$$\frac{dz_0}{dt} = (q - rz_0)v \tag{8.3.26}$$

and find that z_0 is now no longer arbitrary but must be a function that satisfies this differential equation. Introducing $d\alpha = v\,dt$ as before, we find that (8.3.26) takes the same form as the third of (8.3.12) that determines the characteristics. Only if the value of z_0 that we impose is also a solution of this equation does the problem have a solution, and then the integration constant arising in the solution of this equation provides an arbitrary amplitude for this solution, that is, the solution is not unique. As an example, consider the solution of $z_x + v^{-1}z_t + rz = 0$ with v and r constant but with the condition $z(Vt, t) = F(t)$. The parametric solution for this example is obtained from the general solution (8.3.15) as before. We again require $\alpha = 0$ on Γ, which is now the line $x = Vt$. The first two of (8.3.15) yield $\varphi_1(\beta) = V\varphi_2(\beta)$. Choosing $\varphi_1(\beta) = \beta$, we have $x = \alpha + \beta$ and $t = v^{-1}\alpha + V^{-1}\beta$. For $\alpha = 0$, the third of (8.3.15) becomes $z = \varphi_3(\beta) = F(\beta/V)$ so the parametric solution is

$$x = \alpha + \beta, \quad t = v^{-1} + V^{-1}\beta, \quad z = F(\beta/V) \tag{8.3.27}$$

To obtain the solution in terms of x and t, we solve the first two of these equations for α and β and substitute into the solution for z to obtain

$$z(x, t) = F\left(\frac{vt - x}{v - V}\right) e^{-vr(x - vt)/(v - V)} \tag{8.3.28}$$

which agrees with the result obtained in Problem 2.4.3 by the Laplace transform method.

As noted above, the expression of x and t in terms of α and β assumes that the Jacobian of the transformation (8.3.13) is nonvanishing. In the present example a simple calculation shows that

$$J = x_\alpha t_\beta - x_\beta t_\alpha = V^{-1} - v^{-1} \tag{8.3.29}$$

and thus we must require $v \neq V$. For $J = 0$ we return to (8.3.26) and find that on the characteristic we must have

$$\frac{dz_0}{dt} = -vrz_0 \tag{8.3.30}$$

or $z_0 = Ce^{-rvt}$, a result also obtained in Problem 2.4.3c.

The constant coefficient equation considered above is quite simple and, as already noted, can also be solved by the transform techniques introduced in earlier chapters. The advantage of the method of characteristics is that it is still applicable when the coefficients depend upon the independent variables x and

t or even the dependent variable z. In these cases the characteristics are no longer straight lines and various complicating features are found to arise. For example, the characteristics may be confined to certain regions of the xt plane. It is also not possible to specify conditions on a line Γ that intersects a characteristic more than once. To obtain a simple appreciation of these ideas, we now consider two examples.

Example 8.4. Solve $z_x + v^{-1}z_t = 0$ for $v(x) = v_0(1 - x/L)$. Consider the case in which $z(x, 0) = \chi(x) = -H(-x)xe^{x/D}$, which represents a pulse of approximate width $2D$ that initially extends along the negative x axis. At $t = 0$, the pulse begins to propagate in the positive x direction toward the point $x = L$ with a velocity that decreases as it approaches L. Since $v = 0$ at $x = L$, we expect that the pulse will never reach this point but will slow down and compress as it approaches L. Also, on integrating $z_t + v(x)z_x = 0$ over all x and then integrating by parts on the term vz_x, one finds that $A(t) = \int_{-\infty}^{\infty}z(x, t)\, dx$, the area under the pulse, satisfies $dA/dt + v_0A/L = 0$. Hence, $A(t)$ decreases as $e^{-v_0t/L}$.

From (8.3.17) we have

$$\frac{dx}{1} = v_0\left(1 - \frac{x}{L}\right)dt = \frac{dz}{0} \tag{8.3.31}$$

From the first two terms here we obtain a differential equation having the solution $\exp(v_0t/L)(1 - x/L) = c_1$ while the last ratio in (8.3.31) implies $dz = 0$ or $z = c_2$. The initial condition may be recast in the form $t = 0$ and $z - \chi(x) = 0$, corresponding to the pair of surfaces in (8.3.20). One then finds $c_1 = 1 - x/L$ and $c_2 - \chi(x) = \chi[L(1 - c_1)]$, an equation that provides the desired relation between c_1 and c_2. Replacing c_1 and c_2 by u_1 and u_2 at later times, we find that the final solution is

$$z(x, t) = \chi\{L[1 - e^{v_0t/L}(1 - x/L)]\} \tag{8.3.32}$$

For $L \gg x$, v_0t (i.e., before the pulse has an opportunity to sense the presence of the inhomogeneity), we may expand the exponential and keep just the lowest order terms to obtain $z(x, t) = \chi(x - v_0t)$, as expected. A graph of the general result for the initial pulse profile given above is shown in Figure 8.3.

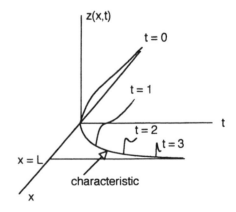

Figure 8.3. Compression of pulse profile as it propagates into inhomogeneous medium.

Continued compression of the pulse will, of course, ultimately lead to a profile that violates the assumption under which the slowly varying envelope approximation was derived. A valid description of the wave profile would then require that one examine the role played by the second derivative terms in (8.3.2). An extension of this type will be considered in Section 8.4.

Finally, it should be noted that this problem could also be solved when $z(0, t)$ is specified, that is, the data are given along the line $x = 0$.

Example 8.5. Solve $z_x + v^{-1}z_t = 0$ for $v(x, t) = v_0(1 - x/L - t/T)$. Consider the initial pulse used in the previous example. In this case, however, the velocity $v(x, t)$ will become negative after a sufficient time. The pulse will then reverse direction and pass the point $x = 0$ at a later time. The characteristic traces will thus bend back so that the information that we impose at $t = 0$ will return at a later time. It is therefore impossible to specify the value of the pulse profile on the line $x = 0$, as could be done in Example 8.4.

From (8.3.17) we have

$$\frac{dx}{1} = v_0 \left(1 - \frac{x}{L} - \frac{t}{T}\right) dt = \frac{dz}{0} \tag{8.3.33}$$

The first two terms provide a differential equation that can be written as

$$\frac{dx}{dt} + \frac{v_0 x}{L} = v_0 \left(1 - \frac{t}{T}\right) \tag{8.3.34}$$

Multiplying by the integrating factor $\exp(v_0 t/L)$, one readily obtains

$$x = L(1 + a - t/T) + Lc_1 e^{-v_0 t/L}, \qquad a \equiv L/v_0 T \tag{8.3.35}$$

From the last term in (8.3.33) we again have the relation $z = c_2$. Setting $z = \chi(x)$ at $t = 0$, we obtain $c_1 = 1 + a - x/L$, with a as specified in (8.3.35), and $c_2 = \chi[L(1 + a - c_1)]$. The solution is then

$$z(x, t) = \chi\{L[1 + a - e^{v_0 t/L}(1 + a - x/L - t/T)]\} \tag{8.3.36}$$

The characteristic traces are shown in Figure 8.4 and show how information specified along the line $x = 0$ at $t = 0$ will return to $x = 0$ at a later time.

The two previous examples involved linear equations. We now consider the quasi-linear equation

$$u_t + uu_x = 0 \tag{8.3.37}$$

which may be obtained from Eq. (8.3.4) by introducing $v = u_0 z$, $r = q = 0$ and then setting $u_0 z = u$. The quantity u thus has dimensions of velocity. Solutions are again found to evolve in such a way that contributions from higher derivatives must eventually be taken into account. While this could be accomplished by reintroducing terms neglected in the slowly varying envelope ap-

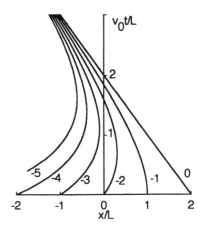

Figure 8.4 Characteristics for the inhomogeneous medium considered in Example 8.5.

proximation, a more customary approach to the association of higher derivatives with (8.3.37) is to consider it to arise from the neglect of the second derivative term in the equation

$$u_t + uu_x = \nu u_{xx} \qquad (8.3.38)$$

known as Burgers' equation. We now consider the "wave breaking" effect associated with solutions of the quasi-linear equation (8.3.37) and in the next section examine Burgers' equation and show how the second derivative term prevents the occurrence of multiple valuedness in the solution. Solution of the quasi-linear equation (8.3.37) by using (8.3.17) yields

$$\frac{dx}{1} = u \, dt = \frac{du}{0} \qquad (8.3.39)$$

Since $du = 0$, we have $u = c_2$ as well as

$$\frac{dx}{dt} = c_2 \qquad (8.3.40)$$

which yields $x - c_2 t = c_1$. Imposing the initial condition $u(x, 0) = \chi(x)$, where χ now has the dimensions of velocity, we have $x = c_1$ and $u = c_2 = \chi(x) = \chi(c_1)$. The solution may be written either in the implicit form

$$u(x, t) = \chi(x - u_0 zt) \qquad (8.3.41)$$

or, by introducing $\xi = x - ut$, in the parametric form

$$u = (x - \xi)/t$$
$$u = \chi(\xi) \qquad (8.3.42)$$

In the latter case the intersection of the two curves in a $u\xi$ plane yields the value(s) of u for a given value of x and t. Both the implicit form (8.1.41) and the parametric form (8.3.42) are seen to reduce to the given initial condition when $t = 0$. For the parametric case, the expected finite value of u at $t = 0$ requires that $x = \xi$ in the first of (8.3.42), and thus the second equation there reduces to $u = \chi(x)$.

Figure 8.5 shows the evolution of the initial profile $\chi(x) = v_0 \text{sech}^2(x/L)$ at the dimensionless times $v_0 t/L = 0, 1.3, 2$. These profiles may be constructed by considering either of the representations introduced above. According to the implicit solution (8.3.41), a point initially at $x = c_1$ with velocity $\chi(c_1)$ moves in the positive x direction during a time t to the new location $x = c_1 + t\chi(c_1)$. The figure shows the displacement of points A, B, and C that have the initial velocities v_0, $0.75\,v_0$, and $0.5\,v_0$, respectively. The corresponding initial values of x are $x_A = L\,\text{sech}^{-1}(1) = 0$, $x_B = L\,\text{sech}^{-1}\sqrt{0.75}$, and $x_C = L\,\text{sech}^{-1}\sqrt{0.5}$, respectively. After the dimensionless time interval $v_0 t/L = 1.3$ these points have moved to

$$x_{A'} = L[0 + 1.3(1)] = 1.30L$$

$$x_{B'} = L[\text{sech}^{-1}\sqrt{0.75} + 1.3(0.75)] = 1.52L \qquad (8.3.43)$$

$$x_{C'} = L[\text{sech}^{-1}\sqrt{0.5} + 1.3(0.5)] = 1.53L$$

At this time the two points $x_{B'}$ and $x_{C'}$ are seen to be nearly equal. For $v_0 t/L > 1.3$, $x_{B'}$ is greater than $x_{C'}$ and the solution becomes multiple valued. Since the model describes a physical situation in which multiple valuedness is meaningless, the solution must be discarded in this region and the neglected term νu_{xx} in (8.3.38) reintroduced into the governing equation. This procedure will be considered in the next section. The solution of the present problem does provide a simple theoretical description of the formation of a shock wave,

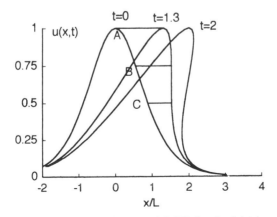

Figure 8.5. Wave breaking predicted by Eq. (8.3.37) for the initial profile $u_0(x, 0) = u_0 \text{sech}^2(x/L)$.

however. The time at which the shock forms may be estimated by identifying it with the earliest time at which the profile $u(x, t)$ becomes vertical, that is, the time at which $u_x = \infty$ for some point on the curve. From (8.3.41) we have the general relation

$$u_x = \frac{\chi'}{1 + t\chi'} \tag{8.3.44}$$

where χ' refers to differentiation with respect to the argument. For the example mentioned above, we have $\chi'(x) = -2(v_0/L)\operatorname{sech}^2 x/L \tanh x/L$. On using the largest negative value of this expression, namely $-(4\sqrt{3}/9)v_0/L$, to obtain the earliest time, we find $v_0 t/L = 1.299$. This result agrees with the numerical value shown in Figure 8.5.

With the parametric form (8.3.42), the intersection of the two curves in the $c_1 c_2$ plane provides the value(s) of $u(x, t)$ for a given value of x, t. For the profile considered above, one finds that, for $v_0 t/L > 1.3$, the two curves intersect in more than one point.

Problems

8.3.1 (a) Solve

$$\frac{\partial x}{\partial x} + \frac{1}{v}\frac{\partial z}{\partial t} = z, \qquad z(0, t) = \varphi(t)$$

where v and r are constant by equating the first and second as well as the first and third members of (8.3.17).

(b) Show that at $t = 0$ the solution reduces to $z(x, 0) = e^{-rx}\varphi(-x/v)$ and, on setting this expression equal to $\chi(x)$, the solution is equivalent to that given in (8.3.16).

8.3.2 Extend Example 8.4 to include damping, that is, solve

$$z_x + v^{-1}z_t + rz = 0$$

and obtain

$$z(x, t) = e^{-rx}\varphi\left\{\frac{L}{v}\ln\left[e^{vt/L}(1 - x/L)\right]\right\}$$

Note that for $x \ll L$ this solution reduces to that of Problem 8.3.1.

8.3.3 The most general form of the differential equation (8.3.17) is

$$\frac{dx}{P} = \frac{dt}{Q} = \frac{dz}{R}$$

where P, Q, and R are functions of x, t, and z. Obtain the relation

$$\frac{\alpha_1 \, dx + \alpha_2 \, dt + \alpha_3 \, dz}{\alpha_1 P + \alpha_2 Q + \alpha_3 R} = \frac{dx}{P} = \frac{dt}{Q} = \frac{dz}{R}$$

where α_1, α_2, and α_3 are arbitrary functions of x, t, and z. Note that if the α_i can be chosen so that the denominator of the first term vanishes, the numerator must also vanish and thus represent a perfect differential $d\varphi$. Thus $\varphi = $ const is an integral of the system.

8.3.4 (a) Apply the method of Problem 8.3.3 to the equation

$$z_x + v^{-1} z_t = f(t - x/V) = \left. \frac{dh(\xi)}{d\xi} \right|_{\xi = t - x/V}$$

(b) Show that by choosing $\alpha_1 = -V^{-1}$, $\alpha_2 = 1$, $\alpha_3 = 0$, the solution may be written

$$z = \frac{vV}{V - v} h(t - x/V)$$

8.3.5 Show that the solution of

$$\frac{\partial z}{\partial x} + \frac{1}{v} \frac{\partial z}{\partial x} + \gamma z = 0, \qquad z(Vt, \, t) = \sin \omega t$$

is

$$z(x, t) = e^{-\gamma v(x - Vt)/(v - V)} \sin \omega_+ (t - x/v)$$

where $\omega_+ = \omega v/(v - V)$.

8.3.6 Derive Eq. (8.3.43).

8.4 BURGERS' EQUATION

The *nonlinear* Burgers' equation (8.3.38) is related to the *linear* diffusion equation by a simple transformation. The origin of this relationship is somewhat subtle, being related to invariants of a general linear partial differential equation, and is summarized in Section 9.10. The relation itself is quite simple and is readily obtained by first writing Burgers' equation in the form

$$\frac{\partial u}{\partial t} = \frac{\partial}{\partial x}\left(v u_x - \frac{1}{2} u^2 \right) \tag{8.4.1}$$

Setting $u = \partial\varphi/\partial x$ and using the equality of mixed second derivatives, we can write

$$\frac{\partial\varphi}{\partial t} = \nu\frac{\partial^2\varphi}{\partial x^2} - \frac{1}{2}\left(\frac{\partial\varphi}{\partial x}\right)^2 \tag{8.4.2}$$

The first and second spatial derivatives of φ occur here in a combination that should remind one of the second derivative of an exponentiated function. When we introduce $\theta(x, t) = e^{-\gamma\varphi(x,t)}$ and examine the constant factors, we are led to set $\gamma = -\frac{1}{2}\nu$ and write

$$\frac{\partial\varphi}{\partial t} = 2\nu^2 e^{\varphi/2\nu}\frac{\partial^2}{\partial x^2}(e^{-\varphi/2\nu}) \tag{8.4.3}$$

Hence, $\theta(x, t) = e^{-\varphi(x,t)/2\nu}$ satisfies the linear diffusion equation

$$\frac{\partial\theta}{\partial t} = \nu\frac{\partial^2\theta}{\partial x^2} \tag{8.4.4}$$

Elimination of the intermediate function φ shows that the solution of Burgers' equation $u(x, t)$ is related to that of the diffusion equation $\theta(x, t)$ through the expressions

$$u(x, t) = -2\nu\frac{\theta_x}{\theta}$$

$$\theta(x, t) = \exp\left(-\frac{1}{2\nu}\int_{x_0}^{x} dx'\, u(x', t)\right) \tag{8.4.5}$$

where the lower limit x_0, effectively a multiplicative integration constant, is irrelevant for our purposes since, according to the first of (8.4.5), θ is used only in a ratio to its first derivative. The lower limit may thus be chosen for convenience.

We may now construct a simple (in principle) prescription for solving Burgers' equation:

1. Choose some initial profile $u(x, 0)$.
2. Determine $\theta(x, 0) = \exp[-\int u(x, 0)\, dx/2\nu]$.
3. Obtain $\theta(x, t)$ by solving the linear diffusion equation (8.4.4).
4. Obtain $u(x, t)$ from the first of (8.4.5).

Unfortunately, the functions $\theta(x, 0)$ that result from the customary expressions for an initial profile $u(x, 0)$ are so cumbersome that the solution for $\theta(x, t)$ and subsequently $u(x, t)$ cannot usually be written down in terms of simple analytical expressions.

However, we expect that the smallness of ν should provide a simplification such that the solution reduces to that of the quasi-linear equation unless u_{xx} is large. To obtain this limiting form, we first write the solution of the diffusion equation (8.4.4) in terms of the Green's function (4.7.6) as

$$\theta(x, t) = \int_{-\infty}^{\infty} dx' \, \theta(x, 0) g(x, t \,|\, x', 0)$$

$$= \frac{1}{2\sqrt{\pi \nu t}} \int_{-\infty}^{\infty} dx' \, \theta(x', 0) e^{-(x'-x)^2/4\nu t} \qquad (8.4.6)$$

The Green's function is immediately available by replacing γ^2 by ν in (4.7.6). For use in (8.4.5) we also determine

$$\theta_x(x, t) = -\int_{-\infty}^{\infty} dx' \, \theta(x', 0) g_{x'}(x, t \,|\, x', 0)$$

$$= -\frac{1}{2\nu} \int_{-\infty}^{\infty} dx' \, u(x', 0)\theta(x', 0)g(x, t \,|\, x', 0) \qquad (8.4.7)$$

where we have used $g_x = -g_{x'}$ and performed an integration by parts. Relating $\theta(x', 0)$ to $r(x', 0)$ by the second of (8.4.5) and writing

$$f(x', x, t) = \frac{1}{2} \int_{x_0}^{x'} dx'' \, u(x'', 0) + \frac{(x'-x)^2}{4t} \qquad (8.4.8)$$

we obtain the solution of Burgers' equation in the form

$$u(x, t) = \frac{\displaystyle\int_{-\infty}^{\infty} dx' \, u(x', 0) e^{-f(x',x,t)/\nu}}{\displaystyle\int_{-\infty}^{\infty} dx' \, e^{-f(x',x,t)/\nu}} \qquad (8.4.9)$$

Since ν plays the role of a small parameter, we may use the approximation described in Problem 1.5.8, namely,

$$\lim_{\nu \to 0} e^{-f/\nu} = \frac{1}{\sqrt{2\pi |f''(x_0)|}} e^{-f(x_0)/\nu} \delta(x - x_0) \qquad (8.4.10)$$

where x_0 is the smallest root of $f'(x) = 0$. From the definition of $f(x)$ given in (8.4.8), we see that x_0 is a solution of $u(x', 0) + (x' - x)/t = 0$, an equation that may be rewritten in the form

$$u(x, t) = u(x_0, 0)$$

$$u(x_0, 0) = (x - x_0)/t \qquad (8.4.11)$$

This result is exactly the same as that obtained for the quasi-linear equation in (8.3.42). In the present instance, however, there is no ambiguity in the choice of the root since the dominant contribution to the integral is provided by the root corresponding to the smallest value of $f(x_0)$.

A more convenient but somewhat restricted approach to the construction of solutions is to choose the function $\theta(x, t)$ or $\theta(x, 0)$ and then evaluate the integral in the first of (8.4.6) to obtain $\theta(x, t)$. The solution $u(x, t)$ is then available from (8.4.5). Since $\theta(x, t)$ appears in the denominator in (8.4.5), solutions $\theta(x, t)$ that vanish should be avoided. This is readily accomplished, however, since $\theta(x, t)$ satisfies a linear equation and hence various solutions may be added to yield additional solutions. Any known solutions $\theta(x, t)$ may be used. Extensive compilations of such solutions are available (e.g., Benton and Platzmann, 1972).

As an example of this procedure, we consider the solution of the diffusion equation that is obtained by using the separation-of-variables technique. Setting $\theta(x, t) = X(x)T(t)$ in the diffusion equation (8.4.4), we obtain $\theta(x, t) = e^{-\alpha(x - \alpha\nu t)}$ where α is the separation constant. (Since θ ultimately appears in a ratio, the divergence of this solution as x, t approach infinity is not the crucial issue that it has been in our previous considerations.) On substituting this solution into (8.4.5), we merely obtain the obvious solution $u(x, t) = $ const for Burgers' equation. To obtain a more interesting result, we need only note that $\theta = $ const is also an immediately available solution of the diffusion equation. Choosing the constant to be unity and adding this solution to the one already obtained by separation of variables, we now have $\theta(x, t) = 1 + Ae^{-\alpha(x - \alpha\nu t)}$ where A is an arbitrary amplitude. Equation (8.4.5) now yields

$$u(x, t) = \frac{2\alpha\nu}{1 + A^{-1}e^{\alpha(x - \alpha\nu t)}} \qquad (8.4.12)$$

To understand the significance of the constants, we set $u(-\infty, t) = 2\alpha\nu = u_0$, $A^{-1} = \exp(-\alpha x_0)$ and write the solution as

$$u(x, t) = \frac{u_0}{1 + e^{(u_0/2\nu)[(x - x_0) - u_0 t/2]}} \qquad (8.4.13)$$

The solution is now seen to have the form of a shelf that propagates in the positive x direction with the velocity $u_0/2$. The shelf drops from the value u_0 at $x = -\infty$ to 0 at $x = +\infty$. The transition takes place in a region of width ν/u_0 centered about $x = x_0$. The relevance of such a solution to the propagation of shock waves is evident. Until a degree of intuition has been developed, however, an approach based upon making a choice for θ clearly leaves one in the position of not knowing what problem has been solved until after the calculation has been completed. We now consider an example in which a choice for $u(x, 0)$ is made.

Example 8.6. Obtain the solution of Burgers' equation for the initial profile $u(x, 0) = A\delta(x)$ by following the four-step procedure outlined below (8.4.5).

Using the second of (8.4.5) at $t = 0$ and setting $R = A/2\nu$, we immediately obtain

$$\theta(x, 0) = \begin{cases} e^R, & x > 0 \\ 1, & x < 0 \end{cases} \tag{8.4.14}$$

From (8.4.6) we then have

$$\theta(x, t) = \int_{-\infty}^{0} dx' \, g + e^{-R} \int_{0}^{\infty} dx' \, g \tag{8.4.15}$$

The integrals are readily expressed in terms of the complimentary error function erfc(x) $= 1 - \text{erf}(x)$ where erf(x) is defined in (2.2.5). Setting $z = x/2\sqrt{\nu t}$ we obtain

$$\theta(x, t) = \tfrac{1}{2}\text{erfc}(z) + e^{-R}[1 - \tfrac{1}{2}\text{erfc}(z)]$$

$$= e^R + \tfrac{1}{2}(1 - e^{-R})\text{erfc}(z) \tag{8.4.16}$$

Determining θ_x by using

$$\frac{d}{dx}\text{erfc}(z_j) = \frac{-1}{\sqrt{\pi \nu t}} e^{-z_j^2} \tag{8.4.17}$$

we find from (8.4.5) that

$$u(x, t) = \sqrt{\frac{\nu}{\pi t}} \frac{(e^R - 1)e^{-z^2}}{1 + \tfrac{1}{2}(e^R - 1)\text{erfc}(z)} \tag{8.4.18}$$

The result obtained in this example provides an opportunity to examine the transition region in the solution $u(x, t)$ that is introduced by the term νu_{xx} in Burgers' equation (8.3.38). To interpret the result, it is convenient to introduce dimensionless variables. The constant A appearing in the amplitude of the delta function in $u(x, 0)$ has dimensions of velocity times length. Introducing a characteristic velocity u_0 and length L, we can set $x = L\xi$ and $t = (L/u_0)\tau$. We also have $R = u_0 L/2\nu$, a ratio reminiscent of a Reynolds number in fluid mechanics. An appropriate magnitude for ν can be expected to be sufficiently small that the term νu_{xx} in Burgers' equation (8.3.38) is of importance only when $u_{\xi\xi}$ is large. More specifically, in dimensionless variables Burgers' equation takes the form $f_\tau + ff_\xi = R^{-1}f_{\xi\xi}$ where $f(\xi, \tau) = u(\xi, \tau)/u_0$. We can thus expect that R will be much greater than unity and can therefore replace $e^R - 1$ by e^R in (8.4.18). In terms of the dimensionless variables ξ and τ, we now find that $u(x, t)$ takes the form

$$\frac{u(\xi, \tau)}{u_0} = \sqrt{\frac{2}{\pi R\tau}} \frac{e^{R(1 - \xi^2/2\tau)}}{2 + e^R\text{erfc}(\sqrt{\xi^2 R/2\tau})} \tag{8.4.19}$$

For $\xi/2\sqrt{\tau} > R^{-1/2}$ the complimentary error function in this result can be approximated by (cf. Problem 2.2.9)

$$\text{erfc}(x) \approx e^{-x^2}/\sqrt{\pi}x, \qquad x \gg 1 \tag{8.4.20}$$

and (8.4.19) is then found to reduce to

$$\frac{u(\xi, \tau)}{u_0} = \frac{\xi}{\tau} \tag{8.4.21}$$

Returning to dimensionless variables, we obtain $u(x, t) = x/t$. Note that this expression is a solution of the quasi-linear equation (8.3.37).

For $\xi/2\sqrt{\tau} \approx 1$, however, the argument of the exponential in the numerator of (8.4.19) becomes zero and for larger values of ξ the exponential decreases very rapidly so that $u(x, t)$ approaches zero. The width of the transition region is determined by setting $\xi/2\sqrt{\tau} = 1 + \Delta$ and then setting the exponent in (8.4.19) equal to -1. The numerator then has a value equal to e^{-1} of its maximum value. We find $\Delta = 1/(2R)$, which is small since R is large.

As noted above, the underlying linearity associated with Burgers' equation enables us to obtain a closed form solution in terms of integrals over the initial profile $u(x, 0)$. As also noted, however, the number of specific initial profiles for which the integrations may be performed analytically is quite limited. One such soluble profile, the delta function $u(x, 0) = A\delta(x)$, has been considered in Example 8.6. The attendant computational procedures are readily extended to treat the case of a sum of delta functions, that is, $u(x, 0) = \Sigma A_i\delta(x - x_i)$, although the resulting expressions do not display the solution in any readily interpreted form. However, since the arbitrary initial profile $u(x, 0)$ enters the general solution only under an integral sign, any such profile may be approximated by a series of delta functions, that is,

$$u(x, 0) = \frac{L}{n+1} \sum_{j=1}^{n+1} u(x, 0)\delta(x - x_j) \tag{8.4.22}$$

where L is the length of the initial profile and n is the number of subsections considered (cf. Fig. 8.6). A numerical evaluation of the resulting solution then provides a clear graphical description of the interplay between shock formation and dissipation for an arbitrary initial profile. We now summarize this solution: Setting $u(x, 0) = u_0 f(x)$ and assuming that the jth region extends from $x_j^- \equiv x_j - \epsilon$ to $x_{j+1}^- \equiv x_{j+1} - \epsilon$ so that the contribution from $\delta(x - x_j)$ is in the jth region, we have

$$\theta(x, 0) = \exp\left[-\frac{R}{n+1} \sum_{j=1}^{n+1} f(x_j)H(x - x_j)\right] \tag{8.4.23}$$

where $R = u_0 L/2\nu$ and H refers to the unit step function. The value of $\theta(x, 0)$ in the jth region thus has the constant value $\theta_j = \exp(-S_j)$ where $S_1 =$

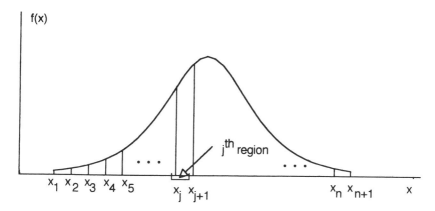

Figure 8.6. Labeling used for subsections of initial pulse profile $u(x, 0) = u_0 f(x)$ in Eq. (8.4.22).

$R f(x_1)/(n + 1)$, $S_2 = R[f(x_1) + f(x_2)]/(n + 1)$, etc. At a later time the value of $\theta(x, t)$ is readily determined by evaluating (8.4.6). One obtains

$$\theta(x, t) = \theta_0(x, t) + \sum_{j=1}^{n} \theta_j(x, t) + \theta_{n+1}(x, t) \qquad (8.4.24)$$

with

$$\theta_0(x, t) = \int_{-\infty}^{x_1^-} dx' \, g, \qquad \theta_{n+1}(x, t) = e^{-S_{n+1}} \int_{x_{n+1}^-}^{\infty} dx' \, g$$

$$\theta_j(x, t) = e^{-S_j} \int_{x_j^-}^{x_{j+1}^-} dx' \, g, \qquad j = 1, 2, 3, \ldots, n \qquad (8.4.25)$$

The integrals over g are immediately expressible in terms of the error function. Introducing the notation

$$z_j = \frac{x - x_j}{2\sqrt{\nu t}} \qquad (8.4.26)$$

we have the first and last terms as

$$\theta_0(x, t) = \tfrac{1}{2} \operatorname{erfc}(z_1)$$

$$\theta_{n+1}(x, t) = e^{-S_{n+1}}[1 - \tfrac{1}{2} \operatorname{erfc}(z_{n+1})] \qquad (8.4.27)$$

while for the intermediate terms we obtain

$$\theta_j(x, t) = \tfrac{1}{2} e^{-S_j}[\operatorname{erfc}(z_{j+1}) - \operatorname{erfc}(z_j)] \qquad (8.4.28)$$

Using the differentiation formula given in (8.4.17) we find

$$\theta_x(x,\, t) = \frac{-1}{2\sqrt{\pi \nu t}} \left[e^{-z_1^2} - e^{-S_{n+1}-z_{n+1}^2} + \sum_{j=1}^{n} e^{-S_j}(e^{-z_{j+1}^2} - e^{-z_j^2}) \right]$$

(8.4.29)

When these expressions for $\theta(x,\, t)$ and $\theta_x(x,\, t)$ are substituted into the solution (8.4.5) and the dimensionless variables $\xi = x/L$ and $\tau = u_0 t/L$ are introduced, we obtain the final result

$$\frac{u(\xi,\, \tau)}{u_0} = \sqrt{\frac{2}{\pi R \tau}} \frac{N}{D}$$

(8.4.30)

where

$$N = e^{-z_1^2} - e^{-S_{n+1}-z_{n+1}^2} + \sum_{j=1}^{n} e^{-S_j}(e^{-z_{j+1}^2} - e^{z_j^2})$$

(8.4.31)

and

$$D = \text{erfc}(z_1) + e^{-S_{n+1}}[2 - \text{erfc}(z_{n+1})]$$

$$+ \sum_{j=1}^{n} e^{-S_j}[\text{erfc}(z_{j+1}) - \text{erfc}(z_j)]$$

(8.4.32)

with $z_j = (\xi - \xi_j)\sqrt{R/2\tau}$ in dimensionless variables. Note that for $n = 0$ the summation is absent and we recover the result given in (8.4.19) for a single delta function. An example using $u(x,\, 0) = u_0 \text{sech}^2 \pi x/L$, with $n = 100$ and $R = 50$, is shown in Figure 8.7.

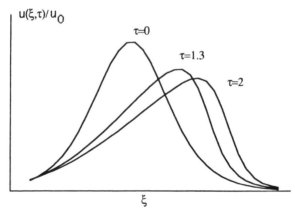

Figure 8.7. Competition between nonlinearity and diffusion in Burgers' equation for initial profile $u(x,\, 0) = u_0 \text{sech}^2 \pi x/L$ as predicted by (8.4.30) for 100 subintervals and $u_0 L/2\nu = 50$.

Problems

8.4.1 Consider the initial profile $u(x, 0) = -(4\nu/L)\tanh x/L$ and show that $\theta(x, 0) = \cosh^2 x/L \, \text{sech}^2 x_0/L$. Show that at later time

$$\theta(x, t) = \frac{1}{2} \text{sech}^2 \frac{x_0}{L} \left(1 + \cosh^2 \frac{x}{L} e^{4\nu t/L^2} \right)$$

and that

$$u(x, t) = \frac{4\nu}{L} \frac{\sinh 2x/L}{\cosh 2x/L + e^{-4\nu t/L^2}}$$

$$\left(\text{Note: } \int_0^\infty dx \, e^{-bx^2} \cosh ax = \frac{1}{2} \sqrt{\frac{\pi}{b}} \, e^{a^2/4b}. \right)$$

8.4.2 Show that the initial value $\theta(x, 0) = A\delta(x)$ leads to $u(x, t) = x/t$.

8.4.3 Show that the initial value $\theta(x, 0) = 1 + A\delta(x)$ leads to

$$u(x, t) = \frac{x}{t} \frac{\theta(x, t) - 1}{\theta(x, 1)}$$

where $\theta(x, t) = 1 + Ae^{-x^2/4\nu t}/2\sqrt{\pi \nu t}$.

8.4.4 Show that the initial velocity profile $u(x, 0) = -Ax/(x^2 + a^2)$ yields

$$u = -2\nu \frac{x}{\nu t + \frac{1}{2}(x^2 + a^2)}$$

8.4.5 Show that the solution obtained in Problem 8.4.4 may be recovered by setting $\theta(x, t) = X(x) + T(t)$ in the diffusion equation and appropriately specifying the constants.

9 Selected Topics

In this final chapter we consider a number of examples that either draw upon various topics from our previous developments or require a somewhat more lengthy treatment of a subject than seemed appropriate when the topic was first being introduced.

9.1 OSCILLATING HEAT SOURCE ON A BEAM

All but the simplest problems dealing with diffusion due to time dependent sources lead to analyses of considerable complexity. As an attempted compromise between problems displaying some physical interest on one hand and tractability on the other, we consider the heating of a beam of length L by a localized source that moves back and forth along the entire beam at a constant frequency Ω. If the frequency of oscillation is increased, one would expect intuitively that the pulsations of heat input received by each point of the beam would eventually blur into some equivalent steady state source. Since the source spends more time near the ends of the beam where it slows down to reverse direction, the equivalent steady state source term is not uniform along the beam. It will be shown that the proper spatial dependence of the steady state source term is readily obtained as a limit to the general solution of this problem.

An adequate theoretical model for the situation envisioned is provided by the one-dimensional diffusion equation (1.7.1) with the source term

$$s(x, t) = q_0 \delta[x - x_0(t)] \qquad (9.1.1)$$

where the source location $x_0(t)$ is given as

$$x_0(t) = \frac{L}{2}(1 + \sin \Omega t) \qquad (9.1.2)$$

and the beam is located in the region $0 \le x \le L$. Following the development of Section 1.7, we express the temperature on the beam $u(x, t)$ as the Fourier series

$$u(x, t) = \sum_{n=1}^{\infty} b_n(t) \sin \frac{n\pi x}{L} \qquad (9.1.3)$$

with the coefficients $b_n(t)$ determined according to (1.7.8) by the coefficients in the corresponding expansion for the source, that is,

$$s(x, t) = \sum_{n=1}^{\infty} s_n(t) \sin \frac{n\pi x}{L} \qquad (9.1.4)$$

Our first task, then, is the determination of the Fourier series representation of the source term. The coefficients are readily seen to be

$$
\begin{aligned}
s_n(t) &= \frac{2}{L} \int_0^L dx \, s(x, t) \sin \frac{n\pi x}{L} \\
&= \frac{2q_0}{L} \int_0^L dx \, \delta[x - x_0(t)] \sin \frac{n\pi x}{L} \\
&= \frac{2q_0}{L} \sin \frac{n\pi x_0(t)}{L} \qquad (9.1.5)
\end{aligned}
$$

From (1.7.8), the coefficients $b_n(t)$ are thus

$$b_n(t) = b_n(0) e^{-n^2 \nu t} + \frac{2q_0}{\gamma^2 KL} \int_0^t dt' \, e^{-n^2\nu(t-t')} \sin\left[\frac{n\pi}{2}(1 + \sin \Omega t')\right] \qquad (9.1.6)$$

Integration of expressions containing terms of the form $\sin[(n\pi/2)\sin \Omega t]$ can frequently be carried out by using expansions in terms of Bessel functions obtainable from the series (D.47). To carry out the evaluation in the present case, it is convenient to introduce the complex quantity

$$F_n(t) = \int_0^t dt' \, e^{-n^2\nu(t-t') + (in\pi/2)(1 + \sin\Omega t')} \qquad (9.1.7)$$

Note that it is the imaginary part of this expression that occurs in (9.1.6).
The expansion coefficients $b_n(t)$ then become

$$b_n(t) = b_n(0) e^{-n^2 \nu t} + \frac{2q_0}{\gamma^2 KL} \, \mathrm{Im}[F_n(t)] \qquad (9.1.8)$$

where Im refers to the imaginary part of the expression in brackets.
Since we will ultimately consider the integral (9.1.7) in the limit $t \to \infty$, it is convenient for convergence purposes to first set $\tau = t - t'$ and write

$$F_n(t) = e^{n\pi i/2} \int_0^t d\tau \, e^{-n^2\nu\tau + i(n\pi/2)\sin\Omega(t - \tau)} \qquad (9.1.9)$$

Using the expansion (D.47) referred to above, we have

$$e^{i(n\pi/2)\sin\Omega(t-\tau)} = \sum_{m=-\infty}^{\infty} J_m\left(\frac{n\pi}{2}\right) e^{im\Omega(t-\tau)} \tag{9.1.10}$$

and we thus obtain $F_n(t)$ in the form

$$F_n(t) = e^{i(n\pi/2)} \sum_{m=-\infty}^{\infty} e^{im\Omega t} J_m\left(\frac{n\pi}{2}\right) \int_0^t d\tau \, e^{-(n^2\nu + im\Omega)\tau} \tag{9.1.11}$$

Although this integral over τ is readily evaluated, the result is somewhat cumbersome. A simpler expression is obtained if we confine attention to the long-time behavior of the temperature dependence. Letting $t \to \infty$ in the upper limit we obtain

$$F_n(t) = e^{i(n\pi/2)} \sum_{m=-\infty}^{\infty} e^{im\Omega t} \frac{J_m(n\pi/2)}{n^2\nu + im\Omega} \tag{9.1.12}$$

Even this result appears rather formidable. However, as Ω increases ($\Omega/\nu \gg 1$) we see that all terms except the term $m = 0$ have a large term containing Ω in the denominator and can thus be neglected in this high-frequency limit. Retaining only the $m = 0$ term, then, we find that at high frequencies ($\Omega/\nu \gg 1$) and after a long time ($\nu t \gg 1$), we have $F_n(t) \approx e^{n\pi i/2} J_0(n\pi/2)/\nu n^2$ and, for use in obtaining $b_n(t)$ from (9.1.8),

$$\text{Im}[F_n(t)] = \frac{\sin(n\pi/2)J_0(n\pi/2)}{\nu n^2} \tag{9.1.13}$$

In the limit $\nu t \gg 1$, the term in (9.1.8) containing the initial condition $b_n(0)$ will have become negligible and the final form of $b_n(t)$ is the time independent expression

$$b_n(t) = \frac{2q_0}{\nu KL}\left(\frac{L}{\pi}\right)^2 \frac{\sin(n\pi/2)J_0(n\pi/2)}{n^2} \tag{9.1.14}$$

The lack of time dependence in $b_n(t)$ is the quantitative manifestation of the steady state situation alluded to in the introduction.

Finally, from (9.1.3), the temperature distribution in the limit $\Omega/\nu \gg 1$, $\nu t \gg 1$ is

$$u(x, t) \to \frac{2q_0 L}{\pi^2 K} \sum_{n=0}^{\infty} \frac{1}{n^2} J_0\left(\frac{n\pi}{2}\right) \sin\frac{n\pi}{2} \sin\frac{n\pi x}{L} \tag{9.1.15}$$

This series was encountered previously in Problem 1.5.9 where it was shown that the steady state temperature distribution due to the source

$$s(x) = \frac{Q_0/\pi}{\sqrt{x(L - x)}}, \qquad 0 \le x \le L \qquad (9.1.16)$$

has precisely the form given here in (9.1.15) if we replace q_0 by Q_0.

It is instructive to examine the physical basis for this equivalence. In a time Δt the oscillating source term in the present problem deposits $q_0 \, \Delta t$ calories in a region of length Δx about a point x. The amount of heat deposited depends upon how rapidly the source is moving past the point x. For locations near $x = 0, L$, where the source is slowing down to reverse direction, more heat is deposited than at the midpoint $x = L/2$, where the source is moving most rapidly. The qualitative correctness of the source strength given in (9.1.16) is thus reasonable. To be more quantitative, note that the time Δt referred to above is related to Δx by $\Delta t = \Delta x/v$ where v is the velocity of the source and is seen from (9.1.2) to be $v = (\Omega L/2)\cos \Omega t$. According to (9.1.2), $\sin \Omega t = (2x/L) - 1$ and thus we have $v = \Omega \sqrt{x(L - x)}$. Hence the amount of heat deposited in a region Δx about x on each complete cycle of the source is

$$H = 2q_0 \, \Delta t = \frac{2q_0 \, \Delta x}{\Omega \sqrt{x(L - x)}} \qquad (9.1.17)$$

where the factor of 2 is required since a point on the beam is passed twice during each complete cycle. After a long time, say N cycles, the amount of heat deposited is

$$NH = \frac{2Nq_0 \, \Delta x}{\Omega \sqrt{x(L - x)}} \qquad (9.1.18)$$

On the other hand, in the steady state problem (1.5.8), the amount of heat (in calories per second) deposited in Δx is

$$s(x) \, \Delta x = \frac{Q_0 \, \Delta x}{\sqrt{x(L - x)}} \qquad (9.1.19)$$

The total amount of heat in, say, calories deposited in the interval Δx about x over the period of time required for the N cycles referred to above is

$$NPs(x) \, \Delta x = \frac{NPQ_0 \, \Delta x/\pi}{\sqrt{x(L - x)}} = \frac{2NQ_0 \, \Delta x}{\Omega \sqrt{x(L - x)}} \qquad (9.1.20)$$

where $P = 2\pi/\Omega$ is the period of the oscillating source.

The amounts of heat in (9.1.18) and (9.1.20) are seen to be equal if we set $q_0 = Q_0$, as was inferred from the comparison of the series expansion in (9.1.15) and Problem 1.5.8.

9.2 TEMPERATURE DISTRIBUTION IN A PIE-SHAPED REGION

In Section 3.7.1 the temperature distribution in a pie-shaped region was determined when an inhomogeneous boundary condition was imposed on the curved rim. When the inhomogeneity is imposed on the straight sides that form the vertex of the region, the analysis is somewhat more complicated. In this section we first develop a method for finding the temperature in an annular pie-shaped region (cf. Fig. 9.1a) and then consider the limiting case in which the inner radius goes to zero. Using polar coordinates, we construct elementary solutions of Laplace's equation for the region $ABCD$ shown in Figure 9.1a in the product form $u = (r, \theta) = R(r)\Theta(\theta)$. The functions R and Θ satisfy

$$r^2 R'' + rR' + v^2 R = 0$$
$$\Theta'' - v^2 \Theta = 0 \tag{9.2.1}$$

where v is the separation constant. The choice of sign for v is determined by the requirement that we be able to impose homogeneous boundary conditions on an oscillatory solution in the radial direction. The radial equation is of Cauchy-Euler form and has solutions $R = r^{\pm iv} = e^{\pm iv \ln r}$. To satisfy the homogeneous boundary conditions at $r = a, b$, we use the combination

$$R(r) = A \cos(v \ln r) + B \sin(v \ln r) \tag{9.2.2}$$

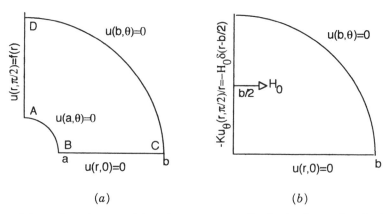

(a)   (b)

Figure 9.1. (a) Boundary conditions for temperature distribution in annular pie-shaped region. (b) Example for the limiting case $a \to 0$.

Imposition of the boundary conditions $R(a) = R(b) = 0$ yields a pair of homogeneous equations in the unknown amplitudes A and B. Requiring that the determinant of this system vanish, we obtain

$$\sin[\nu \ln(b/a)] = 0 \qquad (9.2.3)$$

which yields $\nu = \nu_n = n\pi/\ln(b/a)$ where $n = 1, 2, 3, \ldots$. When the equation resulting from the requirement $R(b) = 0$ is used to eliminate A from (9.2.2), we obtain the radial solution as

$$R_n(r) = C_n \sin(\nu_n \ln r/b) \qquad (9.2.4)$$

where C_n has been written in place of $-B/\cos(\nu \ln b)$.

The appropriate solution of the equation for Θ is $\Theta = \sinh \nu \vartheta$ since it satisfies the homogeneous boundary condition at $\vartheta = 0$. No nonzero solution is obtained by considering the case $n = 0$. Adding these elementary solutions, we have

$$u = \sum_{n=1}^{\infty} C_n \sin\left(\nu_n \ln \frac{r}{b}\right) \sinh \nu_n \theta \qquad (9.2.5)$$

If the temperature on side AD is given as $u(r, \pi/2) = f(r)$, we obtain

$$f(r) = \sum_{n=1}^{\infty} C_n \sin\left(\nu_n \ln \frac{r}{b}\right) \sinh \frac{\nu_n \pi}{2} \qquad (9.2.6)$$

To determine the amplitudes in this expansion, we use the orthogonality of the functions $\sin(\nu_n \ln r/b)$, that is,

$$\int_a^b \frac{dr}{r} \sin\left(\nu_n \ln \frac{r}{b}\right) \sin\left(\nu_m \ln \frac{r}{b}\right) = \frac{1}{2} \ln \frac{b}{a} \delta_{nm} \qquad (9.2.7)$$

which is obtainable either from Eq. (C.20) or by setting $r = be^{-t}$ and transforming to the more familiar Fourier series form. Proceeding in the customary fashion, we obtain

$$C_n \sinh \frac{\nu_n \pi}{2} = \frac{2}{\ln b/a} \int_a^b \frac{dr'}{r'} f(r') \sinh\left(\nu_n \ln \frac{r'}{b}\right) \qquad (9.2.8)$$

With the constants C_n determined, the solution for the annular region is available. We now proceed to take the limit $a \to 0$. The procedure parallels that in Section A.5 where a similar limiting procedure is used to obtain expressions for the Fourier integral from those for the Fourier series. Proceeding as in

Section A.5, we first note that

$$\Delta \nu = \nu_{n+1} - \nu_n = \frac{\pi}{\ln b/a} \tag{9.2.9}$$

Then, replacing C_n in the solution (9.2.4) by the expression in (9.2.8), introducing the identity $\ln(b/a)\,\Delta\nu/\pi = 1$, and replacing the sum by an integral as we let $a \to 0$, we obtain

$$u(r, \theta) = \frac{2}{\pi} \int_0^\infty \frac{d\nu}{\sinh \nu\pi/2} \int_0^b \frac{dr'}{r'} f(r') \sin\left(\nu \ln \frac{r'}{b}\right) \sin \frac{\nu_n r}{b} \sinh \nu\theta \tag{9.2.10}$$

Introducing

$$F(\nu) = \int_0^b \frac{dr}{r} f(r) \sin\left(\nu \ln \frac{r'}{b}\right) \tag{9.2.11}$$

we have the solution in the form

$$u(r, \theta) = \frac{2}{\pi} \int_0^\infty d\nu\, F(\nu) \frac{\sin(\nu \ln r/b)\sinh \nu\theta}{\sinh \nu\pi/2} \tag{9.2.12}$$

When we now set $\theta = \pi/2$ and again write $u(r, \pi/2) = f(r)$, we see that we have an integral transform pair in the form

$$f(r) = \frac{2}{\pi} \int_0^\infty d\nu\, F(\nu) \sin\left(\nu \ln \frac{r}{b}\right)$$

$$F(\nu) = \int_0^b \frac{dr}{r} f(r) \sin\left(\nu \ln \frac{r}{b}\right) \tag{9.2.13}$$

As an example, consider the pie-shaped region $0 \le r \le b$, $0 \le \theta \le \pi/2$ that is maintained at zero temperature on $r = b$ and $\theta = 0$ but is insulated on the side $\theta = \pi/2$ except for a localized heat flow at $r = b/2$. The boundary condition on this side is written

$$J_\theta(r, \theta) = -K \frac{1}{r} \frac{\partial u}{\partial \theta}\bigg|_{\theta = \pi/2} = -H_0\delta\left(r - \frac{b}{2}\right) \tag{9.2.14}$$

The negative sign before H_0 signifies a heat flow into the region and thus a flow in the negative θ direction as indicated in Figure 9.1b.

From (9.2.12) we obtain

$$\frac{\partial u(r, \theta)}{\partial \theta}\bigg|_{\theta = \pi/2} = \frac{2}{\pi} \int_0^\infty d\nu \, F(\nu)\nu \coth \frac{\nu \pi}{2} \sin\left(\nu \ln \frac{r}{b}\right) \quad (9.2.15)$$

From (9.2.14) we have $u_r(r, \pi/2) = (H_0/K)r\delta(r - b/2)$, and by using the inverse relations from (9.2.13), we may write

$$\nu F(\nu) \coth \frac{\nu \pi}{2} = \int_0^b \frac{dr}{r} \frac{H_0}{K} r\delta\left(r - \frac{b}{2}\right) \sin\left(\nu \ln \frac{r}{b}\right)$$

$$= -\frac{H_0}{K} \sin(\nu \ln 2) \quad (9.2.16)$$

From the solution (9.2.12) we now have the following integral representation for the temperature throughout the disk:

$$u(r, \theta) = -\frac{2}{\pi} \frac{H_0}{K} \int_0^\infty \frac{d\nu}{\nu} \frac{\sinh \nu\theta}{\cosh \nu\pi/2} \sin\left(\nu \ln \frac{r}{b}\right) \sin(\nu \ln 2) \quad (9.2.17)$$

By writing the product of two sines as the sum of two cosines, this result may be recast in the form of a tabulated Fourier cosine transform (Table F.4, entry 11). However, by way of obtaining some information that has more useful physical significance, we now determine the amount of heat flowing out through the circular rim at $r = b$ and along the straight edge $\theta = 0$. The sum of the two amounts should, of course, equal the input H_0.

On the circular rim we must evaluate

$$H_{r=b} = \int_0^{\pi/2} b \, d\theta \, J_r = \int_0^{\pi/2} b \, d\theta \left(-K \frac{\partial u}{\partial r}\bigg|_{r=b}\right) \quad (9.2.18)$$

while on the face $\theta = 0$ we require

$$H_{\theta=0} = \int_0^b dr \, J_\theta = \int_0^b dr \left(-\frac{K}{r} \frac{\partial u}{\partial \theta}\bigg|_{\theta=0}\right) \quad (9.2.19)$$

Differentiation of the solution (9.2.17) immediately yields

$$H|_{r=b} = \frac{2H_0}{\pi} \int_0^{\pi/2} d\theta \int_0^\infty d\nu \, \text{sech} \frac{\nu \pi}{2} \sin(\nu \ln 2) \sinh(\nu\theta)$$

$$H|_{\theta=0} = \frac{2H_0}{\pi} \int_0^b \frac{dr}{r} \int_0^\infty d\nu \, \text{sech} \frac{\nu \pi}{2} \sin(\nu \ln 2) \sin\left(\nu \ln \frac{r}{b}\right) \quad (9.2.20)$$

In the integral for $H|_{r=b}$ we can carry out the θ integration to obtain

$$H|_{r=b} = \frac{2H_0}{\pi} \int_0^\infty \frac{d\nu}{\nu} \sin(\nu \ln 2) \left(1 - \operatorname{sech} \frac{\nu\pi}{2} \right)$$

$$= \frac{2H_0}{\pi} \left[\frac{\pi}{2} - \left(2 \tan^{-1} 2 - \frac{\pi}{2} \right) \right]$$

$$= 0.5903 H_0 \tag{9.2.21}$$

In obtaining this result we have used $\int_0^\infty dx \, (\sin ax)/x = \pi/2$ and entry 11 in Table F.3.

When a similar calculation is attempted for $H_{\theta=0}$, we see that the radial integral, when transformed with $r = be^{-t}$, is actually a divergent integral, since we obtain

$$\int_0^b \frac{dr}{r} \sin\left(\nu \ln \frac{r}{b} \right) = - \int_0^\infty dt \, \sin \nu t \tag{9.2.22}$$

We have merely encountered a situation in which the order of integration cannot be interchanged without some special considerations. However, if we carry out the ν integration first, by using entry 9 in Table F.4, and then carry out the radial integration, we obtain the expected result, that is, an amount of heat flow that, when combined with the radial flow, yields the input amount H_0.

We could also recover this result by taking the radial integral only from a lower limit of ϵ out to b and then letting ϵ go to zero at the end of the calculation. Labeling this result as $H(\epsilon)|_{\theta=0}$, we find

$$H(\epsilon)|_{\theta=0} = \frac{2H_0}{K} \int_0^\infty d\nu \, \operatorname{sech} \frac{\nu\pi}{2} \sin(\nu \ln 2) \int_\epsilon^b \frac{dr}{r} \sin\left(\nu \ln \frac{r}{b} \right) \tag{9.2.23}$$

The integral on t in (9.2.22) is now only taken up to $\ln(b/\epsilon)$ and the second of (9.2.20) yields

$$H(\epsilon)|_{\theta=0} = \frac{2H_0}{\pi} \int_0^\infty \frac{d\nu}{\nu} \operatorname{sech} \frac{\nu\pi}{2} \sin(\nu \ln 2) \left[\cos\left(\nu \ln \frac{b}{\epsilon} \right) - 1 \right]$$

$$= \frac{1}{2} \int_0^\infty \frac{d\nu}{\nu} \operatorname{sech} \frac{\nu\pi}{2} \left[\sin\left(\nu \ln \frac{2b}{\epsilon} \right) + \sin\left(\nu \ln \frac{2\epsilon}{b} \right) \right]$$

$$- \left(2 \tan^{-1} 2 + \frac{\pi}{2} \right) \tag{9.2.24}$$

where we have used an addition theorem for the product of trigonometric functions and also employed entry 11 of Table F.3 as in obtaining (9.2.21).

This same entry is used for the two remaining integrals and we obtain

$$H(\epsilon)|_{\theta=0} = \frac{2H_0}{\pi} \left[\tan^{-1} \left(\frac{2b}{\epsilon} \right) + \tan^{-1} \left(\frac{2\epsilon}{b} \right) - 2 \tan^{-1} 2 + \frac{\pi}{2} \right] \quad (9.2.25)$$

Taking the limit $\epsilon \to 0$, we find

$$H(\epsilon)|_{\theta=0} = 2H_0 \left(1 - \frac{4}{\pi} \tan^{-1} 2 \right)$$

$$= -0.4096H_0 \quad (9.2.26)$$

The negative sign merely indicates heat flow in the negative θ direction. The sum of the magnitudes of the two results in (9.2.21) and (9.2.26) equals the heat input H_0.

9.3 BABINET'S PRINCIPLE

In Chapter 4 we considered a number of examples in which the Green's function method provided an expression for the field within a region in terms of an integral over a *known* function multiplied by the Green's function (or its normal derivative) evaluated on the boundary. The solution of such problems was thus reduced to the mere evaluation of an integral. When different parts of a boundary satisfy different boundary conditions, however, the Green's function method leads to an integral containing a product involving Green's function times an *unknown* function. The solution of such problems requires the solution of an integral equation, and the techniques required for the solution of such problems are beyond the framework of the present volume. The procedure used to formulate the integral equation is relatively straightforward, however, and two examples from wave diffraction theory are considered here. The two problems are complimentary in nature. The integral equations obtained in each case are shown to imply, without obtaining their solutions, a general principle of diffraction theory known as *Babinet's principle*.

The problems considered are both two dimensional in their space dependence. One is the scattering of a plane wave by a thin strip of width $2a$ on which the wave field vanishes (Problem I) while the second problem is the passage of a plane wave through a slit of width $2a$ in a thin barrier on which the normal derivative of the wave field vanishes (Problem II). Both situations are depicted in Figure 9.2.

In addition to the surface that produces the wave scattering, an additional surface at infinity is included to provide a closed region for application of Green's theorem as was done in Section 4.8.

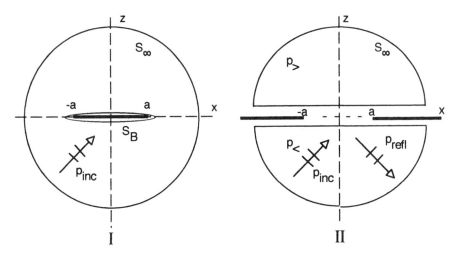

Figure 9.2. Complimentary diffraction problems for Babinet's principle.

For Problem I we have

$$p_1 = \int_{S_\infty + S_B} ds' \mathbf{n}' \cdot (g_F \nabla' p_1 = p_1 \nabla' g_F) \tag{9.3.1}$$

where g_F is the two-dimensional free-space Green's function $g_F(x, z | x', z') = (i/4) H_0^{(1)}(k | \boldsymbol{\rho} - \boldsymbol{\rho}'|)$ given in (4.10.6).

The surface at infinity may be shown (cf. Problem 4.12.6) to yield the incident plane wave $e^{i\mathbf{k} \cdot \mathbf{r}}$, and since $p_1 = 0$ on the barrier S_B, we have

$$p_1 = p_{\text{inc.}}(x, z) + \int_{S_B} ds' \mathbf{n}' \cdot g \nabla' p_1$$

$$= p_{\text{inc.}}(x, z) - \int_{-a}^{a} dx' g \, \partial_{z'} p_1^+ + \int_{-a}^{a} dx' g \, \partial_{z'} p_1^- \tag{9.3.2}$$

where we have used the abbreviated notation ∂_z to represent $\partial/\partial z$ and have also taken account of the fact that $\mathbf{n}$ is always an *outward* normal. The superscripts $\pm$ refer to the side of the barrier on which the field has been evaluated, that is $p^+ = p(x, +\epsilon)$ and $p^- = p(x, -\epsilon)$. Setting $\varphi(x) = p^+ - p^-$, we obtain

$$p_1(x, z) = p_{\text{inc.}}(x, z) - \int_{-a}^{a} dx' g_F(x, z | x', 0) \varphi(x') \tag{9.3.3}$$

and on noting that $p_1 = 0$ on the barrier, we have

$$0 = e^{ik_x x} - \int_{-a}^{a} dx' g_F(x, 0 | x', 0) \varphi(x'), \qquad |x| < a \tag{9.3.4}$$

This result is the integral equation that would have to be solved to complete the solution of this problem.

Turning now to Problem II, we examine the field on opposite sides of the barrier separately and use a Green's function with vanishing normal derivatives on the barrier. By the method of images we have

$$g(\boldsymbol{\rho}|\boldsymbol{\rho}') = \frac{i}{4} [H_0^{(1)}(k\sqrt{(x-x')^2 + (z-z')^2})$$

$$+ H_0^{(1)}(k\sqrt{(x-x')^2 + (z+z')^2})] \quad (9.3.5)$$

Applying Green's theorem to the half space $z < 0$ and noting that $\partial_{n'} p_{2<} = 0$, we obtain

$$p_{2<} = \int_{S_\infty + S_B} ds'\mathbf{n}' \cdot (g\nabla' p_{2<} - p_{2<}\nabla' g) \quad (9.3.6)$$

By a calculation analogous to that of Problem (4.12.6), one can show that the integral over the surface at infinity now yields both an incident and a reflected wave and we have

$$p_{2<}(x, z) = p_{\text{inc.}}(x, z) + p_{\text{refl.}}(x, z) + \int_{S_B} dx'\mathbf{n}$$

$$\cdot (g\nabla' p_{2<} - p_{2<}\nabla' g) \quad (9.3.7)$$

Since $\mathbf{n} \cdot \nabla' g = 0$ on the boundary and $\mathbf{n}' \cdot \nabla p_{2<} = 0$ except in the region $|x| \le a$, we finally obtain

$$p_{2<}(x, z) = p_{\text{inc.}}(x, z) + p_{\text{refl.}}(x, z) + \int_{-a}^{a} dx'g(\boldsymbol{\rho}|x', 0)\partial_{z'}p_{2<}(x', 0) \quad (9.3.8)$$

In the upper half space, the incident and reflected waves are not present and the outward normal on the barrier is in the negative z direction. Thus, a similar procedure applied to the upper half space $z > 0$ yields

$$p_{2>}(x, z) = \int_{-a}^{a} dx'g(\boldsymbol{\rho}|x', 0)\partial_{z'}p_{2>}(x', 0) \quad (9.3.9)$$

In the aperture $y = 0$, $|x| \le 0$ we have $p_{2>} = p_{2<}$, and therefore

$$-\int_{-a}^{a} dx'g(x, 0|x', 0)\partial_{z'}p_{2>}(x', 0)$$

$$= 2e^{ik_x x} + \int_{-a}^{a} dx'g(x, 0|x', 0)\partial_{z'}p_{2<}(x', 0) \quad (9.3.10)$$

Since $\partial_{z'} p_{2<}(x, 0) = \partial_z p_{2>}(x, 0)$ in the aperture as well, and noting that $g(x, 0|x', 0) = 2g_F(x, 0|x', 0)$, we obtain the integral equation

$$0 = 2e^{ik_x x} + 2 \int_{-a}^{a} dx' g_F(x, 0|x', 0)\psi(x'), \qquad |x| \le a \quad (9.3.11)$$

where $\frac{1}{2}\psi(x) = \partial_{z'} p_{2<}(x, 0) = \partial_{z'} p_{2>}(x, 0)$. The integral equation for ψ is exactly the same integral equation as that obtained for $\varphi(x)$ in Problem I. Since $g(x, z|x', 0) = 2G_F(x, z|x', 0)$ and $\partial_{z'} p_{2>}(x, 0) = \frac{1}{2}\psi = \frac{1}{2}\varphi$, we have

$$p_1 = p_{\text{inc}} - \int_{-a}^{a} dx' g_F(\boldsymbol{\rho}|x', 0)\varphi(x') \quad (9.3.12)$$

from (9.3.3) and

$$p_2 = \int_{-a}^{a} dx' g_F(\boldsymbol{\rho}|x', 0)\varphi(x') \quad (9.3.13)$$

from (9.3.9). Adding these two results we have

$$p_1 + p_{2>} = p_{\text{inc}} \quad (9.3.14)$$

which is Babinet's principle.

9.4 COMPARISON OF WAVE MOTION IN ONE, TWO, AND THREE DIMENSIONS—FRACTIONAL DERIVATIVES

Intuitively we might expect that waves in two dimensions would exhibit behavior that, in some sense, is intermediate between that in one and three dimensions. A thorough analysis of this issue is rather subtle, and we shall only give a brief introduction to some basic ideas. Certain aspects of wave motion in various dimensions may be unified by employing a generalization of the concept of differentiation. We introduce this topic first.

Considering integration as an operation inverse to differentiation, we introduce an operator notation for integration by writing[1]

$$\frac{d^{-1}f(x)}{dx^{-1}} = \int_{-\infty}^{x} f(y)\, dy \quad (9.4.1)$$

Defining the second inverse derivative as the nested integral

$$\frac{d^{-2}f(x)}{dx^{-2}} = \int_{-\infty}^{x} dy \int_{-\infty}^{y} f(z)\, dz \quad (9.4.2)$$

[1]Other choices for the lower limit are 0 or an arbitrary constant. For a very readable account of the subject of fractional calculus, see the monograph by Oldham and Spanier (1974).

we find that a change in the order of integration by the method used in Problem B.9 leads to

$$\frac{d^{-2}f(x)}{dx^{-2}} = \int_{-\infty}^{x} dz\, f(z) \int_{z}^{x} dy$$

$$= \int_{-\infty}^{x} dz\, f(z)(x - z) \tag{9.4.3}$$

A sketch of the area involved will show that the same area in the yz plane is covered by the double integral in (9.4.2) and (9.4.3).

When this procedure is applied q times, we obtain

$$\frac{d^{-q}f(x)}{dx^{-q}} = \frac{1}{(q-1)!} \int_{-\infty}^{x} f(y)(x - y)^{q-1}\, dy, \qquad q = 1, 2, 3, \dots \tag{9.4.4}$$

Since this integral does not converge for $q \leq 0$, this result cannot be used directly to calculate the ordinary derivatives that correspond to $q = -1, -2, \dots$. However, if we assume that generalized derivatives satisfy the expected relation

$$\frac{d^{-q}f(x)}{dx^{-q}} = \frac{d^{n}}{dx^{n}} \frac{d^{-q-n}f(x)}{dx^{-q-n}} \tag{9.4.5}$$

then we need merely choose $n > -q > 0$ to obtain the usual result when q is equal to a negative integer. (We will use this same technique later for dealing with negative nonintegral values of q.) As an example, consider $q = -1$ and $n = 2$. We then have

$$\frac{df}{dx} = \frac{d^{2}}{dx^{2}} \frac{d^{-1}f}{dx^{-1}}$$

$$= \frac{d^{2}}{dx^{2}} \int_{-\infty}^{x} f(y)\, dy$$

$$= \frac{df}{dx} \tag{9.4.6}$$

Note that any larger integer value of n would yield the same result.

We now introduce a purely mathematical generalization of this procedure by assuming that it is valid for nonintegral values of q as well. Replacing $(q - 1)!$ by $\Gamma(q)$ in (9.4.4), we have

$$\frac{d^{-q}f(x)}{dx^{-q}} = \frac{1}{\Gamma(q)} \int_{-\infty}^{x} f(y)(x - y)^{q-1}\, dy, \qquad q > 0 \tag{9.4.7}$$

Of particular interest in our later development is the case $q = -\frac{1}{2}$. Choosing $n = 1$ in (9.4.5) we have

$$\frac{d^{1/2}f}{dx^{1/2}} = \frac{d}{dx}\frac{d^{-1/2}f}{dx^{-1/2}} \tag{9.4.8}$$

Applying the general result (9.4.7) with $q = \frac{1}{2}$ and using $\Gamma(\frac{1}{2}) = \sqrt{\pi}$, we obtain

$$\frac{d^{-1/2}f}{dx^{-1/2}} = \frac{1}{\sqrt{\pi}} \int_{-\infty}^{x} f(y)(x - y)^{-1/2}\, dy$$

$$= \frac{2}{\sqrt{\pi}} \int_{-\infty}^{x} f'(y)(x - y)^{1/2}\, dy \tag{9.4.9}$$

where an integration by parts has been performed. For the fractional derivative of order $\frac{1}{2}$ we now find from (9.4.8) that

$$\frac{d^{1/2}f}{dx^{1/2}} = \frac{d}{dx}\left[\frac{2}{\sqrt{\pi}} \int_{-\infty}^{x} f'(y)(x - y)^{1/2}\, dy \right]$$

$$= \frac{1}{\sqrt{\pi}} \int_{-\infty}^{x} f'(y)(x - y)^{-1/2}\, dy \tag{9.4.10}$$

Use of this result will occur later in our consideration of wave motion in two dimensions.

It is of interest to compare a simple function with its $\frac{1}{2}$ and first derivatives. If we choose $f(x) = \delta(x)$, then for $x > 0$ we have

$$\frac{d^{1/2}\delta(x)}{dx^{1/2}} = \frac{1}{\sqrt{\pi}} \int_{-\infty}^{x} \delta'(y)(x - y)^{-1/2}\, dy$$

$$= \frac{-1}{2\sqrt{\pi}} \int_{-\infty}^{x} \delta(y)(x - y)^{-3/2}\, dy$$

$$= \frac{-1}{2\sqrt{\pi}} x^{-3/2} \tag{9.4.11}$$

where an integration by parts has been performed in integrating over the derivative of the delta function. For the first derivative we merely obtain

$$\frac{df}{dx} = \delta'(x) \tag{9.4.12}$$

Thus, both f and df/dx are very localized functions while the derivative of $\frac{1}{2}$ order only decays algebraically as $x^{-3/2}$. It is therefore much less localized than either f or df/dx.

With these mathematical preliminaries introduced, we now turn to a consideration of waves radiated from a point source in one, two, and three dimensions. The wave equation

$$\nabla_N^2 p_N - \frac{1}{c^2} \frac{\partial^2 p_N}{\partial t^2} = -q_N(\mathbf{r}, t), \qquad N = 1, 2, 3 \qquad (9.4.13)$$

where ∇_N^2 refers to the Laplace operator in one, two, or three dimensions, has a solution in an unbounded medium of the form

$$p_N(\mathbf{r}, t) = \int_{-\infty}^{t} dt' \int d^N \mathbf{r}' \, q_N(\mathbf{r}, t) g_N(\mathbf{r}, t | \mathbf{r}', t') \qquad (9.4.14)$$

where $g_N(r, t | r', t')$ is the Green's function in one, two, or three dimensions as obtained in Section 6.7. For a point source at the origin with arbitrary time dependence we have $q_N(\mathbf{r}, t) = f(t) \delta^N(\mathbf{r})$ and thus

$$p_N(\mathbf{r}, t) = \int_{-\infty}^{t} dt' \, f(t') g_N(\mathbf{r}, t | 0, t') \qquad (9.4.15)$$

The function $f(t)$ will be assumed to have a maximum near $t = 0$. In terms of the appropriate Green's function referred to above we have

$$p_1(x, t) = \frac{c}{2} \int_{-\infty}^{t} dt' \, f(t') u \left(t - t' - \frac{|x|}{c} \right)$$

$$p_2(\rho, t) = \frac{1}{2\pi} \int_{-\infty}^{t} dt' \, \frac{f(t')}{\sqrt{(t - t')^2 - \rho^2/c^2}}$$

$$p_3(\mathbf{r}, t) = \frac{1}{4\pi r} \int_{-\infty}^{t} dt' \, f(t') \delta \left(t - t' - \frac{r}{c} \right) \qquad (9.4.16)$$

The integration for p_1 is facilitated by writing $f(t) = dw/dt$. If $w(-\infty)$ is assumed to be zero, we have

$$p_1 = \frac{c}{2} w \left(t - \frac{x}{c} \right) \qquad (9.4.17)$$

The result for p_2 is more readily interpreted if we write $\sqrt{(t - t')^2 - \rho^2/c^2} = \sqrt{(t - t' - \rho/c)(t - t' + \rho/c)}$ and note that for t and ρ/c large but $t' \approx t - \rho/c$ and therefore small (i.e., near the peak of the pulse) the radical is approximately $\sqrt{2}\,\rho/c\,\sqrt{t - t'} - \rho/c$. Using the definition of the fractional derivative from (9.4.10), we then have

$$p_2 = \frac{1}{2\pi}\sqrt{\frac{c}{2\rho}} \int_{-\infty}^{t - \rho/c} \frac{dt'\,w'(t')}{\sqrt{t - \rho/c - t'}} = \frac{1}{2\pi}\sqrt{\frac{c}{2\rho}} \frac{d^{1/2}w}{dt^{1/2}}\bigg|_{t - \rho/c} \qquad (9.4.18)$$

Finally, for p_3 we immediately obtain

$$p_3 = \frac{1}{4\pi r}\frac{dw}{dt}\bigg|_{t - r/c} = \frac{f(t - r/c)}{4\pi r} \qquad (9.4.19)$$

We see that the fractional derivative provides an expression for the pulse shape that is intermediate to those of one and three dimensions. To exhibit this intermediate nature of the $\frac{1}{2}$ derivative, the graphs of a specific initial pulse profile are shown[2] in Figure 9.3 for $w = (t^2 + T^2)^{-1}$. The amplitudes are arbitrary.

The integration associated with the two-dimensional case (9.4.18) provides an opportunity to summarize a number of techniques that are frequently useful for evaluating definite integrals. The integral may be performed by first noting that for the example being considered, namely $w(t) = 1/(t^2 + T^2)$, we may write $dw/dt = (t/T)\,dw/dT$. Also, setting $\tau = t - \rho/c$, we then have

$$2\pi\sqrt{\frac{2\rho}{c}}\,p_2 = \frac{1}{T}\frac{\partial I(\tau, T)}{\partial T} \qquad (9.4.20)$$

where

$$I(\tau, T) = \int_{-\infty}^{\tau} \frac{dt'\,t'}{(t'^2 + T^2)\sqrt{\tau - t'}} \qquad (9.4.21)$$

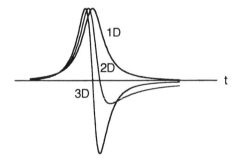

Figure 9.3. Pulse profile for one-, two-, and three-dimensional propagation.

[2]The results of this section have been considered in the context of acoustic wave propagation (Lamb, 1945, p. 524, Lighthill, 1978, p. 17).

On setting $\tau - t' = v^2$, the resulting integral over v may be decomposed to yield

$$I(\tau, t) = \int_0^\infty dv \left(\frac{1}{\tau - v^2 + iT} + \frac{1}{\tau - v^2 - iT} \right)$$

$$= 2 \operatorname{Re} J(\tau, T) \tag{9.4.22}$$

where

$$J(\tau, T) = \int_0^\infty \frac{dv}{\tau - v^2 + iT} \tag{9.4.23}$$

When the denominator in this integrand is replaced by the integral representation

$$\frac{1}{\tau - v^2 + iT} = \frac{1}{i} \int_0^\infty d\omega \, e^{i\omega(\tau - v^2 + iT)} \tag{9.4.24}$$

and the resulting integration over v is carried out using

$$\int_0^\infty dv \, e^{-i\omega v^2} = \frac{1}{2} \sqrt{\frac{\pi}{\omega}} \, e^{i\pi/4} \tag{9.4.25}$$

we find that the T derivative relating J and w is readily performed and yields

$$\frac{\partial J}{\partial T} = \frac{\sqrt{\pi}}{2} e^{i\pi/4} \int_0^\infty d\omega \, \omega^{1/2} e^{-(T - i\tau)\omega} \tag{9.4.26}$$

Setting $\xi = \omega(T - i\tau)$ and recognizing that the resulting integral over ξ is equal to $\Gamma(\frac{1}{2}) = \sqrt{\pi}/2$, we have

$$\frac{\partial J}{\partial T} = \frac{\pi}{4} \frac{e^{i\pi/4}}{(T - i\tau)^{3/4}} \tag{9.4.27}$$

When we rewrite the denominator as an amplitude and phase and subsequently extract the real part, we finally obtain

$$\sqrt{\frac{2\rho}{c}} \, p_2 = \frac{1}{4T^{5/2}} \frac{\cos(3\psi/2 + \pi/4)}{[(\tau/T)^2 + 1]^{3/2}} \tag{9.4.28}$$

where $\psi = \tan(\tau/T)$.

9.5 MODIFIED GREEN'S FUNCTION FOR A SPHERE

In Sections 4.4 and 4.9 it was shown that a Green's function satisfying homogeneous boundary conditions of either Neumann or Dirichlet type on a *plane* surface could be readily constructed by using the method of images. The two cases differ only in the sign of the image term. In Section 5.3 the Green's function satisfying homogeneous Dirichlet conditions within a *spherical* surface was also shown to be expressible in terms of an image source. The corresponding situation for Neumann boundary conditions on a sphere does not bear the simple relation to the Dirichlet case that might be expected on the basis of the result for the plane boundary, however. Imposition of Neumann boundary conditions on the closed spherical surface necessitates the use of a modified Green's function of the type considered for one dimension in Section 4.13. The resultant expression can be put in a form that displays not only a sign change of the single image term expected on the basis of the Dirichlet result for Neumann conditions on a plane, but also an additional contribution that can be interpreted as a *continuous distribution of image sources* extending radially outward to infinity from the location of the expected image source.

On the basis of the result for one dimension obtained in Section 4.13, we expect that the appropriate Green's function is a solution of

$$\nabla^2 G_m = -\delta^3(\mathbf{r} - \mathbf{r}') + \frac{1}{V} \tag{9.5.1}$$

where $V = 4\pi a^3/3$ is the volume of the sphere. As noted in Section 4.13 for one dimension, the solution G_m is expected to have three contributions. The contribution corresponding to the source term $\delta^3(\mathbf{r} - \mathbf{r}')$ has the form of the free-space Green's function as given in (5.1.33), namely,

$$G_1 = \frac{1}{4\pi|\mathbf{r} - \mathbf{r}'|} = \frac{1}{4\pi r_>} \sum_{n=0}^{\infty} \left(\frac{r_<}{r_>}\right)^n P_n(\cos\theta) \tag{9.5.2}$$

where θ is measured from a z axis that passes through the origin and the point $\mathbf{r}'$. The term associated with the uniform background source is a solution of

$$\frac{1}{r^2}\frac{d}{dr}\left(r^2 \frac{dG_2}{dr}\right) = \frac{1}{V} \tag{9.5.3}$$

Integration of this equation immediately leads to

$$G_2 = \frac{r^2}{6V} + \frac{c_1}{r} + c_2 \tag{9.5.4}$$

The final contribution is a solution of the homogeneous equation and can be written in the form

$$G_3 = \sum_{n=0}^{\infty} (A_n r^n + B_n r^{-n-1}) P_n(\cos \theta) \qquad (9.5.5)$$

where, again, θ is measured from a z axis that passes through the origin of the sphere and the point r'. Since the solution must remain finite at the origin, we set $B_n = 0$ in this expression. The three contributions yield an expression for G_m in the form

$$G_m(\mathbf{r}|\mathbf{r}') = \frac{1}{4\pi r} + \frac{r^2}{6V} + \frac{c_1}{r} + c_2 + A_0$$

$$+ \sum_{n=1}^{\infty} \left[\frac{1}{4\pi r} \left(\frac{r'}{r}\right)^n + A_n r^n \right] P_n(\cos \theta) \qquad (9.5.6)$$

where all terms that are independent of the angle θ have been written out explicitly. Imposing homogeneous Neumann boundary conditions at $r = a$, we obtain

$$\left. \frac{\partial G_m}{\partial r} \right|_{r=a} = -\frac{1}{4\pi a^2} + \frac{a}{3V} - \frac{c_1}{a^2}$$

$$+ \sum_{n=1}^{\infty} \left[-\frac{(n+1)r'^n}{4\pi a^{n+2}} + A_n n a^{n-1} \right] P_n(\cos \theta) = 0 \quad (9.5.7)$$

The first two terms cancel identically and since the Legendre polynomials are orthogonal, we obtain $c_1 = 0$ along with

$$A_n = \frac{1}{4\pi a} \left(1 + \frac{1}{n}\right) \left(\frac{r'}{a^2}\right)^n \qquad (9.5.8)$$

The resulting expression for G_m is

$$G_m - c_2 = \frac{1}{4\pi |\mathbf{r} - \mathbf{r}'|} + \frac{r^2}{6V} + \frac{1}{4\pi a} \sum_{n=1}^{\infty} \left(\frac{rr'}{a}\right)^n \left(1 + \frac{1}{n}\right) P_n(\cos \theta)$$

$$(9.5.9)$$

As noted in Section 4.13, the modified Green's function is only determined to within a constant. This constant, c_2, may be specified by imposing some convenient requirement such as the vanishing of the volume integral of G_m. Since only the terms independent of angle survive in the evaluation of such a volume integral, we have

$$0 - c_2 = \frac{1}{r'} \int_0^{r'} r^2 \, dr + \int_r^a r \, dr + \frac{4\pi}{6V} \int_0^a r^4 \, dr \qquad (9.5.10)$$

which leads to

$$c_2 = \frac{r'^2}{8\pi a^3} - \frac{9}{20\pi a} \tag{9.5.11}$$

A final result for G_m is then

$$G_m = -\frac{9}{20\pi a} + \frac{1}{4\pi |\mathbf{r} - \mathbf{r}'|} + \frac{r^2 + r'^2}{8\pi a}$$

$$+ \frac{1}{4\pi a^3} \sum_{n=1}^{\infty} \left(1 + \frac{1}{n}\right) \left(\frac{rr'}{a^2}\right)^n P_n(\cos \theta) \tag{9.5.12}$$

The infinite series appearing in this result can be given an interesting physical interpretation. On setting $a^2/r' = r_i$, the distance from the origin to the image source introduced in Section 5.3, we may write the terms in the sum that do not contain $1/n$ in the form

$$\sum_{1}^{\infty} \left(\frac{rr'}{a^2}\right)^n P_n(\cos \theta) = -1 + r_i \left[\frac{1}{r_i} \sum_{n=0}^{\infty} \left(\frac{r}{r_i}\right)^n P_n(\cos \theta)\right]$$

$$= -1 + \frac{r_i}{|\mathbf{r} - \mathbf{r}_i|} \tag{9.5.13}$$

in which $\mathbf{r}_i = \mathbf{a} a^2/r'$ where $\mathbf{a}$ is a radial unit vector in the direction of the source and its image.

The remaining sum may be rewritten by noting that

$$\frac{1}{nr_i^n} = \int_{r_i}^{\infty} d\lambda \, \lambda^{-n-1}, \qquad n = 1, 2, 3, \ldots \tag{9.5.14}$$

We then have

$$S \equiv \sum_{n=1}^{\infty} \frac{1}{n} \left(\frac{r}{r_i}\right)^n P_n(\cos \theta) = \int_{r_i}^{\infty} d\lambda \, \frac{1}{\lambda} \sum_{n=1}^{\infty} \left(\frac{r}{\lambda}\right)^n P_n(\cos \theta)$$

$$= \int_{r_i}^{\infty} d\lambda \left(\frac{1}{\sqrt{\lambda^2 + r^2 - 2\lambda r \cos \theta}} - \frac{1}{\lambda}\right) \tag{9.5.15}$$

The first term in the integral has the physical significance alluded to in the introduction. It represents the field at the observation point $\mathbf{r}$ due to a source of strength proportional to $d\lambda$ located at a radial distance λ from the origin with these contributions then summed over all distances radially outward from

$\mathbf{r}_i$ to infinity. The second term in the integral is the field produced at the origin by a source of opposite sign and clearly provides convergence for the integral. The integration is readily performed by employing the standard result

$$\int \frac{dx}{\sqrt{x^2 + 2bx + c}} = \ln(x + b + \sqrt{x^2 + 2bx + c}) \qquad (9.5.16)$$

Setting $b = -r\cos\theta$ and $c = r^2$, we have

$$S = \ln \frac{2r_i}{r_i - r\cos\theta + \sqrt{r_i^2 - 2rr'\cos\theta + r^2}}$$

$$= \ln \frac{2a^2}{a^2 - rr'\cos\theta + r'|\mathbf{r} - \mathbf{r}_i|} \qquad (9.5.17)$$

When (9.5.13), and (9.5.17) are substituted into (9.5.12), we have the alternate form of the result

$$G_m + \frac{7}{10\pi a} = \frac{1}{4\pi}\left(\frac{1}{|\mathbf{r} - \mathbf{r}'|} + \frac{1}{|\mathbf{r} - \mathbf{r}_i|} + \frac{r_i/a}{|\mathbf{r} - \mathbf{r}_i|}\right) + \frac{r^2 + r'^2}{8\pi a^3}$$

$$+ \frac{1}{4\pi a}\ln\frac{2a^2}{a^2 - rr'\cos\theta + r'|\mathbf{r} - \mathbf{r}_i|} \qquad (9.5.18)$$

9.6 OSCILLATION OF AN INHOMOGENEOUS CHAIN

Although the standard special functions such as Bessel functions and Legendre polynomials tend to be identified with the separation-of-variables procedure in various coordinate systems, their use in these geometric contexts tends to involve only the simplest properties of these functions. On the other hand, examples of wave motion in inhomogeneous media frequently give rise to other differential equations that, after various transformations, are also found to have solutions expressible in terms of these functions. In these latter contexts, more abstruse properties of such special functions are frequently encountered. As a relatively simple example we consider the small oscillation of an inhomogeneous hanging chain. The oscillation of the *homogeneous* chain, which is expressed in terms of Bessel functions, is considered in Problem 3.7.13. If the density of the chain is chosen to be $\rho(x) = \rho_0 e^{-x/\lambda}$ with x measured upward from the free end to a fixed support a distance L above the free end, then the tension on the string is given by

$$T(x) = g\int_0^x \rho(x)\,dx = \frac{g\rho_0}{\lambda}(1 - e^{-x/\lambda}), \qquad 0 \le x \le L \qquad (9.6.1)$$

where g is the acceleration of gravity. The differential equation governing the oscillation of this chain is

$$\rho(x)\frac{\partial^2 y}{\partial t^2} = \frac{\partial}{\partial x}\left[T(x)\frac{\partial y}{\partial x}\right] \qquad (9.6.2)$$

with $T(x)$ given by (9.6.1). If we consider oscillation at a single frequency ω and write $y(x, t) = f(x)e^{-i\omega t}$, the shape of the chain $f(x)$ is given by the solution of the equation

$$\frac{\partial}{\partial x}\left[(1 - e^{-x/\lambda})\frac{\partial f}{\partial x}\right] + \frac{\lambda\omega^2 e^{-x/\lambda}}{\lambda g}f = 0 \qquad (9.6.3)$$

that satisfies the boundary condition $f(L) = 0$. After making the change of variables[3] $\zeta = 2e^{-x/\lambda} - 1$, we find that $f(\zeta)$ satisfies

$$\frac{d}{d\zeta}\left[(1 - \zeta^2)\frac{\partial f}{\partial \zeta}\right] + \nu(\nu + 1)f = 0, \qquad \nu(\nu + 1) = \frac{\omega^2\lambda}{g} \qquad (9.6.4)$$

which is the Legendre equation. The appropriate solution of this differential equation is $f = P_\nu(\zeta)$. The second solution, $Q_\nu(\zeta)$, is not included since it diverges at $\zeta = 1$ (i.e., at $x = 0$), which is the free end of the chain. Returning to the original variable, x, the shape of the chain is proportional to

$$f(x) = P_\nu(2e^{-x/\lambda} - 1) \qquad (9.6.5)$$

The value of ν is determined by imposing the fixed end boundary condition (i.e., $P_\nu(2e^{-L/\lambda} - 1)) = 0$. For a given value of L/λ there will, in general, be *nonintegral* values of ν that satisfy this equation. For $\lambda < 0$, so that the density increases away from the free end of the chain, the order ν may even be a complex expression that, according to Eq. (9.6.4), is $\nu = -\frac{1}{2} + i\mu$ with $\mu = \sqrt{4\omega^2|\lambda|/g - 1}$. Once these values of ν have been determined, the corresponding oscillation frequencies are obtained from the relation given in (9.6.4). For nonintegral ν the solution $P_\nu(\zeta)$ is no longer a polynomial in ζ and is referred to as a Legendre *function*.

Before addressing the issue of determining appropriate values of ν and the corresponding oscillation frequencies, let us first consider the limit $L/\lambda \ll 1$, which should yield the results for the homogeneous chain obtained in Problem

---

[3]One is led to this choice by first making the obvious substitution $\eta = e^{-x/\lambda}$ and discovering that the transformed equation is $\eta(\eta - 1)f'' + (2\eta - 1)f'' - c\omega^2\lambda(g)f = 0$. The transformation of this equation to the Legendre equation is given by Kamke (1971, p. 463). This volume is extremely useful in such considerations.

3.7.13. In this limit $x/\lambda \ll 1$ and we may write

$$\zeta = 2e^{-x/\lambda} - 1 \approx 2\left(1 - \frac{x}{\lambda}\right) - 1 = 1 - \frac{2x}{\lambda} \qquad (9.6.6)$$

It is clear from the relation between ν and λ given in (9.6.4) that, as λ increases, ν must also become large since ω does not vanish. Then, setting $\nu^2 = \lambda \omega^2/g$ since $\nu \gg 1$, we have

$$f(x) \approx P_\nu\left(1 - \frac{2x}{\lambda}\right) = P_\nu\left(1 - \frac{2x\omega^2}{\nu^2 g}\right) \qquad (9.6.7)$$

The limit of this expression as $\nu \to \infty$ is provided by the result given in Eq. (E.39) where it is shown that

$$\lim_{\nu \to \infty} P_\nu\left(1 - \frac{z^2}{2\nu^2}\right) = J_0(z) \qquad (9.6.8)$$

In the present instance $z = 4x\omega^2/g$ and the limit is

$$f(x) = J_0(2\omega \sqrt{x/g}) \qquad (9.6.9)$$

This result agrees with that obtained for the homogeneous chain in Problem 3.7.13. There it was shown that the lowest frequency of oscillation is equal to $\omega_1 = \frac{1}{2}\alpha_{01}\sqrt{g/L} = 1.202\sqrt{g/L}$.

We now return to a consideration of the oscillation frequencies for arbitrary values of λ. As noted above, a full discussion of this topic involves properties of $P_\nu(\zeta)$ for nonintegral values of ν. In order to avoid such developments, which are outside the scope of this presentation, we use the following indirect approach. By first choosing ν to be an integer, we can then determine those special values of the ratio L/λ that are associated with integer values of ν. The function $P_n(\zeta)$ will then be an algebraic polynomial that can be solved for the values of L/λ. We shall consider only the lowest frequency. This frequency is associated with the largest zero of $P_n(\zeta)$ since it is this zero that is encountered first as we proceed inward from the free end of the chain at $x = 0$ where $\zeta = 1$. (Graphs of the first few Legendre polynomials are shown in Figure E.1.) This root will be labeled ζ_1. From our previous consideration of the limit of a uniform chain, we expect that as increasing integer values of ν are chosen, the lowest frequency will approach that of the uniform chain.

For $\nu = 3$ we require the solution of $P_3(\zeta_1) = \frac{1}{2}(5\zeta_1^3 - 3\zeta_1) = 0$. The largest positive root is at $\zeta_1 = \sqrt{3/5} \approx 0.774597$. Setting $2e^{-L/\lambda} - 1$ equal to this value of ζ_1, we find $\lambda = 8.363L$ and therefore

$$\omega\sqrt{L/g} = \sqrt{\nu(\nu + 1)L/\lambda} = \sqrt{3 \cdot 4/8.363} = 1.1979 \qquad (9.6.10)$$

Results for some other integer values of ν are as follows:

ν	ζ_1	λ/L	$\omega\sqrt{L/g}$
1	0	1.4427	1.1774
2	0.577350	4.2123	1.1935
3	0.774597	8.3630	1.1979
4	0.861136	13.8966	1.1997
5	0.906280	20.8134	1.2006
10	0.973907	76.1463	1.2019

Clearly, as ν increases and the string becomes more homogeneous, the value of $\omega\sqrt{L/g}$ approaches that for the homogeneous chain, that is, $(\frac{1}{2})\alpha_{01} = \frac{1}{2}(2.4048) = 1.2024$, where α_{01} is the first root of J_0.

9.7 POINT SOURCE NEAR THE INTERFACE BETWEEN TWO HALF SPACES

The temperature distribution about a point source in an unbounded medium composed of two half spaces having different thermal conductivities is a problem that exemplifies the use of the Fourier-Bessel transform in a simple way. The conductivity in each half space will be assumed to be constant. The geometry of the problem is shown in Figure 9.4.

The temperature distribution about a point source in a *single* homogeneous medium of infinite extent would, of course, be spherically symmetric about the source. The presence of the plane interface in the problem being considered here gives rise to a temperature distribution that is symmetric about the z axis and thus has only cylindrical symmetry. The problem can therefore be formulated quite naturally in terms of Fourier-Bessel transforms. As is to be expected, the final result can be interpreted in terms of image sources of suitable strengths. We introduce a cylindrical coordinate system ρ, z and write the

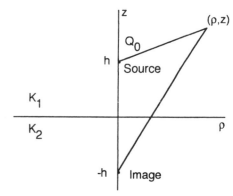

Figure 9.4. Point heat source near interface between two half spaces with different conductivity.

source of strength Q_0 at a distance h above the interface as $s(\mathbf{r}) = Q_0 \delta^2(\mathbf{\rho}) \delta(z - h)$. In the upper half space the temperature $u(\mathbf{\rho}, z)$ therefore satisfies

$$\nabla^2 u_1 = -\frac{Q_0}{K_1} \delta^2(\mathbf{\rho}) \delta(z - h) \qquad (9.7.1)$$

As usual, the solution of this inhomogeneous equation will be composed of a particular solution due to the source term and a solution to the homogeneous equation that enables one to satisfy boundary conditions at the interface $z = 0$ as well as the requirement that the field go to zero at a large distance from the source.

In the lower half space, where there is no source term, the temperature $u(\mathbf{\rho}, z)$ satisfies the homogeneous equation

$$\nabla^2 u_2 = 0 \qquad (9.7.2)$$

At the interface between the two media there is continuity of temperature as well as continuity of heat flow that yield, respectively, the boundary conditions

$$u_1(\rho, 0) = u_2(\rho, 0) \qquad (9.7.3)$$

$$K_1 \frac{\partial u_1}{\partial z}\bigg|_{z=0} = K_2 \frac{\partial u_2}{\partial z}\bigg|_{z=0}$$

The particular solution of (9.7.1) is the spherically symmetric solution associated with a point source, that is, the free-space Green's function given in Eq. (4.9.10). For the present situation it is written

$$u_{1p}(\rho, z) = \frac{Q_0}{4\pi K_1} \frac{1}{|\mathbf{r} - \mathbf{h}|} = \frac{Q_0}{4\pi K_1} \frac{1}{\sqrt{\rho^2 + (z - h)^2}} \qquad (9.7.4)$$

where $\mathbf{r} = \mathbf{\rho} + \mathbf{k}z$ and $\mathbf{h} = \mathbf{k}h$. The homogeneous solution will be expressed in terms of a Fourier-Bessel transform as given in (6.6.3). In the upper half space we thus write the temperature in the form

$$u_1(\rho, z) = \frac{Q_0}{4\pi K_1} \frac{1}{\sqrt{\rho^2 + (z - h)^2}} + \int_0^\infty k \, dk \, J_0(k\rho) A(k) e^{-hz},$$

$$z > 0 \qquad (9.7.5)$$

where the unknown amplitude function $A(k)$ is one of two functions to be determined by the boundary conditions (9.7.3). The solution of Laplace's equation (9.7.2) in the lower half space must decrease as z approaches $-\infty$ and thus contain the exponential dependence e^{kz}. The solution is therefore of the form

$$u_2(\rho, z) = \int_0^\infty k \, dk \, J_0(k\rho) B(k) e^{kz}, \qquad z < 0 \qquad (9.7.6)$$

To proceed with our determination of the unknown coefficients $A(k)$ and $B(k)$ by imposing the boundary conditions at $z = 0$, it will also be necessary to write the particular solution (9.7.4) as a Fourier-Bessel transform. The appropriate expression has already been encountered in (6.5.16). It may also be inferred from the Laplace transform relation

$$\int_0^\infty dt \, e^{-st} J_0(at) = \frac{1}{\sqrt{s^2 + a^2}}, \qquad s > 0 \qquad (9.7.7)$$

Setting $a = \rho$ and $s = |z - h|$ and introducing a factor of unity in the form (t/t) in the integrand of (9.7.7), we obtain

$$u_{1p}(\rho, z) = \frac{Q_0}{4\pi K_1} \frac{1}{\sqrt{\rho^2 + (z - h)^2}}$$

$$= \frac{Q_0}{4\pi K_1} \int_0^\infty d \, dk \, \frac{J_0(k\rho)}{k} e^{-(h-z)k}, \qquad z < h \qquad (9.7.8)$$

The choice $|z - h| = h - z$ has been made since this expression for $u_{1p}(\rho, z)$ will be used at the boundary where $z = 0 < h$.

The temperature distribution in each half space may now be written

$$u_1(\rho, z) = \int_0^\infty k \, dk \, J_0(k\rho) \left[\frac{Q_0}{4\pi K_1} \frac{e^{-(h-z)k}}{k} + A(k) e^{-kz} \right], \qquad z > 0$$

$$(9.7.9)$$

$$u_2(\rho, z) = \int_0^\infty k \, dk \, J_0(k\rho) B(k) e^{kz}, \qquad z < 0 \qquad (9.7.10)$$

Equating temperatures at $z = 0$ we find

$$\int_0^\infty k \, dk \, J_0(h\rho) \left[\frac{Q_0}{4\pi K_1} \frac{e^{-kh}}{k} + A(k) \right] = \int_0^\infty k \, dk \, J_0(h\rho) B(k) \qquad (9.7.11)$$

and since the transforms are unique, we equate the integrands to obtain

$$\frac{Q_0}{4\pi K_1} \frac{1}{k} + A(k) = B(k) \qquad (9.7.12)$$

To impose continuity of heat flow expressed by the second of (9.7.3), we first differentiate (9.7.9) and (9.7.10) with respect to z and again equate integrands after setting $z = 0$. The result is

$$K_1 \left(\frac{Q_0}{4\pi K_1} e^{-kh} - kA \right) = K_2 kB \qquad (9.7.13)$$

Solving (9.7.12) and (9.7.13) for $A(k)$ and $B(k)$, we find

$$A(k) = \frac{K_1 - K_2}{K_1 + K_2} \frac{Q_0}{4\pi K_1} \frac{e^{-kh}}{k}, \qquad B(k) = \frac{2}{K_1 + K_2} \frac{Q_0}{4\pi} \frac{e^{-kh}}{k} \qquad (9.7.14)$$

When these results are introduced into (9.7.5) and (9.7.6), we obtain

$$u_1 = u_{1p} + \frac{K_1 - K_2}{K_1 + K_2} \frac{Q_0}{4\pi K_1} \int_0^\infty k\, dk\, J_0(k\rho) \frac{e^{-k(z+h)}}{k}, \qquad z > 0$$

$$u_2 = \frac{2}{K_1 + K_2} \frac{Q_0}{4\pi} \int_0^\infty k\, dk\, J_0(k\rho) \frac{e^{-k(h-z)}}{k}, \qquad z < 0 \qquad (9.7.15)$$

Since the remaining integrals are in the form of the integral representation for a point source that was used in (9.7.8), we may finally write the solution as

$$u_1 = u_{1p} + \frac{K_1 - K_2}{K_1 + K_2} \frac{Q_0}{4\pi K_1 \sqrt{\rho^2 + (z + h)^2}}$$

$$u_2 = \frac{2}{K_1 + K_2} \frac{Q_0}{4\pi} \frac{1}{\sqrt{\rho^2 + (z - h)^2}} \qquad (9.7.16)$$

In the upper half space the effect of the inhomogeneity is thus seen to be accounted for by placing an image source of strength $Q_0(K_1 - K_2)/(K_1 + K_2)$ in the lower half space at a depth h below the interface. Note also that for $K_1 = K_2$ the image contribution vanishes and both solutions yield the result for the source in free space, u_{1p}.

9.8 WAVES IN AN INHOMOGENEOUS MEDIUM

Solution of the wave equation for a homogeneous string by transformation to characteristic coordinates was first introduced in Section 1.8 and reconsidered in Section 8.1. We now examine a more elaborate case involving an inhomogeneous string. The tension will be assumed constant while the density varies according to $\rho(x) = \rho_0 L^4/(L^2 - x^2)^2$ with the string located in $|x| \le L$. The density thus has the minimum value ρ_0 at its midpoint and increases as $x \to \pm L$ where it becomes infinitely heavy. The wave velocity is given by $v(x) = \sqrt{T/\rho(x)} = c_0(L^2 - x^2)/L^2$ where $c_0 = \sqrt{T/\rho_0}$. The velocity therefore decreases as the ends are approached. We expect that a wave emanating from the center will slow down as it approaches $x = \pm L$. Since $v(\pm L) = 0$, the wave will never actually reach the end points. The inhomogeneity provides a boundary for wave motion on the system and, as time increases, a standing wave develops. While such an example is clearly artificial, it does provide a context for seeing how one's intuitive ideas about wave motion can be displayed in analytical form.

The wave equation for the string takes the form

$$\frac{\partial^2 y}{\partial t^2} - c_0^2 \left(\frac{L^2 - x^2}{L^2}\right)^2 \frac{\partial^2 y}{\partial x^2} = 0 \tag{9.8.1}$$

According to Eq. (8.1.16), the characteristic coordinates are provided by solutions of

$$\frac{dt}{dx} = \pm \frac{1}{v(x)} = \pm \frac{1}{c_0} \frac{L^2}{L^2 - x^2} \tag{9.8.2}$$

Integration yields $c_0 t = \pm L \tanh^{-1}(x/L) + \text{const}$. Representing the integration constants by u and v for the lower and upper choices of sign, respectively, we have

$$u = c_0 t + L \tanh^{-1}(x/L)$$
$$v = c_0 t - L \tanh^{-1}(x/L) \tag{9.8.3}$$

Rather than proceeding directly to a transformation of the wave equation into the coordinates u and v, we first introduce the new length variable $z = L \tanh^{-1}(x/L)$ that is suggested by (9.8.3). Note that as x varies from $-L$ to $+L$, z varies from $-\infty$ to $+\infty$. The spatial derivatives in the original wave equation (9.8.1) may be transformed by first using the chain rule to write $(L^2 - x^2) \partial/\partial x = L^2 \partial/\partial z$. The second derivative expression appearing in (9.8.1) is obtained from the identity

$$(L^2 - x^2) \frac{\partial}{\partial x}\left[(L^2 - x^2)\frac{\partial y}{\partial x}\right] = (L^2 - x^2)^2 \frac{\partial^2 y}{\partial x^2} - 2x(L^2 - x^2)\frac{\partial y}{\partial x} \tag{9.8.4}$$

which is equivalent to

$$L^4 \frac{\partial^2 y}{\partial z^2} = (L^2 - x^2)^2 \frac{\partial^2 y}{\partial x^2} - 2L^3 \left(\tanh \frac{z}{L}\right)\frac{\partial y}{\partial z} \tag{9.8.5}$$

The wave equation (9.8.1) thus takes the form

$$\frac{1}{c_0^2}\frac{\partial^2 y}{\partial t^2} - \frac{\partial^2 y}{\partial z^2} - \frac{2}{L}\tanh\frac{z}{L}\frac{\partial y}{\partial z} = 0 \qquad (9.8.6)$$

As in dealing with a second order ordinary differential equation, it is frequently useful to introduce a transformation that eliminates the first derivative term. A general relation for accomplishing such a transformation is provided by the identity

$$\frac{d^2 y}{dz^2} + f(z)\frac{dy}{dz} = \exp\left[-\frac{1}{2}\int f(z)\,dz\right]\left[\frac{d^2\varphi}{dz^2} - \left(\frac{1}{2}\frac{df}{dz} + \frac{1}{4}f^2\right)\varphi\right] \qquad (9.8.7)$$

where

$$\varphi(z) = y\exp\left[\frac{1}{2}\int f(z)\,dz\right] \qquad (9.8.8)$$

In the present instance $f(x) = (2/L)\tanh(z/L)$ so that $\varphi = y\cosh(z/L)$ and the two terms in (9.8.6) that contain z derivatives become

$$\frac{\partial^2 y}{\partial z^2} + \frac{2}{L}\left(\tanh\frac{z}{L}\right)\frac{\partial y}{\partial z} = \operatorname{sech}\frac{z}{L}\left(\frac{\partial^2\varphi}{\partial z^2} - \frac{2}{L^2}\varphi\right) \qquad (9.8.9)$$

The wave equation (9.8.6) in terms of the new dependent variable φ has the form

$$\frac{1}{c_0^2}\frac{\partial^2\varphi}{\partial t^2} - \frac{\partial^2\varphi}{\partial z^2} + \frac{1}{L^2}\varphi = 0 \qquad (9.8.10)$$

which is an equation with constant coefficients! It is the equation for a homogeneous string embedded in a medium with a constant elastic restoring force and was considered in Problem 4.7.4. The Green's function was found to be

$$g(z, t\,|z', t') = \frac{c_0}{2}J_0\left(\frac{c_0}{L}\sqrt{t^2 - \frac{(z - z')^2}{c_0^2}}\right)H\left(t^2 - \frac{(z - z')^2}{c_0^2}\right) \qquad (9.8.11)$$

In terms of this Green's function, the solution of (9.8.10) may be written as

$$\varphi(z, t) = \int_{-\infty}^{\infty} dz'\,[g(z, t\,|z', 0)\varphi_{t'}(z', 0) - \varphi(z', 0)\,\partial g_{t'}(z, t\,|z', 0)] \qquad (9.8.12)$$

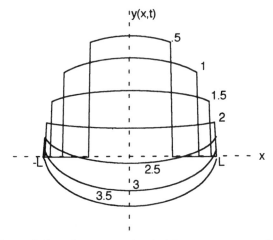

Figure 9.5. Pulse propagation in inhomogeneous medium.

The solution of the original problem is then obtained by retracing our steps through the transformations $z = L \tanh^{-1}(x/L)$ and $\varphi = y \cosh(z/L)$ to obtain

$$y(x, t) = \text{sech}\left(\tanh^{-1}\frac{x}{L}\right) \varphi\left(L \tanh\frac{x}{L}, t\right) \qquad (9.8.13)$$

As an example, consider the case of an initial velocity imposed at $x = 0$. The initial conditions are expressed as $y(x, 0) = 0$, $y_t(x, 0) = A\delta(x)$. In terms of z, we have (using the last property of the delta function listed in (1.5.9))

$$y_t(x, 0) = A\delta\left(L \tanh\frac{z}{L}\right) = A\left(\cosh^2\frac{z}{L}\right)\delta(z) = A\delta(z) \quad (9.8.14)$$

The initial conditions are now $\varphi(z, 0) = 0$ and $\varphi_t(z, 0) = A\delta(z)$. The z' integration in (9.8.12) is now immediate and we obtain

$$y(x, t) = \frac{Ac}{2}\sqrt{1 - \left(\frac{x}{L}\right)^2}\, J_0\left(\frac{c_0}{L}\sqrt{t^2 - \frac{L^2}{c_0^2}\left(\tanh^{-1}\frac{x}{L}\right)^2}\right) \qquad (9.8.15)$$

A graph of the result is shown in Figure 9.5.

9.9 A HYBRID FOURIER TRANSFORM

In Chapter 6 we found that the Fourier sine and cosine transforms provide a convenient approach for solving problems in a semi-infinite region when a boundary condition involving the value of either the function or its derivative is imposed. We now consider a generalization of this method that may be used for solving

$$f'' + k^2 f = \varphi(x), \qquad 0 \le x < \infty \qquad (9.9.1)$$

when the boundary condition

$$f'(0) - af(0) = 0 \qquad (9.9.2)$$

is satisfied.

To determine the appropriate transform pair we first consider a finite region $0 \le x \le L$ and then proceed to the limit $L \to \infty$, as was done for the sine and cosine transform pairs in Section (A.5). To satisfy the homogeneous boundary condition (9.9.2) as well as $f(L) = 0$, the constants in the solution $f(x) = A \cos kx + B \sin kx$ must be related according to

$$aA - kB = 0$$
$$A \cos kL + B \sin kL = 0 \qquad (9.9.3)$$

To obtain nonvanishing solutions of this homogeneous pair of equations, we require the vanishing of the determinant of the system, which yields

$$\begin{vmatrix} a & -k \\ \cos kL & \sin kL \end{vmatrix} = k \cos kL + a \sin kL = 0 \qquad (9.9.4)$$

The solutions of this transcendental equation, which is more conveniently written in the form

$$a \tan kL = k \qquad (9.9.5)$$

yield eigenvalues that will be labeled k_n, $n = 1, 2, 3, \ldots$. The corresponding eigenfunctions are

$$y_n(x) = k_n \cos k_n x + a \sin k_n x \qquad (9.9.6)$$

We now consider the expansion of an arbitrary function $f(x)$ in terms of these functions and write

$$f(x) = \sum_1^\infty c_n y_n(x) \qquad (9.9.7)$$

Use of the results on orthogonality in Appendix C shows that the functions $y_n(x)$ are orthogonal over the interval $0 \le x \le L$ (as long as the constant a does not depend upon n). A standard usage of orthogonality relations leads to

$$c_n = \frac{1}{N_n} \int_0^L d\xi \, f(\xi) \, y_n(\xi) \qquad (9.9.8)$$

where

$$N_n = \int_0^L d\xi \, [\, y_n(\xi)]^2 \tag{9.9.9}$$

We thus obtain the expansion (9.9.7) in the form

$$f(x) = \sum_{n=1}^{\infty} \frac{1}{N_n} \int_0^L d\xi \, f(\xi) \, y_n(\xi) \, y_n(x) \tag{9.9.10}$$

The normalized integral N_n is readily evaluated by using the general result given in Eq. (C.20). One finds

$$N_n = \frac{1}{2k_n} [(1 + aL)\cos k_n L - k_n L \sin k_n L](k_n a \cos k_n L - k_n^2 \sin k_n L) \tag{9.9.11}$$

When (9.9.5) is used to relate $\sin k_n L$ to $\cos k_n L$ and is also used in the form $\sec^2 k_n L = 1 + \tan^2 k_n L = 1 + (k_n/a)$ to express $\cos^2 k_n L$ in algebraic terms, we find

$$N_n = \tfrac{1}{2}[a(1 + aL) + k_n^2 L] \tag{9.9.12}$$

In the limit $aL \gg 1$, we then obtain

$$N_n = \frac{L}{2}(k_n^2 + a^2) \tag{9.9.13}$$

To proceed with consideration of the limit $L \to \infty$, we must develop the prescription for the corresponding decrease in the mode spacing $\Delta k = k_{n+1} - k_n$. From (9.9.4) with $k = k_{n+1}$ we have

$$(k_n + \Delta k)\cos(k_n + \Delta k)L + a \sin(k_n + \Delta k)L = 0 \tag{9.9.14}$$

Expanding this expression and using the identity $k_n \cos k_n L + a \sin k_n L = 0$ from (9.9.4), we have

$$k^{-1}\sin k_n L\{-a \, \Delta k \cos \Delta k \, L - \sin \Delta k \, L[(k_n^2 + a^2) + k_n \, \Delta k]\} = 0 \tag{9.9.15}$$

To determine the value of $\Delta k \, L$ that leads to the vanishing of the term in brackets, we first multiply by L and note that as $L \to \infty$ with $\Delta k \, L$ fixed, we obtain

$$L(k^2 + a^2)\sin \Delta k \, L = 0 \tag{9.9.16}$$

We thus obtain the same relation between mode spacing Δk and length L in this limit as was obtained for the sine and cosine transform, that is, $\Delta k \, L = \pi$. We can now write the normalization constant as

$$N_n^{-1} = \frac{2}{\pi} \frac{\Delta k}{k_n^2 + a^2} \tag{9.9.17}$$

Using this relation in (9.9.10) and replacing the sum by an integral, we obtain

$$f(x) = \frac{2}{\pi} \int_0^\infty dk \, F_H(k) \, y(x, k) \tag{9.9.18}$$

in which $y(k, x) = k \cos kx + a \sin kx$ and where

$$F_H(k) = \frac{1}{k^2 + a^2} \int_0^\infty d\xi \, f(\xi) \, y(\xi, k) \tag{9.9.19}$$

The relations (9.9.18) and (9.9.19) provide the integral transform pair that we are seeking.

When (9.9.19) is introduced into (9.9.18) and the order of integration is interchanged, we obtain

$$f(x) = \frac{2}{\pi} \int_0^\infty d\xi \, f(\xi) \int_0^\infty dk \, \frac{y(x, k) \, y(\xi, k)}{k^2 + a^2} \tag{9.9.20}$$

from which we conclude that

$$\delta(x - \xi) = \frac{2}{\pi} \int_0^\infty dk \, \frac{y(x, k) \, y(\xi, k)}{k^2 + a^2} \tag{9.9.21}$$

For application to the solution of the differential equation (9.9.1), we require the transform of the second derivative. Assuming convergence at infinity as in the case of the sine and cosine transform, we have

$$\frac{1}{k^2 + a^2} \int_0^\infty dx \, \frac{d^2 f}{dx^2} \, y_n(x) = \frac{1}{k^2 + a^2} \int_0^\infty dx \, \frac{d^2 f}{dx^2} (k \cos kx + a \sin kx)$$

$$= -k^2 F_H(k) + \frac{k}{k^2 + a^2} [a f(0) - f'(0)]$$

$$\tag{9.9.22}$$

Note that if $f(x)$ satisfies the boundary condition (9.9.2), the transform of the second derivative reduces to $-k^2 F_H(k)$.

As an example, consider a semi-infinite string excited by a point force of amplitude F_0 at $x = x_0$ and subject to the boundary condition (9.9.2).

We take the transform of

$$\frac{d^2 f}{dx^2} + k_0^2 f = -\frac{F_0}{T} \delta(x - x_0) \tag{9.9.23}$$

by multiplying by $(k^2 + a^2)^{-1} y(x, k)$ and integrating over the length of the system. Using (9.2.22) and the definition of the hybrid transform, we have

$$-k^2 F_H(k) + k_0^2 F_H(k) = -\frac{F_0}{T} \frac{1}{k^2 + a^2} y(x_0, k) \tag{9.9.24}$$

Upon solving for $F_H(k)$ and using the inverse transform, we obtain

$$f(x) = \frac{F_0}{T} \frac{2}{\pi} \int_0^\infty dk \, \frac{y(x, k) \, y(x_0, k)}{(k^2 - k_0^2)(k^2 + a^2)} \tag{9.9.25}$$

which can be rewritten as

$$\begin{aligned} f(x) &= \frac{F_0}{T} \frac{2}{\pi} \int_0^\infty dk \, \frac{y(x, k) \, y(x_0, k)}{k_0^2 + a^2} \left(\frac{1}{k^2 - k_0^2} - \frac{1}{k^2 + a} \right) \\ &= \frac{F_0}{T} \frac{2}{\pi} \frac{1}{k_0^2 + a^2} \int_0^\infty dk \, \frac{y(x, k) y(x_0, k)}{k^2 - k_0^2} - \frac{F_0}{T(k_0^2 + a^2)} \delta(x - x_0) \end{aligned} \tag{9.9.26}$$

where the definition of the delta function given in (9.2.21) has been employed.

Since $y(x, k)$ increases linearly with k, the integrand of the remaining integral increases as k^2 for large k and therefore contains another delta function. The two delta functions are found to enter with equal and opposite amplitude and thus cancel out. When the definition of the eigenfunctions $y(x, k)$ is introduced and trigonometric identities employed, we obtain integrals that can be evaluated by using entries 3 in Table F.3 and 3 in Table F.4 when we replace k_0 by $k_0 + i\epsilon$ as was done in obtaining the Green's function for the semi-infinite string in Section 6.2.1. The result is

$$f(x) = \frac{F_0}{T} \frac{i}{2k_0} \left(e^{ik_0|x - x_0|} + \frac{k_0 - ia}{k_0 + ia} e^{ik_0(x + x_0)} \right) \tag{9.9.27}$$

which is readily interpreted as a wave due to the source term plus that from an appropriate image.

A corresponding problem involving $f'' - k^2 f$ could also be considered. Here, if the constant a in the boundary condition is negative, one finds that in addition to the hybrid transform integrals used above, there is also a localized solution $e^{-|a|x}$ that is not contained in the integral transform. In fact, one can readily show that in this case, where $y(x, k) = k \cos kx - |a| \sin kx$, the localized solution is orthogonal to the solutions $y(x, k)$, that is,

$$\int_0^\infty dx \, e^{-|a|x} y(x, k) = \int_0^\infty dx \, e^{-|a|x} (k \cos kx - |a| \sin kx) = 0 \quad (9.9.28)$$

The integrations are immediately performed as Laplace transforms.

We will not pursue the subject into this stage since it becomes more fruitful to consider it by using complex variable techniques (Friedman, 1990, Ch. 4).

9.10 INVARIANTS OF THE LINEAR PARABOLIC EQUATION

Nearly all of the partial differential equations considered in this volume have been linear. Solutions of nonlinear equations require an understanding of more subtle features of the subject than is required for linear equations. To see briefly how a nonlinear equation can be associated with more subtle mathematical issues, we now consider the nonlinear equation

$$u_{xx} + uu_x + u_y = 0 \quad (9.10.1)$$

and its relation to the general *linear* parabolic equation. The nonlinear equation (9.10.1), with the variable y interpreted as time, was encountered in Section 8.3 where it was referred to as Burgers' equation. We now show how it arises quite incidentally when certain purely mathematical properties of the linear parabolic equation are developed.

Consider the linear equation

$$z_{xx} + 2\alpha z_x + 2\beta z_y + \gamma z = 0 \quad (9.10.2)$$

where α, β, and γ may be functions of x, y. If we transform to a new variable ζ according to $z = \lambda(x, y)\zeta$, then calculation of the various derivatives of z leads to the following equation for ζ:

$$\zeta_{xx} + 2\alpha' \zeta_x + 2\beta' \zeta_y + \gamma' \zeta = 0 \quad (9.10.3)$$

where

$$\beta' = \beta$$

$$\alpha' = \alpha + \frac{1}{\lambda} \frac{\partial \lambda}{\partial x} \quad (9.10.4)$$

$$\gamma' = \gamma + \frac{2\alpha}{\lambda} \frac{\partial \lambda}{\partial x} + \frac{1}{\lambda} \frac{\partial^2 \lambda}{\partial x^2} + \frac{2\beta}{\lambda} \frac{\partial \lambda}{\partial y}$$

We see that the coefficient of the y derivative is unchanged. It is referred to as an invariant of the transformation. We now inquire about the possibility of constructing another invariant by eliminating λ between the two equations relating α' and γ' to α and γ. From the second of (9.10.4) we have

$$\frac{\partial}{\partial x}(\alpha' - \alpha) = \frac{\partial}{\partial \lambda}\left(\frac{1}{\lambda}\frac{\partial \lambda}{\partial x}\right) = -\left(\frac{1}{\lambda}\frac{\partial \lambda}{\partial x}\right)^2 + \frac{1}{\lambda}\frac{\partial^2 \lambda}{\partial x^2}$$

$$= (\alpha' - \alpha)^2 + \frac{1}{\lambda}\frac{\partial^2 \lambda}{\partial x^2} \tag{9.10.5}$$

With this expression for $\partial^2 \lambda/\partial x^2$, we readily find that the third relation in (9.10.4) becomes

$$\gamma' - \frac{\partial \alpha'}{\partial x} - \alpha'^2 = \gamma - \frac{\partial \alpha}{\partial x} - \alpha^2 + \frac{2\beta}{\lambda}\frac{\partial \lambda}{\partial y} \tag{9.10.6}$$

Assuming that $\beta \neq 0$ [for otherwise (9.10.2) would be merely an ordinary differential equation] and then differentiating (9.10.6) with respect to x, we have

$$\frac{\partial}{\partial x}\left[\frac{1}{\beta'}\left(\gamma' - \frac{\partial \alpha'}{\partial x} - \alpha'^2\right)\right] = \frac{\partial}{\partial x}\left[\frac{1}{\beta}\left(\gamma - \frac{\partial \alpha}{\partial x} - \alpha^2\right)\right] = 2\frac{\partial^2 \ln \lambda}{\partial y\,\partial x}$$

$$= 2\frac{\partial}{\partial y}\left(\frac{1}{\lambda}\frac{\partial \lambda}{\partial x}\right) \tag{9.10.7}$$

When the second of (9.10.4) is used to eliminate λ, we finally obtain

$$\frac{\partial}{\partial x}\left[\frac{1}{\beta'}\left(\gamma' - \frac{\partial \alpha'}{\partial x} - \alpha'^2\right)\right] - 2\frac{\partial \alpha'}{\partial y}$$

$$= \frac{\partial}{\partial x}\left[\frac{1}{\beta}\left(\gamma - \frac{\partial \alpha}{\partial x} - \alpha^2\right)\right] - 2\frac{\partial \alpha}{\partial y} \tag{9.10.8}$$

We thus see that a second invariant is

$$J \equiv \frac{\partial}{\partial x}\left[\frac{1}{\beta}\left(\gamma - \frac{\partial \alpha}{\partial x} - \alpha^2\right)\right] - 2\frac{\partial x}{\partial y} \tag{9.10.9}$$

If we wish to begin an investigation of the original equation (9.10.2), we might consider those equations for which this invariant vanishes. Following up this avenue, we see from (9.10.9) that the equation $J = 0$ is equivalent to the requirement that there be a function $\theta(x, y)$ such that

$$\alpha = \frac{\partial \theta}{\partial x} \tag{9.10.10}$$

$$\frac{1}{2\beta}\left(\gamma - \frac{\partial \alpha}{\partial x} - \alpha^2\right) = \frac{\partial \theta}{\partial y}$$

When α and γ in (9.10.2) are written in terms of θ we have

$$\alpha = \frac{\partial \theta}{\partial x}, \qquad \gamma = 2\beta \frac{\partial \theta}{\partial y} + \frac{\partial \alpha}{\partial x} + \alpha^2 \tag{9.10.11}$$

and (9.10.2) becomes

$$z_{xx} + 2\frac{\partial \theta}{\partial x} z_x + \left[\frac{\partial^2 \theta}{\partial x^2} + \left(\frac{\partial \theta}{\partial x}\right)^2\right] z + 2\beta z \frac{\partial}{\partial y}(\ln ze^\theta) = 0 \tag{9.10.12}$$

Setting $w = ze^\theta$ in this equation and multiplying by ye^θ, we obtain

$$\frac{\partial^2 w}{\partial x^2} + 2\beta \frac{\partial w}{\partial y} = 0 \tag{9.10.13}$$

Setting $\beta = \frac{1}{2}$, we obtain the invariant J, which has been set equal to zero, in the form

$$J = 2\left[\frac{\partial}{\partial x}\left(\gamma - \frac{\partial \alpha}{\partial x} - \alpha^2\right) - \frac{\partial \alpha}{\partial y}\right] = 0 \tag{9.10.14}$$

If we restrict ourselves to the case $\gamma = 0$, for which $z = 1$ is a solution and thus $w = e^\theta$, we see that

$$\frac{\partial^2 \alpha}{\partial x^2} + 2\alpha \frac{\partial \alpha}{\partial x} + \frac{\partial \alpha}{\partial y} = 0 \tag{9.10.15}$$

which has the form of Burgers' equation. Recalling that $\alpha = \partial \theta / \partial x$, we have the relationships

$$\frac{\partial^2 \alpha}{\partial x^2} + 2\alpha \frac{\partial \alpha}{\partial x} + \frac{\partial \alpha}{\partial y} = 0$$

$$\alpha = \frac{\partial \theta}{\partial x} = \frac{\partial \ln w}{\partial x}$$

$$\frac{\partial^2 w}{\partial x^2} + \frac{\partial w}{\partial y} = 0 \tag{9.10.16}$$

On setting $w = \theta$, $\alpha = -u/2\nu$, and $y = -\nu t$, we obtain Burgers' equation (8.3.38) and the linear heat equation (8.2.4).

We find, therefore, that the linearizing transformation for Burgers' equation arises in a natural way when we examine a certain invariant possessed by the linear diffusion equation. An extensive discussion of the procedure outlined here may be found in the literature (Forsyth, 1959, Vol. 6, pp. 97 et seq).

APPENDIX A
Fourier Series

The basic notions of Fourier series that are useful for solving partial differential equations as well as the transition from Fourier series to Fourier integral are summarized in this appendix.

A.1 INTRODUCTION

Provided certain conditions that will be considered in Section A.3 are satisfied, a function $f(x)$ having a period $2p$ may be expanded in a series of the form

$$f(x) = \sum_{n=0}^{\infty} \left(a_n \cos \frac{n\pi x}{p} + b_n \sin \frac{n\pi x}{p} \right) \tag{A.1}$$

The coefficients a_n and b_n may be determined by exploiting the fact that over any interval of length $2p$ the functions $\cos n\pi x/p$ and $\sin n\pi x/p$ satisfy the orthogonality relations

$$\int_d^{d+2p} \cos \frac{n\pi x}{p} \cos \frac{m\pi x}{p} \, dx = p\delta_{nm}, \qquad n, m \neq 0$$

$$\int_d^{d+2p} \sin \frac{n\pi x}{p} \sin \frac{m\pi x}{p} \, dx = p\delta_{nm}$$

$$\int_d^{d+2p} \sin \frac{n\pi x}{p} \cos \frac{m\pi x}{p} \, dx = 0, \qquad \text{all } m, n \tag{A.2}$$

where

$$\delta_{nm} = \begin{cases} 1, & n = m \\ 0, & n \neq m \end{cases} \tag{A.3}$$

and d is arbitrary. The first two integrals are thus equal to zero unless $m = n$ and in that special case the integrals have the value p. For $n = m = 0$, the first integral in (A.2) equals $2p$ while the second and third integrals vanish in this case.

To use these relations for determining the coefficients a_n in the series for $f(x)$, we multiply (A.1) by cos $m\pi x/p$ and integrate the resulting equation from d to $d + 2p$. If we assume that the order of integration and summation can be interchanged (cf. Section A.3), the integrals listed above lead to the vanishing of all integrals except the one for which $n = m$. We then obtain

$$\int_d^{d+2p} f(x)\cos \frac{m\pi x}{p} \, dx = pa_m \tag{A.4}$$

while for $n = 0$ we have

$$\int_d^{d+2p} f(x) \, dx = 2pa_0 \tag{A.5}$$

Similarly, multiplication of the series (A.1) by sin $m\pi x/p$ and integration over the same interval yield

$$\int_d^{d+2p} f(x)\sin \frac{m\pi x}{p} = pb_n \tag{A.6}$$

A.2   AN EXAMPLE

For the purpose of displaying the Fourier series method in its simplest form, it is useful to specialize the previous development by setting $d = -\pi$ and $p = \pi$. We then obtain

$$f(x) = \sum_{n=0}^{\infty} (a_n\cos nx + b_n\sin nx) \tag{A.7}$$

with

$$a_0 = \frac{1}{2\pi} \int_{-\pi}^{\pi} f(x) \, dx$$

$$a_n = \frac{1}{\pi} \int_{-\pi}^{\pi} f(x)\cos nx \, dx \tag{A.8}$$

$$b_n = \frac{1}{\pi} \int_{-\pi}^{\pi} f(x)\sin nx \, dx$$

Determination of the coefficients a_n and b_n is largely a computational consideration. As an example, consider the function $f(x) = e^x$ in the region $-\pi < x < \pi$. Then

$$a_0 = \frac{1}{2\pi} \int_{-\pi}^{\pi} e^x \, dx = \frac{\sinh \pi}{\pi}$$

$$a_n = \frac{1}{\pi} \int_{-\pi}^{\pi} e^x \cos nx \, dx = \frac{1}{\pi} \mathrm{Re} \int_{-\pi}^{\pi} e^x e^{inx} \, dx \qquad \text{(A.9)}$$

$$= \frac{1}{\pi} \mathrm{Re} \left. \frac{e^{(1+in)}}{1+in} \right|_{-\pi}^{\pi} = \frac{2(-1)^n}{\pi(1+n^2)} \sinh \pi$$

where Re signifies that only the real part of the subsequent expression is to be taken. This procedure is merely an artifice that enables one to avoid the integration-by-parts computation customarily associated with evaluation of this integral. Also, we have written $\cos n\pi = (-1)^n$. Similarly, we find

$$b_n = \frac{1}{\pi} \mathrm{Im} \int_{-\pi}^{\pi} e^{x+inx} \, dx = -\frac{2(-1)^n n}{\pi(1+n^2)} \sinh \pi \qquad \text{(A.10)}$$

where Im refers to the imaginary part of the subsequent expression.

It should be noted that for large values of n, the coefficients b_n will provide the main contribution to the series since they only decrease as n^{-1} whereas the coefficients a_n decrease as n^{-2} and thus decrease much more rapidly for large n. A graph of the result is shown in Figure A.1. The dotted line is the extension of e^x outside the region $-\pi < x < \pi$ and differs from the Fourier series representation, which is a periodic repetition of the graph that appears in the region $-\pi < x < \pi$. In the use of Fourier series to solve partial differential equations, it usually occurs that the function $f(x)$ is only of interest in the fundamental interval $-\pi < x < \pi$ (or $d < x < d + 2p$ in the general case).

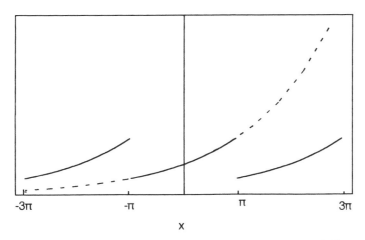

X

Figure A.1. Graph of e^x for $-3\pi < x < 3\pi$ (dashed line) and the periodic repetition of e^x outside the region $-\pi < x < \pi$ provided by the Fourier series.

The fact that the mathematical representation is repeated outside the fundamental interval is then of little concern.

If a function is either odd or even with respect to the origin in the fundamental interval $-\pi < x < \pi$, the Fourier series will contain only sine or cosine terms, respectively. For an expansion in terms of sine functions alone, we have $f(x) = \Sigma_n b_n \sin n\pi x/p$, and since $\sin x = -\sin(-x)$, the function $f(x)$ will satisfy $f(x) = -f(-x)$, that is, it will be an odd function of x. The function $f(x)$ is thus arbitrary for only half of the period $2p$. Similarly, the expansion $f(x) = \Sigma_n a_n \cos n\pi x/p$ implies that $f(x)$ is even, that is, $f(x) = f(-x)$.

If we again consider the exponential function of the previous example, but only in the region $0 < x < \pi$, and then use its odd extension into the region $-\pi < x < 0$, namely,

$$f(x) = \begin{cases} e^x, & 0 < x < \pi \\ -e^{-x}, & -\pi < x < 0 \end{cases} \tag{A.11}$$

we find

$$b_n = \frac{1}{\pi} \left(\int_{-\pi}^{0} (e^{-x}) \sin nx \, dx + \int_{0}^{\pi} e^x \sin nx \, dx \right)$$

$$= \frac{2}{\pi} \int_{0}^{\pi} e^x \sin nx \, dx = \frac{2}{\pi} \frac{n}{1 + n^2} [1 - (-1)^n e^{\pi}] \tag{A.12}$$

A calculation of the coefficients a_n from (A.8) shows that they are zero, as would be expected when one integrates an odd function between symmetric limits. Again, it should be noted that the coefficients decrease as n^{-1} for large values of n. A graph of the function that corresponds to this expansion is shown in Figure A.2.

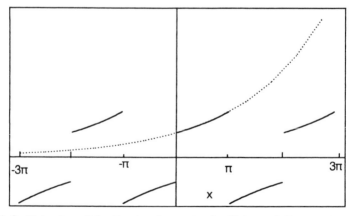

Figure A.2. Extension of the Fourier sine series for $f(x) = e^x$, $0 < x < \pi$, beyond this fundamental interval.

When we consider the *even* extension of e^x into the region $-\pi < x < 0$, that is, the function defined by

$$f(x) = \begin{cases} e^x, & 0 < x < \pi \\ e^{-x}, & -\pi < x < 0 \end{cases} \tag{A.13}$$

or equivalently, $f(x) = e^{-|x|}$, we obtain Figure A.3. The expansion coefficients are now

$$a_0 = \frac{1}{2\pi} \int_{-\pi}^{\pi} e^{|x|}\, dx = \frac{e^{\pi} - 1}{\pi} \tag{A.14}$$

and

$$a_n = \frac{1}{\pi} \int_{-\pi}^{\pi} e^{-|x|} \cos nx\, dx = \frac{2}{\pi} \int_{0}^{\pi} e^x \cos nx\, dx$$

$$= \frac{2}{\pi} \frac{-1 + (-1)^n e^{\pi}}{1 + n^2} \tag{A.15}$$

It should be noted that in this case the coefficients decrease as n^{-2} for large n. The series is thus more rapidly convergent than in the previous instance. A glance at Figure A.3 shows that in the present case there are no discontinuities in the function or its periodic repetition.

In general, we can expect that the smoother a function is, the more easily that function can be described by a Fourier series and consequently the smaller the number of terms that are needed to approximate the function, that is, the more rapidly the series converges.

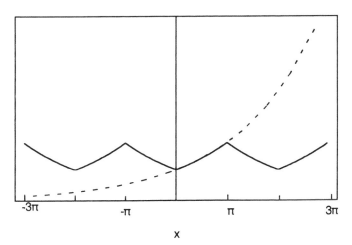

x

Figure A.3. Extension of the Fourier cosine series for $f(x) = e^x$, $0 \le x \le \pi$, beyond this fundamental region.

A.3 CONVERGENCE OF FOURIER SERIES

Most of the results that are summarized in this section will be more readily appreciated if the reader has a few examples in mind. Figure A.4 contains a brief list of Fourier series and approximate sketches of the functions corresponding to them.

In the figure the two sets of three examples each are arranged in order of increasing smoothness of the functions being expanded. Examples (1) and (4) are discontinuous in value, even singular in the case of (4), while (2) and (5) are continuous in value but discontinuous in slope at certain points. Finally, (3) and (6) are continuous in both value and slope. The coefficients in the associated expansions indicate the general result that the smoother the function is, the higher the inverse power of n in the coefficients of the Fourier series

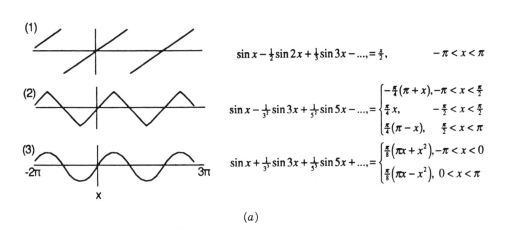

$$\sin x - \tfrac{1}{2}\sin 2x + \tfrac{1}{3}\sin 3x - ..., = \tfrac{x}{2}, \qquad -\pi < x < \pi$$

$$\sin x - \tfrac{1}{3^2}\sin 3x + \tfrac{1}{5^2}\sin 5x - ..., = \begin{cases} -\tfrac{\pi}{4}(\pi + x), & -\pi < x < \tfrac{\pi}{2} \\ \tfrac{\pi}{4}x, & -\tfrac{\pi}{2} < x < \tfrac{\pi}{2} \\ \tfrac{\pi}{4}(\pi - x), & \tfrac{\pi}{2} < x < \pi \end{cases}$$

$$\sin x + \tfrac{1}{3^3}\sin 3x + \tfrac{1}{5^3}\sin 5x + ..., = \begin{cases} \tfrac{\pi}{8}(\pi x + x^2), & -\pi < x < 0 \\ \tfrac{\pi}{8}(\pi x - x^2), & 0 < x < \pi \end{cases}$$

(a)

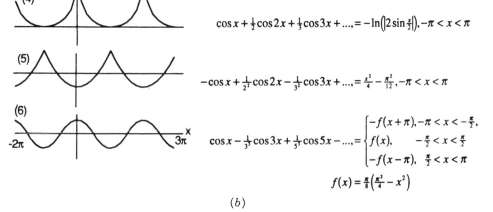

$$\cos x + \tfrac{1}{2}\cos 2x + \tfrac{1}{3}\cos 3x + ..., = -\ln\left(\left|2\sin \tfrac{x}{2}\right|\right), -\pi < x < \pi$$

$$-\cos x + \tfrac{1}{2^2}\cos 2x - \tfrac{1}{3^2}\cos 3x + ..., = \tfrac{x^2}{4} - \tfrac{\pi^2}{12}, -\pi < x < \pi$$

$$\cos x - \tfrac{1}{3^3}\cos 3x + \tfrac{1}{5^3}\cos 5x - ..., = \begin{cases} -f(x+\pi), & -\pi < x < -\tfrac{\pi}{2}, \\ f(x), & -\tfrac{\pi}{2} < x < \tfrac{\pi}{2} \\ -f(x-\pi), & \tfrac{\pi}{2} < x < \pi \end{cases}$$

$$f(x) = \tfrac{\pi}{8}\left(\tfrac{\pi^2}{4} - x^2\right)$$

(b)

Figure A.4.

and thus the more rapid the convergence. Note, for instance, that if we set $x = \pi/2$ in (1) and $x = \pi$ in (4), we obtain the respective results

$$\frac{\pi}{4} = 1 - \frac{1}{3} + \frac{1}{5} - \cdots, \qquad \ln 2 = 1 - \frac{1}{2} + \frac{1}{3} - \frac{1}{4} + \cdots \qquad (A.16)$$

Both of these series are extremely slow in their convergence. On the other hand, setting $x = 0$ in (6) yields

$$\frac{\pi^3}{32} = 1 - \frac{1}{3^3} + \frac{1}{5^3} - \cdots \qquad (A.17)$$

which converges quite rapidly. The sum of the first five terms in the series is 0.9687, while $\pi^3/32 = 0.9689 \ldots$.

There are many instances in which one wishes to apply calculus operations to a Fourier series, such as the differentiation of a series with respect to a parameter appearing in the series or the determination of the value of a series as a parameter appearing in the series approaches some limit. One may question when one is justified in concluding that

$$\frac{d}{dx} \sum a_n \sin nx = \sum n a_n \cos nx$$

$$\lim_{t \to 0} \sum a_n(t) \sin nx = \sum a_n(0) \sin nx \qquad (A.18)$$

For a series with a *finite* number of terms the differentiation or limiting procedure may be carried out in a term-by-term fashion. For an infinite series, however, this term-by-term calculation may not be valid and some criterion for its validity is needed. A somewhat similar situation arises when an infinite series is subjected to *algebraic* operations. In that case it is known (Chrystal, 1964, Vol. 1, p. 141) that a series must be absolutely convergent in order for algebraic operations to be valid. For instance, the sum of the series $1 - \frac{1}{2} + \frac{1}{3} - \frac{1}{4} + \cdots$, which is only conditionally convergent, can be changed at will by grouping the terms in various ways.

Although the function $x/2$ expanded in (1) of Figure A.4a has a derivative equal to $\frac{1}{2}$ in the region $-\pi < x < \pi$, term-by-term differentiation of the series yields $\cos x - \cos 2x + \cos 3x - \cdots$, which does not converge and thus does not, without special considerations (Jeffreys and Jeffreys, 1946, p. 441), represent the value $\frac{1}{2}$. On the other hand, the function $\frac{1}{4}x^2 - (\pi^2/12)$ that is expanded in a Fourier series in (5) of Figure A.4b has the derivative $x/2$ in the interval $-\pi < x < \pi$. The function $x/2$ is expanded in (1) and the term-by-term differentiation of (5) does yield the series of (1) for all values of x except $x = n\pi$ where the periodic figure shown in (5) does not have a derivative. In general, one can expect that the derivative of a Fourier series for $f(x)$ represents $f'(x)$ if the periodic figure of $f(x)$ represented by the series is continuous. This in turn requires that the coefficients in the series decrease at least as rapidly as n^{-2}.

A convergence condition that *guarantees* that calculus operations may be performed on a Fourier series (or on an infinite series in general) in a term-by-term manner in some interval is that of uniform convergence. This is a condition that must be satisfied by the remainder term in a series for any value of x in the interval in which the function is being expanded. Since, in general, the application of this concept to specific cases is quite cumbersome, equivalent but more convenient comparison tests are usually employed. We now consider two of the simpler of these tests.

A.3.1   Weierstrass M-Test

If a function $f(x)$ is expanded in a series $\Sigma_n u_n(x)$ in an interval $a \leq x \leq b$ and after some kth term in the series the functions $u_n(x)$ have the property that, for *all* x in that interval, $|u_n(x)| \leq M_n$ where M_n is a positive constant and ΣM_n converges, then the series $u_n(x)$ converges uniformly in the given interval.

For differentiation of a series, it is the *differentiated* series that must satisfy this condition. For instance, if we use the comparison series $\Sigma M_n = \Sigma_n 1/n^2$, then series (2) satisfies this condition. We can use this result to justify term-by-term differentiation of (6) and it may be noted that the result is equal to (the negative of) (2). As mentioned above, the derivative of (5) yields (1) but the Weierstrass test is not applicable to (1). Here one can use a more subtle test, the Dirichlet test (Carslaw, 1980, p. 151), to show that (1) is uniformly convergent except at the points of discontinuity.

If a series only converges conditionally, as is the case for (1) in Figure A.4a, application of the Weierstrass test will not be possible. A more subtle test that frequently is adequate is Abel's test.

A.3.2   Abel's Test

If Σa_n converges only conditionally in $a \leq x \leq b$ and functions $v_n(t)$ are positive, bounded, and nonincreasing with increasing n for fixed t, then the series $\Sigma a_n v_n(t)$ is uniformly convergent in the interval.

As an example, in Problem 3.6.2 it is shown that the cooling of a sphere initially at a uniform temperature u_0 is given by

$$\frac{u(r, t)}{u_0} = \Sigma (-1)^n e^{-n^2 vt} \frac{\sin(n\pi r/a)}{n\pi r/a} \qquad (A.19)$$

For $t = 0$ this series only converges conditionally and the Weierstrass test is not applicable. Setting $a_n = (-1)^n(\sin n\pi r/a)/(n\pi r/a)$, the exponential term in (A.19) satisfies the conditions on $v_n(t)$ in Abel's test. Thus the series for $u(r, t)$ converges uniformly and one is justified in interchanging the limit and summation procedures to conclude (by using series 23 in Table F.1 with a set equal to 0) that

$$\lim_{t \to 0} \frac{u(r, t)}{u_0} = \sum_{n=1}^{\infty} a_n = 1 \qquad (A.20)$$

that is, the solution does indeed reduce to the expected initial value at $t = 0$.

The Abel test is especially useful in dealing with series solutions that arise in steady state problems in polar coordinates where solutions are expressed in a form such as

$$u(r, \vartheta) = \sum A_n \left(\frac{r}{a}\right)^n \sin n\vartheta \qquad (A.21)$$

As long as $\sum_n A_n \sin n\vartheta$ converges at last conditionally, the function $(r/a)^n$, with r now playing the role of t, corresponds to $v_n(t)$ and one may conclude that

$$\lim_{r \to a} u(r, \vartheta) = \sum A_n \sin n\vartheta = u(a, \vartheta) \qquad (A.22)$$

It should be emphasized that uniform convergence is only a *sufficient* condition to *guarantee* application of calculus operations to a series. It is by no means necessary. As long as one is dealing with physically reasonable functions, there can be a high expectation that the procedures can be applied.

A.4 HALF-RANGE EXPANSIONS—SINE AND COSINE SERIES

A frequent application of Fourier series is in expressing the solution of a partial differential equation over some interval, say $0 \le x \le L$. The boundary conditions satisfied by the solution at the end points impose certain requirements on the type of function that must be used for the expansion. Frequently the functions are either $\sin n\pi x/L$ or $\cos n\pi x/L$. These functions either vanish or have vanishing derivatives at $x = 0, L$ and thus a function expanded in terms of them will satisfy the same boundary conditions. According to the development in Section A.1, a series composed of functions $\sin n\pi x/L$ or $\cos n\pi x/L$ will produce a function that is periodic over a region that is of length $2L$ (say between $-L$ and L). This length is twice the length of interest. We are able to apply the previous results, however, by extending the function beyond the physical region $0 < x < L$ into an unphysical region $-L < x < 0$ in such a way as to satisfy the symmetry requirements imposed by the functions to be used in the expansion, that is, $\sin n\pi x/L$ or $\cos n\pi x/L$. For example, if the expansion is in terms of $\sin n\pi x/L$, then the extension is carried out by requiring that the function satisfy $f(-x) = -f(x)$, that is, $f(x)$ in the entire region $-L < x < L$ must be an odd function. For expansions in terms of $\cos n\pi x/L$, the even extension $f(-x) = f(x)$ is used. Figure A.4 may be viewed as providing a number of examples of functions specified in the region $0 < x < L$ (with $L = \pi$) and extended in an odd (1–3) or even (4–6) manner into the region $-L < x < 0$.

The formulas that are appropriate for calculating the coefficients a_n and b_n are immediately obtained from those in (A.4)–(A.6) by writing $d = -L$ and $p = L$ and also imposing the appropriate symmetry on $f(x)$. In particular, if the expansion functions are $\sin n\pi x/L$, then $f(x)$ will be odd. The coefficients a_n will vanish for all n and the form of b_n given in (A.6) becomes

$$b_n = \frac{1}{L} \int_{-L}^{L} f(x) \sin \frac{n\pi x}{L} \, dx = \frac{2}{L} \int_0^L f(x) \sin \frac{n\pi x}{L} \, dx \qquad \text{(A.23)}$$

Similarly, for expansions in terms of $\cos n\pi x/L$ the extension of $f(x)$ is even. We then have $b_n = 0$ and

$$a_0 = \frac{1}{L} \int_0^L f(x) \, dx$$

$$a_n = \frac{2}{L} \int_0^L f(x) \cos \frac{n\pi x}{L} \, dx, \qquad n \geq 1 \qquad \text{(A.24)}$$

Example A.1. Expand $f(x) = x/2$, $0 < x < \pi$, in terms of the functions $\sin nx$.

Since the expansion is in terms of sine functions in the interval $0 < x < \pi$, we use (A.23) with $L = \pi$ and obtain

$$b_n = \frac{2}{\pi} \int_0^\pi \frac{x}{2} \sin nx \, dx$$

$$= \frac{1}{\pi} \left(-x \frac{\cos nx}{n} \bigg|_0^\pi + \int_0^\pi \frac{\cos nx}{n} \, dx \right)$$

$$= -\frac{\cos n\pi}{n} = \frac{(-1)^{n+1}}{n} \qquad \text{(A.25)}$$

where integration by parts has been used to evaluate the integral. The series obtained is thus

$$\frac{x}{2} = \sin x - \frac{1}{2} \sin 2x + \frac{1}{3} \sin 3x - \cdots \qquad \text{(A.26)}$$

which is listed as (1) in Figure A.4a and also shown in Figure A.5a.

Since the odd extension of the function beyond the original interval gives rise to discontinuities, the expansion coefficients only decrease as n^{-1}.

Example A.2. Expand $f(x) = x/2$, $0 < x < \pi$, in terms of the function $\cos nx$. Setting $L = \pi$ in (A.24) we obtain

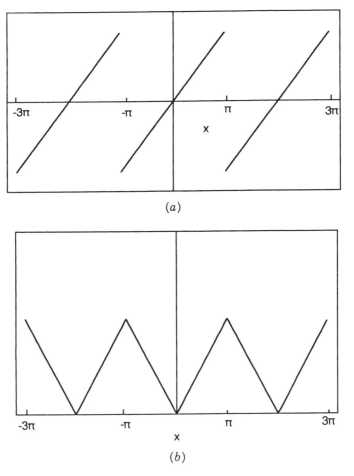

(a)

(b)

Figure A.5. (a) Sine and (b) cosine expansions for $f(x) = x/2$, $0 < x < \pi$.

$$a_0 = \frac{1}{\pi} \int_0^\pi \frac{x}{2} \, dx = \frac{\pi}{4}, \qquad a_n = \frac{2}{\pi} \int_0^\pi \frac{x}{2} \cos nx \, dx = \frac{(-1)^n - 1}{\pi n^2} \qquad (A.27)$$

The figure associated with this result is shown in Figure A.5b.

Since there are no discontinuities in value (only in slope) resulting from this extension, the series converges as n^{-2}.

The method of extending a function must be modified when other combinations of boundary conditions are encountered. For instance, if $f(0) = 0$ while $f'(L) = 0$, then the functions $\sin(2n + 1)\pi x/2L$ are appropriate since they satisfy the same boundary conditions as $f(x)$. We must also determine the additional symmetry requirement imposed on a function expanded in the form

$$f(x) = \sum_{n=0}^{\infty} b_n \sin \frac{(2n + 1)\pi x}{2L} \tag{A.28}$$

Clearly $f(-x) = -f(-x)$ as before, but also $f(x)$ must be an even function with respect to the point $x = L$, that is, $f(L - x) = f(L + x)$ or equivalently $f(2L - x) = f(x)$. We may exhibit this symmetry by writing

$$f(2L - x) = \sum b_n \sin \frac{(2n + 1)\pi(2L - x)}{2L}$$

$$= \sum b_n \sin(2n + 1)\left(\pi - \frac{\pi x}{2L}\right)$$

$$= -\sum b_n \cos(2n + 1)\pi \, \sin(2n + 1)\frac{\pi x}{2L}$$

$$= f(x) \tag{A.29}$$

since $\cos(2n + 1)\pi = -1$. The coefficients b_n are now obtained by considering the expansion of this function in the region $-2L$ to $2L$. Replacing L by $2L$ and n by $2n + 1$ in (A.23) we obtain

$$b_n = \frac{1}{L} \int_0^{2L} f(x) \sin \frac{(2n + 1)\pi x}{2L} \, dx$$

$$= \frac{1}{L}\left[\int_0^L f(x) \sin \frac{(2n + 1)\pi x}{2L} \, dx + \int_L^{2L} f(x) \sin \frac{(2n + 1)\pi x}{2L} \, dx \right]$$

$$\tag{A.30}$$

When we set $z = 2L - x$ in the integral from L to $2L$, it takes the same form as the first integral and we obtain

$$b_n = \frac{2}{L} \int_0^L f(x) \sin \frac{(2n + 1)\pi x}{2L} \, dx \tag{A.31}$$

Example A.3. A function possessing the symmetry considered here is provided by (2) in Figure A.4 when we set $2L = \pi$. The coefficients are given by

$$b_n = \frac{4}{\pi} \int_0^{\pi/2} \left(\frac{\pi}{4} x\right) \sin(2n + 1)x \, dx$$

$$= \frac{\sin(2n + 1)\pi/2}{(2n + 1)^2} \tag{A.32}$$

and the series is that given as (2) in Figure A.4a.

A.5 FOURIER INTEGRAL LIMIT

Thus far we have considered a function in an interval d to $d + 2p$ and its periodic extension outside that interval. We now consider the limiting case in which the interval $2p$ approaches infinity. There is then no periodic repetition of the function. We now show that the series representation considered previously is very naturally converted to an integral representation of the function. To carry out the transition, it is convenient to choose a section of the x axis that is symmetric about the origin. We thus set $d = -L$ and $p = L$ in (A.1) and (A.2) to obtain

$$ f(x) = a_0 + \sum_1^\infty \left(a_n \cos \frac{n\pi x}{L} + b_n \sin \frac{n\pi x}{L} \right) \qquad (A.33) $$

with coefficients given by

$$ a_0 = \frac{1}{2L} \int_{-L}^L f(y)\, dy $$

$$ a_n = \frac{1}{L} \int_{-L}^L f(y)\cos \frac{n\pi y}{L}\, dy, \qquad n = 1, 2, 3, \ldots \qquad (A.34) $$

$$ b_n = \frac{1}{L} \int_{-L}^L f(y)\sin \frac{n\pi y}{L}\, dy, \qquad n = 1, 2, 3, \ldots $$

When these expressions for the coefficients a_n and b_n are inserted into (A.33) and the formula for the cosine of the difference of two angles is used, we find

$$ f(x) = \frac{1}{2L} \int_{-L}^L f(y)\, dy + \frac{1}{L} \sum_1^\infty \int_{-L}^L f(y)\cos \frac{n\pi (x - y)}{L}\, dy \qquad (A.35) $$

Since the cosine is an even function of its argument, we may introduce a sum over negative n and obtain

$$ f(x) = \frac{1}{2L} \sum_{-\infty}^\infty \int_{-L}^L f(y)\cos \frac{n\pi (x - y)}{L}\, dy \qquad (A.36) $$

Furthermore, since the function $\sin n\pi (x - y)/L$ is an *odd* function of the summation variable n, the summation of such a function over all positive and negative values of n will contain terms that cancel in paris and thus will be equal to zero. We therefore make no change in the above expression for $f(x)$ if we add terms $\pm i \sin [n\pi (x - y)/L]$ and write (A.36) in the form

$$ f(x) = \frac{1}{2L} \sum_{-\infty}^\infty \int_{-L}^L f(y) \left[\cos \frac{n\pi (x - y)}{L} \pm i \sin \frac{n\pi (x - y)}{L} \right] dy $$

$$ = \frac{1}{2L} \sum_{-\infty}^\infty \int_{-L}^L f(y)\, e^{\pm in\pi (x - y)/L} \qquad (A.37) $$

Henceforth we shall adopt the convention of using the upper sign in these formulas.

We now consider the process of letting L approach ∞. If we introduce the abbreviation $k_n = n\pi/L$, then the neighboring values k_n and $k_{n+1} = (n+1)\pi/L$ will come closer together as $L \to \infty$. Let us define the difference between these values of k as Δk and write

$$k_{n+1} - k_n = \Delta k = \frac{\pi}{L} \tag{A.38}$$

As $L \to \infty$, the difference Δk approaches zero. It can then be interpreted as a differential dk. Since $L \Delta k/\pi = 1$, we may introduce this form of unity into the above expression for $f(x)$ and obtain

$$f(x) = \frac{1}{2L} \sum_{-\infty}^{\infty} \frac{L \Delta k}{\pi} \int_{-L}^{L} f(y) e^{ik_n(x-y)} \, dy \tag{A.39}$$

When we take the limit $L \to \infty$ and replace k_n by the continuous variable k, we arrive at the integral

$$f(x) = \frac{1}{2\pi} \int_{-\infty}^{\infty} dk \int_{-\infty}^{\infty} f(y) e^{ik(x-y)} \, dy$$

$$= \frac{1}{2\pi} \int_{-\infty}^{\infty} dk \, e^{ikx} \int_{-\infty}^{\infty} f(y) e^{-iky} \, dy \tag{A.40}$$

Setting $F(k) = \int_{-\infty}^{\infty} f(y) e^{-iky} \, dy$ we have the symmetric pair of relations

$$f(x) = \int_{-\infty}^{\infty} \frac{dk}{2\pi} e^{ikx} F(k)$$

$$F(k) = \int_{-\infty}^{\infty} dx \, e^{-ikx} f(x) \tag{A.41}$$

Equations (A.40) are frequently referred to as a complex Fourier transform or Fourier integral pair. The complex exponential is very convenient for evaluating these integrals by complex variable techniques. Since this background is not assumed in the present text, however, we shall use a table of transform pairs as is frequently done for Laplace transforms.

A.6 SINE AND COSINE TRANSFORMS

If $f(x)$ is an even function of x or, as frequently happens, $f(x)$ is only given for $x > 0$ and is then extended to negative values of x by the relation $f(-x) = f(x)$, then $f(x) \sin kx$ is an odd function of x and the integral of this expression over all x will vanish. We then obtain

$$F(k) = \int_{-\infty}^{\infty} f(x) \cos kx \, dx$$

$$= 2 \int_0^{\infty} f(x) \cos kx \, dx \tag{A.42}$$

Since $\cos kx$ is an even function of k, this result shows that $F(k)$ is also an even function of k. Hence, the first of (A.41) may be written

$$f(x) = \frac{1}{\pi} \int_0^{\infty} F(k) \cos kx \, dk \tag{A.43}$$

Writing $F_c(k) = \frac{1}{2}F(k)$, we obtain the relations

$$f(x) = \frac{2}{\pi} \int_0^{\infty} F_c(k) \cos kx \, dk$$

$$F_c(k) = \int_0^{\infty} f(x) \cos kx \, dx \tag{A.44}$$

This pair of relations is known as a *Fourier cosine transform.*

In a similar way, if $f(x)$ is an odd function of x, the term $f(x) \cos kx$ in the second of (A.41) will vanish, and on setting $(i/2)F(k) = F_s(k)$, we obtain the *Fourier sine transform pair*

$$f(x) = \frac{2}{\pi} \int_0^{\infty} F_s(k) \sin kx \, dk$$

$$F_s(k) = \int_0^{\infty} f(x) \sin kx \, dx \tag{A.45}$$

Brief tables of sine and cosine transforms, as well as the exponential transform given by (A.41), are given in Tables F.3–F.5.[1] Note that the two transform pairs are expressed only in terms of the value of $f(x)$ for $x > 0$. The symmetry of $f(x)$ for negative values of x is imposed by the transform that is used.

As an example of these symmetry considerations, consider $f(x) = e^{-ax}$ for $0 < x < \infty$. The cosine transform is seen from the second of (A.44) to be

$$F_s(k) = \int_0^{\infty} e^{-ax} \cos kx \, dx$$

$$= \frac{a}{k^2 + a^2} \tag{A.46}$$

[1]For a more extensive table see Erdelyi et al. (1954).

This result can be obtained either by direct evaluation, or by using the cosine transforms in Table F.4, or due to the structure of this particular integrand from the Laplace transforms in Table F.2. Note that the other relation in the cosine transform pair (A.44) enables us to infer

$$f(x) = \frac{2}{\pi} \int_0^\infty \frac{a}{k^2 + a^2} \cos kx \, dx = e^{-ax}, \qquad x > 0 \qquad (A.47)$$

without actually evaluating the integral at all. Since the integrand in (A.47) is even in x, the value of the integral for all x is $e^{-a|x|}$.

Similarly, e^{-ax}, $0 < x < \infty$, has the sine transform

$$F_s(k) = \int_0^\infty e^{-ax} \sin kx \, dx = \frac{k}{k^2 + a^2} \qquad (A.48)$$

From the first of (A.45) we obtain

$$f(x) = \frac{2}{\pi} \int_0^\infty \frac{k}{k^2 + a^2} \sin kx \, dk = \text{sgn}(x) e^{-a|x|}, \qquad -\infty < x < \infty \qquad (A.49)$$

where the oddness of the result is implied by the symbol $\text{sgn}(x)$, which has the definition

$$\text{sgn}(x) = \begin{cases} +1, & x > 0 \\ -1, & x < 0 \end{cases} \qquad (A.50)$$

A.7 GIBBS' PHENOMENON

We now consider a somewhat subtle issue associated with Fourier series. It occurs whenever a Fourier series is used to represent a function that possesses a discontinuity in value. As an example, consider the expansion

$$\sin x + \tfrac{1}{3} \sin 3x + \tfrac{1}{5} \sin 5x + \cdots = f(x) = \begin{cases} \dfrac{\pi}{4}, & 0 < x < \pi \\[2mm] -\dfrac{\pi}{4}, & -\pi < x < 0 \end{cases} \qquad (A.51)$$

We shall investigate the manner in which the first N terms in the series, $S_N(x)$, approach $f(x)$ as $N \to \infty$. The graph of

$$S_N(x) = \sum_{n=1}^N \frac{\sin(2n-1)x}{2n-1} \qquad (A.52)$$

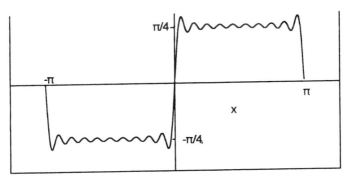

Figure A.6. An example showing the onset of the Gibbs phenomenon.

for $N = 5$ is shown in Figure A.6. In order to examine the oscillations as $N \to \infty$, let us locate the first peak to the right of the origin, determine its height, and then calculate the value of this height as $N \to \infty$. To locate the height we set $dS_n(x)/dx = 0$ and solve for x. The calculation leads to (cf. Problem A.4)

$$\frac{dS_N(x)}{dx} = \sum_{n=1}^{\infty} \cos(2n - 1)x = \frac{\sin 2Nx}{2 \sin x} \tag{A.53}$$

which vanishes at $x = \pi/2N, 2\pi/2N, \ldots, (2N - 1)\pi/2N$. To examine the first peak to the right of the origin, which is located at $x = \pi/2N$, consider the value of the first N terms in the series evaluated at that point, that is,

$$S_N\left(\frac{\pi}{2N}\right) = \sum_{n=1}^{N} \frac{\sin(2n - 1)\pi/2N}{2n - 1}$$

$$= \frac{\pi}{2N} \sum_{n=1}^{N} \frac{\sin(2n - 1)\pi/2N}{(2n - 1)\pi/2N} \tag{A.54}$$

We evaluate this series by first giving it an alternative interpretation. Note that if the integral $\int_0^\pi (\sin \pi x/x) \, dx$ is interpreted as the area under the curve $\sin \pi x/x$ and this area is then approximated by a series of rectangles (as shown in Fig. A.7 for the case $N = 5$), we have

$$\int_0^\pi \frac{\sin x}{x} \, dx = \lim_{N \to \infty} \left[\frac{\sin \pi/2N}{\pi/2N} \frac{\pi}{N} + \frac{\sin 3\pi/2N}{3\pi/2N} \frac{3\pi}{N} \right.$$

$$\left. + \cdots + \frac{\sin(2N - 1)\pi/2N}{(2N - 1)\pi/2N} \frac{\pi}{N} \right]$$

$$= \lim_{N \to \infty} \sum_{n=1}^{N} \frac{\sin(2N - 1)\pi/2N}{(2N - 1)\pi/2N} \frac{\pi}{N}$$

$$= 2 \lim_{N \to \infty} S_N\left(\frac{\pi}{2N}\right) \tag{A.55}$$

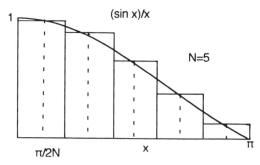

Figure A.7. Riemann sum diagram for replacing sum by integral for evaluating the Gibbs effect.

The area under the curve is thus equal to twice the value of the expression obtained previously in (A.54) for the height of the first peak $S_n(\pi/2N)$. As $N \to \infty$, the height of this peak is thus given in terms of the integral $\int_0^\pi$ (sin x/x) dx, which may be evaluated numerically. One finds $\int_0^\pi$ (sin x/x) dx = 1.85094. The value of the first peak thus overshoots the expected value of $\pi/4 = 0.7854$ and reaches the value

$$\lim_{N \to \infty} S_N(x) = \frac{1}{2} \int_0^\pi \frac{\sin x}{x} \, dx = \frac{1}{2} (1.85094) = 0.92597 \qquad (A.56)$$

The overshoot beyond the expected value of $\pi/4$ is thus $0.92597/(\pi/4)$ = 1.17898, or about 18%. This flange at a discontinuity in the Fourier series representation of a function is known as Gibbs' phenomenon. It should be emphasized that this effect is usually of no practical concern since the area under the flange is equal to zero.

Problems

A.1 A review of the techniques required for evaluating the coefficients in a Fourier series is provided by deriving the series listed in Figure A.4 as well as various entries in Table F.1. Evaluation of the integrals required to obtain (1) in Figure A.4 may be found elsewhere (Carslaw, 1980, p. 241).

A.2 Combine entries 23 and 24 in Table F.1 to yield entry 25, that is, show that

$$\frac{\pi}{4} \frac{\cosh a(\pi/2 - x)}{\cosh \pi a/2} = \sum_{\text{odd} \, n} \frac{n \sin nx}{n^2 + a^2}, \qquad -\pi < x < \pi$$

A.3 To see that a convergent Fourier series may result from integration of a divergent series (Carslaw, 1980, p. 283), consider the expansion

$$\cot x = \Sigma \, a_n \sin nx, \qquad -\pi < x < \pi$$

(a) Determine the coefficients a_n and obtain

$$\cot x = 2 \, (\sin 2x + \sin 4x + \sin 6x \, \ldots)$$

(b) Integrate both sides of the equation from $\pi/2$ to x and obtain series (4) in Figure A.4b.

A.4 Obtain the sum in (A.53) by replacing the series by the real part of Σ_1^N $e^{i(2n-1)x}$ and using the geometric series result $\Sigma_1^N a^n = (1 - a^N)/(1 - a)$.

A.5 Show that

$$\frac{1}{\sqrt{L^2 - x^2}} = \frac{\pi}{2L} + \frac{\pi}{L} \sum_{n=1}^{\infty} J_0(n\pi) \cos \frac{n\pi x}{L}, \qquad -L < x < L$$

Evaluate the coefficients by using the integral representation for the Bessel functions $J_0(x)$ given in Eq. (D.50) along with a trigonometric substitution.

A.6 Show that

$$\frac{1}{\sqrt{x(L - x)}} = \frac{2\pi}{L} \sum_{n=1}^{\infty} J_0\left(\frac{n\pi}{2}\right) \sin \frac{n\pi}{2} \sin \frac{n\pi x}{L}, \qquad 0 < x < L$$

The suggestions in Problem A.5 are again appropriate.

APPENDIX B
Laplace Transform

This appendix contains a brief summary of Laplace transform techniques. Those aspects of the subject that are useful in solving partial differential equations are emphasized.

B.1 INTRODUCTION

A function $f(t)$ has a Laplace transform $F(s)$ defined by

$$F(s) = \int_0^\infty e^{-st} f(t) \, dt \tag{B.1}$$

whenever this integral converges. The requirement of convergence of the integral in (B.1) rules out functions that possess certain types of singularities in the range of integration ($0 \leq t < \infty$). For the types of functions usually encountered in applications the requirements are as follows:

(a) As t approaches zero, an algebraic singularity must be of the form $1/t^\mu$ with $\mu < 1$. For $\mu = 1$ the integral in (B.1) would diverge logarithmically at the origin, and for $\mu > 1$ the divergence would be even stronger. Examples of functions that violate this condition and thus do not possess a Laplace transform are $(\cos t)/t$, $(\sin t)/t^2$, and $e^{\pm t}/t$.

(b) As t approaches infinity, $f(t)$ may at most diverge exponentially, that is, as e^{at}. Then, for $s > a$ the integral in (B.1) will still converge. Functions violating this condition are usually not encountered in applications (at least as functions of time with t approaching infinity). Two such functions are $\exp(t^2)$ and the gamma function, which is introduced in Section B.8.

(c) For $0 < t < \infty$, $f(t)$ can have algebraic singularities of the form $1/(t - \tau)^\mu$ with $\mu \leq 1$. For $\mu = 1$, the integral is evaluated as a principal-value integral (Hildebrand, 1976, Section 10.15). However, such examples will not be encountered in this text.

A singular function that does have a Laplace transform and is frequently encountered is the delta function (cf. Section 1.5). The Laplace transform of this singular function follows from the general properties of the delta function

410

given in Section 1.5. We have

$$\int_0^\infty e^{-st} \delta(t - a) \, dt = \begin{cases} e^{-as}, & a > 0 \\ 0, & a < 0 \end{cases} \tag{B.2}$$

To treat an impulse at $t = 0$ it is convenient to place the impulse at $t = \epsilon$, consider $\delta(t - \epsilon)$, and then let $\epsilon \to 0$ later in the calculation (cf. Problem B.2).

The integral involved in the definition of the Laplace transform, (B.1), is of a type that is known to be uniformly convergent. The implication of uniform convergence for integrals is analogous to that associated with uniform convergence of a series as summarized in Section A.3 in that calculus operations may be freely interchanged with the integration over the variable t. As an example, differentiation of (B.1) with respect to the parameter s may be employed.

It should be noted that our usage of the Laplace transform is usually for the determination of a function that is unknown. When using the method it is implicitly assumed that the function being determined does have a transform. If it does not, then some other method of solution must be employed (cf. Problem B.8). Or, if only some of the possible solutions have a Laplace transform, then only those solutions having a transform may be obtained. For example, the second order ordinary differential equation $ty'' + 2y' + ty = 0$ has the solution $y = c_1 (\sin t)/t + c_2 (\cos t)/t$. This equation will be solved by the Laplace transform method in Example 6 of Section B.3 where it will be found that only the first solution $(\sin t)/t$ is obtained. As noted previously, the second solution does not have a Laplace transform.

The procedure of taking the Laplace transform of a function is frequently abbreviated $L\{f(t)\}$. Equation (B.1) is then written $F(s) = L\{f(t)\}$. Similarly, the determination of the function $f(t)$ when $F(s)$ is known may be denoted by $f(t) = L^{-1}\{F(s)\}$ and $f(t)$ is referred to as the inverse Laplace transform of $F(s)$.

A brief listing of Laplace transform pairs, confined largely to entries used in this text, is contained in Table F.2. Much more extensive tabulations are available (e.g., Erdelyi et al., 1954). Since the relation (B.1) defining the Laplace transform may be differentiated or integrated with respect to the parameter s or parameters contained in the function $f(t)$, many additional examples of functions of time and their associated transforms may be developed from a given transform pair. Some examples are considered in the next section.

B.2 PROPERTIES OF THE LAPLACE TRANSFORM

B.2.1 Transforms of Derivatives

The transform of the first derivative df/dt is related to $F(s)$ by using integration by parts:

$$L\left\{\frac{df}{dt}\right\} = \int_0^\infty e^{-st}\frac{df}{dt}\,dt = e^{-st}f(t)\Big|_0^\infty + s\int_0^\infty e^{-st}f(t)\,dt$$

$$= sF(s) - f(0) \tag{B.3}$$

Repeated partial integration of the nth derivative leads to the general result

$$L\left\{\frac{d^n f}{dt^n}\right\} = s^n F(s) - s^{n-1}f(0) - s^{n-2}f'(0) - \cdots - f^{(n-1)}(0) \tag{B.4}$$

where $f^{(n-1)}(0)$ signifies $d^{n-1}f(t)/dt^{n-1}\big|_{t=0}$.

Differentiating under the integral sign, we have

$$\frac{dF(s)}{ds} = \frac{d}{ds}\int_0^\infty e^{-st}f(t)\,dt = -\int_0^\infty e^{-st}tf(t)\,dt \tag{B.5}$$

Consequently $L\{tf(t)\} = -dF(s)/ds$. In general

$$L\{t^k f(t)\} = (-1)^k \frac{d^k F(s)}{ds^k}, \qquad k = 0, 1, 2, \ldots \tag{B.6}$$

Combining this result with (B.4), the expression for the transform of an nth derivative, we have the general result

$$L\left\{t^k \frac{d^n f}{dt^n}\right\} = (-1)^k \frac{d^k}{ds^k}\left[s^n F(s) - s^{n-1}f(0) - \cdots - f^{(n-1)}(0)\right] \tag{B.7}$$

This result is useful for solving certain types of differential equations with variable coefficients.

B.2.2   Transforms of Integrals

Provided $\int_0^t f(\tau)\,d\tau$ diverges no more rapidly than e^{at}, an integration by parts yields

$$L\left\{\int_0^t f(\tau)\,d\tau\right\} = \int_0^\infty e^{-st}\left[\int_0^t f(\tau)\,d\tau\right]dt$$

$$= -\frac{1}{s}e^{-st}\int_0^t f(\tau)\,d\tau\Big|_{t=0}^{t=\infty} - \int_0^\infty \left(-\frac{1}{s}\right)e^{-st}f(t)\,dt$$

$$= \frac{1}{s}F(s), \qquad s > a \tag{B.8}$$

Note that the integrated term in (B.8) vanishes at both limits.

It is well known that the process of integration can introduce functions that are more complicated (transcendental) than the functions occurring in the integrand [e.g., integrals with respect to s of s^{-1} and $(s^2 + 1)^{-1}$ lead to $\ln s$ and $\tan^{-1} s$, respectively]. Consequently a relation involving the integral of a transform can be expected to provide transform pairs that involve more elaborate functions of s. Consider, for example, the result of integrating the defining integral, (B.1), with respect to s and subsequently interchanging the order of the t and s integrations. We then obtain

$$\int_0^\infty f(t) \int_s^\infty e^{-st}\, ds = \int_s^\infty F(s)\, ds \qquad (B.9)$$

provided the integrals exist. [Note that for $F(s) = 1/s$, the integral over s would not exist.] Carrying out the s integration we obtain

$$\int_0^\infty e^{-st} \frac{f(t)}{t}\, dt = \int_s^\infty F(s)\, ds \qquad (B.10)$$

Thus, if $L\{f(t)\} = F(s)$, then $L\{f(t)/t\} = \int_s^\infty F(s)\, ds$.

B.2.3 Limits

We now summarize some relations between the transform $F(s)$ and $f(t)$ for either large or small values of s or t as well as relations involving the integral of $f(t)$ over all time. On occasion, such results are the only information about $f(t)$ that is of interest in a problem, and consequently they may be obtained *directly from the transform itself* without ever determining $f(t)$.

The Laplace transform of $f(t)$ may be interpreted as the area between the positive t axis and the curve $e^{-st}f(t)$ for positive values of t. If s is increased, the exponential factor decreases more rapidly so that the total area under $e^{-st}f(t)$ diminishes. In the limit $s \to \infty$ we may expect this area to approach zero. We can also expect that the value of the integral in this limit will depend only on the value of $f(t)$ for small values of t. This notion may be made more precise by first rewriting (B.3) as

$$sF(s) = f(0) + \int_0^\infty e^{-st} \frac{df}{dt}\, dt \qquad (B.11)$$

Then, always assuming that the indicated derivatives exist, we may apply repeated partial integration to this result and develop the series

$$sF(s) = f(0) + \frac{1}{s} f'(0) + \frac{1}{s^2} f''(0) + \cdots \qquad (B.12)$$

which now yields

$$\lim_{s \to \infty} [sF(s)] = f(0) \qquad (B.13)$$

and provides a quantitative relation between $F(s)$ for large s and $f(0)$. With $f(0)$ thus determined, we may continue this procedure to evaluate

$$\lim_{s \to \infty} [s^2 F(s) - sf(0)] = f'(0) \tag{B.14}$$

which is obtained from (B.12) by first multiplying by s. The procedure may be continued to higher derivatives (cf. Example 4 in Section B.3).

As an example of what happens when $f(t)$ or higher derivatives do *not* exist at $t = 0$, note that $L\{t^{-1/2}\} = \sqrt{\pi}/s$ (cf. Table F.2, entry 3). Hence $t^{-1/2}$ does not possess an expansion in integral inverse powers of s. Note that a byproduct of these considerations is the fact that a function of s cannot be a possible Laplace transform unless it vanishes as $s \to \infty$.

Similarly, from (B.11), provided the integral converges,

$$\lim_{s \to 0} [sF(s)] = f(0) + \int_0^\infty \frac{df}{dt} \, dt$$

$$= f(0) + f(\infty) - f(0)$$

$$= f(\infty) \tag{B.15}$$

The value of the transform for small values of s is thus related to $f(t)$ for large values of t. Note that there is no possibility of applying this theorem if s cannot approach 0 due to a restriction that it be larger than some positive value (e.g., entry 7 in Table F.2).

Finally, note that the integral of $f(t)$ over all t is

$$\int_0^\infty f(t) \, dt = \lim_{s \to 0} \int_0^\infty e^{-st} f(t) \, dt = F(0) \tag{B.16}$$

provided, of course, that the transform integral exists when $s = 0$.

Equations (B.13), (B.15), and (B.16) are extremely useful since they provide information about $f(t)$ when only $F(s)$, and not $f(t)$, itself is known.

B.2.4   Shift Theorems

If $f(t)$ is of the form $f(t) = g(t - t_0) H(t - t_0)$ where $H(t - t_0)$ is the unit step function

$$H(t - t_0) = \begin{cases} 1, & t > t_0 \\ 0, & t < t_0 \end{cases} \tag{B.17}$$

then

$$L\{f(t)\} = \int_0^\infty e^{-st} g(t - t_0) H(t - t_0) \, dt$$

$$= \int_{t_0}^{\infty} e^{-st} g(t - t_0) \, dt \tag{B.18}$$

Setting $u = t - t_0$ we then have

$$L\{f(t)\} = \int_0^{\infty} e^{-s(u + 0)} g(u) \, du = e^{-st_0} G(s) \tag{B.19}$$

where $G(s)$ is the Laplace transform of $g(t)$. We thus have the *first shift theorem of Laplace transforms*

$$L\{g(t - t_0) H(t - t_0)\} = e^{-st_0} G(s) \tag{B.20}$$

On the other hand, if $f(t)$ is of the form $f(t) = e^{at} g(t)$, then

$$L\{f(t)\} = \int_0^{\infty} e^{-st} e^{at} g(t) \, dt = \int_0^{\infty} e^{-(s - a)t} g(t) \, dt$$

$$= G(s - a) \tag{B.21}$$

This result is known as the *second shift theorem of Laplace transforms*

$$L\{e^{at} g(t)\} = G(s - a) \tag{B.22}$$

B.2.5 The Convolution Theorem

A problem of frequent occurrence when using the Laplace transform is the determination of an inverse transform for a function $F(s)$ that has the product form $F_1(s) F_2(s)$ where the inverse transforms of each of the functions $F_1(s)$ and $F_2(s)$ is already known. The inverse transform of $F(s)$ may be written

$$L^{-1}\{F(s)\} = L^{-1}\{F_1(s) F_2(s)\}$$

$$= L^{-1}\left\{ \int_0^{\infty} e^{-st_1} f_1(t_1) \, dt_1 \int_0^{\infty} e^{-st_2} f(t_2) \, dt_2 \right\}$$

$$= \int_0^{\infty} f_1(t_1) \, dt_1 \int_0^{\infty} f_2(t_2) \, dt_2 \, L^{-1}\{e^{-s(t_1 + t_2)}\} \tag{B.23}$$

Since the inverse transform of an exponential is a delta function [cf. (B.2)], we have

$$L^{-1}\{F(s)\} = \int_0^{\infty} f_1(t_1) \, dt \int_0^{\infty} f_2(t_2) \delta[t - (t_1 + t_2)] \, dt_2 \tag{B.24}$$

Integrating over t_2 and noting that the argument of the delta function will vanish at $t_2 = t - t_1$, we obtain

$$L^{-1}\{F(s)\} = \int_0^\infty f_1(t_1)f_2(t - t_1)H(t - t_1)\,dt_1 \qquad (B.25)$$

The step function is necessary in order to insure that the point $t_2 = t - t_1$ is actually in the range of the t_2 integration (i.e., $t_2 > 0$). Since the effect of the step function is to cut off the range of integration at $t_1 = t$, we finally have

$$L^{-1}\{F_1(s)F_2(s)\} = \int_0^t f_1(t_1)f_2(t - t_1)\,dt_1 \qquad (B.26)$$

This result is actually symmetric in f_1 and f_2, although it may not appear so at first glance, since the change of integration variable to $u = t - t_1$ yields

$$L^{-1}\{F_1(s)F_2(s)\} = \int_0^t f_1(t - u)f_2(u)\,du \qquad (B.27)$$

The replacement of the lower limit in this result by $-\infty$ to obtain steady state solutions will be considered in Example 3 in the next section.

B.3   EXAMPLES

We now summarize some calculations that exhibit the usefulness of many of the above topics when treating ordinary differential equations. They will also be found useful when dealing with partial differential equations.

1. *Green's Function for an Oscillator.* The motion of a damped mechanical oscillator that receives an impulse I at $t = t_0$ is determined by solving the equation $mx'' + Rx' + kx = I\delta(t - t_0)$ where the prime indicates a time derivative and m, R, and k are the mass, resistance, and spring constant of the oscillator, respectively. Setting $x(t) = Ig(t)/m$ and writing $2\gamma = R/m$, $\omega^2 = k/m$, we find that $g(t)$ satisfies

$$g'' + 2\gamma g' + \omega^2 g = \delta(t - t_0) \qquad (B.28)$$

For the case $g(0) = g'(0) = 0$, the Laplace transform of the equation for $g(t)$ is

$$(s^2 + 2\gamma s + \omega^2)G(s) = e^{-st_0} \qquad (B.29)$$

where (B.2) and (B.4) have been employed. Solving for $G(s)$ and completing the square, we have

$$G(s) = \frac{e^{-st_0}}{(s + \gamma)^2 + \Omega^2}, \qquad \Omega^2 = \omega^2 - \gamma^2 \qquad (\text{B.30})$$

Since this result contains both an exponential in s and a displacement in s, the inverse transform is obtained with the help of the two-shift theorems developed in the previous section. Using (B.20) and (B.21), in conjunction with entry 10 in Table F.2, we obtain

$$g(t) = \frac{1}{\Omega} e^{-\gamma(t - t_0)} \sin \Omega(t - t_0) H(t - t_0) \qquad (\text{B.31})$$

This result is the response (amplitude) at time t of an oscillator of unit mass when it is struck by a unit impulse at time t_0. The solution is frequently written $g(t - t_0)$ or $g(t|t_0)$ and is known as the Green's function for the oscillator (cf. Section 4.5).

2. *More General Driving Force—Convolution Theorem.* If the oscillator has initial amplitude and velocity x_0 and v_0, respectively, and is driven by the external force $f(t)$, then the equation governing its displacement, $x(t)$, is

$$x'' + 2\gamma x' + \omega^2 x = \frac{1}{m} f(t), \qquad x(0) = x_0, \qquad x'(0) = v_0 \quad (\text{B.32})$$

The Laplace transform of this equation is

$$s^2 X - sx(0) - x'(0) + s\gamma[sX - x(0)] + \omega^2 X = \frac{1}{m} F(s) \quad (\text{B.33})$$

Solving for $X(s)$, we obtain

$$X(s) = \frac{sx_0 + (2\gamma x_0 + v_0)}{(s + \gamma)^2 + \Omega^2} + \frac{1}{m} \frac{F(s)}{(s + \gamma)^2 + \Omega^2} \qquad (\text{B.34})$$

where again $\Omega^2 = \omega^2 - \gamma^2$. Carrying out the inversion of the first term with the help of entries 10 and 11 of Table F.2 plus the shift theorems and use of the convolution theorem in the second term, we have

$$x(t) = e^{-\gamma t}[x_0 \cos \Omega t + \Omega^{-1}(2\gamma x_0 + v_0) \sin \Omega t]$$

$$+ \frac{1}{m} \int_0^t \left\{ \frac{1}{\Omega} e^{-\gamma(t - \tau)} \sin \Omega(t - \tau) \right\} f(\tau) \, d\tau$$

$$= e^{-\gamma t}[x_0 \cos \Omega t + \Omega^{-1}(2\gamma x_0 + v_0) \sin \Omega t] + \frac{1}{m} \int_0^t g(t|\tau) f(\tau) \, d\tau$$

$$(\text{B.35})$$

where we have used the notation introduced above in connection with Example 1 since the expression in brackets under the integral in (B.35) is just the function $g(t|\tau)$ obtained in that example.

3. *Steady State Solutions.* In the previous example the first two terms in the solution damp out in a time on the order of γ^{-1}. After this initial period of time, only the integral over $f(t)$ will remain. It is frequently referred to as the steady state solution. A convenient way of treating this integral in the steady state regime is to first consider an external force $f(t) = F(t)H(t - t_0)$ where H is the unit step function. The integral in (B.35) is then $\int_{t_0}^{t} g(t - \tau)F(\tau) \, d\tau$. Although this result has been obtained by using the Laplace transform method, which implies positive values of t, it is actually a general result that is valid for arbitrary initial time t_0. There is therefore no reason for not allowing t_0 to recede to $-\infty$. The initial transient regime may then be thought of as having taken place in the remote past and we have only the steady state response expressed in terms of the integral

$$x_{ss}(t) = \frac{1}{m} \int_{-\infty}^{t} g(t - \tau)F(\tau) \, d\tau$$

$$= \frac{1}{m} \int_{0}^{\infty} g(u)F(t - u) \, du \tag{B.36}$$

where the second form of the integral follows from the first by setting $u = t - \tau$.

4. *Solution of an Integrodifferential Equation.* The present example not only gives an indication of the use of the Laplace transform for solving equations containing a certain type of integral expression, but also gives an example of the type of information on $f(t)$ that may be obtained from the transform $F(s)$ without actually determining $f(t)$. We consider the transform of the equation

$$y'(t) = -\omega^2 \int_{0}^{t} e^{-2\gamma(t - \tau)} y(\tau) \, d\tau, \qquad y(0) = y_0 \tag{B.37}$$

and use the transform $Y(s)$ to determine the additional initial information $y'(0)$ and $y''(0)$ as well as $\int_{0}^{\infty} y(t) \, dt$ before actually determining the solution $y(t)$. Taking the transform with the help of the convolution theorem, we obtain

$$sY(s) - y_0 = -\omega^2 \frac{1}{s + 2\gamma} Y(s) \tag{B.38}$$

The solution for $Y(s)$ is

$$Y(s) = y_0 \frac{s + 2\gamma}{s^2 + 2\gamma s + \omega^2} \tag{B.39}$$

From (B.14) we have

$$y'(0) = \lim_{s \to \infty} [s^2 Y(s) - sy(0)] = \lim_{s \to \infty} \frac{s\omega^2}{s^2 + 2\gamma s + \omega^2} = 0 \quad \text{(B.40)}$$

This result is also evident if the equation being solved, (B.37), is evaluated at $t = 0$. In addition, by extending the procedure used in obtaining (B.14), we find

$$y''(0) = \lim_{s \to \infty} [s^3 Y(s) - s^2 y(0) - sy'(0)]$$

$$= \lim_{s \to \infty} \frac{-y_0 \omega^2 s^2}{s^2 + 2\gamma s + \omega^2} = -\omega_0^2 y_0 \quad \text{(B.41)}$$

This is the expected result since differentiation of the original equation yields $y'' + 2\gamma y' + y = 0$ and thus $y''(0) = -\omega^2 y(0)$ since $y'(0) = 0$.

The value of $\int_0^\infty y(t)\, dt$ may be obtained by noting that, provided the integral exists,

$$\int_0^\infty y(t)\, dt = \lim_{s \to \infty} \int_0^\infty e^{-st} y(t)\, dt = \lim_{s \to \infty} Y(s) = Y(0) \quad \text{(B.42)}$$

Evaluating $Y(0)$ in the present example, we obtain

$$\int_0^\infty y(t)\, dt = \frac{2\gamma y_0}{\omega^2} \quad \text{(B.43)}$$

Finally, the inverse transform of $Y(s)$ in (B.39) yields

$$y(t) = y_0 e^{-\gamma t} \left(\cos \Omega t + \frac{\gamma}{\Omega} \sin \Omega t \right), \qquad \Omega^2 = \omega^2 - \gamma^2 \quad \text{(B.44)}$$

from which the previous results for $y''(0)$ and $\int_0^\infty y(t)\, dt$ may be obtained directly.

5. *Solution of a Boundary Value Problem.* Solve $y'' - y = 0$, $y(0) = 1$, $y(2) = 0$. Since $y'(0)$ is not initially known, the Laplace transform of the differential equation will contain an unknown initial condition. This unknown must be carried through the solution procedure and determined at a later stage in the calculation. The transform of the differential equation is

$$s^2 Y(s) - sy(0) - y'(0) - Y(s) = 0 \quad \text{(B.45)}$$

Then

$$Y(s) = \frac{s + y'(0)}{s^2 - 1} \tag{B.46}$$

and (cf. entries 14 and 15 in Table F.2)

$$y(t) = \cosh t + y'(0)\sinh t \tag{B.47}$$

We may now determine the unknown initial condition by applying the boundary condition at $t = 2$. We have

$$y(2) = \cosh 2 + y'(0)\sinh 2 = 0 \tag{B.48}$$

so that $y'(0) = -\coth 2$. The solution is thus $y(t) = \cosh t - \coth 2 \sinh t$, which may be rewritten

$$y(t) = \operatorname{csch} 2 \sinh(2 - t) \tag{B.49}$$

6. *Solution of* $ty'' + 2y' + ty = 0$. Before proceeding, note that upon setting $y = tu$, the equation for u is found to be $u'' + u = 0$, an equation with constant coefficients that has the solution $u = c_1\cos t + c_2\sin t$. The solution of the equation for y is thus $y = c_1(\cos t)/t + c_2(\sin t)/t$. Note that only the second of these solutions possesses a Laplace transform.

Using (B.7), the transform of the equation for y is

$$-\frac{d}{ds}[s^2 Y(s) - sy(0) - y'(0)] + 2[sY(s) - y(0)] - \frac{dY}{ds} = 0 \tag{B.50}$$

Simplification yields

$$\frac{dY}{ds} = -\frac{y_0}{s^2 + 1} \tag{B.51}$$

Integrating from s to ∞ and recalling that $Y(\infty) = 0$, we obtain

$$Y(s) = y_0 \int_s^\infty \frac{ds'}{s'^2 + 1} \tag{B.52}$$

Since $L^{-1}\{1/(s^2 + 1)\} = \sin t$, the relation (B.10) yields

$$y(t) = \frac{\sin t}{t} \tag{B.53}$$

Note that only the solution that has a Laplace transform has been obtained.

An alternative procedure for obtaining $y(t)$ that is sometimes useful in more complicated cases is to integrate (B.51) directly and obtain

$$Y(s) = -y_0 \tan^{-1} s + c \tag{B.54}$$

To determine the integration constant c, we introduce the series expansion for $\tan^{-1} s$ that is valid for large s and write

$$sY(s) = s \left[c - y_0 \left(\frac{\pi}{2} - \frac{1}{s} + \frac{1}{3s^2} - \cdots \right) \right] \tag{B.55}$$

Using (B.15), we have

$$y(0) = s \left(c - y_0 \frac{\pi}{2} \right) + y_0 \tag{B.56}$$

Since $y(0)$ must be independent of s, we obtain $c = y_0 \pi/2$. The transform of the solution is then

$$Y(s) = y_0 \left(\frac{\pi}{2} - \tan^{-1} s \right) = y_0 \tan^{-1} \left(\frac{1}{s} \right) \tag{B.57}$$

From a table of transforms (cf. Table F.2, entry 12) one may now recover a result in agreement with (B.53).

7. *A Laplace Transform Occurring in Diffusion Theory.* The Laplace transforms listed in Figure 2.3 may be obtained by first considering the integral

$$I(a, b) = \int_0^\infty e^{-(a^2 u^2 + b^2/u^2)} \, du \tag{B.58}$$

This integral may be evaluated by differentiating under the integral sign and then setting $v = b/au$. We obtain

$$\frac{\partial I}{\partial b} = -2b \int_0^\infty e^{-(a^2 u^2 + b^2/u^2)} \frac{du}{u^2}$$

$$= -2a \int_0^\infty e^{-(a^2 v^2 + b^2/v^2)} \, dv$$

$$= -2aI \tag{B.59}$$

The general solution of this differential equation for I is $I(a, b) = Ce^{-2ab}$. The integration constant C is determined by noting that for $b = 0$ the integral reduces to $I(a, 0) = \int_0^\infty \exp(-a^2 u^2) \, du = \sqrt{\pi}/2a = C$. Consequently,

$$\int_0^\infty e^{-(a^2u^2 + b^2/u^2)}\, du = \frac{\sqrt{\pi}}{2a} e^{-2ab} \tag{B.60}$$

The second entry in the table in Figure 2.3 is directly related to this result. On replacing the transform variable t by u^2, we obtain

$$L\left\{\frac{e^{-k^2/4t}}{\sqrt{t}}\right\} = \int_0^\infty \frac{e^{-st - k^2/4t}}{\sqrt{t}}\, dt$$

$$= 2\int_0^\infty e^{-(su^2 + k^2/4u^2)}\, du$$

$$= 2I(\sqrt{s}, k/2)$$

$$= \sqrt{\frac{\pi}{s}}\, e^{-k\sqrt{s}} \tag{B.61}$$

The first entry in the table is obtained by differentiating (B.61) with respect to k while the third and fourth entries follow from successive integration of (B.61) with respect to k from k to ∞.

B.4 THE GAMMA FUNCTION

The transform of t^n given in entry 2 of Table F.2 may be generalized to include the case of nonintegral values of n. The integral involved is then

$$L\{t^\nu\} = \int_0^\infty dt\, e^{-st} t^\nu, \qquad \nu > 1 \tag{B.62}$$

Setting $st = \zeta$, we have

$$L\{t^\nu\} = \frac{1}{s^{\nu+1}} \int_0^\infty d\zeta\, e^{-\zeta} \zeta^\nu \tag{B.63}$$

The resulting integral, which converges for $\nu > -1$, may be expressed in terms of the tabulated function

$$\Gamma(z) = \int_0^\infty d\zeta\, e^{-\zeta} \zeta^{z-1} \tag{B.64}$$

known as the gamma function. By setting $z - 1 = \nu$, we have

$$L\{t^\nu\} = \frac{\Gamma(\nu + 1)}{s^{\nu+1}} \qquad (B.65)$$

For $z = 1, 2, 3, \ldots$ the integral defining $\Gamma(z)$ may be evaluated by continued integration by parts to yield $\Gamma(n) = (n - 1)!$. For nonintegral n the integral is evaluated numerically. Certain special nonintegral values may also be evaluated analytically, for example,

$$\Gamma(\tfrac{1}{2}) = \sqrt{\pi} \qquad (B.66)$$

The gamma function retains the expected property of the factorial. Thus, for nonintegral n, the relation $n! = n(n - 1)!$ is replaced by $\Gamma(\nu + 1) = \nu\Gamma(\nu)$. As an example, $\Gamma(\tfrac{5}{2}) = \tfrac{3}{2}\Gamma(\tfrac{3}{2}) = \tfrac{3}{2} \cdot \tfrac{1}{2} \cdot \Gamma(\tfrac{1}{2}) = \tfrac{3}{2} \cdot \tfrac{1}{2} \cdot \sqrt{\pi}$.

The gamma function has many other properties that will not be required in this text. Some of them may be found elsewhere (Hildebrand, 1976). A satisfactory appreciation of the properties of the gamma function is only obtained when it is examined in the complex ν plane (Whittaker and Watson, 1969, Chapter 12).

Problems

B.1 Use Laplace transform methods to solve the following differential equations:

(a) $y'' - y = 1$, $y(0) = 1$, $y'(0) = 0$
(b) $y'' - y = 0$, $y(0) = 1$, $y(\infty) = 0$
(c) $y'' + y = \delta(t - 1)$, $y(0) = 0$, $y'(0) = 0$

B.2 (a) Solve $y'' + \omega^2 y = 0$, $y(0) = 0$, $y'(0) = v_0$.
(b) Solve $y'' + \omega^2 y = (I/m)\delta(t - \epsilon)$, $y(0) = 0$, $y'(0) = 0$, and show that, as ϵ goes to zero, the result of part (a) is recovered when one sets $I = mv_0$.

B.3 Solve $y'' + 4y' + 5y = a\delta(t - 2)$, $y(0) = 0$, $y'(0) = 1$. Determine the constant a so that $y(t) = 0$ for $t > 2$. Evaluate $\int_0^2 y(t)\,dt$ by using the Laplace transform $Y(s)$.

B.4 Solve $ty'' - (t + 1)y' + y = 0$.

B.5 Solve $y'' + 4y' + 5y = e^{-2t}\cos t$, $y(0) = 1$, $y'(0) = 2$. (Note that the Laplace transform method automatically takes care of the fact that the inhomogeneous term is a solution of the homogeneous equation.)

(a) Show that $\lim_{s \to \infty} [sY(s)] = y(0)$ and $\lim_{s \to \infty} [s^2 Y(s) - y(0)] = y'(0)$.

(b) Show that $Y(0) = \int_0^\infty y(t)\,dt$.

B.6 Solve $y'' + 4y' + 5y = \cos tH(t - t_0)$, $y(0) = y'(0) = 0$. Use the convolution theorem to obtain the steady state solution.

B.7 Show that the Laplace transform of $ty'' + y' + ty = 0$ leads to the differential equation

$$\frac{dY}{ds} = y_0(1 + s)^{-1/2}$$

Expand the radical, integrate term by term, and show that term-by-term inversion leads to

$$y(t) = y_0 \left(1 - \frac{t^2}{2^2} + \frac{t^4}{2^4(2!)^2} - \cdots \right)$$

$$= y_0 J_0(t)$$

B.8 Consider the Laplace transform solution of $y' - 2ty = 0$. Show that $Y(s)$ diverges as $s \to \infty$ so that $y(t)$ does not possess a Laplace transform. (The solution is $y = c \exp t^2$.)

B.9 Derive the convolution theorem (B.27) by writing the product of two transforms as

$$F_1(s)F_2(s) = \int_0^\infty dt_1\, f_1(t_1) \int_0^\infty dt_2\, f_2(t_2) e^{-s(t_1 - t_2)}$$

Now introduce new integration variables t and τ as

$$t = t_1 + t_2, \qquad \tau = t_1$$

and note that for a fixed value of $t_1 = \tau$, t varies from t_1 to ∞ as t_2 varies from 0 to ∞. Thus

$$F_1(s)F_2(s) = \int_0^\infty d\tau\, f_1(\tau) \int_\tau^\infty dt\, f_2(t - \tau) e^{-st}$$

Interpret the result as a double integral that covers the shaded region indicated in Figure B.1. The order of integration may be reversed and the same area of the τt plane still be covered by writing

$$F_1(s)F_2(s) = \int_0^\infty dt\, e^{-st} \int_0^t d\tau\, f_1(\tau) f_2(t - \tau)$$

that is, for a fixed value of t_1 the τ integration proceeds from $\tau = 0$ to $\tau = t$.

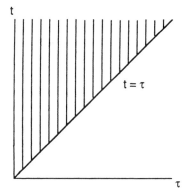

t = τ

τ   **Figure B.1.**

The coefficient of e^{-st} in the t integrand, namely,

$$\int_0^t d\tau\, f_1(\tau) f_2(t - \tau)$$

is now seen to have the Laplace transform $F_1(s)\, F_2(s)$ as required by the convolution theorem.

APPENDIX C
Sturm-Liouville Equations

Some of the concepts that arose in the consideration of Fourier series in Appendix A are here shown to occur in a more general context.

The orthogonality relation

$$\int_0^{\pi} \sin mx \sin nx \, dx = 0, \qquad m \neq n \tag{C.1}$$

may be obtained by a direct evaluation of the integral. However, the result may also be derived in a more general way that enables one to show the vanishing of certain other integrals that arise in more complicated situations. We first introduce this general method by showing how it may be used to recover the result quoted above.

Labeling the two terms in the integral in (C.1) as $y_m(x) = \sin mx$ and $y_n(x) = \sin nx$, we note that they are solutions, respectively, of the differential equations plus boundary conditions

$$y_m'' + m^2 y_m = 0, \qquad y_m(0) = y_m(\pi) = 0$$
$$y_n'' + n^2 y_n = 0, \qquad y_n(0) = y_n(\pi) = 0 \tag{C.2}$$

We now multiply the first of these equations by y_n and the second by y_m and subtract the resulting equations. The result is

$$y_n y_m'' - y_m y_n'' + (n^2 - m^2) y_m y_n = 0 \tag{C.3}$$

If this expression is now integrated over x from 0 to π, we obtain

$$\int_0^{\pi} (y_m y_n'' - y_n y_m'') + (n^2 - m^2) \int_0^{\pi} y_m y_n \, dx = 0 \tag{C.4}$$

Note that the second integral is the one occurring in (C.1). The first integral in (C.4) is readily shown to vanish. To see this, we perform an integration by parts and rewrite that first term as

$$\int_0^\pi (y_m y_n'' - y_n y_m'') \, dx = y_m y_n' \Big|_0^\pi - \int_0^\pi y_m' y_n' \, dx$$

$$- \left[y_n y_m' \Big|_0^\pi - \int_0^\pi y_n' y_m' \, dx \right] \qquad \text{(C.5)}$$

Since y_n and y_m vanish at the end points, the integrated terms vanish. The two remaining integrals are equal and hence cancel. We have thus shown that the first integration (C.4) is zero and we have

$$(n^2 - m^2) \int_0^\pi y_m y_n \, dx = 0 \qquad \text{(C.6)}$$

Hence for $n \neq m$ this integral must vanish and we have obtained the orthogonality relation (C.1). Note that this conclusion has been reached without any actual evaluation of the integral. The method tells us nothing about the value of the integral when $n = m$, however. This issue will be addressed subsequently.

The technique outlined above can also be used to demonstrate the vanishing of other integrals as well. Note that the integrated term in (C.5) would also vanish if the derivatives $y_n'(x)$ and $y_m'(x)$ were to vanish at $x = 0$ and π. This would be the case if we chose $y_m = \cos mx$ and $y_n = \cos nx$. Hence the method can also be used to show that

$$\int_0^\pi \cos mx \cos nx \, dx = 0, \qquad m \neq n \qquad \text{(C.7)}$$

In general, we can expect orthogonality for any pair of solutions of $y'' + n^2 y = 0$ that have boundary conditions leading to the vanishing of the integrated terms in (C.5).

The procedure used above may be applied to the more general equation

$$\frac{d}{dx} \left[p(x) \frac{dy}{dx} \right] + [q(x) + \lambda s(x)] y = 0, \qquad a \leq x \leq b \qquad \text{(C.8)}$$

which is known as the Sturm-Liouville equation. Note that if we set $p = s = 1$, $q = 0$, and $\lambda = n^2$, m^2 in (C.8), we obtain the equations listed in (C.2). For the most part, differential equations of Sturm-Liouville type arise in two distinct ways. For inhomogeneous systems such as heat flow on a beam with variable conductivity as considered in Problem 1.6.4 or vibration of an inhomogeneous string considered in Problem 1.8.5, the governing differential equation is of this form. The Sturm-Liouville equation also arises when the separation-of-variables procedure is applied to the Laplace operator in various curvilinear orthogonal coordinate systems (cf. Section 3.5). For the inhomogeneous systems, the functions p and s have the physical significance of being quantities such as density, conductivity, or tension, quantities that cannot vanish or be negative. When these equations arise in the separation-of-variables procedure, however, the functions p and s may become zero or infinite.

If we consider two different values of λ, say λ_m and λ_n, with corresponding solutions y_m and y_n, then the equations for y_m and y_n may be combined as was done above in obtaining (C.4) to yield

$$\int_a^b \left[y_m(py_n')' - y_n(py_m')' \right] dx + (\lambda_n - \lambda_m) \int_b^a s(x) y_n y_m \, dx = 0 \quad (C.9)$$

Integration by parts may again be performed on the first integral and cancellation of the resulting integrals again takes place. We are then left with

$$\left[y_m(py_n') - y_n(py_m') \right]\Big|_{x=a}^{x=b} + (\lambda_n - \lambda_m) \int_a^b s(x) y_n y_m \, dx = 0 \quad (C.10)$$

When the term evaluated at the end points a and b is equal to zero, the remaining integral may be interpreted as expressing the orthogonality of $y_n(x)$ and $y_m(x)$ *with respect to the weight function* $s(x)$ over the interval $a < x < b$. The terms evaluated at the end points can vanish in various ways, for example,

(a) $c_1 y'(a) + c_2 y(a) = 0$ (C.11)
    $c_3 y'(b) + c_4 y(b) = 0$
    where the c_i are constants independent of λ (cf. Problem C.6).
(b) $p(a) = p(b) = 0$
(c) Periodic boundary conditions, that is, $y(a) = y(b)$, $y'(a) = y'(b)$.

For any situation in which the integrated terms vanish we have the orthogonality relation

$$\int_a^b s(x) y_m(x) y_n(x) \, dx = 0, \quad m \neq n \quad (C.12)$$

The orthogonality integral obtained here can be used to expand a function in a series of the form

$$f(x) = \sum_{n=1}^{\infty} a_n y_n(x) \quad (C.13)$$

The procedure is analogous to that used for Fourier series. On multiplying (C.13) by $s(x) y_p(x)$ and integrating over x in the region $a \leq x \leq b$, all terms in the series vanish except for $n = p$ and one obtains

$$a_n = \frac{\displaystyle\int_a^b dx\ s(x)\ y_n(x)\ f(x)}{\displaystyle\int_a^b dx\ s(x)\ [\ y_n(x)]^2} \tag{C.14}$$

from which the coefficients a_n may, in principle, be calculated. Evaluation of the integral in the numerator of (C.13) can be quite formidable since the function $y_n(x)$ is a solution of (C.8), an equation that may have variable coefficients. The integral in the denominator of (C.13) may be carried out in general, however, and we now consider this subject.

C.1   NORMALIZATION INTEGRAL

The previous considerations of orthogonality give no indication of the value of the orthogonality integral when $n = m$. We now consider this topic. Note that (C.10) may be written

$$\int_a^b sy_m y_n\ dx = \frac{p(y_m y_n' - y_n y_m')\ \big|_b^a}{\lambda_n - \lambda_m} \tag{C.15}$$

The ratio becomes indeterminate for $\lambda_n = \lambda_m$, and this situation should suggest consideration of a derivative with respect to λ. Hence we return to the original Sturm-Liouville equation (C.8), with no boundary conditions imposed upon the solution and thus with λ still a continuously varying parameter, and differentiate the equation with respect to λ. Since p, q, and s are independent of λ, we obtain

$$(py_x)_{x\lambda} + qy_\lambda + sy + \lambda sy_\lambda = 0 \tag{C.16}$$

We now combine this expression with the original Sturm-Liouville equation by the multiply-and-subtract procedure. Multiplying (C.8) by y_λ (C.16) by y and subtracting, we obtain

$$y(py_x)_{x\lambda} - y_\lambda(py_x)_x + sy^2 = 0 \tag{C.17}$$

Integrating with respect to x over the interval $a \le x \le b$ and then integrating by parts, we have

$$[\,y(py_x)_\lambda - y_\lambda(py_x)]\,\big|_a^b - \int_a^b dx\ [\,y_x(py_x)_\lambda - y_{\lambda x}py_x] + \int_a^b sy^2\ dx = 0 \tag{C.18}$$

Since p is independent of λ, the integrand in the second term vanishes. Now letting λ become one of the eigenvalues λ_n, we have

$$\int_a^b sy_n^2(x)\,dx = p(y_{\lambda_n}y_x) - yy_{x\lambda_n}\Big|_a^b \tag{C.19}$$

For $\lambda = k^2$ we have $\partial/\partial\lambda = (\tfrac{1}{2}k)\partial/\partial k$ and the result is

$$\int_a^b sy_n^2(x)\,dx = \frac{p}{2k_n}\left(\frac{\partial y_n}{\partial k_n}\frac{\partial y_n}{\partial x} - y_n\frac{\partial^2 y_n}{\partial x\,\partial k_n}\right)\Big|_a^b \tag{C.20}$$

Since the $y_n(x)$ satisfy boundary conditions at the end points a and b, this result usually simplifies considerably when specific examples are treated.

Example C.1. Use the general result in (C.20) to obtain the value of the integral in (C.1) when $n = m$.

We consider the differential equation $y'' + n^2 y = 0$. Here $p = s = 1$ and $\lambda = n^2$. We have $k = n$ and the various terms in (C.20) are

$$y_n = \sin nx$$

$$\frac{\partial y_n}{\partial n} = x\cos nx$$

$$\frac{\partial y_n}{\partial x} = n\cos nx$$

$$\frac{\partial^2 y_n}{\partial x\,\partial n} = \cos nx - nx\sin ux \tag{C.21}$$

Substitution into (C.20) yields

$$\int_0^\pi \sin^2 x\,dx = \frac{1}{2n}(nx - \sin nx\cos nx)\Big|_0^\pi = \frac{\pi}{2} \tag{C.22}$$

as expected. (Strictly speaking, of course, one differentiates with respect to a continuous variable ν and then sets $\nu = n$ after the differentiation.)

Example C.2. The radially symmetric modes of a vibrating membrane may be described by solutions of the zero order Bessel equation (cf. Section 3.7 and Appendix D):

$$xy_{xx} + y_x + k^2 xy = 0, \qquad y(0) = y(a) = 0 \tag{C.23}$$

In Sturm-Liouville form this equation is

$$(xy_x)_x + k^2 xy = 0 \tag{C.24}$$

Here $p = x$, $q = 0$, $s = x$, and $\lambda = k^2$. From the solution $y = J_0(kx)$ we calculate the derivatives

$$y_x = kJ_0'(kx)$$

$$y_k = xJ_0'(kx)$$

$$y_{kx} = J_0'(kx) + kxJ_0''(kx) \tag{C.25}$$

where a prime denotes differentiation with respect to the argument kx. When these results are introduced into (C.20) and the second derivative is eliminated by using the equation for $J_0(kx)$ in the form $k^2xJ_0'' + kJ_0' + k^2xJ_0 = 0$, we obtain

$$\int_0^a x \, dx \, J_0^2(kx) = \frac{x^2}{2} \left\{ [J_0'(kx)]^2 + [J_0(kx)]^2 \right\} \Big|_0^a \tag{C.26}$$

Now setting $k = k_n$ where $J_0(k_n a) = 0$ and using the general result $J_0'(ka) = -J_1(ka)$, we obtain the final expression for the normalization integral as

$$\int_0^a x \, dx \, J_0^2(kx) = \tfrac{1}{2} a^2 J_1^2(k_n a), \qquad J_0(k_n a) = 0 \tag{C.27}$$

C.2 THE WRONSKIAN

If y_1 and y_2 are two linearly independent solutions of the *same* Sturm-Liouville equation, that is,

$$(py_1')' + (q + \lambda s) y_1 = 0$$

$$(py_2')' + (q + \lambda s) y_2 = 0, \qquad a \le x \le b \tag{C.28}$$

the multiply-and-subtract procedure leads to

$$y_1(py_2')' - y_2(py_1')' = 0 \tag{C.29}$$

Carrying out the differentiation and regrouping terms, we obtain

$$p(y_2 y_1'' - y_1 y_2'') + p'(y_2 y_1' - y_1 y_2') = 0 \tag{C.30}$$

When the quantity $py_2'y_1'$ is added and subtracted to the term multiplying p, a perfect derivative of the quantity $W = y_1 y_2' - y_2 y_1'$ is obtained and (C.28) has the form

$$p \frac{dW}{dx} + p'W = 0 \tag{C.31}$$

Integration of this perfect derivative yields $pW = c$ where c is a constant of integration. We finally obtain

$$W[y_1(x), y_2(x)] \equiv y_1(x) y_2'(x) - y_2(x) y_1'(x) = \frac{c}{p(x)} \qquad \text{(C.32)}$$

Since this expression holds for *all* values of x, the constant c can frequently be determined by evaluating the solutions y_1 and y_2 in some limit, usually for $x \gg 1$ or $x \ll 1$, where y_1 and y_2 may assume a simple analytical form (cf. Problem C.3). Note that when $p(x)$ is a constant, the Wronskian is also a constant.

A simple example of a Sturm-Liouville equation with variable coefficients is provided by the equation that results from a transformation of the equation

$$\frac{d^2 y}{d\vartheta^2} + \nu^2 y = 0, \qquad 0 \le \vartheta \le \pi \qquad \text{(C.33)}$$

Before introducing the transformation to a new independent variable, we first summarize the solution in terms of ϑ. The equation has solutions $y = u_\nu = \cos \nu\vartheta$ and $y = v_\nu = \sin \nu\vartheta$. If ν is an integer, we can rewrite these solutions as polynomials in $\cos \vartheta$ (or $\sin \vartheta$) by using the standard trigonometric multiple angle formulas. For the first few values of $\nu = n$ we obtain

$$
\begin{array}{llll}
n = 1: & u_1 = \cos \vartheta & v_1 = \sin \vartheta = \sqrt{1 - \cos^2 \vartheta} \\
n = 2: & u_2 = \cos 2\vartheta & v_2 = \sin 2\vartheta \\
& \quad = 2\cos^2 \vartheta - 1 & \quad = 2\sqrt{1 - \cos^2 \vartheta} \cos \vartheta \\
n = 3: & u_3 = \cos 3\vartheta & v_3 = \sin 3\vartheta \\
& \quad = 4\cos^3 \vartheta - 3\cos \vartheta & \quad = \sin(-1 + 4\cos^2 \vartheta)
\end{array}
$$

$$\text{etc.} \qquad\qquad\qquad\qquad\qquad\qquad\qquad\qquad\qquad\qquad\qquad \text{(C.34)}$$

Equation (C.33) and its solutions (C.34) can now be transformed so as to provide a simple context for examining an equation with variable coefficients. When we introduce the transformation $z = \cos \vartheta$, the solutions u_n in (C.34) will take the form of polynomials in z while the other solutions v_n will be $\sqrt{1 - z^2}$ times a polynomial in z. Let us now change variables to $z = \cos \vartheta$ in the original differential equation and examine the form of that equation when expressed in terms of z. Since $z = \cos \vartheta$, we have

$$\frac{d}{d\vartheta} = \frac{dz}{d\vartheta}\frac{d}{dz} = -\sin \vartheta \frac{d}{dz} = -\sqrt{1 - z^2}\frac{d}{dz}$$

$$\frac{d^2}{d\vartheta^2} = \frac{d}{d\vartheta}\left(-\sqrt{1 - z^2}\frac{d}{dz} \right) = \sqrt{1 - z^2}\frac{d}{dz}\left(\sqrt{1 - z^2}\frac{d}{dz} \right)$$

$$\text{(C.35)}$$

and the original equation (C.33) can be written

$$\frac{d}{dz}\left(\sqrt{1 - z^2}\, \frac{dy}{dz}\right) + \frac{n^2}{\sqrt{1 - z^2}}\, y = 0, \qquad -1 \le z \le 1 \qquad \text{(C.36)}$$

This equation is seen to have the form of a Sturm-Liouville equation with $p = \sqrt{1 - z^2}$, $q = 0$, $\lambda = n^2$, and $s = (1 - z^2)^{-1/2}$. We have thus encountered one of the situation mentioned earlier, namely $p(z)$ vanishes at the ends of the interval and s is infinite at these points. Since $\vartheta = \cos^{-1}z$, the solution may be written in terms of z as

$$u_n = \cos(n \cos^{-1}z), \qquad v_n = \sin(n \cos^{-1}z) \qquad \text{(C.37)}$$

As noted above, the solutions u_n are polynomial in z while the solutions v_n are $\sqrt{1 - z^2}$ times a polynomial in z. The polynomials u_n are known as Tschebycheff polynomials[1] and are frequently written $T_n(z)$. Writing $v_n = S_n(z)$, we have, for the first few values of n,

$$
\begin{array}{ll}
T_0 = 1 & S_0 = 0 \\
T_1 = z & S_1 = \sqrt{1 - z^2} \\
T_2 = 2z - 1 & S_2 = \sqrt{1 - z^2}\,2z \\
T_3 = 4z^3 - 3z & S_3 = \sqrt{1 - z^2}(4z^2 - 1)
\end{array}
\qquad \text{(C.38)}
$$

$$\text{etc.}$$

Since $p(z)$ vanishes at the end points $z = \pm 1$, we know that the orthogonality relation (C.12) is satisfied. Thus we can write

$$\int_{-1}^{1} \frac{T_n(z)\, T_m(z)}{\sqrt{1 - z^2}}\, dz = 0, \qquad n \ne m \qquad \text{(C.39)}$$

and similarly for the function $S_n(z)$. Perhaps the easiest way to evaluate the normalization integral for $n = m$ is to revert to the variable ϑ and obtain

$$\int_{-1}^{1} \frac{[T_n(z)]^2\, dt}{\sqrt{1 - z^2}} = \int_{0}^{\pi} [T_n(\cos \vartheta)]^2\, d\vartheta$$

$$= \int_{0}^{\pi} \cos^2 n\vartheta \, d\vartheta = \frac{\pi}{2} \qquad \text{(C.40)}$$

Evaluation of the orthogonality integral by using the general expression (C.20) is considered in Problem C.2.

---

[1]When the differentiation in (C.36) is carried out, this equation may be written $(1 - z^2)\,y'' - zy' + n^2 y = 0$, which is the standard form for the equation defining these polynomials.

Another simple instance of a Sturm-Liouville equation is provided by the Cauchy-Euler equation. An example is considered in Problem C.5.

Problems

C.1 Show that if one solution of the Sturm-Liouville equation (C.8) is y, then a second solution may be written $y_2 = vy$, where v satisfies the equation $py_1 v'' + (2py_1 + p'y_1)v' = 0$. Show that

$$v = \int \frac{dx}{p(x)\, y_1^2}$$

C.2 Obtain the normalization integral for Tschebycheff polynomials given in (C.40) by using the general relation (C.20).

C.3 Evaluate the Wronskian (C.32) for Bessel functions by using both the large- and small-argument approximations given in (D.25) and (D.26). Show that

$$W[J_0(x),\, Y_0(x)] = \frac{2}{\pi x}, \qquad W[H_0^{(1)}(x),\, H_0^{(2)}(x)] = -\frac{4i}{\pi x},$$

$$W[I_0(x),\, K_0(x)] = -\frac{1}{x}$$

C.4 Obtain the normalization constant given in (C.20) by setting $\lambda_n = \lambda$, $\lambda_m = \lambda + \Delta\lambda$, and $y_m = y_n + (\partial v_n/\partial\lambda)\,\Delta\lambda$ and then using the definition of a derivative.

C.5 Consider the Cauchy-Euler equation $x^2 y'' + \lambda y = 0$ with boundary conditions $y(a) = y(b) = 0$.

(a) Show that $y(x) = \sqrt{x}\, \sin[\mu_n \ln(x/a)]$, $\mu_n = n\pi/\ln(b/a) = \sqrt{\lambda_n - \tfrac{1}{4}}$.

(b) Obtain the normalization constant by using (C.20).

C.6 **(a)** Show that the integrated term in (C.10) vanishes when condition (a) of Eq. (C.11) is satisfied.

(b) Consider the case in which the constants c_i depend upon n (i.e., λ) and show that the integrated term in (C.10) no longer vanishes. The eigenfunctions are thus no longer orthogonal. Note that in the case of the vibrating string, the boundary condition $Ty_x = -Ry_t = i\omega Ry$, which is appropriate for a string with a resistive support, would lead to such a situation. In this case, since $T = \rho c^2$, we have

$$\rho c y_x - ikRy = 0$$

C.7 Expand unity in a Fourier-Bessel series in the region $0 < x < 1$ by determining the coefficients A_n in the expansion

$$1 = \sum_{n=1}^{\infty} A_n J_0(\alpha_{0n} x), \qquad J_0(\alpha_{0n}) = 0$$

Show that for $x = 0$ the result reduces to

$$\frac{1}{2} = \sum_{n=1}^{\infty} \frac{1}{\alpha_{0n} J_0(\alpha_{0n})}$$

APPENDIX D
Bessel Functions

The role played by Bessel functions in describing circular waves is more readily appreciated if it is compared with the much simpler situation for spherical waves that is described in Section 3.6. It is suggested that the reader be familiar with the wave equation for spherical waves that is described there before considering Bessel functions.

When the Laplacian is written in polar coordinates, the wave equation

$$\nabla^2 w - \frac{1}{c^2}\frac{\partial^2 w}{\partial t^2} = 0 \tag{D.1}$$

becomes (cf. Section 3.5)

$$\frac{1}{r}\frac{\partial}{\partial r}\left(r\frac{\partial w}{\partial r}\right) + \frac{1}{r^2}\frac{\partial^2 w}{\partial \vartheta^2} - \frac{1}{c^2}\frac{\partial^2 w}{\partial t^2} = 0 \tag{D.2}$$

To introduce the subject in the simplest possible manner, we consider only waves with circular symmetry, that is, waves having no angular dependence. We therefore set $\partial^2 w/\partial \vartheta^2 = 0$ in (D.2). We will also consider waves at a single frequency ω and thus write $w(r, t) = w(r)e^{-i\omega t}$. With these specializations, (D.2) reduces to

$$\frac{1}{r}\frac{d}{dr}\left(r\frac{dw}{dr}\right) + k^2 w = 0 \tag{D.3}$$

where $k = \omega/c$. After performing the indicated differentiation we obtain

$$r\frac{d^2 w}{dr^2} + \frac{dw}{dr} + k^2 rw = 0 \tag{D.4}$$

which is the simplest case of an equation known as Bessel's equation. Before discussing the solutions of this equation, known as Bessel functions, we first consider certain properties that should be expected of the solution on the basis of simple physical considerations. If w represents the amplitude of a circular wave radiating outward in two dimensions from a source point, then since there

436

is no dissipation included in the original wave equation [i.e., no term proportional to $\partial w/\partial t$ in (D.1)], we expect that a constant amount of energy in the wave, E, will be spread out over a circle with ever-increasing circumference $C = 2\pi r$. Then, the energy density $\mathcal{E} = E/C$ should decrease as r^{-1}. Furthermore, at large radii ($r \gg \lambda$), the wave front is nearly plane and energy density in a plane wave is proportional to the square of the wave amplitude, that is, $\mathcal{E} = w^2$ (cf. Section 1.8). We therefore expect w to be proportional to $r^{-1/2}$. If we factor out this expected dependence on r and write the solution of (D.4) as $w(r) = r^{-1/2}f(r)$, we may then examine the remaining radial dependence in $f(r)$. This is done by substitution into (D.4) to obtain the equation satisfied by $f(r)$. The required derivatives are

$$w' = r^{-1/2}(f' - \tfrac{1}{2}r^{-1}f)$$

$$w'' = r^{-1/2}(f'' - r^{-1}f' + \tfrac{3}{4}r^{-2}f) \tag{D.5}$$

and (D.4) becomes, after dividing out a common factor of $r^{1/2}$,

$$f'' + \left(k^2 + \frac{1}{4r^2}\right)f = 0 \tag{D.6}$$

For values of r that are large enough so that $1/(2r) \ll k$, that is, $kr \gg \tfrac{1}{2}$, the term $\tfrac{1}{4}r^2$ in this last equation may be neglected in comparison to k^2. We then obtain the more familiar equation $f'' + k^2f = 0$, which has solutions $\cos kr$ and $\sin kr$. For $kr \gg \tfrac{1}{2}$ we thus expect *each* of the two *exact* solutions for w to reduce to some linear combination of these two solutions and be of the form

$$w \cong \frac{1}{\sqrt{r}}(A\cos kr + B\sin kr) \tag{D.7}$$

Before considering the appropriate values of A and B, we return to an examination of the exact equation for w.

D.1 SERIES SOLUTION

Perhaps the most elementary method for obtaining a solution of an equation with variable coefficients is to try to express it in the form of an infinite series and write

$$w = a_0 + a_1r + a_2r^2 + a_2r^3 + \cdots \tag{D.8}$$

We then substitute this expression into Bessel's equation (D.4). The two derivatives of w that are required are

$$w' = a_1 + 2a_2r + 3a_3r^2 + 4a_4r^3 + \cdots$$

$$w'' = 2a_2 + 3 \cdot 2a_3r + 4 \cdot 3a_4r^2 + \cdots \qquad \text{(D.9)}$$

Substituting into (D.4) and grouping terms according to powers of r, we find

$$a_1 + (4a_2 + k^2a_0)r + [(3 \cdot 2 + 3)a_3 + k^2a_1]r^2$$

$$+ [(4 \cdot 3 + 4)a_4 + k^2a_2]r^3 + \cdots = 0 \qquad \text{(D.10)}$$

Since this result is to hold for *all* values of r, we set the coefficient of each power of r equal to zero separately and obtain

$$a_1 = 0$$

$$a_2 = -\frac{k^2}{4}a_0 = -k^2\frac{a_0}{2^2}$$

$$a_3 = -\frac{k^2a_1}{3 \cdot 3} = 0$$

$$a_4 = -\frac{k^2a_2}{4(3 + 1)} = \frac{k^4a_0}{4^2 \cdot 2^2} = \frac{k^4}{(2!)^2 2^4}$$

$$\cdots \qquad \text{(D.11)}$$

In general, all coefficients a_n with an odd value of n are proportional to a_1 and thus vanish. An inspection of the coefficients that arise for even values of n leads to a simple generalization of the expression given above for a_4. We find

$$a_{2n} = \frac{(-1)^n k^{2n}}{(n!)^2 2^{2n}}a_0 \qquad \text{(D.12)}$$

A solution of (D.4) is thus

$$w(r) = a_0 \sum_{n=0}^{\infty} (-1)^n \frac{1}{(n!)^2}\left(\frac{kr}{2}\right)^{2n} \qquad \text{(D.13)}$$

This series is customarily referred to as $J_0(kr)$. If we set $kr = z$, the first few terms are

$$J_0(z) = 1 - \frac{1}{1!}\left(\frac{z}{2}\right)^2 + \frac{1}{(2!)^2}\left(\frac{z}{2}\right)^4 - \cdots \qquad \text{(D.14)}$$

The constant a_0 in (D.13) plays the role of an integration constant.

Since Bessel's equation is a second order differential equation, it can be expected to possess two linearly independent solutions. The second solution was not obtained above since it is not of the form of the series expansion assumed in (D.8). To motivate the choice of a different form for the solution, we first note that the series solution obtained above contains k and r in the dimensionless form kr. In fact, if we set $z = kr$ in Bessel's equation, it takes the form $z\, d^2w/dz^2 + dw/dz + zw = 0$, which is independent of k. The form of the solution for small z may thus be examined by considering either small k or small r in (D.4). For the extreme case $k = 0$ in the original form of Bessel's equation (D.3) we would obtain

$$\frac{1}{r}\frac{d}{dr}\left(r\frac{dw}{dr}\right) = 0 \tag{D.15}$$

which may be integrated directly to yield

$$w = a + b \ln r \tag{D.16}$$

This result may be considered to be a linear combination of the *two* solutions of Bessel's equation (D.4) in the limit of small kr. The series solution obtained previously in (D.13) approaches the constant value a_0 as r tends to zero and can be considered to correspond to the constant a obtained here in (D.16). The $\ln r$ term must be included in any starting point for a second solution to Bessel's equation. Due to the singular nature of this second solution, however, it will be discarded in all problems that require a finite value for the solution w at $r = 0$. The situation is the same as that encountered for spherical waves in Section 3.6. Since a detailed use of the second solution will not be encountered here, we shall only list the final result, which is usually labeled either $Y_0(kr)$ or $N_0(kr)$ and has the form (Hildebrand, 1976, p. 144)

$$Y_0(z) = \frac{2}{\pi}\left[\left(\ln\frac{z}{2} + \gamma\right)J_0(z) + \frac{z^2}{2^2} - \frac{z^4}{2^4(2!)^2}\left(1 + \frac{1}{2}\right)\right.$$
$$\left. + \frac{z^6}{2^6(3!)^2}\left(1 + \frac{1}{2} + \frac{1}{3}\right) - \cdots\right] \tag{D.17}$$

where $\gamma = \lim_{n\to\infty}(1 + \frac{1}{2} + \frac{1}{3} + \cdots 1/n - \ln n) = 0.577216\ldots$, the Euler-Mascheroni constant.

D.2 LARGE-ARGUMENT FORMS

For large values of z it is found that the exact solution given for $J_0(z)$ in (D.14) and for $Y_0(z)$ in (D.17) will match the approximate large-argument form given

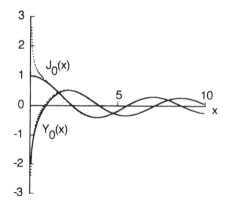

Figure D.1. Graphs of the exact (solid) and large-argument approximate (dotted) expressions for $J_0(x)$ and $Y_0(x)$.

in (D.7) if the constants A and B are chosen so as to yield[1]

$$J_0(z) \approx \sqrt{\frac{2}{\pi z}} \cos \left(z - \frac{\pi}{4} \right)$$

$$Y_0(z) \approx \sqrt{\frac{2}{\pi z}} \sin \left(z - \frac{\pi}{4} \right) \qquad \text{(D.18)}$$

These results can be obtained when the solutions of Bessel's equation are analyzed by complex variable techniques. A comparison between the exact and approximate solution is given in Figure D.1.

D.3 ZEROS OF $J_0(x)$

The exact location of each zero of $J_0(x)$, labeled α_{0n}, may be obtained from numerical study of the series solution given in (D.14).[2] It is evident from Figure D.1 that, at least for qualitative purposes, the location of each zero is given quite accurately by the zeros of the approximate forms given in (D.18), which are at $z - \pi/4 = \pi/2,\ 3\pi/2,\ 5\pi/2,\ \ldots$. A comparison between exact and approximate values is given in Table D.1. The use of these results to describe *standing waves* in cylindrical geometry is considered in Section 3.7.

To obtain a representation of *traveling waves* with circular symmetry, first recall the consideration of the geometric attenuation of spherical waves in

---

[1]This result requires $A = \sqrt{(2/\pi)}\cos \pi/4$, $B = \sqrt{(2/\pi)}\sin \pi/4$ for J_0 and $A = -\sqrt{(2/\pi)}\sin \pi/4$ and $B = \sqrt{(2/\pi)}\cos \pi/4$ for Y_0.
[2]For an extensive analysis of the theory of zeros of Bessel functions see Watson (1944, Chapter 15).

Table D.1 Exact and Approximate Values of First Five Zeros of $J_0(x)$

n	$\pi/4 + (2n - 1)\pi/2$	Exact
1	2.356	2.4048256
2	5.498	5.5200781
3	8.639	8.6537279
4	11.781	11.7915344
5	14.922	14.9309177

Section 3.6. On the basis of considerations just developed for cylindrical waves we might expect that at least at large distances from the source, outgoing cylindrical waves are of the form $r^{-1/2}f(r - ct)$ while incoming waves are functions $r^{-1/2}g(r + ct)$. In conjunction with a time dependence $e^{-i\omega t}$ we would then expect outgoing and incoming cylindrical waves to be of the form $r^{-1/2}e^{ikr}$ and $r^{-1/2}e^{-ikr}$, respectively. From (D.18) we see that the combinations of J_0 and Y_0 that provide such expressions are $J_0 + iY_0$ since

$$J_0 \pm iY_0 \cong \sqrt{\frac{2}{\pi kr}} \left[\cos \left(kr - \frac{\pi}{4} \right) \pm i \sin \left(kr - \frac{\pi}{4} \right) \right]$$

$$\cong \sqrt{\frac{2}{\pi kr}} e^{\pm i(kr - \pi/4)} \tag{D.19}$$

These combinations of J_0 and Y_0 are known as Hankel functions of the first and second kind, respectively, and for all values of r are written

$$H_0^{(1)}(kr) = J_0(kr) + iY_0(kr)$$

$$H_0^{(2)}(kr) = J_0(kr) - iY_0(kr) \tag{D.20}$$

D.4 HIGHER ORDER BESSEL FUNCTIONS

When angular dependence is retained in the two-dimensional wave equation (D.2), waves at a single frequency ω satisfy the equation

$$\frac{1}{r} \frac{\partial}{\partial r} \left(r \frac{\partial w}{\partial r} \right) + \frac{1}{r^2} \frac{\partial^2 w}{\partial \vartheta^2} + k^2 w = 0, \qquad k = \frac{\omega}{c} \tag{D.21}$$

A solution in the factored form $w(r, \vartheta) = R(r)\Theta(\vartheta)$ leads to the separated equation

$$\frac{r}{R}(rR')' + k^2 r = -\frac{\Theta''}{\Theta} = m^2 \tag{D.22}$$

where m^2 is the separation constant. As in Section 3.7 we choose m to be an integer to guarantee continuity of the solution as ϑ increases from 0 through 2π. The equation for R is now found to be

$$r^2 R'' + rR' + (k^2 r^2 - m^2)R = 0 \tag{D.23}$$

This equation may be analyzed in the same manner as done already for $m = 0$. In particular, the substitution $R(r) = r^{-1/2} f(r)$ leads to the equation

$$f'' + \left[k^2 + \frac{1}{r^2} \left(\frac{1}{4} - m^2 \right) \right] f = 0 \tag{D.24}$$

from which the large-argument form of the solution may again be inferred. The results are

$$J_m(z) \cong \sqrt{\frac{2}{\pi z}} \cos(z - \varphi_m)$$

$$Y_m(z) \cong \sqrt{\frac{2}{\pi z}} \sin(z - \varphi_m) \tag{D.25}$$

where $\varphi_m = (2m + 1)\pi/4$. For $z \ll 1$ one finds (Hildebrand, 1976, p. 148)

$$J_m(z) \approx \frac{1}{2^m m!} z^m$$

$$Y_m(z) \approx -\frac{2^m (m - 1)!}{\pi} z^{-m}, \qquad m \neq 0$$

$$Y_0(z) = \frac{2}{\pi} \ln z \tag{D.26}$$

Graphs of $J_m(x)$ for $m = 0, 1, 2, 3$ are shown in Figure D.2. Again, outgoing and incoming cylindrical waves are expressed, respectively, by $H_m^{(1)}(z)$ and

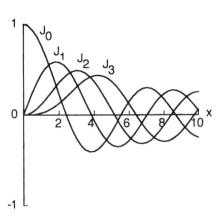

Figure D.2. Graphs of $J_m(x)$ for $m = 0,$ 1, 2, 3.

$H_m^{(2)}(z)$ and are defined in a manner completely analogous to that given in (D.20) for zero order Hankel functions.

It should be emphasized that these large-argument forms are only valid as long as $m \ll kr$. When both order and argument are large, other expressions, obtained by using complex variable techniques, are required (Watson, 1944, Chapter 8).

D.5 ORTHOGONALITY

As shown in Section 3.7, satisfaction of boundary conditions for the case of cylindrical symmetry ($m = 0$) in a region of radius a leads to consideration of a series expansion of the form

$$f(r) = \sum_{n=1}^{\infty} A_n J_0(k_n r) \tag{D.27}$$

where $k_n = \alpha_{0n}/a$. The first few of the constants α_{0n}, the roots of $J_0(\alpha_{0n}) = 0$, are listed in Table D.1. Since Bessel's equation in the form given in (D.3) is in the form of a Sturm-Liouville equation, that is,

$$\frac{d}{dr}\left(r \frac{dw}{dr} \right) + k^2 w = 0 \tag{D.28}$$

it is possible to develop orthogonality relations that enable one to determine the constants A_n in (D.27). From the discussion of Sturm-Liouville theory in Appendix C, one finds that a coefficient A_p in (D.27) is obtained by multiplying that equation through by $r J_0(\alpha_{0p} r/a)$ and integrating over r from 0 to a. Replacing x by r in relation (C.14) and setting $s(r) = r$ as well as using limits of integration appropriate for the present situation, we have

$$A_n = \frac{\displaystyle\int_0^a dr\ rf(r)J_0(\alpha_{0n}r/a)}{\displaystyle\int_0^a dr\ r[J_0(\alpha_{0n}r/a)]^2} \tag{D.29}$$

Examples of the use of this result are contained in Section 3.7.

D.6 FUNCTIONS $I_n(z)$ AND $K_n(z)$

We have seen that there is a close correspondence between solutions $J_0(kr)$ and $Y_0(kr)$ of Bessel's equation and the solutions $\cos kx$ and $\sin kx$ of the equation $y'' + k^2 y = 0$. It should be recalled that in the separation-of-variables procedure in rectangular coordinates one also encounters the equation $y'' - k^2y = 0$ with associated solutions $e^{\pm kx}$ or, equivalently, $\cosh kx$ and $\sinh kx$. Note that both the equation and the solutions are obtainable by replacing k by ik in the equation $y'' + k^2y = 0$. Similarly, there are corresponding solutions to the equation

$$r^2R'' + rR' + (-k^2r^2 - m^2)R = 0 \tag{D.30}$$

which is obtained by replacing k by ik in Bessel's equation. Just as the exponential and hyperbolic functions are nonoscillatory, so also the solutions to (D.30) are nonoscillatory. The solutions that are usually employed are

$$I_n(kr) = i^{-n}J_n(ikr)$$

$$K_n(kr) = \frac{\pi}{2} i^{n+1}H_n^{(1)}(ikr) \tag{D.31}$$

Graphs of I_0 and K_0 are shown in Figure D.3. The factors of i enter in the definitions in (D.31) in such a way that all I_n and K_n are real.

Since J_0 corresponds to $\cos kx$ and $H_0^{(1)}$ to e^{ikx}, it is seen that a linear combination of the solutions in (D.31) (for $n = 0$) is analogous to replacing k by ik in the linear combination $a \cos kx + be^{ikx}$, which is a solution of $y'' + k^2y = 0$. This choice is not as curious as might be thought at first and is motivated by the occurrence of the singularity of the Bessel function Y_0 at the origin. From Figure D.3 we see that the choice of I_0 and K_0 yields one solution that remains finite at the origin and one that decays to zero for large kr. This is the pair of solutions that proves useful in solving most problems.

For $kr \gg 1$ one finds

$$I_n(kr) \cong \frac{e^{kr}}{\sqrt{2\pi kr}}, \qquad K_n(kr) \cong \sqrt{\frac{\pi}{2kr}}\, e^{ikr} \tag{D.32}$$

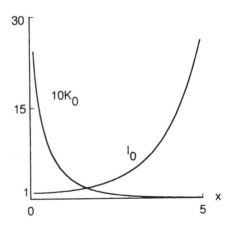

Figure D.3. Graphs of I_0 and K_0.

while for $kr \ll 1$

$$I_n(kr) \approx \frac{1}{2^n n!}(kr)^n$$

$$K_n(kr) \approx 2^{n-1}(n-1)!(kr)^{-n}, \qquad (n \neq 0) \tag{D.33}$$

$$K_0(kr) \approx -\ln kr$$

D.7 RECURRENCE RELATIONS

Bessel's functions of neighboring order are interrelated in ways that prove to be extremely useful. Some of these relations are listed below (Hildebrand, 1976, p. 149) where $Z_n(x)$ refers to one of the Bessel functions J, Y, $H^{(1)}$, $H^{(2)}$, I, or K:

$$\frac{d}{dx}[x^n Z_n(kn)] = \begin{cases} kx^n Z_{n-1}(kx), & Z = J, Y, I, H^{(1)}, H^{(2)} \\ -kx^n Z_{n-1}(kx), & Z = K \end{cases} \tag{D.34}$$

$$\frac{d}{dx}[x^{-n} Z_n(kx)] = \begin{cases} -kx^{-n} Z_{n+1}(kx), & Z = J, Y, K, H^{(1)}, H^{(2)} \\ kx^{-n} Z_{n+1}(kx), & Z = I \end{cases} \tag{D.35}$$

$$\frac{d}{dx} Z_n(kx) = \begin{cases} kZ_{n-1}(kx) - \dfrac{n}{x} Z_n(kx), & Z = J, Y, I, H^{(1)}, H^{(2)} \\ \\ -kZ_{n-1}(kx) - \dfrac{n}{x} Z_n(kx), & Z = K \end{cases} \tag{D.36}$$

$$\frac{d}{dx} Z_n(kx) = \begin{cases} -kZ_{n+1}(kx) + \dfrac{n}{x} Z_n(kx), & Z = J, Y, K, H^{(1)}, H^{(2)} \\ \\ kZ_{n+1}(kx) + \dfrac{n}{x} Z_n(kx), & Z = I \end{cases} \tag{D.37}$$

$$2\frac{d}{dx} Z_n(kx) = k[Z_{n-1}(kx) - Z_{n+1}(kx)], \qquad Z = J, Y, H^{(1)}, H^{(2)} \tag{D.38}$$

$$2\frac{d}{dx} I_n(kx) = k[I_{n-1}(kx) + I_{n+1}(kx)] \tag{D.39}$$

$$2\frac{d}{dx} K_n(kx) = -k[K_{n-1}(kx) + K_{n+1}(kx)] \tag{D.40}$$

$$Z_{n+1}(kx) = \frac{2n}{kx} Z_n(kx) - Z_{n-1}(kx), \qquad Z = J, Y, H^{(1)}, H^{(2)} \quad \text{(D.41)}$$

$$I_{n+1}(kx) = -\frac{2n}{kx} I_{n+1}(kx) + I_{n-1}(kx) \tag{D.42}$$

$$K_{n+1}(kx) = \frac{2n}{kx} K_n(kx) + K_{n-1}(kx) \tag{D.43}$$

Note that if kx happens to correspond to a zero of some Bessel function, then a simpler relation is obtained. For instance, from (D.36) and (D.37), if $kx = \alpha_{nm}$ so that $J_n(\alpha_{nm}) = 0$, we have

$$\frac{d}{dx} J_n(kx)\Big|_{kx=\alpha_{nm}} = \frac{\alpha_{nm}}{a} J_{n-1}(\alpha_{nm}) = -\frac{\alpha_{nm}}{a} J_{n+1}(\alpha_{nm}) \tag{D.44}$$

Note also that this relation holds for *all* values of the argument if $n = 0$. In particular,

$$\frac{d}{dx} J_0(kx) = -kJ_1(kx) \tag{D.45}$$

D.8   GENERATING FUNCTION

In dealing with special functions (i.e., solutions of Sturm-Liouville equations), it is often convenient to have a function of two variables, say $g(x, z)$, such that the special functions are the coefficients in the McLauren expansion of $g(x, z)$. By using the method outlined in Problem D.1, one may show that

$$e^{(x/2)(z-1/z)} = \sum_{n=-\infty}^{\infty} z^n J_n(x) \tag{D.46}$$

As an example of the usefulness of this expression, note that on setting $z = e^{i\vartheta}$ one obtains

$$e^{ix\sin\vartheta} = \sum_{n=-\infty}^{\infty} J_n(x) e^{in\vartheta} \tag{D.47}$$

Both sides of this expression have period 2π in ϑ. (It is an example of a complex Fourier series.) Multiplying (D.47) by $e^{-im\vartheta}$ and integrating over a full period, we obtain

$$J_m(x) = \frac{1}{2\pi} \int_{-\pi}^{\pi} e^{i(x\sin\vartheta - m\vartheta)} d\vartheta \tag{D.48}$$

The limits of integration could equally well be taken from 0 to 2π since the integrand has period 2π. This last relation proves to be an extremely useful representation for $J_m(x)$. Note that when $m = 0$ the result may be written

$$J_0(x) = \frac{1}{2\pi} \int_0^{2\pi} e^{ix\cos\varphi} \, d\varphi \tag{D.49}$$

where ϑ has been replaced by $\varphi + \pi/2$ and, again, since the range of integration is a complete circle, it has been written as 0 to 2π. If, on the other hand, the range of integration is taken between the symmetric limits $-\pi \leq \vartheta \leq \pi$, then since only the even part of the integrand contributes, we obtain

$$J_0(x) = \frac{1}{\pi} \int_0^{\pi} \cos(x \cos \varphi) \, d\varphi$$

$$= \frac{2}{\pi} \int_0^{\pi/2} \cos(x \cos \varphi) \, d\varphi \tag{D.50}$$

Similar considerations applied to $J_m(x)$ in (D.48) yield

$$J_m(x) = \frac{1}{\pi} \int_0^{\pi} \cos(x \sin \varphi - m\varphi) \, d\varphi \tag{D.51}$$

Finally, since $J_m(x) = (-1)^m J_{-m}(x) = e^{-im\pi} J_{-m}(x)$, we may add the integral for $J_m(x)$ in (D.48) to that obtained by replacing m by $-m$ and φ by $\theta + \pi/2$ to obtain

$$2J_m(x) = \frac{1}{2\pi} e^{-im\pi/2} \int_{-\pi}^{\pi} d\theta \, e^{ix\cos\theta} (e^{im\theta} + e^{-im\theta}) \tag{D.52}$$

and therefore

$$J_m(x) = \frac{i^{-m}}{2\pi} \int_{-\pi}^{\pi} d\theta \, e^{ix\cos\theta} \cos m\theta \tag{D.53}$$

Problems

D.1 The generating function introduced in (D.46) may be obtained by beginning with the expression

$$g(x, z) = \sum_{n=-\infty}^{\infty} z^n J_n(x)$$

(a) Obtain $\partial g/\partial x$ and use recurrence relations, as well as a relabeling of summation indices, to obtain

$$\frac{\partial g}{\partial x} = \frac{x}{2}\left(z - \frac{1}{z}\right) g$$

(b) Solve this differential equation and note that $g(0, z) = 1$ to obtain

$$g(x, z) = e^{(x/2)(z - 1/z)}$$

D.2 (a) By expanding the exponent in the integrand in (D.49) and integrating term by term, show that the resulting series expansion is the series for $J_0(z)$ given in (D.14), and therefore,

$$J_0(z) = \frac{1}{2\pi} \int_0^{2\pi} e^{iz\cos\theta} \, d\theta$$

Note: $\int_0^{2\pi} \cos^{2k} \theta \, d\theta = (\pi/2)[1.3.5 \cdots . (2k - 1)]/(2.4.6 \cdots . 2k)$.

(b) Show by direct substitution into Bessel's equation that the integral is a solution of that equation.

D.3 Show by direct substitution that $y = J_n(e^z)$ is a solution of $y'' + (-n^2 + e^{2z})y = 0$.

D.4 (a) By rewriting Bessel's equation in the form $n^2 J_n(x)/x = xJ_n(x) - [xJ_n'(x)]'$, and similarly for $J_m(x)$ and using the multiply-and-subtract procedure, show that

$$\int \frac{J_n(x)J_m(x)}{x} \, dx = \frac{x(J_n J_m' - J_m J_n')}{n^2 - m^2}$$

and from this result conclude that

$$\int_0^\infty \frac{J_n(x)J_m(x)}{x} \, dx = \frac{2}{\pi} \frac{\sin(n - m)\pi/2}{n^2 - m^2}$$

(b) Since n and m need not be integers for this result to be valid, use L'Hospital's rule to obtain

$$\int_0^\infty \frac{J_n^2(x) \, dx}{x} = \frac{1}{2n}, \qquad n \neq 0$$

D.5 (a) Show that $x^2 y'' + axy' + (b + cx^{2d})y = 0$ has the solution $y = x^{(1-a)/2} Z_p(\sqrt{c} \, x^d/d)$ where $p = d^{-1}\sqrt{(1 - a)^2/4 - b}$ and Z refers to any Bessel function.

(b) Show that $4y'' + 9xy = 0$ has solutions $x^{1/2} Z_{\pm 1/3}(x^{3/2})$.

D.6 Express $e^{ikr\cos\theta}$ as a Fourier series by writing

$$e^{ikr\cos\theta} = a_0 + \sum_{n=1}^{\infty} a_n\cos n\theta$$

and then using (D.53) to obtain

$$e^{ikr\cos\theta} = \sum_{n=0}^{\infty} \epsilon_n i^n J_n(kr)\cos n\theta$$

where ϵ_n is 1 for $n = 0$ and 2 for $n = 1, 2, 3, \ldots$.

APPENDIX E
Legendre Polynomials

It is shown in Section 3.5 and in Chapter 5 that, when rotational symmetry is present in spherical coordinates (i.e., no φ dependence), the separation-of-variables procedure leads to a differential equation of the form

$$\frac{1}{\sin \theta} \frac{d}{d\theta} \left(\sin \theta \frac{d\Theta}{d\theta} \right) + k^2 \Theta = 0, \qquad 0 \le \theta \le \pi \qquad (E.1)$$

where k is a separation constant. We now consider the solutions of this equation that are of importance in those developments.

E.1 NATURE OF THE SOLUTION—LEGENDRE POLYNOMIALS

In analyzing this equation it is convenient to introduce a new variable μ defined as $\mu = \cos \theta$. Then

$$\sin \theta \frac{d}{d\theta} = \sin \theta \frac{d\mu}{d\theta} \frac{d}{d\mu} = -\sin^2 \theta \frac{d}{d\mu} = -(1 - \mu^2) \frac{d}{d\mu} \qquad (E.2)$$

and the original equation becomes

$$\frac{d}{d\mu} \left[(1 - \mu^2) \frac{d\Theta}{d\mu} \right] + k^2 \Theta = 0, \qquad -1 \le \mu \le 1 \qquad (E.3)$$

Some idea of the nature of the solutions possessed by this equation may be obtained by considering the case $k = 0$. A first integration immediately yields

$$\frac{d\Theta}{d\mu} = \frac{c_1}{1 - \mu^2} \qquad (E.4)$$

where c_1 is a constant of integration. After a second integration is carried out, the general solution is seen to be

$$\Theta = c_1 \ln \frac{1 + \mu}{1 - \mu} + c_2 \qquad (E.5)$$

The solution proportional to c_1 is infinite at $\mu = \pm 1$ (i.e., at $\vartheta = 0, \pi$, which correspond to the north and south poles of the sphere, respectively). In any problem in which the solution must remain finite at these points, we must reject this solution and set $c_1 = 0$.

For $k \neq 0$, the customary power series solution yields two series, neither of which converge at $\mu = \pm 1$ unless k^2 is of the form $k^2 = n(n + 1)$, $n = 0, \pm 1, \pm 2, \ldots$, in which case one series terminates after a finite number of terms (cf. Problem E.1). Note that replacement of n by $-n - 1$ in the expression for k^2 yields $(-n - 1)(-n) = n(n + 1)$. Hence, no additional values of k^2 are provided by negative values of n. Accordingly, we shall only consider the positive values $n = 0, 1, 2, 3, \ldots$ in the following development. With this choice for k^2, (E.3) then has the form

$$\frac{d}{d\mu}\left[(1 - \mu^2)\frac{d\Theta}{d\mu}\right] + n(n + 1)\Theta = 0 \tag{E.6}$$

and for k^2 equal to one of these values $(0, 2, 6, 12, \ldots)$, one solution of (E.6) is finite at $\mu = \pm 1$. The other solution is still singular at $\mu = \pm 1$. The divergent solution has logarithmic singularities of the type encountered above for the case $n = 0$. For $n = 0, 1, 2, \ldots$, the solution that remains finite is thus a polynomial of order n. It is known as a Legendre polynomial of order n and is denoted by $P_n(\mu)$. We shall now show that this polynomial, with the integration constant chosen so that $P_n(1) = 1$, can be written in the form

$$P_n(\mu) = \frac{1}{2^n n!}\frac{d^n}{d\mu^n}(\mu^2 - 1)^n \tag{E.7}$$

which is known as Rodrigues' formula.

To see that this expression is a solution of (E.6), we first consider the second order differential equation satisfied by $w = (\mu^2 - 1)^n$. Two derivatives of w yield

$$\frac{dw}{d\mu} = 2n\mu(\mu^2 - 1)^{n-1}$$

$$\frac{d^2w}{d\mu^2} = 2n(\mu^2 - 1)^{n-1} + 4n(n - 1)\mu^2(\mu^2 - 1)^{n-2} \tag{E.8}$$

Multiplying the second derivative by $\mu^2 - 1$ and also using the expression for $dw/d\mu$, we obtain

$$(\mu^2 - 1)\frac{d^2w}{d\mu^2} - 2(n - 1)\mu\frac{dw}{d\mu} - 2nw = 0 \tag{E.9}$$

Differentiating this equation m times, we find (cf. Problem E.9)

$$(\mu^2 - 1) \frac{d^{m+2}w}{d\mu^{m+2}} - 2(n - m - 1)\mu \frac{d^{m+1}w}{d\mu^{m+1}}$$

$$- (m + 1)(2n - m) \frac{d^m w}{d\mu^m} = 0 \tag{E.10}$$

Choosing $m = n$ and writing $u = d^n w/d\mu^n$, we arrive at

$$(1 - \mu^2) \frac{d^2 u}{d\mu^2} - 2\mu \frac{du}{d\mu} + n(n + 1) = 0 \tag{E.11}$$

which is an equivalent form of the Legendre's equation (E.6). A solution of this equation is thus

$$u(\mu) = c \frac{d^n w}{d\mu^n} = c \frac{d^n (\mu^2 - 1)^n}{d\mu^n} \tag{E.12}$$

where c is an arbitrary constant. Since $(\mu^2 - 1)^n$ is a polynomial of order $2n$, n derivatives of this polynomial yield a polynomial of order n. Hence $u(\mu)$ is a polynomial in μ of order n. As noted above, the polynomial is Legendre polynomial $P_n(\mu)$ when the constant c is chosen so that $u(1) = 1$ for each n. The appropriate value of c for each n may be determined by noting that $\mu^2 - 1 = (\mu + 1)(\mu - 1)$ and that near $\mu = 1$ we may approximate the product $(\mu + 1)(\mu - 1)$ by $2(\mu - 1)$. Thus, for μ near unity we have

$$u(\mu) \cong c2^n \frac{d^n (\mu - 1)^n}{d\mu^n}$$

$$\cong c2^n n(n - 1)(n - 2) \ldots 2.1 = c2^n n! \tag{E.13}$$

To obtain $u(1) = 1$ we thus set $c = 1(2^n n!)$ and obtain the expression given in (E.7). When the corresponding approximation is applied near $\mu = -1$, one finds that $P_n(-1) = (-1)^n$.

For the first seven values of n the polynomials are

$$P_0(\mu) = 1$$

$$P_1(\mu) = \mu$$

$$P_2(\mu) = \tfrac{1}{2}(3\mu^2 - 1)$$

$$P_3(\mu) = \tfrac{1}{2}(5\mu^3 - 3\mu)$$

$$P_4(\mu) = \tfrac{1}{8}(35\mu^4 - 30\mu^2 + 3)$$

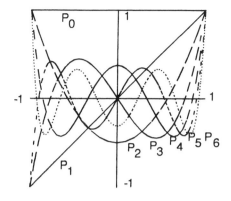

Figure E.1. Graphs of $P_n(x)$ for $n = 0$, ..., 6.

$$P_5(\mu) = \tfrac{1}{8}(63\mu^5 - 70\mu^3 + 15\mu)$$

$$P_6(\mu) = \tfrac{1}{16}(231\mu^6 - 315\mu^4 + 105\mu^2 - 5) \qquad (E.14)$$

Graphs of these expressions are shown in Figure E.1. As noted earlier, the constant n occurs in the defining equation (E.6) in the form $n(n + 1)$, and replacement of n by $-n - 1$ yields the same value for this product. This result provides the identity $P_{-n-1}(\mu) = P_n(\mu)$. As noted above, no additional solutions are provided by negative values of n and we consider only $n = 0, 1, 2, 3, \ldots$.

The second solution, which is usually labeled $Q_n(\mu)$, is infinite at $\mu = \pm 1$, as was seen above for $k = 0$, which corresponds to $n = 0$. The solution given in (E.5) is equal to $Q_0(\mu)$ when we set $c_1 = \tfrac{1}{2}$ and $c_2 = 0$. We will not have further occasion to use these second solutions.

E.2 ORTHOGONALITY, RECURRENCE RELATIONS, AND GENERATING FUNCTION

Equation (E.6) defining the Legendre polynomial $P_n(\mu)$ is of the form of a Sturm-Liouville equation (cf. Appendix C) with $p = \sqrt{(1 - \mu^2)}$, $q = 0$, $s = 1$, and $\lambda = n(n + 1)$. The vanishing of p at $\mu = \pm 1$ leads to the vanishing of the integrated term in (C.10) and provides orthogonality of the Legendre polynomials over the interval $-1 \le \mu \le 1$. The value of the normalization integral is determined in Problem E.7. One finds

$$\int_{-1}^{1} P_n(\mu)P_m(\mu)\, d\mu = \begin{cases} 0, & n \ne m \\ \dfrac{2}{2n + 1}, & n = m \end{cases} \qquad (E.15)$$

Note that the general expression for the normalization constant given in (C.20) is difficult to use in this instance since for n not equal to an integer, the solutions are no longer polynomials and their properties are more complicated. One must differentiate the Legendre function with respect to order.

As was the case for Bessel functions (cf. Appendix D), the Legendre polynomials satisfy recurrence relations and have a generating function. The recurrence relations are

$$nP_n(\mu) = (2n - 1)\mu P_{n-1}(\mu) - (n - 1)P_{n-2}(\mu)$$

$$(\mu^2 - 1)\frac{dP_n}{d\mu} = n[\mu P_n(\mu) - P_{n-1}(\mu)]$$

$$= \frac{n(n + 1)}{2n + 1}[P_{n+1}(\mu) - P_{n-1}(\mu)] \tag{E.16}$$

The generating function is (cf. Problem E.6)

$$g(\mu, h) = \frac{1}{\sqrt{1 - 2h\mu + h^2}} = \sum_{n-0}^{\infty} h^n P_n(\mu) \tag{E.17}$$

Evaluating the generating function at $\mu = 0$, we obtain

$$\frac{1}{\sqrt{1 + h^2}} = \sum_{n=0}^{\infty} h^n P_n(0) \tag{E.18}$$

Expanding the left-hand side by the binomial theorem and equating equal powers of h, we have

$$P_{2n+1}(0) = 0, \qquad P_{2n}(0) = (-1)^n \frac{1.3.5 \ldots (2n - 1)}{2^n n!} \tag{E.19}$$

Since only even powers of h appear in (E.18), it may be written as

$$\frac{1}{\sqrt{1 + h^2}} = \sum_{n=0}^{\infty} h^{2n} P_{2n}(0) \tag{E.20}$$

This expansion is useful in obtaining closed form expressions for various series that occur when solving problems in terms of Legendre polynomials. For instance, if (E.20) is integrated over h from 0 to ζ, where $0 \le \zeta \le 1$, we obtain

$$\ln(\zeta + \sqrt{1 + \zeta^2}) = \sum_{n=0}^{\infty} \frac{\zeta^{2n+1}}{2n + 1} P_{2n}(0) \tag{E.21}$$

Also, after first multiplying (E.20) by h and then integrating, we find

$$\sqrt{1 + \zeta^2} - 1 = \sum_{n=0}^{\infty} \frac{\zeta^{2n+2}}{2n+2} P_{2n}(0), \qquad 0 \le \zeta \le 1 \qquad \text{(E.22)}$$

Clearly, many other such expressions may be developed.

E.3 GEOMETRIC INTERPRETATION OF THE GENERATING FUNCTION

If the distance from an origin $(0, 0, 0)$ to a field point (x, y, z) is written r, then the distance from (x, y, z) to a point $(0, 0, a)$ on the z axis is

$$R = \sqrt{x^2 + y^2 + (z - a)^2} = \sqrt{r^2 - 2az + a^2}$$

$$= r\sqrt{1 - 2az/r^2 + (a/r)^2}, \qquad r > a \qquad \text{(E.23)}$$

$$= a\sqrt{1 - 2z/a + (r/a)^2}, \qquad r < a$$

Setting $z = r \cos \vartheta$, $\mu = \cos \vartheta$, and $h = a/r$, we have

$$\frac{r}{R} = \frac{1}{\sqrt{1 - 2\mu h + h^2}} \qquad \text{(E.24)}$$

which is just the generating function introduced in (E.17). In general, we may write

$$\frac{1}{R} = \frac{1}{r} \sum_{n=0}^{\infty} \left(\frac{a}{r}\right)^n P_n(\mu), \qquad r > a$$

$$\frac{1}{R} = \frac{1}{a} \sum_{n=0}^{\infty} \left(\frac{r}{a}\right)^n P_n(\mu), \qquad r < a \qquad \text{(E.25)}$$

These relations may be used to obtain another expression for the Legendre polynomials. Consider R to be a function of a and write

$$\frac{1}{R(a)} = f(a) = \frac{1}{\sqrt{x^2 + y^2 + (z - a)^2}} \qquad \text{(E.26)}$$

This expression may be developed in powers of a as the Maclaurin expansion

$$f(a) = f(0) + a \left.\frac{\partial f}{\partial a}\right|_{a=0} + \frac{a^2}{2!} \left.\frac{\partial^2 f}{\partial a^2}\right|_{a=0} \cdots \qquad \text{(E.27)}$$

Noting that $\partial f / \partial a = -\partial f / \partial z$, we may write the expansion in terms of z derivatives and take the limit $a \rightarrow 0$. Since $f(0) = 1/r$, we have

$$f(a) = \frac{1}{R} = \frac{1}{r} - a \frac{\partial}{\partial z}\left(\frac{1}{r}\right) + \frac{a^2}{2!} \frac{\partial^2}{\partial z^2}\left(\frac{1}{r}\right) - \cdots \qquad (E.28)$$

On comparing coefficients of equal powers of a in the first of (E.25) and (E.28), we find

$$\frac{1}{r^{n+1}} P_n(\mu) = \frac{(-1)^n}{n!} \frac{\partial^n}{\partial z^n} \frac{1}{r} \qquad (E.29)$$

The significance of this result in terms of axial multipoles is considered in Section 5.2.

E.4   RELATION BETWEEN LEGENDRE FUNCTIONS AND BESSEL FUNCTIONS

Spherical coordinates are composed of spheres $r = $ const, cones $\vartheta = $ const, and planes $\varphi = $ const that intersect orthogonally. Consider the situation for $\vartheta \approx 0$ (near the z axis) at a large distance above the origin. In this region the spheres are approximately horizontal planes, the cones are approximately cylinders, and the planes $\varphi = $ const remain unchanged. The system of intersecting surfaces thus resembles the planes and cylinders of cylindrical coordinates. In this limit we should expect that there would be a relation between a horizontal distance expressed in terms of the cylindrical coordinate ρ and the corresponding distance $r \sin \vartheta$ expressed in spherical coordinates. The relation takes the form of a limit in which the Legendre polynomial reduces to a Bessel function, namely

$$\lim_{n \to \infty} P_n\left(\cos \frac{z}{n}\right) = J_0(z) \qquad (E.30)$$

We now outline a way in which the plausibility of this result may be deduced.

By adding and subtracting the quantity $\mu^2 h^2$ under the radical defining the generating function in (E.17), we have

$$g(\mu, h) = \frac{1}{\sqrt{(1 - \mu h)^2 - h^2(\mu^2 - 1)}} \qquad (E.31)$$

Using a standard evaluation of the definite integral

$$\frac{1}{\pi} \int_0^\pi \frac{d\varphi}{a + b \cos \varphi} = \frac{1}{\sqrt{a^2 - b^2}}, \qquad a > b > 0 \qquad (E.32)$$

we can rewrite the generating function in the integral form

$$g(\mu, h) = \frac{1}{\pi} \int_0^\pi \frac{d\varphi}{1 - \mu h + h\sqrt{\mu^2 - 1}\,\cos\varphi} \tag{E.33}$$

where we have set $a = 1 - \mu h$ and $b = h\sqrt{\mu^2 - 1}$ in (E.32). Expanding the denominator in the integrand, we have

$$g(\mu, h) = \frac{1}{\pi} \int_0^\pi d\varphi \sum_{n=0}^\infty (\mu - \sqrt{\mu^2 - 1}\,\cos\varphi)^n h^n \tag{E.34}$$

Since, according to (E.17), $P_n(\cos\vartheta)$ is the coefficient of h^n in the expansion of the generating function, we may equate corresponding coefficients of h^n and obtain (setting $\mu = \cos\vartheta$)

$$P_n(\cos\theta) = \frac{1}{\pi} \int_0^\pi (\mu - \sqrt{\mu^2 - 1}\,\cos\varphi)^n \, d\varphi \tag{E.35}$$

On using the identity $z^n = \exp(n \ln z)$, we have

$$P_n(\cos\theta) = \frac{1}{\pi} \int_0^\pi d\varphi \, e^{n\ln(\cos\theta - i\sin\theta\cos\varphi)} \tag{E.36}$$

For small values of θ we may use the first term in the Taylor expansion for $\cos\theta$ and $\sin\theta$. The argument of the logarithm is then approximated by

$$\ln(\cos\theta - i\sin\theta\cos\varphi) \approx \ln\left(1 - \frac{\theta^2}{2} - i\theta\cos\varphi\right)$$

$$\approx -i\theta\cos\varphi \tag{E.37}$$

If we now consider the small angle θ to be expressed as $\theta = z/n$ where $n \gg 1$, then for fixed z we have

$$\lim_{n\to\infty} P_n\left(\cos\frac{z}{n}\right) = \frac{1}{\pi} \int_0^\pi d\varphi \, e^{-iz\cos\varphi} = J_0(z) \tag{E.38}$$

which is the relation referred to above. Note that since $z/n \ll 1$ we may write $\cos z/n \approx 1 - z^2/2n^2$ and obtain the equivalent form

$$\lim_{n\to\infty} P_n\left(1 - \frac{z^2}{2n^2}\right) = J_0(z) \tag{E.39}$$

An oscillation problem that uses this relation is considered in Section 9.6.

Problems

(An extensive sequence of problems involving Legendre polynomials may be found in Edwards, 1930, Vol. II, pp. 931–939.)

E.1 Obtain two series solutions for the Legendre equation (E.11) by substituting the series form $u(\mu) = \Sigma_0^\infty a_k \mu^k$.
   (a) Show that the recurrence relation is

$$a_{k+2} = -\frac{(n-k)(k+n+1)}{(k+2)(k+1)} a_k, \qquad k = 0, 1, 2, \ldots$$

   (b) Show that the solution is $u(\mu) = a_0 u_0(\mu) + a_1 u_1(\mu)$ where

$$u_0 = 1 - \frac{n(n+1)}{2!}\mu^2 + \frac{n(n-2)(n+1)(n+3)}{4!}\mu^4 + \cdots$$

$$u_1 = \mu - \frac{(n-1)(n+2)}{3!}\mu^3$$

$$+ \frac{(n-1)(n-3)(n+2)(n+4)}{5!}\mu^5 + \cdots$$

E.2 If the parameter n in Eq. (E.11) is not an integer, both of the series obtained in Problem E.1 have an infinite number of terms. The radius of convergence of each series may be determined by considering the ratio of two successive terms in each series, that is,

$$\frac{|a_{n+2}|}{|a_n|}\mu^2$$

This ratio must be less than unity to guarantee convergence of the series. Use the results obtained in Problem E.1 to show that the limit of this ratio is equal to μ^2 for either series. Thus both series converge for $|\mu| < 1$.

   For $|\mu| = 1$ the test fails. Show that in this case the kth term in each series decreases as $1/k$ for $k \gg n$ and thus each series diverges logarithmically unless n is an integer, in which case one series terminates to provide the convergent solution.

E.3 Since μ^n is the highest power of μ in $P_n(\mu)$, the expansion of μ^n in terms of Legendre polynomials will contain no polynomials of order higher than n. Show that

$$\mu^2 = \tfrac{1}{3}[P_0(\mu) + 2P_2(\mu)]$$

$$\mu^3 = \tfrac{1}{5}[3P_1(\mu) + 2P_3(\mu)]$$

$$\mu^4 = \tfrac{1}{5}[P_0(\mu) + \tfrac{20}{7}P_2(\mu) + \tfrac{8}{7}P_4(\mu)]$$

E.4 Show that

$$\frac{\sin 3\theta}{\sin \theta} = \frac{1}{3} + \frac{8}{3}P_2(\cos \theta), \qquad \frac{\sin 4\theta}{\sin \theta} = \frac{4}{5}P_1(\cos \theta) + \frac{16}{5}P_3(\cos \theta)$$

$$\sin^4 \theta = \frac{8}{15}P_0(\cos \theta) - \frac{16}{21}P_2(\cos \theta) + \frac{8}{35}P_4(\cos \theta)$$

E.5 Integrate the Legendre equation from $\mu = \mu_0$ to $\mu = 1$ and obtain the relation

$$\int_{\mu_0}^{1} P_n(\mu)\, d\mu = \frac{1}{n + 1}[P_{n-1}(\mu_0) - \mu_0 P_n(\mu_0)]$$

$$= \frac{1}{(2n + 1)}[P_{n-1}(\mu_0) - P_{n+1}(\mu_0)]$$

where the second form is obtained by using (E.16).

E.6 The generating function for the Legendre polynomials is a function $g(\mu, h)$ such that

$$g(\mu, h) = \sum_{n=0}^{\infty} h^n P_n(\mu)$$

By differentiating this series term by term and using the first recurrence relation (E.16), show that

$$\frac{\partial g}{\partial h} = (2\mu h - h^2)\frac{\partial g}{\partial h} + (\mu - h)g$$

Integrate this ordinary differential equation (by separating variables g and h) and obtain

$$g(\mu, h) = \frac{c(\mu)}{\sqrt{1 - 2\mu h + h^2}}$$

where $c(\mu)$ is a constant of integration. Show that $g(0, \mu) = 1$ and thus that

$$g(\mu, h) = \frac{1}{\sqrt{1 - 2\mu h + h^2}}$$

E.7 Obtain the orthogonality relation for the Legendre polynomials by evaluating

$$\int_{-1}^{1} \frac{d\mu}{\sqrt{1 - 2h\mu + h^2}\sqrt{1 - 2k\mu + k^2}}$$

$$= \int_{-1}^{1} \sum_{n=0}^{\infty} \sum_{m=0}^{\infty} h^n k^m P_n(\mu) P_m(\mu)\, d\mu$$

Write the integral over μ in the form

$$\frac{1}{2\sqrt{hk}} \int_{-1}^{1} \frac{d\mu}{\sqrt{(a - \mu)(b - \mu)}}$$

where $a = (1 + h^2)/2h$, $b = (1 + k^2)/2k$. Evaluate the integral by using the substitution $u = \sqrt{a - \mu} + \sqrt{a + \mu}$ and obtain

$$\frac{1}{2\sqrt{hk}} \tanh^{-1}\sqrt{hk} = \sum_{n=0}^{\infty} \sum_{m=0}^{\infty} h^n k^m \int_{-1}^{1} P_n(\mu) P_m(\mu)\, d\mu$$

Since only equal powers of hk occur in the expansion of $\tanh^{-1}\sqrt{hk}$, and $\tanh^{-1}x = \Sigma x^{2n+1}/(2n + 1)$, conclude that

$$\int_{-1}^{1} P_n(\mu) P_m(\mu)\, d\mu = \begin{cases} 0, & n \neq m \\ \dfrac{1}{2n + 1}, & n = m \end{cases}$$

Note that since $P_0(\mu) = 1$, we may interpret the integral $\int_{-1}^{1} P_n(\mu)\, d\mu$ as $\int_{-1}^{1} P_0(\mu) P_n(\mu)\, d\mu$, which vanishes for $n \neq 0$, that is,

$$\int_{-1}^{1} P_n(\mu)\, d\mu = \int_{0}^{\pi} P_n(\cos\theta)\sin\theta\, d\theta = \begin{cases} 2, & n = 0 \\ 0, & n > 0 \end{cases}$$

E.8 Show that

$$\sum_{k=1}^{\infty} \frac{P_{2k}(0)}{k} = 2 \int_{0}^{1} \frac{dh}{h}\left(\frac{1}{\sqrt{1 + h^2}} - 1\right) = 2 \ln\left(2\sqrt{2} - 2\right)$$

E.9 A simple procedure for carrying out the successive differentiations of (E.9) to obtain (E.10) is to first set $\varphi_m = d^m w/d\mu^m$. Then use a prime to denote a derivative with respect to μ and show that (E.9) and its first two derivatives become

$$(\mu^2 - 1)\varphi_0'' - 2\mu(n - 1)\varphi_0' - 2n\varphi_0 = 0$$

$$(\mu^2 - 1)\varphi_1'' - 2\mu(n - 2)\varphi_1' - 2(2n - 1)\varphi_1 = 0$$

$$(\mu^2 - 1)\varphi_2'' - 2\mu(n - 3)\varphi_2' - 2(2n - 1 - 2)\varphi_2 = 0$$

Use the relation $1 + 2 + 3 + \cdots + m = m(m + 1)/2$ to obtain the general result given in (E.10).

APPENDIX F
Tables of Sums and Integral Transforms*

Table F.1 Summation of Series

1. $1 - \frac{1}{2} + \frac{1}{3} - \frac{1}{4} + \cdots = \ln 2$

2. $1 - \frac{1}{3} + \frac{1}{5} - \frac{1}{7} + \cdots = \pi/4$

3. $1 + \frac{1}{2^2} + \frac{1}{3^2} + \frac{1}{4^2} + \cdots = \pi^2/6$

4. $1 - \frac{1}{2^2} + \frac{1}{3^2} - \frac{1}{4^2} + \cdots = \pi^2/12$

5. $1 + \frac{1}{3^2} + \frac{1}{5^2} + \frac{1}{7^2} + \cdots = \pi^2/8$

6. $1 - \frac{1}{3^3} + \frac{1}{5^3} - \frac{1}{7^3} + \cdots = \pi^3/32$

7. $1 + \frac{1}{2^4} + \frac{1}{3^4} + \frac{1}{4^4} + \cdots = \pi^4/90$

8. $1 - \frac{1}{2^4} + \frac{1}{3^4} - \frac{1}{4^4} + \cdots = 7\pi^4/720$

9. $1 + \frac{1}{3^4} + \frac{1}{5^4} + \frac{1}{7^4} + \cdots = \pi^4/96$

10. $1 + \frac{1}{2^6} + \frac{1}{3^6} + \frac{1}{4^6} + \cdots = \pi^6/945$

11. $1 - \frac{1}{2^6} + \frac{1}{3^6} - \frac{1}{4^6} + \cdots = 31\pi^6/42 \cdot 6!$

12. $1 + \frac{1}{3^6} + \frac{1}{5^6} + \frac{1}{7^6} + \cdots = \pi^6/960$

13. $1 + \frac{1}{3^8} + \frac{1}{5^8} + \frac{1}{7^8} + \cdots = 17\pi^8/4 \cdot 8!$

14. $\Sigma(-1)^{n-1}/[(2n)^2 - 1] = \pi/4 - \frac{1}{2}, \; n = 1, 2, 3, \ldots$

15. $\Sigma\, 1/(n^2 - 4) = 0, \; n = 1, 3, 5, \ldots$

Bessel Functions, $J_0(\alpha_{0i}) = 0, \; i = 1, \ldots, \infty$

16. $\Sigma[1/(\alpha_{0i})^2] = \frac{1}{4}$

17. $\Sigma[1/(\alpha_{0i})^4] = \frac{1}{32}$

18. $\Sigma[1/(\alpha_{0i})^6] = \frac{1}{192}$

Algebraic Sums

19. $x - x^2/2 + x^3/3 - x^4/4 + \cdots = \ln(1 + x), \; -1 < x \le 1$

20. $x + x^3/3 + x^5/5 + x^7/7 + \cdots = \tanh^{-1}x, \; |x| < 1$

21. $x - x^3/3 + x^5/5 - x^7/7 + \cdots = \tan^{-1}x, \; -1 < x \le 1$

22. $\Sigma[1/(n^2 + a^2)] = (\pi/4a)\tanh(\pi a/2), \; n = 1, 3, 5, \ldots$

*More extensive tables of the material contained in this appendix may be found in Gradshteyn and Ryzhik, 1965; Hansen, 1975; Jolley, 1961 and Magnus and Oberhettinger, 1954.

Table F.1 (*Continued*)

Trigonometric Sums[a]

23. $\Sigma[(n \sin nx)/(n^2 + a^2)] = (\pi/2)\sinh a(\pi - x)/\sinh \pi a, \ 0 < x < 2\pi$

24. $\Sigma[(-1)^{n-1}(n \sin nx)/(n^2 + a^2)] = (\pi/2)\sinh ax/\sinh \pi a, \ -\pi < x < \pi$

25. $\Sigma[(n \sin nx)/(n^2 + a^2)] = (\pi/4)\cosh a(\pi/2 - x)/\cosh a\pi/2, \ n = 1, 3, 5, \ldots,$
$\quad -\pi < x < \pi$

26. $\Sigma(a^n/n)\sin nx = \frac{1}{2}\tan^{-1}[(2a \sin x)/(1 - a^2)], \ n = 1, 3, 5, \ldots, \ |a| < 1,$
$\quad x \neq n\pi$

27. $\Sigma(a^n/n)\cos nx = -\frac{1}{2}\ln(1 - 2a \cos x + a^2), \ |a| < 1$

28. $\Sigma(a^{-n}/n)\cos nx = \ln a - \frac{1}{2}\ln(1 - 2a \cos x + a^2), \ |a| > 1$

29. $\Sigma[(\sin^2 nx)/n^2] = \pi x/4, \ n = 1, 3, 5, \ldots$

30. $\Sigma[(\sin na \sin nb)/n^2] = a(\pi - b)/2, \ -b \le a \le b$
$\qquad\qquad\qquad\qquad\quad = b(\pi - a)/2, \ b \le a \le (2\pi - b)$

31. $\Sigma[(\cos na \cos nb)/n^2] = [a^2 + (b - \pi)^2 - \pi^2/3]/4, \ 0 \le a < b$
$\qquad\qquad\qquad\qquad\quad = [b^2 + (a - \pi)^2 - \pi^2/3]/4, \ b < a \le \pi$

Hyperbolic Functions

32. $\Sigma(-1)^{(n-1)/2}(1/n)\text{sech}(n\pi/2) = \pi/8, \ n = 1, 3, 5, \ldots$

33. $\Sigma(1/n^3)\tanh(n\pi/2) = \pi^3/32, \ n = 1, 3, 5, \ldots$

[a]Sums are over all integers, $1, \ldots, \infty$, unless indicated otherwise.

Table F.2 Laplace Transform

$f(t)$	$F(s) = \int_0^\infty e^{-st}f(t)\,dt$	s
1. 1	$1/s$	>0
2. t^n	$n!/s^{n+1}$	>0
3. $1/\sqrt{\pi t}$	$1/\sqrt{s}$	>0
4. $\sqrt{t/\pi}$	$1/(2s^{3/2})$	>0
5. $(t^2 - a^2)^{-1/2}, \ t > a$	$aK_0(as)$	>0
$0, \ t < a$		
6. $t^\nu, \ \nu > -1$	$\Gamma(\nu + 1)\,s^{-\nu-1}$	>0
7. e^{at}	$1/(s - a)$	$>a$
8. $t^{-1/2}\exp(-a^2/4t)$	$\sqrt{\pi/s}e^{-a\sqrt{s}}$	>0
9. $a/(2\sqrt{\pi}\ t^{3/2})\exp(-a^2/4t)$	$e^{-a\sqrt{s}}$	>0
10. $\sin at$	$a/(s^2 + a^2)$	>0
11. $\cos at$	$s/(s^2 + a^2)$	>0
12. $(\sin at)/t$	$\tan^{-1}(a/s)$	>0
13. $(\sin at \sin bt)/t$	$\frac{1}{4}\ln\{[s^2 + (a + b)^2]/[s^2(a - b)^2]\}$	>0
14. $(\sin at \cos bt)/t$	$\frac{1}{2}\tan^{-1}[2as/(s^2 - a^2 + b^2)]$	>0
15. $\sinh at$	$a/(s^2 - a^2)$	$>a$

Table F.2 (*Continued*)

$f(t)$	$F(s) = \int_0^\infty e^{-st}f(t)\,dt$	s
16. $\cosh at$	$s/(s^2 - a^2)$	$> a$
17. $\mathrm{erf}(at)$	$s^{-1}\exp(s^2/4a^2)\mathrm{erfc}(s/2a)$	> 0
18. $\mathrm{erfc}(a/2\sqrt{t})$	$s^{-1}e^{-a\sqrt{s}}$	> 0
19. $\sqrt{t}\,\exp(-a^2/4t) - \pi^{1/2}a/2\,\mathrm{erfc}(a/2\sqrt{t})$	$(\tfrac{1}{2})\pi^{-1/2}s^{-3/2}e^{-a\sqrt{s}}$	> 0
20. $J_0(at)$	$1/\sqrt{s^2 + a^2}$	> 0
21. $J_1(at)/t$	$(\sqrt{s^2 + a^2} - s)/a$	> 0
22. $J_0(a\sqrt{t^2 - b^2}),\ t > b$	$\exp(-b\sqrt{s^2 + a^2})/\sqrt{s^2 + a^2}$	> 0
$\quad 0,\ t < b$		
23. $\delta(t - a),\ a > 0$	e^{-as}	> 0

Table F.3 Fourier Sine Transform

$$f(x) = \frac{2}{\pi}\int_0^\infty F_s(k)\sin kx\,dk \qquad\qquad F_s(k) = \int_0^\infty f(x)\sin kx\,dx$$

1. $1/x$	$\tfrac{1}{2}\pi$
2. $x^{-1/2}$	$\sqrt{\pi/2k}$
3. $x/(x^2 + a^2)$	$\tfrac{1}{2}\pi e^{-ak}$
4. $\dfrac{1}{x(x^2 + a^2)}$	$\dfrac{\pi}{2a^2}(1 - e^{-ak})$
5. e^{-ax}	$k/(k^2 + a^2)$
6. $xe^{-a^2x^2}$	$\dfrac{\sqrt{\pi}}{4a^3}e^{-k^2/4a^2}$
7. $\dfrac{e^{-a^2x^2}}{x}$	$\dfrac{\pi}{2}\,\mathrm{erf}\left(\dfrac{k}{2a}\right)$
8. $\sin ax$	$\tfrac{1}{2}\pi\delta(k - a)$
9. $\dfrac{\sin ax}{x}$	$\dfrac{1}{2}\ln\left|\dfrac{k + a}{k - a}\right|$
10. $(\cos ax)/x$	$\tfrac{1}{2}\pi H(k - a)$
11. $(\mathrm{sech}\,ax)/x$	$2\tan^{-1}(e^{\pi k/2a}) - \tfrac{1}{2}\pi$

Table F.4 Fourier Cosine Transform

$$f(x) = \frac{2}{\pi} \int_0^\infty F_c(k)\cos kx \, dk \qquad F_c(k) = \int_0^\infty f(x)\cos kx \, dx$$

1. 1	$\pi\delta(\kappa)$
2. $x^{-1/2}$	$\sqrt{\pi/2k}$
3. $\dfrac{1}{x^2 + a^2}$	$\dfrac{\pi}{2a}e^{-ak}$
4. e^{-ax}	$\dfrac{a}{k^2 + a^2}$
5. $e^{-a^2x^2}$	$\dfrac{\sqrt{\pi}}{2a}e^{-k^2/4a^2}$
6. $\cos ax$	$\frac{1}{2}\pi\delta(k - a),\, a > 0$
7. $(\sin ax)/x$	$\frac{1}{2}\pi H(a - k)$
8. $\dfrac{\sin a\sqrt{b^2 + x^2}}{\sqrt{b^2 + x^2}}$	$\frac{1}{2}\pi J_0(b\sqrt{a^2 - k^2})H(a - k)$
9. $\operatorname{sech} ax$	$(\pi/2a)\operatorname{sech}(\pi k/2a)$
10. $\dfrac{\sinh ax}{\sinh bx},\, a < b$	$\dfrac{\pi}{2b}\dfrac{\sin(\pi a/b)}{\cosh(\pi k/b) + \cos(\pi a/b)}$
11. $\dfrac{\sinh ax}{x\cosh bx},\, a < b$	$\dfrac{1}{2}\ln\dfrac{\cosh(\pi k/2b) + \sin(\pi a/2b)}{\cosh(\pi k/2b) - \sin(\pi a/2b)}$
12. $K_0(ax)$	$\frac{1}{2}\pi(k^2 + a^2)^{-1/2}$

Table F.5 Exponential Fourier Transform

$$f(x) = \int_{-\infty}^{\infty} \frac{dk}{2\pi} e^{ikx} F(k) \qquad F(k) = \int_{-\infty}^{\infty} e^{-ikx} f(x) \, dx$$

1. 1	$2\pi\delta(x)$
2. e^{iax}	$2\pi\delta(k - a)$
3. $1/(x^2 + a^2)$	$\pi e^{-|k|a}$
4. $\operatorname{sech} ax$	$\dfrac{\pi}{a}\operatorname{sech}\dfrac{\pi k}{2a}$

Table F.6 Fourier-Bessel Transform

$$f(r) = \int_0^\infty k \, dk \, J_0(kr)F_B(k) \qquad F_B(k) = \int_0^\infty r \, dr \, J_0(kr)f(r)$$

1. $r^{-1}e^{-\gamma r}$	$1/(k^2 + \gamma^2)^{1/2}$
2. $e^{-\gamma r}$	$\gamma/(k^2 + \gamma^2)^{3/2}$
3. $e^{-\gamma^2 r^2}$	$\dfrac{1}{2\gamma^2}e^{-k^2/4\gamma^2}$
4. $H(a - r)/(a^2 - r^2)^{1/2}$	$(\sin ka)/k$
5. $\dfrac{\sin(a\sqrt{b^2 + r^2})}{\sqrt{b^2 + r^2}}$	$\dfrac{\cos(b\sqrt{a^2 - k^2})}{\sqrt{a^2 - k^2}}H(a - k)$

References

E. R. Benton and G. W. Platzmann, *Q. Appl. Math.*, 30, 195–212 (1972).

H. S. Carslaw, *An Introduction to the Theory of Fourier's Series and Integrals*, 3rd ed., Dover, New York, 1980.

G. Chrystal, *Algebra*, 2 vols., Chelsea, New York, 1964.

H. T. Davis, *Introduction to Nonlinear Differential and Integral Equations*, Dover, New York, 1960.

J. Edwards, *Integral Calculus*, 2 vols., Macmillan, London, 1930.

A. Erdélyi, et al., *Tables of Integral Transforms*, 2 vols., McGraw-Hill, New York, 1954.

B. Friedman, *Principles and Techniques of Applied Mathematics*, Dover, New York, 1990.

A. R. Forsyth, *Theory of Differential Equations*, 6 vols., Dover, New York, 1959.

I. S. Gradshteyn and I. M. Ryzhik, *Tables of Integrals, Series and Products*, 4th ed., Academic, New York, 1965.

E. R. Hansen, *A Table of Series and Products*, Prentice-Hall, Englewood Cliffs, N.J., 1975.

F. B. Hildebrand, *Advanced Calculus for Applications*, 2nd ed., Prentice-Hall, Englewood Cliffs, N.J., 1976.

H. J. Jeffreys and B. S. Jeffreys, *Methods of Mathematical Physics*, Cambridge University Press, New York, 1946.

L. B. W. Jolley, *Summation of Series*, 2nd ed., Dover, New York, 1961.

E. Kamke, *Differentialgleichungen, Lösungsmethoden und Lösungen*, Vol. 1, Chelsea, New York, 1971.

H. Lamb, *Hydrodynamics*, 6th ed., Dover, New York (1945).

M. J. Lighthill, *Waves in Fluids*, Cambridge University Press, Cambridge, 1978.

W. Magnus and F. Oberhettinger, *Formulas and Theorems for the Function of Mathematical Physics*, Chelsea, New York, 1954.

P. M. Morse and H. Feshbach, *Methods of Theoretical Physics*, 2 vols., McGraw-Hill, New York, 1953.

K. B. Oldham and J. Spanier, *The Fractional Calculus*, Academic, New York, 1974.

J. W. Strutt (Lord Rayleigh), *The Theory of Sound*, 2 vols. Dover, New York, 1945.

G. N. Watson, *A Treatise on the Theory of Bessel Functions*, 2nd ed., Cambridge University Press, Cambridge, 1944.

H. F. Weinberger, *Partial Differential Equations*, Wiley, New York, 1965.

E. T. Whittaker and G. N. Watson, *A Course of Modern Analysis*, Cambridge University Press, Cambridge, 1969.

INDEX

467